☑ Y0-AQW-199

1BEL7965

SPRINGER
LAB MANUAL

Springer

Berlin
Heidelberg
New York
Barcelona
Budapest
Hong Kong
London
Milan
Paris
Singapore
Tokyo

Martin Clynes (Ed.)

Animal Cell Culture Techniques

With 76 Figures

 Springer

DR. MARTIN CLYNES

National Cell & Tissue Culture Center
Bioresearch Ireland
Dublin City University
Glasnevin
Dublin 9, Ireland

A91.
QH
585
.2
.A566
1998

ISBN 3-540-63008-2 Springer-Verlag Berlin Heidelberg New York

Library of Congress Cataloging-in-Publication Data
Animal cell culture techniques / Martin Clynes (ed.). – Berlin ;
Heidelberg ; New York : Springer, 1998
   (Springer lab manual)
   ISBN 3-540-63008-2

This work is subject to copyright. All rights are reserved, whether the whole or part of the material is concerned, specifically the rights of translation, reprinting, reuse of illustrations, recitation, broadcasting, reproduction on microfilm or in any other way, and storage in data banks. Duplication of this publication or parts thereof is permitted only under the provisions of the German Copyright Law of September 9, 1965, in its current version, and permissions for use must always be obtained from Springer-Verlag. Violations are liable for prosecution under the German Copyright Law.

© Springer-Verlag Berlin Heidelberg 1998
Printed in Germany

The use of general descriptive names, registered names, trademarks, etc. in this publication does not imply, even in the absence of a specific statement, that such names are exempt from the relevant protective laws and regulations and therefore free for general use.

Product liability: The publisher cannot guarantee the accuracy of any information about dosage and application thereof contained in this book. In every individual case the user must check such information by consulting the relevant literature.

Cover Design: design & production GmbH, D-69121 Heidelberg
Typesetting: Mitterweger Werksatz GmbH, S-68723 Plankstadt
SPIN 10515510   31/3133   5 4 3 2 1 0 – Printed on acid free paper

Contents

Introduction

This book arose out of a request from Springer-Verlag to produce a laboratory manual based on two lecture and practical courses, "Introduction to Animal Cell Culture for Biochemistry", sponsored by FEBS (Federation of European Biochemical Societies), which took place at the National Cell and Tissue Culture Centre (NCTCC) at Dublin City University, Ireland, in June 1994 and June 1996. The chapters have been written mainly by scientists working at the NCTCC but also by guest lecturers who contributed to the two courses. There are also several additional chapters by experts in areas not covered in the courses.

There are a number of good texts available on basic cell culture technique, but books covering a wide range of techniques are less readily available. The classic, *Tissue culture – methods and applications*, by P.F. Kruse and M.K. Patterson, is the model to which all such texts should aspire. Although this excellent volume would be hard to equal, it obviously does not cover a number of important techniques developed over the past 25 years. I hope that this book will, to some extent, fill this gap in the literature.

I would like to draw the reader's attention to some important aspects of working with animal cells that are sometimes overlooked in laboratories in which cell culture is not a major focus, but instead is used, perhaps intermittently, as a practical tool. Important issues exist in relation to safety, cell stability, cell cross-contamination and cell characterization/authentication. If these issues are ignored, experimental results may be meaningless.

Safety

Working with animal cells in culture has, in general, turned out to be a low-risk procedure. Nevertheless, there have been fatal incidents arising from work with viruses, and work with primary monkey cells which harbored filoviruses (e.g., Marburg virus). Many animal cells contain C-type particles, which may be retrovirus-related. Particular care should be taken when working with human and other primate cells. Generally, good "housekeeping" and working in a class II downflow recirculating laminar flow cabinet will provide a safe working environment. When working with viruses, or cultures suspected of containing infective viruses, however, specialist information on safety and containment should be sought.

In designing housekeeping or "good laboratory practice" guidelines, it is important to remember that safety issues occur at several levels:

- Safety of operator

- Safety of others in the laboratory

- Safety of others, e.g., cleaning personnel

- Safety of the general public

- Environmental safety (including possible infection of domestic or wild animals)

When working with any cell line or primary culture, make the assumption that it may possibly contain infective virus. Work in a downflow recirculating class II laminar flow cabinet, never in a horizontal flow cabinet (which can, even in the absence of viruses, possibly increase exposure to allergens). Make sure that any cuts, especially on the hands, are covered. Wearing gloves (with cotton inserts, if you have sensitive hands) is recommended but not essential. Thorough washing of hands before and after cell culture work is, of course, mandatory. If working with primary human material, immunization against hepatitis B is advisable. Laboratory coats (preferably buttoning up to the neck) are essential. Face masks are recommended if dangerous aerosols are anticipated, but are not generally necessary. Waste media, flasks, cotton wool, pipettes, etc. – anything that might have been in contact with cells or serum-containing medium – should be stored securely in sealed bags or containers and adequately autoclaved before disposal. Broken glass must be stored separately and safely.

Chemical sterilization is an option, but requires careful monitoring since the sterilizing agents have a limited half-life in solution, so that autoclaving is preferable, certainly for high- or medium-risk material.

Consideration must also be given to containment of viruses – escape of mammalian or even insect viruses which would have no consequence for human health might have undesirable ecological consequences.

Stability of Phenotypes of Cells in Culture

Cell lines can change properties – drug resistance, growth rate, antigen expression, enzyme profile, etc. – quite rapidly, sometimes even within five to ten passages. It is essential to set up a large frozen stock of each cell line and to work only within a defined number of cell doublings (or passages at defined dilution/split ratio). You will need to define this period for the cell lines and phenotypes of interest to you – some cells/ phenotypes are very stable, others are unstable. As a rough guideline, a ten-passage period is satisfactory for many cell lines and properties.

Pedigree of Cell Lines

Get your cell lines from a reputable source (preferably the laboratory of origin or an established cell repository). Then set up your own frozen stocks. Cross-contamination of cell lines is disturbingly commonplace (for a thriller-style account of how this was first discovered, see Michael Gold's, *A Conspiracy of Cells*, State University of New York Press, 1986).

As an example, MCF-7, a breast cancer line very widely used in cancer research, has a different patient of origin in stocks from different laboratories (Osborne et al. 1987).

Cell Contamination and Authentication

Cross-contamination of cell lines is a surprisingly frequent event, even in well run laboratories. Many of the earlier cell lines deposited at the American Type Culture Collection Catalogue are contaminated by HeLa, the first human cell line to be established, and a very rapidly growing cell type. Stock cultures of two cell lines should never be worked on at the same time in a laminar flow cabinet, and between work with different cell lines a thorough cleaning of surfaces and a 15-min "clearing" gap should occur. These procedures minimize the chance of aerosol contamination. Bear in mind that aerosol droplets, containing cells, can be formed by simple manipulations such as pipetting and centrifugation. Use of pipettes, medium bottles, etc. for more than one cell line is another possible source of cross-contamination and should never occur in a cell culture laboratory. Simple errors in labeling may also be a common cause.

When setting up a large frozen stock line, aliquots should be thawed to test for viability, growth and absence of contamination (including mycoplasma). They should also be characterized by some appropriate criteria, both for comparison to the parent cell line, and as a standard for comparison of future stocks. DNA fingerprinting is probably the best method for individual cell identification, and if such techniques are not available in your laboratory, commercial fingerprinting is available at a reasonable cost (e.g., Cellmark Diagnostics, address below). Cytogenetic analysis is also useful for this purpose and (like isoenzyme analysis) is an easy method for detecting interspecies contamination.

DNA fingerprinting will not, of course, distinguish between variants of a cell line – in this case specific differences, e.g., antigenic markers or drug resistance, may have to be investigated.

The methods necessary to authenticate your cell lines will depend on the properties of your cells, how many cell types you are culturing, and the identity of likely contaminants. A simple characterization protocol, which can be implemented without excessive effort, but which provides the necessary degree of assurance, is more likely followed in the long term than an exhaustive but time-consuming set of procedures. Cell stocks should be mycoplasma-free; this is dealt with in the text.

Some Useful Addresses

- American Type Culture Collection (ATCC): wide range of animal cells, vectors, viruses, etc., and manuals on cell characterization. 12301 Parklawn Drive, Rockville, Maryland 20852, USA. Phone: 301–881–2600; fax: 310–816–4361; e-mail: sales@atcc.org; /www.atcc.org/general.html.
- European Collection of Animal Cell Cultures (ECACC): CAMR, Salisbury, SP4 OJG, UK. Phone: 44–1–980–612512; fax: 44–1–980–611315; e-mail:etcs@ecacc.demon.co.uk.
- European Tissue Culture Society (ETCS) – more research based – and European Society for Animal Cell Technology – more industrially oriented – can both be contacted via their Secretariats c/o ECACC.
- Animal Cell Technology Industrial Platform (ACTIP): a working group of companies involved in animal cell culture. c/o Scientific Writing and Consultancy, P.O. Box 23161, 3001 KD Rotterdam, The Netherlands. Phone: 31–10–4363725; fax: 31–10–4361004.
- Cellmark Diagnostics. Blacklands Way, Abingden Business Park, Abingden, Oxfordshire OX14 2DY, UK. Phone:+44–1–235–5286609; fax: +44–1–235–528141.

– Society for In Vitro Biology (formerly Tissue Culture Association). 8815 Centre Park Drive, Suite 210, Columbia, Maryland 21045, USA. Phone: +1–410–9920946; fax: +1–410–9920949.

Some Useful Journals

– *Cytotechnology* and *Methods in Cell Science*, published by Kluwer Academic, Division of Biosciences, PO Box 17, 330 AA Dordrecht, The Netherlands.
– *In Vitro Cellular and Developmental Biology* (included in membership fees for Society for In Vitro Biology, see above).

Acknowledgements. I would like to thank all who contributed to this book, Dr. Jutta Lindenborn of Springer-Verlag who organized publication, and especially Yvonne Reilly and Mairéad Callan who helped put the book together. Finally, I would like to dedicate the book to my parents, Katie and Jack, and to Kate, Eugene, Mary, Honor, Aedin, Peter, Isolde and Róisín.

Dublin, March 1998   MARTIN CLYNES

References

Kruse PF, Patterson MK (1973) Tissue culture – methods and applications. Academic Press, New York
Osborne CK, Hobbs K, Trent JM (1987) Biological differences among MCF-7 human breast cancer cell lines from different laboratories. Breast Cancer Res Treat 9:111–121

Part I

General Methods

Chapter 1

Cloning Animal Cells

SHIRLEY MCBRIDE[*,1,2], MARY HEENAN[1], and MARTIN CLYNES[1]

Introduction

The word "clone" is used to describe a cell population that has derived from a single cell. Such single cells are termed "clonogenic" and only a small proportion of normal cells and probably a higher proportion of tumor cells have the proliferative capabilities required to give rise to clones. Both normal and neoplastic stem cells may be considered clonogenic as they have the ability to renew themselves and produce descendants which differentiate.

Clonogenic cells are detected in vitro by their ability to form individual colonies following isolation from the host tissue or from an established culture. Most detection methods employ microscopic inspection to identify colonies, and cells therefore must be seeded on average a few millimeters apart in order to discriminate between colonies. These are extremely artificial conditions compared with those in vivo as intercellular communications are effectively lost and many cells with clonogenic potential may not survive or proliferate at such low densities. Therefore, it is not assumed that clonogenic cells detected under in vitro conditions are entirely representative of stem cell populations present in vivo.

Colony-forming efficiency (CFE) is a measure of the ability of a cell population to form colonies (colonies formed/total cells seeded $\times 100\%$) and is often used as a means of comparing the tumorigenic potential of different cell populations. However, the CFE of a population may vary

* *Correspondence to* Shirley McBride: phone +353–1–705401; fax +353–1–7045484; e-mail smcbride@serv
[1] Shirley McBride, Mary Heenan, Martin Clynes, National Cell and Tissue Culture Center, Dublin City University, Glasnevin, Dublin 9, Ireland
[2] Shirley McBride, Current address: The University of Edinburgh, Department of Pathology, The University of Edinburgh Medical School, Teviot Place, Edinburgh EH8 9AG

depending on the assay conditions used. Cells with low CFEs may, for example, require feeder layers, whereas those with higher CFEs may be cloned using a limiting dilution assay or cloning rings, all of which are described below.

The isolation of clones from parental populations has several applications. Cytogenetic and phenotypic heterogeneity is an important feature of tumor development and progression and has significant implications for the diagnosis and treatment of cancer. The ability to study individual tumor cells and descendants derived from a heterogeneous population is invaluable to the understanding of tumor cell biology. Cloning cells from cell lines following exposure regimes with agents such as mutagens and toxins allows the establishment of useful variant populations including, for example, biological and genetic variants. Cloning is also carried out on cells which have been transfected with DNA to yield genetically homogenous populations and on hybridoma cells after fusion.

1.1
Preliminary Determination of Colony-Forming Efficiency

Note: All manipulations described in this chapter are carried out aseptically in a laminar flow cabinet except when cells/colonies are being counted.

Seal all plates with Parafilm when incubating or inspecting microscopically to reduce contamination risks.

To determine which cloning procedure is most appropriate for a particular cell type, a preliminary estimation of CFE is first carried out.

▨ Materials

reagents
- Suspension of actively growing cells obtained by standard subculture procedures
- Growth medium appropriate for cells
- Tissue culture flasks, petri dishes or multiwell plates

equipment
- Laminar flow cabinet
- Humidified CO_2 incubator
- Inverted microscope

Procedure

1. Count the cell suspension, prepare dilutions containing 1×10^2, 1×10^3 and 1×10^4 cells per ml and inoculate the tissue culture vessel of choice with appropriate volumes of cell dilutions.   **estimation of CFE**

2. Incubate at 37 °C, 5 % CO_2, in a humidified incubator.

3. After 4 days, inspect the cells microscopically for colony formation (a shorter or longer incubation time may be required to observe colonies, depending on the growth rate of the particular cells). Mark the colonies on the underside of the vessel.

4. Continue to mark colonies every 2–3 days as they appear until no further growth is apparent.

5. Determine the CFE (see above) and select the appropriate cloning method from the following procedures. If the CFE is very low (0–1 %), feeder layers may be required in the cloning assays (see below).

1.2
Isolation of Clonal Populations Using Cloning Rings

Note: This procedure is suitable for cells with a CFE < 5%. The isolation of clonal populations using cloning rings is achieved by segregating specific colonies of cells. The objective of the cloning procedure is to selectively remove individual colonies of cells which grew from a single cell and are at a distance from any surrounding colonies, and maintain them in an isolated environment. This cloning method can only be used in association with strongly adherent cells which will not readily detach from the surface and reattach elsewhere.

Materials

– Cloning rings (stainless steel, glass or Teflon, 0.5–0.8 cm inside diameter and approx. 1 cm high; The base of the ring must be smooth and flat.)
– Silicone grease (Fisons Scientific Equipment; cat. no. LUB-450-U)

Procedure

1. Sterilize the silicone grease and the cloning rings by baking at 120 °C overnight in separate glass petri dishes.

2. Subculture the cells of interest and resuspend in a growth medium known to support growth of the cell line. The cells must be in a single cell suspension at this stage. If doublets or clumps are present, this may give rise to nonclonal colonies.

3. Dilute the cell suspension to a concentration which results in cells attaching in a dispersed manner and which allows cell growth. This optimum concentration varies for each cell line and is dependent on the CFE of the cells. Cells with high CFEs should be diluted to a higher amount than cells with low CFEs. The objective is to plate the cells at a density whereby the cells will attach in a singular and dispersed manner, allowing for the formation of colonies from single cells and which are well separated from surrounding colonies.

4. Incubate the plates at 37 °C, 5 % CO_2 in a humidified atmosphere and allow the cells to attach overnight. Examine the plates. Attached cells are noted (and marked on the underside of the petri dish for easy relocation). Areas where cells have attached as two or more cells are also noted, so that colonies arising from these cells can be ignored.

5. The cells are allowed to grow with regular monitoring and refeeding, if required. **Note:** Care should be taken to ensure colonies are not beginning to grow in areas where cells did not originally attach. This would indicate an ability by the cells to detach and reattach at a distant point. In these cases, migration of cells to another colony and mixing of the individual colonies may be occurring, resulting in the growth of nonclonal colonies of cells.

6. When the colonies have reached approximately 50 cells per colony, individual colonies are subcultured and transferred to a well of a 96-well plate. The location of the colony being subcultured and the nearest surrounding colonies are determined. The subculturing of the specific colony may be performed during constant microscopic monitoring, but if the locations of all the colonies of interest are clearly marked on the plate, this may not be necessary.

7. Remove the medium from the petri dish and wash the cells with 5 ml sterile PBS.

8. Using sterile forceps, dip one end of the cloning ring in the silicone grease, ensuring the grease is evenly dispersed. Place the cloning ring around the colony of interest and with the edge of the forceps, press the cloning ring down to ensure a good seal between the bottom of the cloning ring and the petri dish. The plate should be monitored at this stage to ensure that the cloning ring encircles only the colony of interest.

9. Add approximately 100 μl trypsin solution to the cloning ring and incubate the dish at 37 °C until the cells begin to detach (the exact time length will depend on the cell line being handled).

10. When the cells begin to detach, gently pipette the trypsin solution up and down a number of times to detach the remaining cells and transfer the solution to a well of a 96-well plate. Add 100 μl medium to the cloning ring, gently pipette up and down a number of times and remove the medium to the same well of the 96-well plate. Care must be taken in the pipetting procedure to ensure that the cells are not damaged. A micropipette may be used for the subculturing step, which should not touch against the sides of the cloning ring, to prevent disturbing the position of the ring on the plate.

11. The cells in the 96-well plate are allowed to attach overnight and then are refed with 200 μl fresh medium.

12. The cells are allowed to grow until the well is at least 80 % confluent. At this stage the cells are subcultured into a larger dish. The number of scale-up steps depends on the growth rate and CFE of the cells. If the cells have a high CFE, the cells may be transferred from the well of a 96-well plate, to a 24-well plate and then to a 25 cm^2 flask; however cells with low CFEs may need intermediate steps (from a 96- to a 48-well plate and subsequently to 24- and 6-well plates before transfer to 25 cm^2 flasks).

13. Master stocks of the individual cell lines should be stored in liquid nitrogen as soon as possible. Obviously, to keep the new cell line as identical to the original clones as possible, the lowest possible number of subculturing steps is preferred.

Comments

- Cells which have low CFEs and/or poor growth rates may benefit from growing in conditioned medium throughout the isolation of the clones. Conditioned medium may by obtained by growing an 80 % confluent culture of the original cell line in medium for 24 h. The medium is collected, centrifuged at 120 g for 5 min to remove any cells and filtered through a 0.2 μm low protein binding filter. The conditioned medium is aliquoted and stored at 4 °C. This medium may contain growth promoters and attachment factors secreted by the cell line, and therefore may help promote cell growth. A mixture of this conditioned medium may be used to supplement the fresh growth medium (50/50 v/v).

- Because of the large number of manipulation steps encountered in the cloning of the cells and the length of time the cells are growing in an open dish, it may be necessary to grow the cells in antibiotic containing medium, if a problem with microbial contamination is encountered. This is not recommended, but if it is required, the cells should be regrown in antibiotic-free medium, as soon as it is feasible to do so.

- When the colonies of cells are being subcultured and during the subsequent maintenance of the cells, they should be treated as separate cell lines. This includes using a separate plastics supply for each clonal cell line and keeping a designated medium and reagents supply for a particular cell line.

- A clonal cell line may become more heterogeneous with sequential passaging. Therefore, in order to keep the cell line as identical as possible to the originally isolated colony of cells, it is important to generate frozen master stocks as soon as possible. The cell line should only be used up to a predesignated passage number, at which stage a fresh vial of cells should be revived from frozen stocks.

1.3
Cloning Procedure with the Limiting Dilution Assay

Note: This cloning procedure is only suitable for adherent cells with a CFE>5 %.

▨ Materials

– Equipment and reagents as above.

▨ Procedure

cloning
procedure

1. Dilute a cell suspension to produce a density of 10 cells per 3 ml medium.

2. Plate 100 μl of suspension into each well of a 96-well plate and incubate at 37 °C, 5 % CO_2 for 4 days. If a large number of clones are required, multiple plates should be used as usually only 5–10 clones will result from each plate.

3. Inspect the plate microscopically and mark those wells which contain single cells.

Note: If cells grow particularly well, inspect after 2–3 days to ensure detection of single cells.

If cells are small, it may be necessary to incubate for longer until colonies begin to appear.

Pay particular attention to the edge of wells during the inspection as cells can be more difficult to see here due to the curvature of the wells and may be overlooked. Conditioned medium may be used to enhance the growth of slow growing colonies as outlined in the previous procedure .

4. Continue to inspect the plate every 2–3 days to ensure marked wells contain single colonies. The cells may need to be fed occasionally. If the medium contains a pH indicator, the medium should be replaced following a change of color. Otherwise, replace approximately once a week.

5. When the marked wells are almost confluent, harvest the cells by trypsinization and seed into 6-well plates and then up into 25 cm² and larger flasks.

6. Create frozen stocks of the clones and store in liquid nitrogen as quickly as possible.

See notes for cloning rings procedure.

1.4
Feeder Layers

As mentioned above, if the CFE of particular cells is very low (0–1 %), feeder layers may be employed to enhance proliferation. The feeder layer principle involves the destruction, by irradiation (Melioli et al. 1994) or exposure to cell proliferation inhibitors, of the reproductive integrity of cells from the population that is to be cloned. Such cells retain their structural integrity and continue to metabolize until the test cells have begun to proliferate, thus acting as a supplement to the clones. The method for the establishment of feeder layers using mitomycin C is now outlined.

▪ Materials

– As for previous cloning procedures
– Mitomycin C

▪ Procedure

establishing feeding layers

1. A limiting dilution assay is set up in a 96-well plate as described above, using the cells to be cloned, and those wells containing single cells are identified.
   At this stage, the feeder layer cells are prepared separately as follows:

2. Inoculate tissue culture vessels with cells and grow to approximately 50 % confluency.

3. Add 2 µg/ml mitomycin C to the medium and incubate for 24 h.

4. Remove the medium from the cells, add fresh medium plus 2 µg/ml mitomycin C and incubate for a further 24 h.

5. Harvest the cells, resuspend in fresh medium (without mitomycin C) at a concentration of 5×10^5 cells/ml and inoculate 100 µl per well into the 96-well plates containing the single cells from step 1. Include (and examine regularly) control wells (feeders only) to ensure that no viable cells remain in the feeder population. This is very important.

6. Incubate the plates and continue from step 5 in the limiting dilution assay procedure above.

1.5
Semi-solid Media Cloning

This technique can be used to clone cells which grow in suspension, as well as many cells which are adherent under normal conditions. Agar or methylcellulose is added to the normal growth medium to produce a semi-solid medium which supports growth in suspension. Colonies are again detected microscopically and isolated using a micropipette. Following isolation, clones may be maintained in semi-solid medium or grown as monolayers.

▩ Materials

- Growth medium   reagents
- Agar medium (see procedure below)
- Agar (Bacto Difco)
- 35 mm sterile petri dishes
- Micropipette

- Equipment as above   equipment
- Water bath

▩ Procedure

The semi-solid agar is prepared as follows: cloning in
                                                             semi-solid agar
1. Dissolve 0.77 g agar in 50 ml ultrapure water and autoclave.

2. Immediately before use, melt the agar. (Use a Bunsen burner or microwave oven. Careful! "Bumping" may occur in agar which can cause serious burns). Incubate at 44 °C in a water bath.

3. Prepare agar medium as follows and equilibrate to 44 °C: 50 ml 2x DMEM (20 ml 10x DMEM, 76 ml ultrapure H_2O, 4 ml 1 mM Hepes; adjust to pH 7.4 with 1 M NaOH); 2 ml 1 mM Hepes; 1 ml $NaHCO_3$; 1 ml pen/strep; 14 ml normal growth medium. Immediately prior to use, add the thermolabile components: 2 mM L-glutamine and 10 ml fetal calf serum.

4. Add the melted agar to the agar medium, mix well using a swirling motion, return to the water bath and reduce the temperature to 41 °C.

5. The cells to be assayed are harvested and resuspended in medium without serum. It is important that a single cell suspension is obtained, so resuspend the cells well.

6. Dilute the cells to give a concentration of 2×10^4 cells per ml in a total of 5 ml. **Note:** The cell density used may need to be adjusted for different cell lines to optimize colony formation.

7. Add 5 ml agar (41 °C) to each 5 ml of cell suspension, mix well and quickly dispense 3 ml into each petri dish.

8. Place the petri dishes onto trays which contain a small volume of sterile water to prevent the agar drying out and incubate at 37 °C, 5 % CO_2 for 10 days.

9. Inspect the petri dishes microscopically for colony formation. Bring a microscope into a laminar flow cabinet and remove individual colonies using a micropipette. The colonies may be cultured individually in soft agar or in monolayer cultures.

References

Heenan M, O'Driscoll L, Cleary I, Connolly L, Clynes M (1997) Isolation and characterisation of multiple clonal subpopulations of an MDR human lung cell line which exhibits significantly different levels of resistance. Int J Cancer 71: 907–915

Meliolo G, Ratto G, Guastella M, Meta M, Biassoni R, Semino C, Casartelli G, Pasquetti W, Catrullo A, Moretta L (1994) Isolation and in vitro expansion of lymphocytes infiltrating non-small cell lung carcinoma: functional and molecular characterisation for their use in adoptive immunotherapy. Eur J Cancer 30A: 97–102

Chapter 2

Cell Culture and Diagnostic Virology

EDWIN O'KELLY*

Introduction

Since the discovery by Enders (1949) that polioviruses could be cultured successfully in nonneural tissue, cell culture has become a very useful and convenient method for isolating viruses in vitro. Even though more modern diagnostic virological techniques such as polymerase chain reaction (PCR), enzyme immunoassay (EIA) and immunofluorescence (IF) have become increasingly popular recently, viral isolation in cell culture still remains the "gold standard" for many cultivable viruses (Schmidt 1989). A single cell culture can be used to cultivate a broad spectrum of viral agents. Viral culture also facilitates the production of high titered virus used in antibody testing, viral characterization or molecular analysis. In the past many diagnostic virology laboratories initiated and propagated their own cell lines. Nowadays, it is more convenient and indeed common practice to buy cells, due to the abundance of commercial sources available. This is also true for the growth media in which the cells are propagated. Irrespective of how the cells or media are acquired, it is important for the diagnostic virology laboratory to maintain their own appropriate quality control procedures to ensure that cells and media are free of contaminants and remain susceptible to viral challenge.

Most diagnostic virology laboratories use monolayer cell cultures to propagate viruses. The main advantage of using monolayer cultures is the ease with which the infected cultures can be monitored microscopically. Many viruses present themselves in cell culture by producing degenerative changes in the cells, the so-called cytopathic effect (CPE).

* Edwin O'Kelly, Virus Reference Laboratory, University College Dublin, Belfield, Dublin 4, Ireland; phone +353–1–2691214; fax +353–1–2697611; e-mail eokelly@-ollamh.ie

The CPE is often characteristic of a specific virus and this allows the experienced observer to make a presumptive diagnosis based on the type of CPE present on the monolayer. A more definitive viral diagnosis is carried out by further testing of the viral isolate. This can be achieved by performing a viral neutralization assay on the isolate in fresh cell cultures. A useful alternative to viral neutralization is the application of immunoassay techniques such as IF staining of infected cells, EIA or nucleic acid hybridization. The application of these techniques are particularly useful for detecting specific viral replication in cultures in the absence of a CPE. Not all viruses will produce a CPE and some viruses are slow to grow. Immunoassay techniques can also allow early detection of viral replication prior to the formation of a CPE and allow more rapid viral diagnosis. The availability of specific and sensitive monoclonal antibodies directed against viral antigen has greatly enhanced the use of these techniques in viral diagnosis.

This chapter will outline some of the current methods used for detecting viruses in cell culture in a busy diagnostic virology setting.

▨ Materials

reagents
- Tissue culture flasks (Costar, Bibby Sterilin, Nunc, Falcon)
- Microtiter plates, cell clusters (Costar, Bibby Sterilin, Nunc, Falcon)
- U-bottomed cell culture microtiter plate (Costar)
- Dram/shell vials with coverslips (Bibby Sterilin)
- Multichannel pipette (BCL)
- Glass tissue culture tubes (Bibby Sterilin)
- Silicon rubber bungs (BDH)
- Cell counting chamber (Becton Dickinson)
- Kova Glasstic disposable counting chamber (Hycor)
- Cell cultures (Biowhittaker, ATCC, ECACC)

The following reagents can be obtained from Gibco BRL, Biowhittaker.

- Minimum essential medium (MEM) with Earle's salts without L-glutamine, sodium bicarbonate (10x)
- L-glutamine 200 mM (100x)
- MEM nonessential amino acids (NEAA) (100x)
- Trypsin-EDTA (1x)
- Penicillin-streptomycin solution (10 000 Units–10 000 µg)
- Gentamicin sulfate (10 mg/ml)
- Fungizone (250 µg/ml)

- Sodium bicarbonate
- Fetal bovine serum
- Nystatin (100x), 10 000 Units/ml
- Hanks balanced salt solution (10x)
- PBS tablets (Dulbecco formula A)

Virus viability is crucial for successful isolation in cell culture. Enveloped viruses are particularly labile. Viruses such as respiratory syncytial virus, influenza and the herpes viruses, to mention but a few, may lose infectivity if they are not adequately protected. Rapid transportation to the laboratory under the proper conditions can greatly enhance effective isolation. Viruses should be transported to the laboratory in the appropriate transport medium (viral transport medium) which can be bought commercially or made up in-house. **transport medium**

- Viral transport medium (500 ml): made up as described and aseptically dispensed into 2 ml aliquots

Hanks balanced salt solution (10x)	50 ml
Sterile reverse osmosis(RO) purified water (Millipore)	415 ml
Fetal bovine serum	10 ml
Penicillin/streptomycin	10 ml
Gentamicin sulfate	2.5 ml
Fungizone	2.5 ml
Few drops of sodium hydroxide 1 M	pH 7.2 –7.4
Hepes buffer	10 ml

- Sucrose phosphate (2SP) *Chlamydia* transport medium (1 liter)

Although *Chlamydia trachomatis* is not classified as a virus, it has always been traditional to cultivate this organism in the virology laboratory as it grows readily in cell culture. Prepare the medium as follows:

Sucrose	68.46 g
Dipotassium hydrogen phosphate	2.088 g
Potassium dihydrogen phosphate	1.088 g

Dissolve in 928 ml RO water. Adjust to pH 7.0 with NaOH and autoclave at 10 lbs for 15 min. Add 10 ml streptomycin sulfate (100 Units/ml final conc.); 2 ml vancomycin (20 Units/ml final conc.); 2.5 ml fungizone (250 µg/ml); 5 ml phenol red (10 mg/ml); 50 ml fetal bovine serum.

- Agar: prepared as a 3 % solution using sterile reverse osmosis water. The agar is dissolved by boiling on a heating block for 5–10 min. In **media** order to keep the agar in a molten state it must be maintained at a temperature greater than 44 °C.

- 2X EMEM maintenance medium (100 ml): 20 ml 10x MEM; 68 ml sterile RO water; 2 ml penicillin/streptomycin; 0.4 ml nystatin (50 U/ml). Just before use add: 8 ml sodium bicarbonate and 2 ml fetal calf serum.
- Growth medium (500 ml stock): prepare stock solution as described and store at 4 °C until required. Sodium bicarbonate and fetal bovine serum are not added to the stock solution until it is about to be used.

MEM (10x)	50 ml
Sterile RO water	433 ml
NEAA (100X)	5 ml
L-glutamine	5 ml
Penicillin-streptomycin	5 ml
Fungizone	2 ml
Sodium bicarbonate 7.5 %	Use at 2–3 ml per 100 ml stock
Fetal bovine serum	Use at 10 %

- Maintenance medium (MM) (500 ml stock)

MEM (10x)	50 ml
Sterile RO water	438 ml
NEAA (100x)	5 ml
Penicillin-streptomycin	5 ml
Fungizone	2 ml
Sodium bicarbonate 7.5 %	Use at 4 ml per 100 ml stock
Fetal bovine serum	Use at 1 %

2.1
Isolation of Viruses in Cell Culture

The ability to culture viruses successfully in the laboratory depends on a number of important factors. These include the sensitivity of the cells used, the viability of the virus, the type of specimens sent to the laboratory and the way they are processed, the culture conditions and the stage of the patients illness when the specimen is taken (Schmidt and Emmons 1989). Even when all these considerations are taken into account it must be remembered that not all viruses can be cultured and there are certain viruses that are very difficult to grow or require very specialized culture conditions (Table 2.1). However most of the more common human

Table 2.1. Cultivation of human viral pathogens

Virus family	Easily isolated viruses	Sensitive cell culture systems for isolation	Associated disease	Preferred clinical specimens for viral isolation
Herpes-viridae	Herpes simplex virus type 1 and 2	African green monkey kidney (VERO); primary monkey kidney (PMK)	Cold sores; genital herpes; congenital infections	Skin swab; genital swab
Adeno-viridae	Adenovirus	Human larynx carcinoma (HEp-2); human embryonic kidney (HEK)	Respiratory infections; eye infections; gastroenteritis	Throat swab; feces; conjunctival swab
Picorna-viridae	Echovirus	Human diploid lung fibroblast (MRC-5); PMK; rhabdomyosarcoma (RD)	Aseptic meningitis; rash; myocarditis; pericarditis	Cerebrospinal fluid (CSF); stools
	Coxsackievirus type B	PMK, VERO, RD	Myocarditis; pericarditis	Pericardial fluid; stools
	Coxsackievirus type A7,A9	PMK, VERO, RD	Aseptic meningitis; rash	Cerebrospinal fluid; feces
	Polioviruses	MRC-5, VERO	Paralysis	Cerebrospinal fluid; stools
Orthomyxo-viridae	Influenza A,B,C	Madin Darby canine kidney (MDCK) PMK	Respiratory infections	Throat swab; nasal swab; nasopharyngeal aspirate
	Difficult viruses to culture			
Paramyxo-viridae	Parainfluenza 1,2,3,4	African green monkey kidney (LLCMK2) PMK, MDCK	Respiratory infections; croup	Nasopharyngeal aspirate; nasal swab
	Mumps	PMK	Parotitis, orchitis	Saliva
	Measles	Primary human embryonic kidney (PHEK)	Masculopapular rash	

Table 2.1. (Continue)

Virus family	Easily isolated viruses	Sensitive cell culture systems for isolation	Associated disease	Preferred clinical specimens for viral isolation
	Respiratory syncytial virus	Hep-2	Bronchiolitis; pneumonia	Nasopharyngeal aspirate
Herpesviridae	Cytomegalovirus	MRC-5	Generalized illness Congenital infections	Urine; throat swab; Buffy coat; bronchiolar lavage
	Varicella zoster	MRC-5	Vesicular rash; chickenpox/ shingles	Vesicular fluid
Reoviridae	Rotavirus	African green monkey kidney (BSC-1) Intestinal epithelium (CACO$_2$)	Infantile gastroenteroitis	Stools
Picornaviridae	Rhinovirus	MRC-5	Common cold	Nasal swab; nasopharyngeal aspirate
Togaviridae	Rubella	Rabbit kidney (RK13)	German measles; congenital Infections	Blood
	Viruses requiring specialized techniques/ facilities			
Hepadenoviridae	Hepatitis A	Fetal rhesus kidney(FRK)	Hepatitis/ jaundice	Stools
Retroviridae	Human immunodeficiency viruses	T lymphocytes	AIDS	Blood
Herpesviridae	Epstein-Barr virus	B lymphocytes	Glandular fever	Blood
Bunyaviridae	Hantaviruses	Subclone of African green monkey kidney cells (Vero E6)	Hemhorragic fever with renal syndrome; pulmonary syndrome	Biopsy material

Table 2.1. (Continue)

Virus family	Easily isolated viruses	Sensitive cell culture systems for isolation	Associated disease	Preferred clinical specimens for viral isolation
Rhabdo-viridae	Rabies virus	Hamster kidney cells; chick embryo cells	Convulsions; paralysis	Urine; CSF; saliva; brain tissue
	Viruses never been isolated in cell culture			
Hepadeno-viridae	Hepatitis B virus		Liver disease	
Flaviviridae	Hepatitis C virus		Liver disease	
Parvoviridae	Parvovirus B19		Damage to fetus; congenital infections	
Caliciviridae	Norwalk -like viruses		Gastroenteritis	
Papovaviri-dae	Human papillo-mavirus		Genital warts	
Filoviridae	Ebola virus		Hemhorragic fever	

pathogenic viruses can be cultured relatively easily provided the proper conditions are satisfied. A wide variety of virus-sensitive cell lines are available either commercially or through one of the national or international cell bank collections such as the American Type Culture Collection (ATCC; Rockville, Maryland, USA) or the European Collection of Animal Cell Cultures (ECACC; Salisbury, Wiltshire, UK).

Cell cultures are normally grown in $25 \, cm^2$, $75 \, cm^2$ or $150 \, cm^2$ plastic tissue culture plates depending on the volume of cells required (Fig. 2.1). These can be seeded into shell vials, tissue culture tubes or microtiter plates for viral isolation. When seeding cells it is necessary to ensure that the appropriate split ratio is used. This depends on the seeding efficiency of the culture. Slow growing cells will have a low split ratio. Split ratios for some commonly used cell lines are shown in Table 2.2.

Fig. 2.1. Commonly used tissue culture vessels

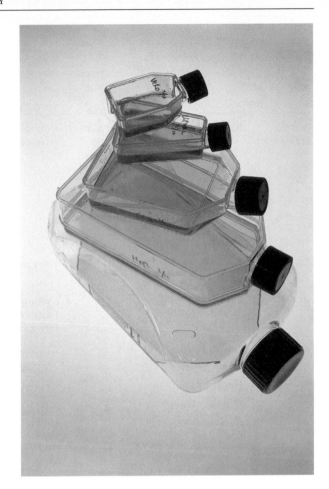

Table 2.2. Split ratio

Cell type	Split ratio
African green monkey kidney (Vero)	1:3
Fetal rhesus kidney (FRK)	1:2
Human epithelium (Hep-2)	1:5
Human embryonic lung (MRC-5)	1:2
Human rhabdomyosarcoma (RD)	1:2
Madin Darby canine kidney (MDCK)	1:4

Anchorage-dependent cells are dispersed from the monolayers using equal volumes (10 ml) of a chelating agent, versene (EDTA), and a proteolytic enzyme, trypsin. Cells may also be removed by scraping with a disposable cell scraper or by using sterile glass beads. Some cells, if seeded at too low a cell density, fail to monolayer. Cell seeding densities may vary depending on the cell type but generally cells should not be seeded at concentrations lower than 1×10^4 cells/ml.

Procedure

Subculturing of Cells

1. Examine the condition of the cell monolayer using an inverted microscope and ensure that the cells are healthy and confluent.

2. Discard the spent growth medium (GM) from the vessel and wash the monolayer twice with 10 ml of prewarmed versene/trypsin wash solution leaving the solution on the monolayer for 20 s with each wash.

3. Discard the wash solution and incubate the flask at 37 °C for 2–5 min. Some cells are difficult to detach from the monolayer and require a

subculturing
procedure

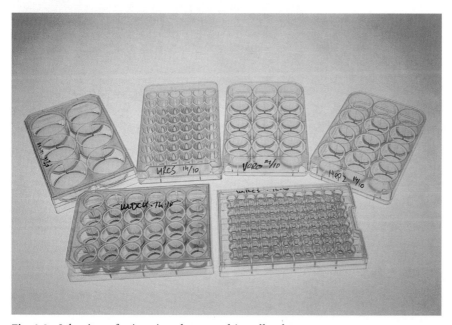

Fig. 2.2. Selection of microtiterplates used in cell culture

more vigorous removal routine. In such cases, the cells may be incubated for longer intervals with the versene/trypsin solution left on.

4. As the cells detach add fresh prewarmed GM to the flask and pipette the cell suspension several times to break up the cell clumps.

5. Count the cells and seed into fresh flasks, tubes or microtiter plates (Figs. 2.2, 2.3).

Fig. 2.3. Stationary rack of inoculated tubes

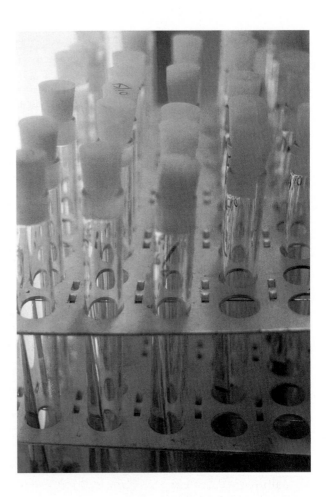

Standard Virus Isolation

1. Seed 1 ml cell suspensions at a concentration of approx. 10^5 cells/ml into standard culture tubes using freshly made GM.
   For viral isolation it is usual to prepare at least three different cell types for inoculation to increase the chances of isolation (Leland and French 1988). While some cell lines have a broad range of viral susceptibility, no single cell line is sensitive to every virus.

2. Incubate the culture tubes in stationary racks at 37 °C and allow to monolayer.

3. When the monolayers are 90 % confluent, discard the GM and replace with 4 ml MM per tube.

4. Label the tubes accordingly and inoculate 0.2 ml freshly prepared specimen into each tube in duplicate.

5. Keep at least two negative control tubes per rack.

6. Incubate at 37 °C. Some viruses (influenza, parainfluenza) need to be cultured at 33 °C with trypsin but without fetal calf serum in the MM.

7. Examine the cultures daily for CPE.

standard isolation

Microtiter Method of Virus Isolation

This method represents an enhancement of conventional monolayer isolation techniques (Fong and Landry 1991; Brumback et al. 1995). Using this method six cell lines are seeded in suspension on microtiter plates thereby improving the sensitivity of virus isolation (O'Neill et al. 1996). Up to four specimens can be inoculated with each plate. The cell lines selected for microtiter plate work should represent a broad range of viral susceptibility. The cells used may be varied accordingly, depending on the focus of isolation. Such a cell line range is shown in Table 2.3.

1. Grow the selected cells in 75 cm^2 plastic cell culture flasks in standard culture medium supplemented with 10 % fetal bovine serum and buffered with 7.5 % sodium bicarbonate.

2. Disperse confluent monolayers of cells by washing twice with equal volumes of a preheated versene/trypsin mixture.

microtiter method

Table 2.3. Recommended seeding densities

Cell line type	Origin	Seeding density/ml
HEL/MRC-5	Human embryonic lung fibroblasts	4×10^5
Hep-2	Human epithelium	6×10^5
E6-VERO	African green monkey	6×10^5
RMK	Rhesus monkey kidney	6×10^5
RD	Human rhabdomyosarcoma	1×10^6
FRK-4r	Fetal rhesus kidney	1×10^6
MDCK	Madin Darby canine kidney	6×10^5

Table 2.4. Scheme for dividing plate into four sections

Rows A–D	Columns 1–6
Rows A–D	Columns 7–12
Rows E–H	Columns 1–6
Rows E–H	Columns 7–12

3. Resuspend in 5 ml prewarmed GM, count the cells using a counting chamber and adjust, if necessary, to the appropriate cell concentration (Table 2.3).

4. Divide the plate (Fig. 2.4) as shown in Table 2.4.

5. Add 90 µl serum-free medium to each well of a U-bottomed cell culture microtiter plate.

6. Add 10 µl of the first specimen to row A wells 1–6.

7. Similarly add 10 µl of the second, third and fourth specimens to row A wells 7–12, row E wells 1–6, and wells 7–12, respectively.

8. Make serial 10-fold dilutions of each specimen from row A to row D and from row E to row H using 10 µl volumes and a multichannel pipette.

Fig. 2.4. Microtiter template

9. Add 25 µl of each different cell suspension to paired columns 1 and 7; 2 and 8; 3 and 9; 4 and 10; 5 and 11 and 6 and 12, respectively. Cover the plate with a plastic lid and incubate at 37 °C in a humidified, 5 % CO_2 atmosphere.

10. Monitor the plates daily for CPE using an inverted microscope.

2.2
Identification of Virus Isolates

Development of characteristic CPE in cell culture is often useful in making a presumptive identification of the viral isolate (Fig. 2.5). This identification would also be based on the specimen source and the cell type in which the virus has grown. However, final identification of the viral isolate needs to be confirmed.

▦ Procedure

Virus Neutralization

Serial dilutions of the isolate are prepared and challenged with specific neutralizing anti-serum. The test is conveniently performed using microtiter plates.

1. Prepare cell cultures in flat bottomed tissue culture microtiter plates by seeding the wells with 200 µl of cell suspension containing 1×10^5 cells/ml.

   neutralization procedure

2. Cover the plate and incubate at 37 °C in a moist CO_2 atmosphere.

3. When the cells are confluent, replace GM with 100 µl MM.

4. Make 10-fold serial dilutions of the virus isolate in clean sterile glass containers.

5. Reconstitute the specific anti-serum to a working concentration.

6. Mix equal volumes (100 µl each) of virus and anti-serum and incubate at 37 °C for 1 h. Make a positive control with virus and diluent only.

7. Add 100 µl serum-virus mixture to each well and incubate at 37 °C in a moist CO_2 atmosphere.

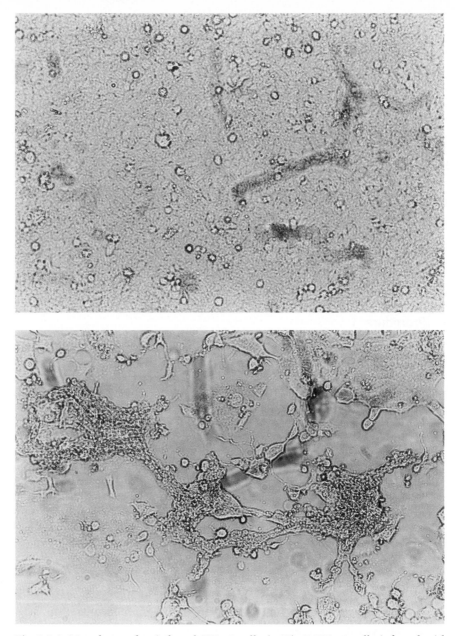

Fig. 2.5.A Monolayer of uninfected HEp 2 cells (×20); **B** HEp 2 cells infected with respiratory syncytial virus showing typical syncytial cytopathic effect (×40)

8. After a number of days examine the plate using an inverted microscope. A confirmed identification of the viral isolate is made when
   development of CPE has been effectively inhibited by the specific anti-
   serum.

Immunofluorescence (IF)

This procedure is very useful for confirming cell culture isolates and has
been applied to great effect in identifying a wide variety of viral antigens. It can also be used directly to detect viruses in clinical specimens.
Two methods are used, direct and indirect (Figs. 2.6–2.8). In the direct
assay, infected cells are harvested, washed and fixed onto the wells of a
Teflon coated glass slide (Fig. 2.6). The fixed cells are stained with a specific anti-viral monoclonal antibody which is conjugated to a fluorescein
dye (FITC). Unbound antibody is washed off and the slide is read using
a fluorescent microscope. The indirect method involves using an extra
antibody and incubation step (Fig. 2.8). The procedure for the direct
method is as follows.

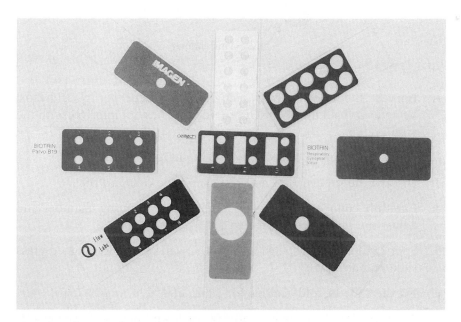

Fig. 2.6. Selection of teflon coated immunofluorescence slides

Fig. 2.7. Direct immunofluorescence

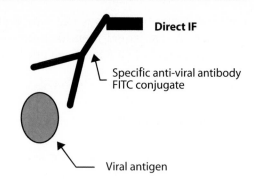

Direct IF

Specific anti-viral antibody
FITC conjugate

Viral antigen

Fig. 2.8. Indirect immunofluorescence

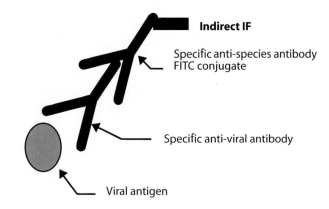

Indirect IF

Specific anti-species antibody
FITC conjugate

Specific anti-viral antibody

Viral antigen

**immuno-
fluorescence**

1. Harvest the infected cells using a cell scraper or glass beads. Versene-trypsin may be used provided it does not interfere with the viral antigen.

2. Spin the harvested cells gently at 300–400 g for 5 min.

3. Discard the supernatant and resuspend the cells in PBS.

4. Carefully count the cells using a disposable counting chamber and adjust the cell concentration to 1×10^6 cells/ml.

5. Spot 15 µl of cell suspension onto the well of a Teflon coated glass slide and air dry.

6. Fix the slide in cold acetone for 5 min at 4 °C.

7. After fixation the slide is relatively stable and may be stored at 4 °C for up to 24–48 h or frozen in a sealed container at −20 °C to await staining at a later date.

8. Add 20 µl specific conjugate (monoclonal antibody tagged to FITC) to the well of the slide and incubate in a moist chamber at 37 °C for 30 min.

9. Wash the slide thoroughly for 10 min by immersing in a staining trough containing PBS.

10. Allow the slide to air dry, mount in buffered glycerol and cover with a clean plastic coverslip.

11. Read with blue fluorescent light at 20×–40× magnification.

Shell Vial Assay

The isolation of many viruses which would normally take some time to grow in culture can be greatly enhanced by this method (Grundy 1988; Stirk and Griffiths 1988). The use of shell vials was first described by Gleaves (1984). The procedure employs a spin amplification technique to facilitate entry by the virus into the cell. After a short incubation period (24–48 h) intracellular viral antigen may be detected before the appearance of a CPE, using IF, thereby resulting in earlier viral detection than with standard cell culture.

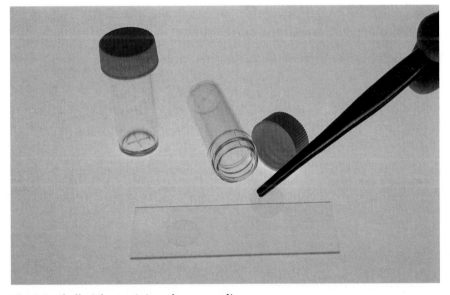

Fig. 2.9. Shell vial containing glass coverslip

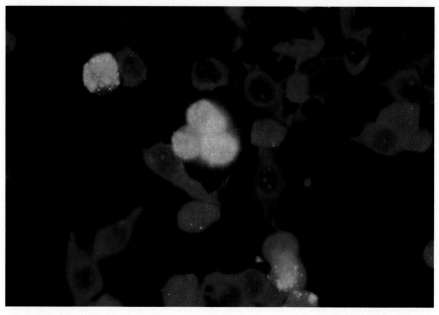

Fig. 2.10. Direct immunofluorescence (IF) of parainfluenza type 2 infected FRK cells showing cytoplasmic and wholecell immunofluorescence (×40)

assay

1. Prepare the coverslips (Fig. 2.9) by seeding the shell vials at 1×10^5 cells/ml with 2 ml of cell suspension.

2. Allow the cells to monolayer on the coverslip.

3. When a healthy monolayer has formed discard the GM and inoculate the shell vial with 0.2 ml prepared specimen.

4. Spin the vial at 700 g for 1 h.

5. Add 2 ml of freshly prepared MM to the vial and incubate at 37 °C in a moist 5 % CO_2 atmosphere.

6. After 24–48 h fix the cells on the coverslip with acetone and perform a direct IF assay (see Fig. 2.10 for example).

2.3
Quantitation of Viruses

Many approaches are available to determine the concentration of viruses in a given tissue. Infectivity assays, chemical assays and direct counting of virus particles using electron microscopy are among those used. Perhaps the most widely utilized approach is the use of infectivity assays. These enable the virologist to calculate the number of infectious viral particles per unit volume, called the "virus titer." Two types of infectivity assays using cell culture systems are described. Both rely on the virus being infectious and capable of producing a CPE in a sensitive cell culture.

▨ Procedure

TCID$_{50}$ (Tissue Culture Infectious Dose)

This is a quantal assay which determines the dilution of virus required to infect or cause CPE in 50 % of inoculated cell cultures. The assay can be carried out in culture tubes or 96-well microtiter plates.

1. Make up a 50 ml cell suspension containing 1×10^5 cells/ml.   **tube cultures**

2. Seed 1 ml of cell suspension into each of 50 sterile cell culture tubes and allow to monolayer by incubating at 37 °C.

3. When 90 % of a cell monolayer has formed, the cultures are ready for viral inoculation. Prepare 10-fold (log) serial dilutions of virus suspension in MM diluent (10^{-1}–10^{-8}).

4. Remove the GM from each tube and replace with 1 ml of viral suspension. Inoculate five culture tubes for each virus dilution. Set up five control tubes which will contain MM diluent only.

5. Incubate at 37 °C.

6. Monitor the cell cultures daily for signs of CPE.

Note: Be aware that some viruses grow quickly while others may take some time to culture.

The development of a CPE may be scored according to the following regime shown in Table 2.5. As soon as a CPE becomes established in the

Table 2.5. Grading scheme for CPE

Grade of CPE	Percentage of cell monolayer infected
–	No CPE formed
±	<30%
+	30–50%
++	≤75%
+++	>75% of cell monolayer infected

Table 2.6. Example of data used to calculate TCID$_{50}$

Virus Dilution	Infected cultures	Total cultures Infected (X) (summing up)	Cumulative non-infected cultures (summing down) (Y)	$\dfrac{X}{X+Y}$	Percent infected
10^{-1}	5/5	21	0	21/21	100
10^{-2}	5/5	16	0	16/16	100
10^{-3}	4/5	11	1	11/12	91.7
10^{-4}	4/5	7	2	7/9	77.7
10^{-5}	2/5	3	5	3/8	37.5
10^{-6}	1/5	1	9	1/10	10
10^{-7}	0/5	0	14	0/14	0
10^{-8}	0/5	0	19	0/19	0

tubes and no further CPE appears to be developing, the TCID$_{50}$ may be calculated.

Calculate the TCID$_{50}$ by determining the dilution of virus causing CPE in 50% of the inoculated cell cultures (50% end point dilution). The appearance of any grade of CPE in the cell monolayer is indicative of infection. The data from Table 2.6 is used by way of example to enable calculation of the TCID$_{50}$.

Calculation of TCID$_{50}$

calculation   The calculation is made according to the Reed-Muench method. It is clear from Table 2.6 that the 50% end point lies between the virus dilution of 10^{-4} (77.7%) and 10^{-5} (34.5%). To find out exactly where the end point lies the proportionate distance between these two dilutions is first calculated.

$$\text{Proportionate distance} = \frac{(\% \text{ positive above} 50\%) - 50\%}{(\% \text{ positive above } 50\%) - (\% \text{ positive below } 50\%)}$$

$$= \frac{77.7 - 50\%}{77.7 - 37.5\%} = 0.7$$

The 50 % end point dilution is now calculated as follows:

Log $TCID_{50}$ = (log dilution above 50 %) + (proportionate distance×log dilution*)

(*Note: 10-fold dilutions were used)

$$= -4 + (0.7 \times -1) = -4.7$$

$$TCID_{50} = 10^{-4.7}$$

Since the virus inoculation was 1 ml per tube, the viral titer is therefore $10^{4.7}$ $TCID_{50}$/ml.

Plaque Assay

This is a focal assay which is based on the ability of infectious virus particles to form small areas of cell lysis or foci of infection on the cell monolayer. This is achieved by first adsorbing the virus onto a confluent cell monolayer and then overlaying the monolayer with agar or some nonviscous medium (methylcellulose). The overlay medium restricts the spread of secondary infection so that only areas of the cell monolayer adjacent to the initially infected cells will become infected and form plaques or small areas of CPE (Fig. 2.11). These plaques can then be counted and the viral titer calculated. Plaque assays can be carried out in 24-well cell cluster plates or petri dishes.

1. Seed 2 ml of cell suspension containing 1×10^5 cells per ml into the wells of a 24-well cell cluster plate and allow to form a healthy, confluent monolayer (2–3 days).

2. Prepare overlay medium by combining 2x MM and agar.

3. Equilibrate in a 50 °C water bath and ensure no clumping occurs.

4. Prepare 10-fold (log) serial dilutions of virus suspension in MM diluent (10^{-1}–10^{-8}).

plaque assay

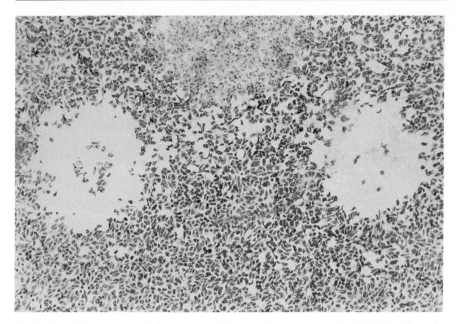

Fig. 2.11. Plaques of parainfluenza type 3 in VERO cells stained with 2 %(v/v) methylene blue (×20)

5. Remove the GM from the wells of the cluster plate.

6. Inoculate duplicate wells with 0.2 ml virus suspension (two wells for each dilution) and adsorb for 1 h at 37 °C . Tilt the plate at 15-min intervals to prevent the monolayer from drying out and to ensure an even distribution of inoculum.

7. Wash the infected monolayers with prewarmed, sterile PBS to remove unadsorbed virus.

8. Remove the overlay medium from the water bath and allow to cool before pouring onto the monolayer. Ensure the overlay medium is not allowed to overcool or it will solidify prematurely.

9. Carefully add 2 ml overlay into each well and incubate at 37 °C to allow plaque formation.

10. When plaques have formed, fix cell monolayers with 30 % formaldehyde for 20 min.

11. Carefully remove the overlay and stain the monolayer with 1 % methylene blue solution.

12. Remove the stain and count the number of plaques.

13. Calculate the viral titer.

Calculation of Virus Titer from Plaque Assay

The infectivity of the virus is expressed as plaque-forming units (PFU) per ml and is calculated from the number of plaques observed at the appropriate dilutions.

Count the total number of plaques for each dilution, i.e., for each pair of wells since they have been inoculated in duplicate. From Table 2.7 it is clear that there were too many plaques to count at the lower dilutions. The first countable plaques were therefore observed at a dilution of 10^{-3}. Since the initial inoculum per well was 0.2 ml and the wells were inoculated in duplicate it follows: *calculation*

0.4 ml	of a 10^{-3} dilution contains	121 PFU
0.04 ml	of a 10^{-3} dilution contains	66 PFU
0.004 ml	of a 10^{-3} dilution contains	24 PFU
0.0004 ml	of a 10^{-3} dilution contains	8 PFU
0.4444 ml	of a 10^{-3} dilution contains	219 PFU

Therefore 0.4444 ml of undiluted virus contains 219×10^3 PFU $= 2.19 \times 10^5$ PFU. Therefore 1 ml of undiluted virus contains $(1/0.4444) \times 2.19 \times 10^5$ PFU $= 4.93 \times 10^5$ PFU/ml.

Table 2.7. Example of data used to calculate plaque titer

Dilution	Count 1	Count 2	Total count
10^{-1}	>100	>100	>200
10^{-2}	>100	>100	>200
10^{-3}	49	72	121
10^{-4}	30	36	66
10^{-5}	15	9	24
10^{-6}	3	5	8
10^{-7}	0	0	0
10^{-8}	0	0	0

References

Enders JF, Weller TH, Robbins FC (1949) Cultivation of the Lansing strain of polio-myelitis virus in cultures of various human embryonic tissues. Science 109: 85–87

Grundy JE, Super M, Griffiths PD (1988) Symptomatic cytomegalovirus infection in seropositive kidney recipients: reinfection with donor virus rather than reactivation of recipient virus. Lancet ii: 132–135

Leland DS, French MLV (1988) Virus isolation and identification. In: Lennette EH, Halonen P, Murphy FA (eds) Laboratory diagnosis of infectious diseases; principles and practice. Springer, Berlin, Heidelberg, New York, pp 39–59

Stirk PR, Griffiths PD (1988) Comparative sensitivity of three methods for the detection of cytomegalovirus lung infections. J Virol Methods 20: 133–135

Brumback BG, Cunningham DM, Morris MV, Villavicencio JL (1995) Rapid culture for influenza virus types A and B in 96 well plates. Clin Diagn Virol 4: 251–256

Fong CKY, Landry ML (1991) Advantages of multiple cell culture systems for the detection of mixed – virus infections. J Virol Methods 33: 283–290

O'Neill HJ, Russell JD, Wyatt DE, McCaughey C, Coyle PV (1996) Isolation of viruses from clinical specimens in microtitre plates with cells inoculated in suspension. J Virol Methods 62: 169–178

Gleaves CA, Smith TF, Shuster EA, Pearson GR (1984) Rapid detection of cytomegalovirus in MRC-5 cells inoculated with urine specimens by low -speed centrifugation and monoclonal antibody to an early antigen. J Clin Microbiol 19: 917–919

Schmidt NJ (1989) Cell culture procedures for diagnostic virology. In: Schmidt NJ, Emmons RW (eds). Diagnostic procedures for viral, rickettsial and chlamydial infections. American Public Health Association, Baltimore, pp 51–100

Schmidt NJ, Emmons RW (1989) General principles of laboratory diagnostic methods for viral, rickettsial and chlamydial infections. In: Schmidt NJ, Emmons RW (eds). Diagnostic procedures for viral, rickettsial and chlamydial infections. American Public Health Association, Baltimore, pp 1–36

Reed LJ, Muench HA (1938) A simple method of estimating fifty percent endpoints. Am J Hyg 27: 493–497

Suppliers

Costar Corp.
Cambridge MA 01720 U.S.
Tel. 1 617 868 6200
Fax. 1 617 868 0355
Toll Free 1 800 492 1110

Nunc. A/S
Kemstrupvej 90
Roskilde DK 4000 Denmark
Tel. 45 42 359065
Fax. 4542 350165

Bibby Sterilin Ltd, Tilling Drive
Stone, Staffordshire ST15OSA UK
Tel. +44(0) 1785–812121
Fax. +44(0) 1785–813748

BDH Laboratory Supplies
Poole Dorset UK BH151TD
Tel. (01202) 660–444
Fax. (01202) 666–856

Becton Dickinson European Division
Chemin des Sources
BP 37
Meylan CEDEX, France
76 416464 (phone)

Hycor Biomedical Inc.
18800Von Karman Ave.
Irvine, California 92715–1517, USA
Tel. 714–440–2000
Fax. 714–440–2222

Biowhittaker,Inc.
Biggs Ford Rd.
Walkersville,Maryland 21793, USA
Toll Free 800 638 8174
Fax. 301 845 8338

American Type Culture Collection, (ATCC)
Rockville, Maryland, U.S.A.
European Collection of Animal Cell Cultures (ECACC) Salisbury,
Wiltshire, UK

Life Technologies Ltd.(Gibco BRL)
Fountain Drive
Inchinnan Business Park
Paisley PA4 9RF
Tel. 0141 814 6100
Fax. 0141 814 6317
Free phone orders 0800 269 210

Chapter 3

Screening for Mycoplasma Contamination in Animal Cell Cultures

MARY HEENAN[*][1] and MARTIN CLYNES[1]

Introduction

Mycoplasma is the trivial name of the class *Mollicutes*, a group of minute wall-less bacteria comprising over 120 species. Their lack of a cell wall distinguish them from classical bacteria. They are the smallest self-replicating prokaryotic organisms and carry the smallest genomes recorded in prokaryotes, ranging from about 600 to 1700 kb, depending on the species and strain. They are very fastidious organisms, requiring complex growth medium and many species have not been cultivated as yet *in vitro*. As a consequence of their small genome, mycoplasma lack many biosynthetic pathways and are dependent on their artificial culture media or host cells for their supply of many essential nutrients.

Mycoplasma are common and serious contaminants of cell cultures and remain one of the major problems encountered in biological research and diagnosis and in the biotechnology industry, where cell cultures are used. Mycoplasma contamination is widespread, with the incidence varying depending on the cell line, detection method used and the quality control practised by the laboratory. An infection of cell cultures may persist for an extended time without apparent cell damage and can affect virtually every parameter within a cell culture system (McGarrity et al. 1985) such as changes in cell metabolism, alterations in cell karyotype and retardation of cell growth (Boyle et al. 1981; Sasaki et al. 1984; McGarrity et al. 1984).

Mycoplasma are parasites and many are pathogens of a variety of animal hosts. In humans, mycoplasma are mostly surface parasites, colonising the epithelial lining of the respiratory and urinogenital tracts; how-

* *Correspondence to* Mary Heenan: phone +353–1–7045691; fax +353–1–7045484; e-mail heenanm@dcu.ie
[1] Mary Heenan, Martin Clynes, National Cell and Tissue Culture Centre/BioResearch Ireland, Dublin City University, Glasnevin, Dublin 9, Ireland

ever several human pathogenic mycoplasma have also been identified, including *M. pneumoniae, M. pulmonis* and *M. gallisepticum*. Although mycoplasma are generally membrane parasites, avidly adhering to the outer surface of the host cell membrane, some have been shown to penetrate into the host cells (Lo et al. 1991).

Mycoplasma are pleomorphic, but are generally coccoid forms of approximately 0.3 μm in diameter. Because of their small size and pliability, mycoplasma can pass through 0.22 μm pore size membrane filters usually used in cell culturing (Hay et al. 1989). Although the class *Mollicutes* consists of more than 120 species (Tully 1992), only approximately 20 species have been isolated from cell cultures. It is estimated that between 5 % and 35 % of all cell lines are infected with mycoplasma, with the species *A. laidlawii, M. arginini* (bovine), *M. hyorhinis* (porcine), *M. fermentans* and *M. orale* (human) representing the majority (95 %) of contaminating isolates (Hay et al. 1989). The main sources of contamination of clean cultures are thought to be previously infected cultures (Barile 1979).

A number of methods exist for the detection of mycoplasma contamination of cell cultures. They include direct culture procedures, DNA fluorochrome stains, immunodetection methods, ELISA procedures, electron microscopic analysis and molecular biology tests, including PCR. A number of extensive reviews on mycoplasma contamination in cell cultures have been published, including detailed mycoplasma detection and identification protocols (Kahane and Adoni 1993; Tully and Razin 1983). Only the culture method is a direct method of mycoplasma detection. All other procedures are indirect assay methods. However, each method has its advantages and limitations. *M. hyorhinis* is difficult to grow on agar (Uphoff et al. 1992). The DNA fluorochrome method cannot be used for direct identification of contamination. PCR does not differentiate dead from viable mycoplasma and the range of mycoplasma species detectable is limited by the PCR probes used.

This chapter describes methods of mycoplasma detection by DNA fluorochrome staining and by the direct culture method. These are the two most widely used techniques, can be routinely used in all cell culture laboratories and identify mycoplasma contamination by a broad range of mycoplasma species.

Collection of Samples for Mycoplasma Testing

The sample of medium collected from the cell culture for mycoplasma testing should be in the best possible condition to allow detection of the microorganism. Therefore, it should have the maximum possible number of mycoplasma present in the sample, it should not contain inhibitors of mycoplasma growth and it should be stored in a way that facilitates detection of the mycoplasma.

- Prior to removing a sample for mycoplasma analysis, cells should be passaged a minimum of three times after thawing to facilitate the amplification of low-level infection.

- Although mycoplasma display absolute resistance to penicillin, other antibiotics may mask low-level infection. Therefore, cells should be subcultured for three passages in antibiotic-free medium before testing the culture for mycoplasma infection.

- The optimum conditions for harvesting supernatant for mycoplasma testing occur when the culture is in log phase, near confluency, and the medium has not been renewed in 2–3 days.

- At least 5 ml of supernatant should be removed from the culture for analysis. Ideally, samples containing cells should be removed for analysis in addition to centrifuged supernatant, as mycoplasma adhere to cells and therefore the concentration of mycoplasma on the cell surface may be greater than that found in the growth medium. Cells may be removed from adherent cultures by decanting all but 2–3 ml of supernatant and scraping some of the monolayer into the liquid with a cell scraper.

- Samples of supernatant should preferably be analyzed immediately; however samples may be stored at −70 °C until analyzed.

3.1
DNA Fluorochrome Stain

This indirect method used for the detection of mycoplasma infection is based on the method of Chen (1977). It involves growing a mycoplasma-free indicator cell line in medium which has previously been exposed to the cell line under investigation. The indicator cell line is then fixed and stained using a fluorescent stain which is specific for DNA. The nucleus of the indicator cells fluoresce when viewed under UV light and any extranuclear fluorescence is an indication of possible mycoplasma contamination.

This is a simple, inexpensive and sensitive test for mycoplasma contamination. It will detect a broad range of species, but in general it is most effective in detecting the cytoadsorbing mycoplasma. However, it is an indirect test and the presence of extranuclear fluorescence is not diagnostic for mycoplasmas, because it may represent other prokaryotic DNA and so further characterization may be necessary. In general, extranuclear mycoplasma DNA will appear as regular small spherical dots or chains of fluorescence and is much smaller than other microbial cell contamination. It is important in the analysis of the slides that a range of fields are viewed, in order to detect low level contamination and also because, in some mycoplasmas infections, only a small number of cells have adsorbed mycoplasma.

In this procedure, the indicator cell line used is NRK, a normal rat kidney line. Vero cells (an African green monkey kidney cell line) and NIH 3T3 or 3T6 (mouse fibroblast cell lines) are also commonly used as the indicator cells. The main requirements of the indicator cells are that they are mycoplasma free, they are relatively robust and their cellular integrity is maintained during the fixing process (if cells and nuclei are ruptured, nuclear DNA may give false positive results) and that they possess a good nucleus to cytoplasma ratio, so that extranuclear fluorescence is easily determined.

The two DNA-binding stains commonly used in mycoplasma testing are bisbenzimidazole (Hoechst 33258) and 4'-6-diamidino-2-phenyl-indole (DAPI) (Russell et al. 1975). This procedure uses Hoechst 33258 as the DNA stain.

Materials

- NRK cells in log phase (ATCC; CRL 1570)
- 35 mm petri dishes
- Microscope slides
- Coverslips
- Sterile PBS
- Hoechst 33258
- Mounting medium
- *Mycoplasma orale* (ATCC; 45539)

– *Mycoplasma hyorhinis* (ATCC; 23234)
– Carnoy's fixative
– Carnoy's/PBS
– Fluorescent microscope

preparation of
solutions and
equipment

Hoechst 33258 Stock

Note: This stain is carcinogenic. Care should be taken when handling the stain, gloves should be worn and waste must be disposed of according to local regulations.

1. Weigh out 1 mg of powder and dissolve in 20 ml of sterile PBS by stirring on a magnetic stirrer for 30 min.

2. Filter sterilize with a 0.22 µm filter.

3. Dispense into 1.0 ml aliquots, cover in aluminium foil and store in the dark at −20 °C.

4. A fresh stock should be prepared every 6 months, so an expiration date should be written on the aliquots.

5. Prepare a working stock (50 ng/ml) of Hoechst 33258 by thawing a vial of 50 µg/ml stock solution and diluting it in sterile PBS to 50 ng/ml. Cover with aluminum foil and use immediately.

Mounting Medium

1. Autoclave or filter sterilize 22.2 ml of 0.1 M citric acid and 27.8 ml of 0.2 M disodium phosphate.

2. Mix with 50 ml of glycerol.

3. Adjust the pH to 5.5 with 1 M HCl.

4. Filter sterilize through a 0.22 µm filter, and refrigerate until required.

Note: Check pH perodically, as it is critical for optimal fluorescence of the DNA stain.

Preparation of Coverslips and Slides

1. Soak the required number of coverslips and slides in detergent for 1 h.

2. Rinse thoroughly in tap water for 5 min and in deionized water for a further 5 min.

3. Rinse the coverslips in methanol and wipe dry with lint-free tissue. The slides can be refrigerated in methanol until required (use a spark-proof refrigerator). Then, they are also wiped with lint-free tissue.

4. Place the cleaned coverslips in a large glass petri dish with a lid. Cover the dish in aluminium foil and sterilize in a hot air oven at 160 °C for at least 1 h. Allow to cool before use.

Carnoy's Fixative

1. Mix 1 part glacial acetic acid to 3 parts methanol.

2. Place at −20 °C for 30 min before use and use within 4 h of preparation.

Carnoy's/PBS

1. Make up 1:1 (v:v) Carnoy's fixative: PBS

2. Refrigerate for 30 min before use and use within 4 h of preparation.

Procedure

1. Sterilize a forceps by dipping in 70 % alcohol and flaming, and place coverslips into 35 mm petri dishes (one into each). Prepare two coverslips per sample (including positive and negative controls).

staining for mycoplasma infection

2. Trypsinize a growing culture of NRK cells. Resuspend at a concentration of 2×10^3 cells/ml in DMEM +5 % FCS +1 % L-glutamine. Add 1 ml of cell suspension to each coverslip. Incubate at 37 °C in a CO_2 incubator overnight to allow cell attachment.

3. Thaw one aliquot of *M. orale* positive culture and dilute to the concentration required for the positive control. This concentration of

the mycoplasma stock should have been previously tested to ensure that the infection can be detected by the indirect test procedures in use. The procedure should be repeated for *M. hyorhinis.*

4. Check the coverslips microscopically to ensure that the cells have attached. Add 1 ml of conditioned medium from the test samples to each of two labeled coverslips. Suitable codes should be used to differentiate between duplicate samples.

5. Set up duplicate positive and negative controls. The positive controls are samples of *M. orale* and *M. hyorhinis* (two common mycoplasma contaminants of cell cultures).The negative control is NRK growth medium. **Note:** The presence of positive and negative controls are essential to monitor the validity of the test system.

6. Incubate samples in a CO_2 incubator until the indicator cells reach approximately 50 % confluency (approximately 5 days).

7. Remove the waste medium from the plates. Do not allow the coverslips to dry out before completing fixation. Wash the coverslips twice with sterile PBS. Wash once with Carnoy's/PBS.

8. Add 2 ml of Carnoy's fixative to each coverslip and allow to fix for 10 min. Remove and dispose of the fixative and allow all the samples to air dry for at least 1 h. (Samples may be left for up to a week before staining).

9. Add 2 ml of Hoechst 33258 working stock (50 ng/ml) to each sample (exercise caution when handling the stain). Only stain approximately ten coverslips at a time to prevent over-exposure to the stain. Allow the samples to stain for 10 min. (The stain is light sensitive, therefore staining should be performed in very low light.) After 10 min, carefully remove the stain and wash each coverslip thoroughly with 3×2 ml of sterile distilled water. **Note:** Do not use a salt solution in this rinsing step as it can increase background fluorescence and/or cause artifacts.

10. Mount the coverslips onto the prelabeled slides with a drop of the mounting medium. Dry off the bottom of each coverslip carefully with lint-free tissue before mounting top side down on the slide. Place each of the preparations into a slide folder and store in the dark until ready to analyze. Preferably, the samples should be analyzed on the day of staining, but may subsequently be stored in the dark for further analysis if necessary.

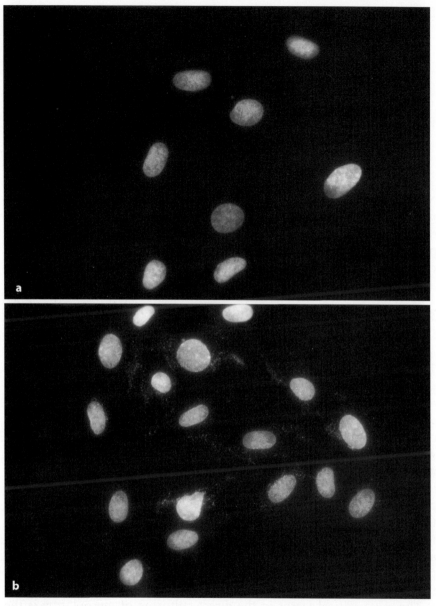

Fig. 3.1a, b. Hoechst 33258 stained NRK indicator cells, used in the analysis of a mycoplasma-free sample (**a**) and a sample contaminated with *Mycoplasma hyorhinis* (**b**)

11. Examine samples by flourescent microscopy for mycoplasma contamination. Optimum fluorescence excitation of the Hoechst 33258 stain occurs at approximately 360 nm, whereas emission peaks at approximately 475 nm. Therefore, excitation is performed with illumination from a high-pressure mercury lamp. Emission is best viewed through filter combinations that exclude light below 460 nm.

12. The test results should be deemed invalid if the positive or negative controls produce unexpected results. A sample of a DNA fluorochrome stain of mycoplasma-free and mycoplasma-infected cells is shown in Fig. 3.1a and 3.1b, respectively.

Note: Fading of the Hoechst 33258 stain may occur under the conditions required for analysis of the slides. However, the fluorescence half-life of the stain can be increased by the presence of the antioxidant p-phenylenediamine (Sigma; P6001) in the mounting medium at a concentration between 0.005 % and 0.1 % (w/v) (Battaglia et al. 1994).

3.2
Direct Culture Procedure

This direct method of mycoplasma detection involves growing the mycoplasma on a selective agar and observing the colonies by microscopical examination. Mycoplasma can reproduce in a cell-free medium where, on agar, they exhibit a characteristic colony morphology, with the center of the colony embedded beneath the surface. Mycoplasma colonies in agar are viewed by microscope. Most mycoplasma colonies appear as round colonies (approximately 100–1000 µm in diameter) with a dense center and a less dense periphery, giving the appearance of a fried egg (Fig. 3.2). Mycoplasma colonies have been isolated, however, that do not conform strictly to this appearance. They may appear to lack a distinct periphery and to be totally embedded in the agar. These colonies are usually very small and appear granular. Common errors made when attempting to identify mycoplasma colonies on agar are usually attributed to air bubbles trapped when the plate is poured or tissue culture cells added to the agar in sample.

The most critical factor in the direct culturing of mycoplasma is the standardization and pretesting of medium components and growth conditions. The quality of components may vary with manunfacturers and batches, so each component must be pretested and found satisfactory for mycoplasma growth promotion before use. A number of variations to the mycoplasma culture agar and broths are cited in the literature

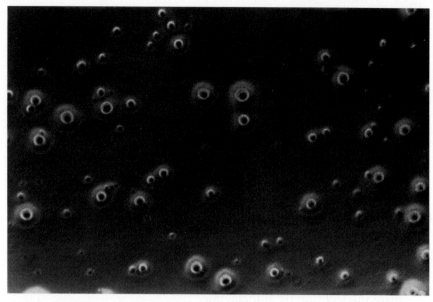

Fig. 3.2. *Mycoplasma orale* colonies. The center of the colony is embedded in the agar, giving a "fried egg" type morphology

(Kahane and Adoni 1993; Tully and Razin 1983). However, as long as the agars and broths are shown to support the growth of a range of common cell culture-infecting mycoplasma and the internal assay controls are as expected, any of these formulations are appropriate. The protocol outlined here is that used within this laboratory and which has been shown to support the growth of both *M. arginini* and *M. orale*. Variations to the procedure include the addition of penicillin to the agar and broth (to decrease the risk of microbial contamination) and the inclusion of DNA and different types of serum (horse serum and pig serum in both the standard and heat inactivated forms).

Materials

- Mycoplasma broth base (Becton-Dickinson BBL; 11458)
- Mycoplasma agar base (Becton-Dickinson BBL; 11456)
- Yeast extract (Gibco BRL; 20047–056)
- *Mycoplasma orale* (American Type Culture Collection (ATCC); 45539)
- *Mycoplasma arginini* (ATCC; 23243)
- Anaerobic incubator or jar at 36±1 °C

Preparation of Stock Solution

1. To 225 ml distilled water add 12.5 g dextrose and 2.5 g L-arginine HCl.

2. Mix the ingredients at 37 °C until dissolved. Bring the final volume up to 250 ml with distilled water.

3. Sterilize the solution by filtration using a 0.22 µm filter.

4. Dispense into 25 ml aliquots and store at −70 °C until required.

Mycoplasma Broth Medium

1. Add 3.7 g mycoplasma broth base and 0.005 g phenol red to 150 ml distilled water. Heat to dissolve.

2. Sterilize the solution by autoclaving.

3. Allow the broth mixture to cool to room temperature.

4. Aseptically add 50 ml serum, 25 ml yeast extract (15 % w/v) and 25 ml stock solution. Mix the solution completely. The final pH should be adjusted to 7.2–7.4.

5. Dispense 5 ml aliquots of the broth medium into sterile test tubes and store at 4 °C. Use within 3–4 weeks.

Mycoplasma Agar Medium

1. Add 5.95 g mycoplasma agar base to 150 ml water. To dissolve, bring the solution to the boil.

2. Sterilize the solution by autoclaving.

3. Place the sterilized medium in a water bath at 50 °C. Place 50 ml serum, 25 ml yeast extract (15 % w/v) and 25 ml stock solution in a water bath at 37 °C. Allow the components to equilibrate at these respective temperatures.

4. Aseptically add the serum, yeast extract and stock solution to the medium. Mix well.

5. Proceed immediately to dispense 10 ml aliquots into 60 mm petri dishes. Add the fluid as quickly as possible in order to eliminate the problem of the agar solidifying before the medium is completely dispensed. The plates may be stored at 4 °C for up to 4 weeks.

Procedure

1. Samples, including positive and negative controls, should be set up in duplicate. Agar plates should be labeled on both the top and base, and it should be possible to differentiate between replica plates.

microscopy for mycoplasma contamination

2. Thaw one aliquot of *M. orale* positive culture and dilute to the desired concentration. The chosen concentration should have been pretested to ensure that the infection can be detected by the direct culture procedure. The procedure should be repeated for *M. arginini.*

3. Inoculate agar plates with 100 µl of test samples and the relevant positive and negative controls. Incubate anaerobically in a 5 % CO_2/95 % N_2 atmosphere, at 36±1 °C for 3 weeks (36 °C is the optimum growth temperature for mycoplasma). The anaerobic environment may be achieved by using Becton-Dickinson BBL Mycrobiology Systems gas packs (catalogue number 70304), catalysts (catalogue number 70303) and jars. After approx. 7 days and again after 3 weeks, examine agar plates microscopically for evidence of mycoplasma contamination.

4. Inoculate broths with 0.25 ml of test sample and the relevant positive and negative controls. Incubate broths for 5–7 days. (Arginine-hydrolyzing mycoplasma release ammonia, alkalizing the broth and the shift in pH causes the red color to be intensified. Glucose-fermenting mycoplasma acidify the broth and the red color changes to yellow. Therefore, visual examination of the broths may facilitate the detection of mycoplasma and it allows amplification of low-level mycoplasma contamination prior to transfer to agar plates.)

5. Transfer 100 µl of broth culture to an agar plate. Incubate plates anaerobically in a gas jar for 3 weeks, as in step 3 above. After approximately 7 days and again after 3 weeks, examine agar plates microscopically for evidence of mycoplasma contamination.

6. A result should be deemed invalid if the positive or negative controls give unexpected results. A typical fried-egg type mycoplasma colony is shown in Fig. 3.2.

Results

Identification of Mycoplasma Colonies

Mycoplasma colonies can be distinguished from other bacterial colonies and debris by either disrupting the surface of the agar or by staining with Dienes' stain.

- If the surface of the agar is gently disrupted using a cell scraper, the mycoplasma colonies (which are embedded in the agar) should not be disturbed, while other microbial colonies and debris should be moved.

- Dienes' stain (Dienes et al. 1948) may also be used to distinguish between mycoplasma colonies and pseudocolonies/debris. Dienes' stain is prepared by dissolving 2.5 g of methylene blue (Sigma M9140), 1.25 g of azure II (Sigma A2507), 10 g of maltose and 0.25 g of sodium carbonate in 100 ml of distilled water. The stain may be stored for up to 12 months. An area of the agar adjacent to the suspect colony is swabbed with a cotton swab soaked in Dienes' stain. The stain will diffuse to the colony and the colony is then microscopically examined. Stained mycoplasma colonies have dense blue centers and light blue peripheries. Bacterial colonies are also stained but are decolorized in 30 min, in contrast to the mycoplasma colonies, which do not become decolorized. A possible source of error could be a colony of stained microorganisms composed of dead bacteria (which cannot decolorize the stain). However, the morphology of a bacterial colony can usually be distinguished from that of a mycoplasma colony, even without Dienes' stain.

Positive Control Stocks

1. Stocks of mycoplasma may be grown, stored and used as positive control stocks in future work. Mycoplasma may be grown as broth or agar cultures (see above "Procedure").

2. In cases where the mycoplasma is grown as a broth culture, the broth may be aliquoted into appropriate volumes and stored at −70 °C. When the broth is thawed, it is diluted to a predetermined volume, which will result in mycoplasma growth in the relevant test protocol.

3. When agar stocks are grown, the portion of the agar containing mycoplasma colonies is aseptically cut into small subsections and these

individual fractions stored at −70 °C. Upon revival of frozen stocks, the agar section is added to a volume of either broth or cell growth medium and the suspension mixed, to allow mycoplasma to detach into the solution. The solution may then be diluted to the predetermined concentration, appropriate for use in the test protocols.

Comments

Eradication of Mycoplasma

There is no rapid, simple and universally effective method in eradicating mycoplasma from contaminated cells. The elimination treatment is prolonged, generally toxic to the cells and may result in the selection of a subpopulation of the cell culture which do not retain all the characteristics of the original culture. It is often unsuccessful and poses a risk of infection of other cultures. Therefore, elimination of mycoplasma should be a last resort and only used when a cell culture cannot be replaced. Instead, preventative procedures should be employed to impede introduction of mycoplasma into the culture, routine examination of the culture for mycoplasma contamination should be performed and all mycoplasma contaminated cultures should be discarded, to prevent cross-contamination of clean cultures.

When a culture cannot be replaced and must be treated, the treatments most widely used involve exposure of the cultures to antibiotics which are toxic to the mycoplasma. Although all species exhibit an absolute resistance to penicillin, many strains are inhibited by a range of antibiotics, including tetracyclines, ciprofloxacin, tylosine, kanamycin, novobiocin, gentamycin, quinolones and minocyclin (Schmidt and Erfle 1984; Ter Laak et al. 1991; Schmitt et al. 1988; Gignac et al. 1992; Coronato et al. 1994). The concentration of antibiotic used in the elimination of mycoplasma should be that which falls between the effective dose in killing mycoplasma and the toxic dose for the cell culture. A number of commercial antibiotic-based mycoplasma eradicating preparations are available, with instructions for their use available from the manufacturers. These include Ciprofloxacin (Bayer), BM-cycline from Boehringer Mannheim (catalogue number 799050) and Mycoplasma Removal Agent from ICN Biomedicals Ltd. (catalogue number 30–500–44). These treatments have been used to various effects by different researchers in the eradication of mycoplasma from cell stocks (Fleckenstein et al. 1994; Somasundaram et al. 1992; Hlubinova et al. 1993; Nissen 1995; Drexler et al. 1994).

Antibiotic treatment of mycoplasma contamination should last until the contamination is completely removed. This should be determined by negative results in mycoplasma tests, after the cultures have been grown in treatment-free medium for at least 4 weeks. This allows low level mycoplasma infection to regrow and be easily detected by standard test methods. If mycoplasma contamination persists, the elimination process should be repeated, using either a higher concentration of the initial treatment or a different treatment (mycoplasma can develop resistance to antibiotics) and the cultures retested at the end of this treatment.

References

Barile M (1979) Mycoplasma-tissue cell culture interactions. In: Tully J, Whitcomb G (eds) The mycoplasmas, vol 2, Academic, New York, pp 425–474

Battaglia M, Pozzi D, Grimaldi S, Parasassi T (1994) Hoechst 33258 staining for detecting mycoplasma contamination in cell cultures: a method for reducing fluorescence photobleaching. Biotech Histochem 69:152–156

Boyle J, Hopkins J, Fox M, Allen T, Leach R (1981) Interference in hybrid clone selection caused by *Mycoplasma hyorhinis* infection. Exp Cell Res 132:67–72

Chen TR (1977) In situ detection of mycoplasma contamination in cell cultures by fluorescent Hoechst 33258 stain. Exp Cell Res 104:255–262

Coronato S, Vullo D, Coto C (1994) A simple method to eliminate mycoplasma from cell cultures. J. Virol Methods 46:85–94

Dienes L, Ropes W, Smith W, Madoff S, Bauer W (1948) The role of pleuropneumonia-like organisms in genitourinory and joint diseases. NEJM 238:509–515

Drexler H, Gignac S, Hu Z-B, Hopert A, Fleckenstein E, Voges M, Uphoff C (1994) Treatment of mycoplasma contamination in a large panel of cell cultures. In Vitro Cell Dev Biol 30A:344–347

Fleckenstein E, Uphoff C, Drexler H (1994) Effective treatment of mycoplasma contamination in cell lines with enrofloxacin (Baytril). Leukemia 8:1424–1434

Gignac S, Uphoff C, MacLeod R, Steube K, Voges M, Drexler H (1992) Treatment of mycoplasma contaminated continuous cell lines with mycoplasma removal agent (MRA). Leukemia Res 16:815–822

Hay R, Macy M, Chen T (1989) Mycoplasma infection of cultured cells. Nature 339:487–488

Hlubinova K, Feldsamova A, Prachar J (1994) Evaluation of two methods for elimination of mycoplasma. In Vitro Cell Dev Biol 30A:21–22

Kahane I and Adoni A (1993) Rapid diagnosis of mycoplasma. Plenum, New York.

Lo S-C, Hayes M, Wang R, Pierce P, Kotani H, Shih J (1991) Newly discovered mycoplasma isolated from patients infected with HIV. Lancet 338:1415–1418

McGarrity G, Sarama J, Vanaman V (1995) Cell culture techniques. Am Soc Microbiol 51:170–183

McGarrity G, Vanaman V, Sarama J (1984) Cytogenetic effects of mycoplasmal infection of cell cultures: a review. In Vitro Cell Dev Biol 20:1–18

Nissen E (1995) Treatment of mycoplasma contamination. In Vitro Cell Dev Biol 31:260.

Russell W, Newman C, Williamson D (1975) A simple cytochemical technique for demonstration of DNA in cells infected with mycoplasmas and viruses. Nature 253:461–462

Sasaki T, Shintani M, Kihara K (1984) Inhibition of growth of mammalian cell cultures by extracts of arginine utilizing mycoplasmas. In Vitro Cell Dev Biol 20:369–374

Schmidt J, Erfle V (1984) Elimination of mycoplasma from cell cultures and establishment of mycoplasma-free cell lines. Exp Cell Res 152:565–570

Schmitt K, Daubener W, Bitter-Suermann D, Hadding U (1988) A safe and efficient method for elimination of cell culture mycoplasmas using ciprofloxacin. J Immunol Methods 109:17–25

Somasundaram C, Matzku S, Nicklas W (1992) Use of ciprofloxacin and BM-cycline in mycoplasma decontamination. In Vitro Cell Dev Biol 28A:708–710

Ter Laak E, Pijpers A, Noordergraaf J, Schoevers S, Verheijden J (1991) Comparison of methods for in vitro testing of susceptibility of porcine mycoplasma species to antimicrobial agents. Antimicrob Agents Chemotherap 35:228–233

Tully J (1992) Mollicutes. In: Lederberg J (ed) Encyclopedia of microbiology, vol 3. Academic, New York, pp 181–191

Tully J, Razin S (1983) Methods in mycoplasmology, vol. IV. Academic, New York

Uphoff C, Gignac S, Drexler H (1992) Mycoplasma contamination in human leukemia cell lines. I. Comparison of various detection methods. J Immunol Methods 149:43–53

Suppliers

Bayer: Diagnostics Division, Evans House, Hamilton Close, Basingstoke, Hampshire, RG21 27E, UK

Boehringer Mannheim :Diagnostics and Biochemicals, Bell Lane, Lewes, East Sussex, BN7 1LG, UK

ICN Biomedicals Ltd, Eagle House, Peregrine Business Park, Gomm Road, High Wycombe, Buckinghamshire, HP13 7DL, UK

Chapter 4

Serum-Free Media

Joanne Keenan*[1], Paula Meleady[1], and Martin Clynes[1]

▨ Introduction

The development and routine use of serum-free media (SFM) is a high priority for cell culture from both an industrial and scientific point of view. With the extent of specific nutritional and supplemental requirements exhibited by different cell types and the possible combinations that can be used, the task of finding a selection of factors that can replace the complex mixture of factors in serum for a particular cell line can be daunting.

SFM generally consist of a basal medium and additional supplements. The basal medium provides most of the nutrients required by the cell including: amino acids (essential and nonessential); vitamins (especially B group); nucleic acids; lipids (essential fatty acids, glycerides, etc.); inorganic salts (as buffering agents, co-enzymes and cofactors) and an energy source (usually glucose or fructose). Many of the basal media are already optimized for a given cell line, so further optimization may be required for a particular cell line.

The additional supplements include: growth promoters, such as insulin, insulin-like growth factors (IGFs), epidermal growth factor (EGF), platelet-derived growth factor (PDGF), fibroblast growth factor (FGF), estradiol, and dexamethasone; attachment factors, such as collagen, fibronectin and laminin; transport proteins and detoxifying agents, such as albumin and transferrin; trace elements, such as Fe, Cu, Sn, Co Mn, Mo, Va, Ni, Zn and Se; and lipids, such as essential fatty acids, phospholipids and triglycerides.

* *Correspondence to* Joanne Keenan: phone +353–1–7045721; fax +353–1–7045484; e-mail keenanJ@ccmail.dcu.ie

[1] Joanne Keenan, Paula Meleady, Martin Clynes, National Cell and Tissue Culture Center, Dublin City University, Glasnevin, Dublin 9, Ireland

Alternatively, commercially available SFM may be used. These are produced by companies including Sigma, Boehringer Mannheim, Costar, Flow Laboratories, Hyclone and New Brunswick. However with many commercial SFM, a period of adaptation to the SFM is required and the proprietary state of some products means that the consumer does not know the full composition of the SFM.

There are many advantages to growing cells in serum-free conditions:

- Defined nature of SFM: interpretation of data not hampered by unknown factors

- Selective growth of one cell type over another, e.g., in primary cultures

- Increased productivity (Yabe et al. 1986)

- No growth inhibitors, such as chalones, transforming growth factor-β (TGF–β), and glucocorticoids, as in serum

- Reduced batch-to-batch variation as compared to serum

- Less possibility of bacterial, viral and mycoplasma contamination

- Control of cellular differentiation (Rahenchilla et al. 1989; Najar et al. 1990).

- Lower protein levels, resulting in fewer problems in downstream processing

Not only are most of these SFM totally defined, many are also relatively cheap, usually comprising low molecular weight components and defined polymers with few or no added proteins. However, there are also some problems with SFM.

- Different cell types exhibit varying requirements for cell growth. In general, transformed cell lines have simpler requirements than untransformed cells. Hybridomas and myelomas are generally less fastidious and will grow in SFM with minimal additions and little dependence on growth factors. For anchorage-dependent cells, the development of SFM has had varied success (Hutchings and Sato; Zirvi et al. 1986; Taub et al. 1979). Many other cell lines require the introduction of growth factor combinations in addition to the other more usual components, thus resulting in more expensive alternatives to serum.

- Growth is often slower in SFM (Ahearn et al. 1992).

- Use of ill-defined components to supplement media, e.g., Pedersens fetuin and bovine pituitary extract, has compromised the defined nature of some SFM.

- Adaptation of cells to growing in SFM is often required.

- Changes in phenotype sometimes occurs in SFM, e.g., differentiation or loss of some specific cellular activity.

Materials

General Tips for Handling Cells in Serum-Free Media

water Highly pure water should be used for SFM to maintain batch to batch repeatability, e.g., ultra-pure water prepared by ion exchange followed by reverse osmosis. The resistance of the water should be 1.5–2 mOhms.

chemicals The purity of the chemicals should be checked. Impurities may be growth promoting, e.g., Se, Cr, V, As, Sn, or growth inhibitory, e.g., Hg. Chemicals produced for cell culture should be used where possible. Remember to keep a record of the batch or lot of each chemical used. Caution should be exercised with chemicals or reagents that are reconstituted in solvents, as the solvents themselves may be detrimental to cells unless suitably diluted, e.g., 1 mg dexamethasone should be reconstituted in 1 ml ethanol and diluted 50-fold in basal medium.

basal media Check the expiration date on basal media and try to ensure a 2 month leeway so that media components should not have broken down and resulted in a loss of activity. Store in the dark where possible, to prevent the formation of free radicals which could be detrimental to cells under serum-free conditions and to prevent the degradation of light sensitive vitamins (B group). When filter-sterilizing, ensure that detergent-free filters are used to avoid introducing harmful detergents into the SFM. Plastic containers are preferred as glass tends to leach more potentially harmful components into the medium. If glass is to be used, it should be dedicated for serum-free work from the beginning. Thoroughly clean glass to ensure that residues from cleaning, which are harmful to cells grown in SFM, are removed.

supplements Supplements should be reconstituted as recommended by the supplier and further diluted and aliquoted so that any of the solutions need to be

Table 4.1. Reconstitution of bovine insulin, transferrin and selenite

Product	Reconstitution	Assay range
Hybrimax insulin (Sigma England, cat. no. I4011)	Reconstitute in 100 µl glacial acetic acid and dilute in sterile PBS A initially to 5 mg/ml. Subsequent dilutions in basal medium.	0.1 –50 µg/ml
Bovine transferrin (Bayer Diagnostic, cat. no. 82–0425–02)	Reconstitute in basal medium and dilute so that stocks are at 5 mg/ml. Subsequent dilutions in basal medium.	0.1–50 µg/ml
Sodium Selenite (Sigma England, cat. No. S5261)	Reconstitute in basal medium and dilute so that stocks are at 20 µg/ml. Subsequent dilutions in basal medium.	1–100 nm

thawed only once or twice. Most cells respond to insulin (or IGF-I), transferrin, selenium and a lipid combination such as fraction V albumin or lipoprotein (e.g., Pentex Ex-cyte from Bayer Diagnostic, England). Such supplements will lose activity if a stock is repeatedly frozen and thawed. Complete medium should only be made up just before use. Table 4.1 details the reconstitution of bovine insulin, transferrin and selenite.

Most anchorage-dependent cells are capable of making the components of an extracellular matrix (ECM) including collagen and fibronectin, but precoating of a plate or providing attachment factors in the medium increases the rate of ECM deposition. Laminin (epithelial, endothelial, some tumors and hepatocytes), fibronectin (epithelial, mesenchymal, neuronal and fibroblasts) and collagen (epithelial, endothelial, muscle and nerve cells) are most commonly used $(1-10 \,\mu g/cm^2)$. These can be added to the plates in a minimum volume and allowed to air-dry in a laminar flow hood. Alternatively, they may be incorporated into the medium $(1-50 \,\mu g/ml)$.

addition of attachment factors

Scale-up of cell numbers in SFM requires several considerations. Cost of SFM containing growth factors in large scale operations can be as, or more, expensive than serum. When the product is other than cell mass, low cost maintenance medium (little or no growth factors) may be used to keep high cell densities viable and maintain high production rates. It is also important when scaling-up cell numbers to monitor the integrity of the cells and the production rate. In addition, the loss of serum in

scaling-up

large-scale operations leaves mammalian cells more susceptible to damage from high aeration and agitation rates. To protect against fluid mechanical damage, media additives such as pluronic F68, polyvinylpyrrolidone or methylcellulose are included.

4.1
Trypsinizing Cells in Serum-Free Medium

When trypsinizing cells in serum-free conditions, serum should not be used to inactivate trypsin as residues from the serum may be left in the cell suspension. To overcome this, trypsin inhibitor is used. The following procedure describes the trypsinization process. All manipulations with cells should be carried out in a laminar flow cabinet. For best results, cells should still be in the log growth phase, i.e., not more than 80 % confluent.

■ Materials

reagents
- Trypsin versene (2.5 mg/ml in PBS A; Gibco cat. no. 043–0509H)
- Serum-free medium
- Trypsin inhibitor (TI; Sigma cat. no. T6522, where an equal volume of TI will inhibit trypsin (see suppliers instructions))

equipment
- Centrifuge
- Inverted microscope
- Hemocytometer
- Sterile flasks
- Universal
- Pipettes and tips

■ Procedure

trypsinizing
1. The day before trypsinization, cells should be fed with fresh medium.

2. On the day of trypsinization, thaw trypsin, TI (in basal medium) and make up fresh stock of SFM. Do not warm trypsin solution.

3. Remove medium from flask.

4. Rinse cells with a small amount of trypsin and remove trypsin.

5. Add in sufficient trypsin to cover the surface of the flask, a maximum of 0.5 ml per 25 cm^2 flask.

6. Allow flask to sit on bench for several minutes, monitoring the state of detachment regularly. As soon as the cells have rounded up (the length will depend on the cell type), gently tap the end of the flask to dislodge the cells.

7. Add in an equal volume of TI.

8. Remove cell suspension to a sterile universal.

9. Wash flask with 5–10 ml sterile medium to remove residual cells and add to cell suspension.

10. Centrifuge at 1000 rpm for 5 min. If a pellet of cells has not formed, it may be necessary to centrifuge the cells for an additional 5 min.

11. Very gently, remove the supernatant and resuspend the cells in 1 ml sterile medium. Add 9 ml medium.

12. Centrifuge at 1000 rpm for 5 min.

13. Very gently, remove the supernatant and resuspend cells in 5 ml SFM.

14. Determine the cell count and re-seed cells immediately.

4a. Cover cells with trypsin and immediately remove as much as possible. Close flask to prevent evaporation. **alternative to steps 4–14**

5a. Leave flask on ice or at 4 °C while cells are detaching.

6a. When cells partially loosen, add 5 ml chilled SFM to flask and shake gently to resuspend cells. Pipette up and down to gently remove clumps.

7a. If making a cell count, keep cells on ice (or in refrigerator at 4 °C) until count is completed.

4.2
Long-Term Storage of Cells in Serum-Free Medium

Cells routinely stored for indefinite periods of time in liquid nitrogen are normally frozen using a combination of the cryopreservative dimethyl sulfoxide (DMSO) with medium containing 10 % – 20 % serum. Cryopreservatives may be penetrative or nonpenetrative. Penetrative cryopreservatives, such as DMSO and glycerol, protect the cells against freezing damage caused by intracellular ice crystals and osmotic effects. Nonpenetrative cryopreservatives, such as serum, protect the cells from damage by extracellular ice crystals. To freeze cells under serum-free conditions, other nonpenetrative cryopreservatives are employed (Ashwood-Smith 1987; Yoshida and Takeuchi 1991; Ohno et al. 1988; Merten et al. 1995), e.g., polyvinylpyrrolidone (PVP) or methylcellulose (MC).

▨ Materials

reagents
- Polyvinylpyrrolidone at 360 000 MW (Sigma cat. no. PVP-360)
- Methylcellulose at 4000 cp (Sigma cat. no. M0512)
- Dimethyl sulfoxide (Sigma cat. no. D-5879)
- PBS A

equipment
- Cryovials
- Hemocytometer
- Inverted microscope
- Centrifuge
- Sterile universals
- Sterile flasks
- Pipettes and tips

▨ Procedure

the day before freezing
1. Ensure that cells are not confluent and that they will be in the log phase of growth for the next day. Feed with fresh SFM.

2. Prepare 1 % (w/v) MC (10X) or 30 % (w/v) PVP (10X) in PBS A and autoclave.

3. Dilute DMSO to 50 % (w/v) in basal medium and filter sterilize.

4. Prepare 2X freezing medium: Add 1 ml 1 % MC or 1 ml 30 % PVP to 2 ml SFM. Add in 2 ml 50 % DMSO. **on the day of freezing**

5. Trypsinize cells as described in previous section. Obtain a cell concentration of about 1×10^7 cells/ml.

6. **Slowly,** in a dropwise manner, add an equal volume of freezing medium to the cells, while at the same time gently swirling the cell suspension to allow the cells to adapt to the presence of DMSO (toxic to cells if added too quickly!).

7. Aliquot cells into cryovials and slowly freeze in the vapor phase of liquid nitrogen for 3 h and then into the liquid phase ($-196\,^{\circ}$C) for an indefinite period of time.

8. Check viability: Thaw cells rapidly at $37\,^{\circ}$C and immediately transfer cells to a universal containing 9 ml pre-warmed SFM.

9. Centrifuge cell suspension at 800 rpm.

10. Seed cells into 25 cm^2 flasks containing 9 ml pre-warmed SFM.

11. Feed cells with fresh SFM on the following day.

4.3
Designing Your Own Serum-Free Medium

There are many detailed reviews which outline the development of SFM (Barnes and Sato 1980; Mather 1984; Hewlett et al. 1991; Miyazaki et al. 1984; Bjare 1992; Sandstrom et al. 1994). A number of different approaches to replace fetal calf serum have been undertaken. The amount of serum in the medium can be reduced and gradually replaced by defined components. This method of adaption helps to maintain the original cell phenotype. Alternatively, the cells may be cultured directly into SFM and any resulting clonal populations isolated.

In order to design/develop a SFM for a particular cell line, consider the cell type and the origin of the cell. It may be possible to use an existing SFM for a cell line of similar origin or cell type and then adapt the medium to the cell line in question (Table 4.2). Where there is no existing SFM, it may be necessary to adapt the cell line or select clonal subpopulations. When optimizing a SFM, remember that high growth does not always correspond to high rates of product synthesis. Also, as the adaptation process may take some time, it would be useful to first try and grow the cells in a simple SFM, as some cell types have autocrine mechanisms which are turned on in SFM.

Table 4.2. Composition of some published serum-free media

Reference	Cell line	Basal medium	Ins (μg/ml)	Tf (μg/ml)	BSA (mg/ml)	SEL (nM)	Growth promoters	Others
Iscove and Melchers (1978)	Fo, NS-1, Sp2/0	IMDM	–	1	0.5	100	–	50 μM β-ME
Chang et al. (1980)	Ag8	MEM	5	5	–	–	–	NEAA
Chiang et al. (1985)	NIH 3T3	ATCC	10	25	–	–	10 ng/ml EGF	5 μg/ml F, 25 μg/ml HDL
Hosoi et al. (1991)	Bovine granulosa	ATCC	5	–	25	–	500 ng/ml aprotinin 10 ng/ml HBGF-2	25 μg/ml Lp
De Boer et al. (1989)	Splenocytes	DMEM	5	36	2.5	–	20 μM EA	1 μg/ml LA, O, Pal
Monette et al. (1990)	Erythroid stem cells	MEMα	10	–	10	–	2 U/ml EPO, 50 U/ml IL-3	0.1 mM βME, NEAA
Johnson et al. (1992)	Human keratinocytes	DMEM	5	5	0.02	53	0.4 μg/ml HC, 20 pM T	0.1 mM EA,0.1 mM PEA, SnCl$_2$
Hutchings and Sato (1978)	HeLa	Ham's F12	5	5	–	15	30 ng/ml EGF, 30 mM HC	Ni$_2$SO$_4$, MnCl$_2$, NH$_4$M o$_2$O$_7$, SnCl$_2$, He$_2$SO$_4$, CdSO$_4$ Na$_2$SiO$_3$, NaHVO$_4$,
Darlington et al. (1987)	Human hepatoma	MEM/MAB (3:1)	10	10	–	30	1 μg/ml Gi, GH, 1 μM T, Dex	1 μg/ml LA

Table 4.2. (Continue)

Reference	Cell line	Basal medium	Ins (µg/ml)	Tf (µg/ml)	BSA (mg/ml)	SEL (nM)	Growth promoters	Others
Branchaud et al. (1990)	Human trophoblasts	RPMI	15	10	0.7	–	–	1–2 µg/ml LDL
Briand et al. (1987)	Human primary mammary epithelial	ATCC	0.25	10	–	15	100 ng/ml EGF, 0.1 nM bE, 1 mM HC, 5 µg/ml Prl	2 mM L-glut, CIV
Gazdar and Oie (1986)	Human primary NSCLC	RPMI-1640 or ATCC	20	10	2.0	25	100 ng/ml EGF	2 mM L-Glut, CIV
Allen et al. (1985)	Satellite cells	MCDB-104	10^{-6} M	5	1	30	0.5 mg/ml Fu, 100 ng/ml FGF	100 nM Dex
Taub et al. (1979)	Kidney cells	ATCC	5	5	–	10	25 ng/ml PGE	50 pM T, 50 nM HC

Abbreviations: *Ins*, insulin; *Tf*, transferrin; *BSA*, bovine serum albumin; *Sel*, selenium; *NEAA*, nonessential amino acids; *GF*, growth factor; *β-ME*, β-mercaptoethanol; *EA*, ethanolamine; *LA*, linoleic acid; *O*, oleic acid; *P*, pyruvate; *Gl*, glucagon; *Pu*, putrescine; *HC*, hydrocortisone; *Dex*, dexamethasone; *Pal*, palmitic acid; *PEA*, phosphoethanolamine; *T*, triiodothreonine; *Fu*, fetuin; *EPO*, erythropoietin; *Il*, interleukin; *ATCC*, 50:50 vol/vol DME:Ham's F12; *F*, fibronectin; *HBGF-2*, heparin binding growth factor-2; *Lp*, lipoprotein; *EGF*, epidermal growth factor; *CT*, cholera toxin; *PDGF*, platelet-derived growth factor; *FGF*, fibroblast growth factor; *CIV*, collagen type IV; *Prl*, prolactin.

Procedure

adaptation

1. Reduce the serum content and allow the cells to grow for several passages in order to adapt., e.g., 5 % → 2 % → 1 %. It may be necessary to increase the inoculation density initially, until the cells adapt. As adaptation occurs, the inoculation density may then be reduced. Freeze samples of the cells and check the viability when thawed.

2. Select the most suitable basal medium at low serum concentrations, e.g., Hams' F12 , DMEM, Iscoves DMEM, Hams' F12: DMEM and RPMI-1640.

3. Reduce to 0.1 % serum. At this stage it may be necessary to add in some supplements. The most universally used supplements to SFM are insulin (1–10 µg/ml) and transferrin (1 –10 µg/ml) and selenium (1–100 nM). It may be necessary to increase the inoculation density again until the cells adapt.

4. Remove serum completely and test the effect of a panel of growth factors, lipids and other trace elements. If cells do not grow, they may require an attachment factor. It may be necessary to increase the inoculation density until the cells adapt.

5. Modification of basal medium: In order to modify the basal medium for a particular cell line, it is necessary to determine the optimal concentration for each component (amino acids, vitamins, minerals and inorganic salts). When modifying the medium, growth should be distinctly suboptimal in order to observe differences in growth promotion.
   - Start with one component: test at tenfold differences from the initial concentration in the basal medium, e.g., 0.1x, 1x and 10x. If there is a significant growth improvement, carry out a complete titration of this component. Choose a concentration at the midpoint in the log-phase as the new concentration. By choosing such a concentration it is less likely to become rate-limiting as the optimization proceeds. The new concentration should then be incorporated into the medium and the next component titrated.

6. Long-term culture: Test the behavior of cells over an extended period, usually ten passages. When serum carryover has been eliminated, other factors may be required. Monitor glucose and lactate production and amino acid kinetics.

References

Ahearn GS, Daehler CC, Majumdar SK (1992) SFM for murine erythroleukemia cells still not as good as serum-supplemented medium. In Vitro 28a:227–229

Allen RE, Dodson MV, Luiten LS, Boxhorn LK (1985) A SFM that supports the growth of cultured skeletal muscle satellite cells. In Vitro 21:636–640

Ashwood-Smith MJ (1987). Mechanisms of cryoprotectant action. In: Bowler K and fuller BJ (eds) Temperature and animal cells, Symposia of the Society for Experimental Biology, no 41. The Company of Biologists, Cambridge, UK, pp 395–406

Barnes D, Sato G (1980) Methods for growth of cultured cells in SFM. Anal Biochem 102:255–270

Bjare U (1992) Serum-free cell culture. Pharmac Ther 53:355–374

Branchaud CL, Goodyear CG, Guyda HJ, Lefebvre Y (1990) A serum-free system for culturing human placental trophoblasts. In Vitro 26:865–870

Briand P, Petersen OW, Van Deure B (1987) A new diploid non-tumorigenic human breast epithelial cell line isolated and propagated in chemically-defined medium. In Vitro 23:181–188

Chiang LC, Silnutser J, Pipas JM, Barnes D (1985) Selection of transformed cells in SFM. In Vitro 21:707–712

Darlington GJ, Kelly JH, Buffone GJ (1987) Growth and hepatospecific gene expression of human hepatocyte cells in defined medium. In Vitro 23:349–354

De Boer M, Ossendorp F, Van Duijn G, Ten Voorde G, Tager J (1989) Optimum conditions for the generation of monoclonal antibodies using primary immunization of mouse splenocytes in vitro under serum-free conditions. J Immunol Methods 121:253–260

Gazdar AF, Oie HK (1986) Correspondence re: Brower et al. , Growth of cell lines and clinical specimens of human non-small cell lung cancer in serum-free defined medium. Cancer Res 46:6011–6012

Hewlett G (1991) Strategies for optimising serum-free media. Cytotechnology 5:3–14

Hosoi S, Miyaji H, Satoh M, Kurimoto T, Mihara A, Fujiyoshi N, Itoh S, Sato S (1991) Optimisation of cell culture conditions for production of biologically active proteins. Cytotechnology 5: S17–34

Hutchings S, Sato G (1978) Growth and maintenance of HeLa cells in SFM supplemented with hormones. Pro Natl Acad Sci USA 75:901–904

Iscove N, Melchers F (1978) Complete replacement of serum by albumin, transferrin and soybean lipid in cultures of lipopolysaccharide reactive B lymphocytes. J Exp Med:147, 923–933

Johnson EW, Meunier SF, Roy CJ, Parenteau NL (1992) Serial cultivation of normal human keratinocytes: a defined system for studying the regulation of growth and differentiation. In Vitro 28a:429–435

Mather JP (1984) Mammalian cell culture: the use of serum-free supplemented media. Plenum, New York

Merten OW, Peters S, Couve E (1995) A simple serum-free freezing medium for serum-free cultured cells. Biologicals 23: 185–189

Miyazaki K, Masui H, Sato G (1984) Growth and differentiation of human bronchogenic carcinoma cells in SFM. In: Barnes DW, Sirbasku DA, Sato GH (eds) Chapter 2: Cell culture methods for molecular and cell biology. Alan R Liss, NY, pp 83–94

Monette FC, Hartwell R, Wu D, Sigounas G (1990) Erythroid stem cell culture in serum-depleted medium. J Tissue Culture Methods 13:69–75

Najar HM, Bru-Capdeville A, Gieseler R, Peters J (1990) Differentiation of human monocytes into accessory cells at serum-free conditions. Eur J Biol 51: 339–346

Ohno T, Kurita K, Abe S, Eimori N, Ikawa Y (1988) A simple serum-free freezing medium for serum-free cultured cells. Cytotechnology 1:257–260

Ranchilla F, Moover C, Wille J (1989) Biosynthesis of proteoglycans by proliferating and differentiating normal human keratinocytes cultured in SFM. J Cell Physiol 140: 98–106

Sandstrom CE, Miller WM, Papoutsakis ET (1994) SFM for culture of primitive and mature haematopoietic cells. Biotech Bioeng 43: 706–733

Taub M, Chuman L, Saier MH, Sato G (1979) Growth of MDCK epithelial cell line in hormone supplemented SFM. Proc Natl Acad Sci USA 76:3338–3342

Yabe N, Kato M, Matsuga Y, Yamane I, Takada M (1986) Enhanced formation of mouse hybridomas without HAT treatment in SFM. In Vitro 22: 363–368

Yoshida T, Takeuchi M (1991) Primary culture of mouse astrocytes under serum-free conditions. Cytotechnology 5: 99–106

Zirva KA, Chee DO, Hill GH (1986) Continuous growth of human tumor cell lines in SFM. In Vitro 22:369–374

Futher Reading

Barnes DW, Sirbasku DA, Sato GH (1984) Cell culture methods for molecular and cell biology. vol. 1. Methods for the preparation of media, supplements and substrata for serum-free animal cell culture. Alan R. Liss, NY

Mather JP (1984) Mammalian cell culture: the use of serum-free supplemented media. Plenum, New York

Suppliers

Sigma, Fancy Road, Poole, Dorset, BH12 4QH, UK

Gibco, Life Technologies Ltd. 3 Fountain Drive, Inchinnan Business Park, Paisley PA4 9RF

Chapter 5

Screening Culture Media and Sera for Endotoxin Testing

JOANNE KEENAN*[1] and MARTIN CLYNES[1]

Introduction

Endotoxin, part of the cell wall of all gram-negative bacteria, contains a highly biologically active lipopolysaccharide which causes a pyrogenic reaction in warm-blooded animals and is a direct result of bacterial contamination. Therefore, the presence of endotoxin in pharmaceutical products or medical devices is a cause of great concern. Once present, pyrogenic endotoxin is difficult or impossible to remove, depending on the product.

The lipopolysaccharide molecule consists of three regions: the o-polysaccharide which imparts hydrophobicity; the lipid A region which is the hydrophilic, biologically active portion and the core region, linking the o-polysaccharide and lipid A regions.

Development of Limulus Amoebocyte Lysate

Originally the medical industry relied solely on the pyrogenic response of rabbits to detect an unacceptably high level of endotoxin. The limulus amoebocyte lysate (LAL) test system offers a good alternative to the pyrogen tests in rabbits as it is comparatively inexpensive, rapid, reliable and convenient.

LAL is derived from *Limulus polyphemus*, the horseshoe crab, which is found off the eastern coast of the United States of America. These crabs possess a simple specific defense against gram-negative bacteria. The blood of the horseshoe crab contains only one cell type called amoebo-

* *Correspondence to* Joanne Keenan: phone +353–1–7045721; fax +353–1–7045484; e-mail Keenanj@ccmail.dcu.ie
[1] Joanne Keenan, Martin Clynes, National Cell and Tissue Culture Center, Dublin City University, Glasnevin, Dublin 9, Ireland

cytes and when the crab sustains an injury, the amoebocytes seal off the injury by releasing clotting enzymes which cause the formation of a clot. Every summer, horseshoe crabs are bled without injury. The blood is centrifuged and then lysed into pyrogen-free water to release the clotting enzymes, referred to as the lysate.

The complex set of reactions invoked by the LAL is only partly understood. A cascade of enzyme reactions results in the activation of a clotting enzyme. The clotting enzyme then cleaves coagulogen to coagulin which in turn activates the clot formation.

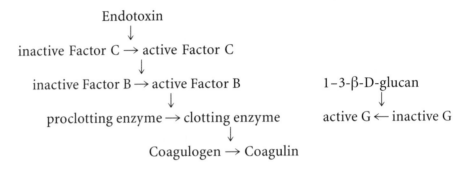

A second set of reactions may activate coagulin through the G factor. A few forms of glucose, including glucans (especially β-1,3-D-glucans), have been shown to be LAL reactive when present in sufficiently high concentrations. Their proposed mechanism is to activate factor G which can then act on coagulogen. This becomes particularly important when testing samples from cell culture, as there are likely to be LAL reactive glucans (LRG) present in the supernatant of cells. Some lysates have been designed to take the possibility of such false positives into account. Zwittergent (BioWhittaker, Inc.) contains sulfobetaine which in some way, inhibits factor G, thereby only showing coagulin activation by endotoxin. However, as glucans are also produced by yeast and fungi, it may be useful to use both systems in order to see the effect of yeast and fungal as well as bacterial contamination.

Outline

There are three methods of testing for endotoxin currently available based on the LAL, gel clot, chromogenic and turbidometric techniques.

The gel clot method involves incubating the testing sample with the lysate at 37 °C for 1 h. The tube is then gently inverted through 180° to see if a clot forms. This is the simplest method and requires the least capital investment. However, it is only semi-quantitative, with at best a sensitivity of 0.03 EU/ml.

The chromogenic technique involves a shorter incubation time. The lysate contains a chromogen, normally a paranitroaniline (pNA)-tetrapeptide, which is cleaved by the activated clotting enzyme from the amino acid chain to give pNA. The amount of pNA produced is proportional to the degree of lysate activation. This assay system is quantitative and easy to perform, has a sensitivity as low as 0.01 EU/ml, and suffers less from factors which may be present in the sample that can inhibit clot formation.

In the turbidometric method, incubation occurs in a turbidometer where the optical density of the mixture can be measured. This quantitative method has a sensitivity as low as 0.001 EU/ml. However, the capital costs are quite high.

As the gel clot method is easy to perform and requires little or no capital investment, this is the procedure that will be described in detail. Technician and sample validation will also be outlined. The chromogenic technique will also be described.

Materials

Tips for Trouble-Free Limulus Amoebocyte Lysate Testing

- Use LAL reagent grade water for reconstitution and dilution. As water is used as the diluent, it is the most common source of contamination. The levels of endotoxin present in the water should be low enough so that it does not interfere with the test. USP sterile water need only pass a level of 0.25 EU/ml. Clearly, this would interfere if the sensitivity of the lysate was 0.125 EU/ml or lower. Once opened, the bottle should only be used on the one day. The rubber stopper should not be reused. The bottle may be covered with Parafilm using the side facing the backing, provided that the backing is intact.

- Store all reagents at correct conditions.

- Follow manufacturer's/supplier's instructions for the reconstitution of the reference standard endotoxin (RSE) or the control standard endotoxin (CSE). Observe expiration dates.

- Use pyrogen-free consumables, e.g., tips, glass vials.

reagents

- Use a fresh pipette tip for each liquid transfer.

- Ensure to use the correct type of lysate for the samples being tested.

glass

- It is often easier for beginners to use disposable glassware as this reduces the risks of contamination. If recycling glassware, **do not** use detergents of any kind as residues of these may interfere with clot formation. Instead, rinse out in distilled water. Dry before wrapping and depyrogenating.

- Soda glass contains sodium ions which enhance clot formation so borosilicate should be used.

incubation chamber

- Use a dry heater block with holes wide enough to allow lysate tubes to be comfortably inserted. Dry heater blocks are recommended as they are more stable and have less chance of contamination than a water bath. If using a water bath, **do not** use a circulating water bath and avoid putting the lid on the water bath as condensation may result in contamination.

- The incubator should be checked to ensure that it is reading the correct temperature, i.e., $37 \pm 1\,°C$.

- The incubator should be placed on a vibration-free table. **Do not** vortex samples on the same bench that the samples are incubated on.

- Use aseptic technique without a flame (as heat may affect the adsorption characteristics of the endotoxin).

- Avoid passing hand over the open bottles.

choosing a lysate

The type of lysate used will depend on the sensitivity required, the number of samples to be tested and on the type of samples being tested. For samples from cell culture, the lysate should be able to show only the effect of endotoxin and not LAL reactive material. Where large amounts of samples are tested, multi-test vials are desirable. If however, only a small number of vials are to be tested, or for inexperienced testers of LAL, the single test vials are recommended. Most suppliers offer advice on lysates.

sample collection

Collect samples aseptically in endotoxin-free tubes. It may be necessary to filter the sample if there are cells present. If the sample is not to be tested immediately, it should be frozen to prevent any (additional!) contamination.

5.1
Procedure for Gel Clot Test for Endotoxin

The gel clot test will give only a positive or negative result. In order to
semi-quantify the concentration, it is necessary to carry out serial dilutions and test these as described below to get a range.

▨ Materials

- Endotoxin standard RSE or purchased CSE with certificate of analysis
  from LAL supplier; store at 2–8 °C
- LAL single test vials, e.g., Pyrotell (ACC Inc.), 0.125 EU/ml; store at
  −20 °C
- LAL 50 test vial, e.g., Pyrotell (ACC Inc.), 0.125 EU/ml; store at −20 °C
- LAL single test vials, e.g., Pyrogent (BioWhittaker), 0.125 EU/ml; store
  at 2–8 °C
- LAL reagent water (LRW)

- Heater block for LAL test tubes including insert blocks (e.g., Techne   *equipment*
  Dri-block) or noncirculating water bath
- Pipettes, e.g., p100, p1000
- Sterile pyrogen-free pipette tips for p100 and p1000
- Test tubes, e.g., Greiner 120161
- Sodium silicate glass reaction tubes, 75×9/10 mm
- Timer
- Vortex

▨ Procedure

1. 50 LAL test vials: Reconstitute the contents of each vial as described   *preparation of*
   by the supplier using LRW. Pipette specific volumes (as prescribed by   *reagents*
   supplier) into pyrogen-free glass reaction tubes. These are used in
   the same way as the single test vials. Unused vials may be stored at
   the appropriate temperature for later use.

2. Sample pretreatment: Detection of endotoxin in samples containing
   serum is complicated by inhibition or enhancement. Factors which
   interfere with clot formation include serine proteases, glucans, trypsin and other LAL reactive material (LRM). Various methods of

overcoming interference include heating, dilution, solvent extraction and acid treatment. A combination of diluting and heat treatment is recommended by the ACC. The amount of dilution required will depend on the sample and on the type of serum present. Neat serum should be diluted by one fourth, heated at 75 °C for 10 min or by boiling for 1–2 min. If the sample congeals, larger dilutions should be made before heating.

3. Standard spike solution: Dissolve the contents of one flask of RSE in 5.0 ml of LRW. This will give a concentration of 2000 EU/ml. Mix for 25–30 min. Aliquot into endotoxin-free tubes (about 20) and store for a maximum of 2 weeks at 2–8 °C.

on the day of testing
To prepare the spike solution for controls, the dilutions to be made will depend on λ (the sensitivity of the lysate). The sensitivity of the lysate is validated within a twofold range of λ. For this reason, the spiked controls should contain enough endotoxin to cause a clot in this range. So as not to dilute the sample, only a small volume of the spike endotoxin solution should be added. For example, if using vials with a sensitivity of 0.125 EU/ml, the final concentration of the sample should be 0.25 EU/ml. If, however, only 25 µl of a spike solution is to be added to 500 µl of the sample, the spike solution should be at 5.25 EU/ml. Dilutions of a 1/19 followed by a 1/20 of RSE stock (2000 EU/ml) will give a suitable spike solution. Remember to mix each dilution for not less than 30 s before proceeding to the next dilution.

Alternatively, CSE can be used. For reconstitution, follow the suppliers instructions. Units are in ng/ml. It is therefore necessary to determine the potency of CSE by comparison with the RSE. This is achieved by performing quadruplicate series of dilutions (as indicated on certificate of analysis). The endpoint for this assay is defined as the lowest concentration of endotoxin to give a positive test. The sensitivity of the CSE can then be compared to that of the RSE to determine the potency. The stock CSE solution can be stored for no more than 4 weeks at 2–8 °C.

on the day of use
The CSE should be vortexed for 3 min. The solution should be subsequently diluted to a concentration with equivalent potency to the USP RSE.

4. Place the heater block on a vibration-free table and ensure that the temperature is 37 ± 1°C.

5. Dilute samples: Where there is an estimate of the endotoxin concentration for a given sample that is greater than the sensitivity of the lysate, dilutions of the sample are required. Remember for each dilution:
   - Use sterilized pyrogen-free pipette tips, LRW and pyrogen-free reactions tubes for dilutions.
   - At each stage of the dilution, the samples should be vortexed for not less than 30 s.
   - The final diluted sample should be 1 ml volume. Aliquot 500 µl to a separate endotoxin-free vial for the positive sample control.

6. Prepare controls: For each test carried out, a positive and negative control are required. For each sample tested, the sample and a positive sample control are needed. See Table 5.1.   controls

7. Gently tap the end of each lysate vial on the bench to bring all the contents to the bottom. Remove the rubber stopper carefully, to avoid contamination.

8. Using a timer, vortex each test sample for 30 s.

9. Add the prescribed volume of test sample to the lysate.

10. Gently mix the contents together and place the tube in a preheated heater block.

11. Incubate (with a timer) for 60 ± 2 min at 37 °C.

12. After incubation, take the tube out of the heater block and carefully turn it through 180° in one movement. If a gel has formed and remains attached to the end of the vial, the test is positive and the concentration of endotoxin in the sample is greater than the sensitivity of the lysate used. If no gel has formed or if a gel has formed but is not retained, the endotoxin content of the sample is smaller than the sensitivity of the lysate used.

Table 5.1. Preparation of controls

Control	LRW (µl)	Spike (µl)	Test sample (µl)
Negative control	500	–	–
Positive control	500	25	–
Test sample	–	–	500
Positive sample control	–	25	500

Results

For each test:

- The negative control should not clot. If a clot forms, the LRW tubes, pipette tips, the lysate or the operator is the source of the endotoxin. If this clots the test in not validated and needs to be repeated, taking steps to prevent contamination of reagents, tips, etc.

- The positive control should clot. If this fails to clot, there is a problem with the spike (either interference from LRW or the spike solution has lost activity). If there is no clot, the test is not validated and the assay needs to be repeated.

- The positive sample control should clot. If this fails to clot, then the sample is interfering with the formation of a clot and further dilutions are required.

Troubleshooting

Part of the requirements of the FDA for the validation of endotoxin testing for any product is to ensure that the product is not interfering with possible clot formation.

- High salt, high pH, ethanol and proteins interfere with the formation of a clot. If the positive sample control fails to clot, inhibition is occurring and it will be necessary to dilute the samples further and retest. The dilution at which clotting of the positive sample control, but no clotting of the negative control, occurs is the minimum dilution in which no interference with clot formation is taking place. Such samples should be diluted to this concentration before retesting. This may present problems if the dilution required to eliminate the inhibition is greater than the permissible FDA dilution for that product, the maximum valid dilution.

If the pH of the sample is not between 6 and 8, it may be necessary to adjust the pH to be within this range. The lysate itself has some buffering capacity, so it is important to measure the pH of the sample and lysate mixture. A change in pH may by achieved by addition of endotoxin-free acid or base. However, use of acid or base is not recommended as the ionic strength of the solution may be affected if the pH is overshot several times, so as to inhibit LAL functionality. Many suppliers recommend the use of endotoxin-free Tris buffers which will bring the pH back into the neutral range.

- Another source of interference with the clot formation is the presence of chelating molecules such as heparin and EDTA. Divalent ions are important for the chain reactions involved in generating the clot. If these ions are bound up by a chelating agent, the formation of a clot may be impaired. This may be overcome by using a solution of $MgCl_2$ in the normal dilutions.

A simple qualitative test for differentiating LRG from endotoxin consists of treating the sample with 0.1 N NaOH. Samples containing LRG will not lose LAL reactivity in the presence of 0.1 N NaOH, while those with endotoxin will have diminished reactivity due to mild hydrolysis.

5.2
Confirmation of Lysate Sensitivity

For each individual who tests endotoxin, or for each new batch of lysate vials, it is recommended to test the sensitivity (λ) of the lysate, i.e., to test the label claim provided by the manufacturer. This is achieved as follows:

Procedure

Prepare standard endotoxin concentrations in LRW that bracket the   sensitivity labeling sensitivity, i.e., 4 λ, 2 λ, 1 λ, 0.5 λ, 0.25 λ. For example, if the sensitivity of the lysate is 0.125 EU/ml, the concentration range is 0.5, 0.25, 0.125, 0.063, 0.0315 EU/ml.

Perform the LAL test four separate times for each set of dilutions. The endpoint for this assay is defined as the least concentration of endotoxin to give a positive test. The lysate sensitivity is calculated by determining the geometric mean of the endpoint. Each endpoint value is converted to log_{10}. The individual log_{10} values are averaged and the lysate sensitivity is taken as the anti-log_{10} of this average log value. If the lysate sensitivity is $\lambda \pm 2$-fold dilution, to the labeled sensitivity, then the lysate or the individual is confirmed. The examples in Tables 5.2 and 5.3 show how operator A is validated for a lysate with 0.125 EU/ml sensitivity, while operator B is not.

Table 5.2. Results for validation of operator A using a lysate with a sensitivity of 0.125 EU/ml

Assay	0.5	0.25	EU/ml 0.125	0.063	0.0315	$\log_{10}$ endpoint
Assay 1	+	+	+	−	−	−0.903
Assay 2	+	+	−	−	−	−0.602
Assay 3	+	+	+	−	−	−0.903
Assay 4	+	+	−	−	−	−0.602
Mean $\log_{10}$ endpoint						−0.752
Anti-$\log_{10}$ endpoint						0.176

Table 5.3. Results for validation of operator B using a lysate with a sensitivity of 0.125 EU/ml

Assay	0.5	0.25	0.125	0.063	0.0315	$\log_{10}$ endpoint
Assay 1	+	+	−	−	−	−0.602
Assay 2	+	−	−	−	−	−0.301
Assay 3	+	+	−	−	−	−0.602
Assay 4	+	−	−	−	−	−0.301
Mean $\log_{10}$ endpoint						−0.451
Anti-$\log_{10}$ endpoint						0.353

sample validation

In order to validate a product or sample solution, it is necessary to carry out a preliminary test in which a series of product dilutions are tested, both with and without added endotoxin spike. This will show if the product is inhibiting clot formation. When the dilution in which no interference to clot formation is found, this dilution will be used to carry out a label claim. By carrying out the label claim, the results will show if there is enhancement of clot formation by the product.

preliminary test

Set up a series of 2-fold (or 10-fold if the product is thought to be extremely inhibitory) dilutions. A duplicate set should be set up with only the sample, while a second set should be set up containing endotoxin spike. The dilution at which the spiked samples clot and the unspiked samples do not is the minimum dilution at which there is no inhibition caused by the sample. In the example below (Table 5.4), a 1/16 dilution is required to remove clot inhibition.

Table 5.4. Dilution series to identify the dilution at which no inhibition of clot occurs

Sample/dilution	1:2	1:4	1:8	1:16	1:32	1:64	1:128
Sample A	–	–	–	–	–	–	–
Sample A	–	–	–	–	–	–	–
Sample A + spike	–	–	–	+	+	+	+
Sample A + spike	–	–	–	+	+	+	+

Using the 1/16 dilution of the sample as the diluent, a label claim is now performed in duplicate. A duplicate label claim is also performed using LRW. If there is enhancement of clot formation, the label claim will not be obtained. The label claim will be tested from 2λ to 0.25λ. The above results will be used as an example.

1. Prepare a solution of CSE (or RSE) at 4λ to the label claim in LRW.

2. Prepare a 1/8 and 1/16 dilution of the product.

3. Mix 4 ml 4λ CSE to 4 ml 1/8 dilution. This gives a 1/16 dilution of the product with 2λ CSE.

4. Dilute the 2λ with the 1/16 dilution of the product.

5. Remember to include negative controls.

Results

The results in Table 5.5 show that there was no inhibition or enhancement caused by the product at 1/16 dilution. This means, for a 1/16 dilution the product is validated.

If enhancement was taking place, clots would have been observed in the $1/2\lambda$ or $1/4\lambda$ product samples. This would necessitate diluting further to perhaps a 1/64 or a 1/128 and repeating the label claim.

Table 5.5. Determination of label claim using product and LAL reagent water (LRW)

	LRW				Product				
Neg[a]	2λ	1λ	$1/2\lambda$	$1/4\lambda$	2λ	1λ	$1/2\lambda$	$1/4\lambda$	Neg[a]
–	+	+	–	–	+	+	–	–	–
–	+	+	–	–	+	+	–	–	–

[a] Refers to negative control.

5.3
Chromogenic Technique

The chromogenic technique will vary from supplier to supplier so the following methodology is general.

▣ Materials

– Chromogenic test kit, e.g., Endochrome-K (Endosafe) QCL-1000 or Kinetic-QCL kinetic chromogenic LAL test kit (BioWhittaker)

equipment   – As before but also a multichannel pipette

▣ Procedure

chromogenic test

1. Prepare a standard curve using USP RSE or CSE, e.g., 1.0, 0.8, 0.6, 0.4 and 0.2 EU/ml. The RSE or CSE should be prepared as previously described, ensuring that at each dilution the solution should be vortexed vigorously for at least 1 min. It is recommended to have triplicate repeats of each standard dilution.

2. Warm samples, standards and blanks (LRW) up to 37 °C.

3. Pipette into a pre-warmed 96-well plate.

4. Reconstitute lysate as instructed by the manufacturer and immediately, using a timer, apply the indicated amount of lysate to each well using a multichannel pipette. Gently move the plate in a swirling motion to mix the contents of each well. Incubate for the appropriate time.

5. During incubation, prepare the substrate as indicated by the manufacturer and warm to 37 °C.

6. When the incubation period is over, add in the required volume of substrate per well, using a multichannel pipette. Gently move the plate in a swirling motion to mix the contents of each well. Incubate for the appropriate time

7. After the second incubation, the reaction is stopped, usually by the addition of acetic acid. Again, mix the contents of the wells and read at the relevant wave length (usually about 450 nm).

8. With the absorbances obtained, construct a standard curve. The unknown samples may be read off the standard curve. If the sample absorbance is outside the standard curve, the sample has to be diluted and tested again.

Comments

The gel clot method using LAL provides a quick and easy method to semi-quantify the presence of endotoxin in samples, so long as the procedure is followed correctly, and the suitable controls are incorporated. This is suitable for cell culture supernatent. Where sera or solutions with high concentrations of protein are to be tested, the chromogenic method can offer a better alternative as it suffers less from sample inhibition than the gel clot method. In all cases, it must be remembered that the technician who carries out the tests and the sample/s should be validated. For more detailed information, the FDA guidelines are recommended.

References

United States Pharmacopoeia XXII, Biological Tests/Bacterial Endotoxin test. 85 pp 1493–1495

US Department of Health and Human Services, Public Health Service, Food and Drug Administration (1987) Guideline on validation of the limulus amoebocyte lysate test as an end-point endotoxin test for human and animal parenteral drugs, biological products medical devices

Berzofsky RN (1989) LAL-RM. LAL review. BioWhittaker, Walkersville, MD, 5:1–4

Donova MA (1994) Overcoming your inhibitions. LAL review BioWhittaker, Walkersville, MD, 5:1–5

Friberger P, Knos M and Mellstam (1982) Endotoxin and their detection with limulus amoebocyte lysate test. Alan R. Liss, New York, pp 195–206.

Rolansky PF and Novitsky TJ (1991) Sensitivity of limulus amoebocyte lysate (LAL) to LAL-reactive glucans. J Clin Micro 29: 2477–2482

Suppliers

Below are the addresses of three major companies supplying lysate and associated products. By contacting these, the local distributor may be found:

Associates of Cape Cod Inc., P.O. Box 224, Woods hole, Massachusetts 02543, USA

Biowhittaker, Inc. 8830 Biggs Ford Road, Wakersville, Maryland 21793–0127, USA

Endosafe, 1023 Wappoo road, Suite 43-B, Charleston, South Carolina 29407, USA

Abbreviations

LAL Limulus amoebocyte lysate

LRG LAL reactive glucans

LRW LAL reagent grade water

RSE Reference standard endotoxin

CSE Control standard endotoxin

Glossary

Lambda (λ) = Concentrations of lysates are given as sensitivity or λ. The sensitivity of a particular lysate is the minimum concentration of endotoxin needed to activate the formation of a clot which stays on the end of the vial when fully inverted.

Units = The references (RSE or CSE) which are used to validate the sensitivity of lysates and act as spikes for controls may have measurements of endotoxin units (EU), international endotoxin units (IU) or ng/ml. All concentrations are measured against the United States Pharmacopoeia reference standard endotoxin (USP RSE). Information on the concentrations are supplied on the certificate of analysis that should accompany every lysate and reference. The conversion factor from ng/ml activity to EU/ml will depend on the batch of endotoxin in question. This is achieved by carrying out serial dilutions of the sample in ng/ml and the RSE in EU/ml and comparing the endpoints for both.

Endotoxin limit = The pyrogenic threshold for humans has been determined as 5 EU/ml per kg body weight. So for an average person, weighing 70 kg, an endotoxin limit of 350 EU/h should not be exceeded. The FDA has a constantly increasing list of therapeutics for which an endotoxin limit has been determined. The limits set are normally much lower than that required to give a pyrogenic response.

MVD = The maximum valid dilution that can be carried out for a particular product so that, when tested, the results are still within the endotoxin limit for that product. Dilutions may be necessary to overcome any inhibition/enhancement that the sample is causing. MVD is defined as the endotoxin limit (in EU/ml)/λ. Thus, by increasing the sensitivity of the lysate, it is possible to increase the MVD.

MVC = The minimum valid concentration and is the product concentration at the MVD. MVC is defined as the product concentration/MVD.

Chapter 6

Generation of Monoclonal Antibodies and Characterization of Novel Antibodies by Western Blotting and Immunocytochemistry

Elizabeth Moran[*][1], Annemarie Larkin, Allan Masterson, and Martin Clynes[1]

Introduction

The development by Kohler and Milstein (1975) of murine monoclonal antibodies by somatic cell fusion techniques (i.e., the production of hybridomas) has resulted in the routine production and availability of these antibodies for research and medical purposes. The two most commonly used cells in this hybridization technique are antibody-secreting B cells from the immunized animal (which provide functional immunoglobulin genes) and mouse myeloma cells which provide the correct genes for continued cell division in culture. Any material that can elicit a humoral response, e.g., viruses, whole cells, cellular components, purified proteins, synthetic peptides, carbohydrates, can be used for the production of monoclonal antibodies. Using carefully defined conditions for efficient fusions and selection of antibody-producing hybridomas from unfused cells, monoclonal antibodies specific for a wide range of antigens have been produced. Monclonal antibodies have many advantages over polyclonal antibodies, e.g., their specificity of binding, their homogeneity (antibody produced by all the decendants of one hybridoma cell is identical) and their ability to be produced in unlimited quantities (as cell culture supernatant or ascitic fluid). A major disadvantage is that the production of monclonal antibodies is more time consuming and usually more costly than polyclonal antibody production and often a panel of monoclonal antibodies to different epitopes on the same antigen is required to give similar binding affinity as a polyclonal antibody.

* *Correspondence to* Elizabeth Moran: phone +353–1–7045701; fax +353–1–7045484; e-mail moranl@ccmail.dcu.ie

[1] Elizabeth Moran, Martin Clynes, National Cell and Tissue Culture Centre, Dublin City University, Glasnevin, Dublin 9, Ireland

Outline for Production and Characterisation of Monoclonal Antibodies

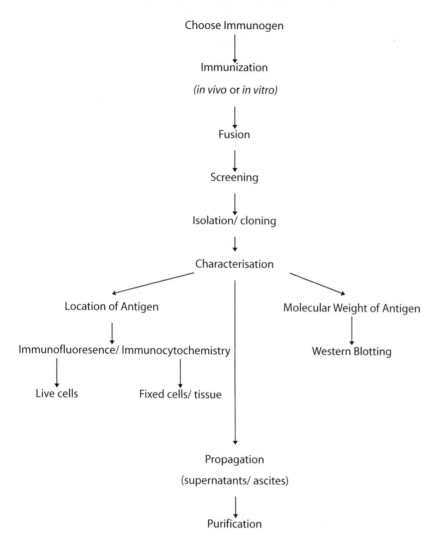

Fig. 6.1. Outline for production and characterization of monoclonal antibodies

Outline

This chapter will discuss the production of monoclonal antibodies to cellular markers associated with multiple drug resistance (MDR) and their characterization, i.e., epitope location by immunofluorescence on live cells and immunocytochemistry on fixed cells and determination of molecular weight of antigen by western blotting (Fig. 6.1).

6.1
Generation of Monoclonal Antibodies

Materials

Choice of Immunogen for Monoclonal Antibody Production

A range of immunogens have been successfully employed in our laboratory to produce monoclonal antibodies to specific target antigens (i.e., MDR markers). These include whole cells (particulate antigen), lysed cells, isolated cell membranes and synthetic peptides (soluble antigens). Soluble cellular antigens are usually administered in the presence of an adjuvant. Adjuvants are non-specific stimulators of the immune response which, when used appropriately, increase the immune response greatly. Whole cells generally make excellent antigens as they are readily phagocytosed by the antigen processing cells of the immune system. They are used for immunization at 10^5–10^7 cells/0.5 ml injection, usually in a neutral physiological buffer such as phosphate-buffered saline (PBS). Whole cells are not normally administered with an adjuvant and are usually administered intraperitoneally (they should not be administered intravenously). However, very small cells or cells with fragile cell membranes which are readily disrupted are often administered with an adjuvant. Synthetic peptides, normally coupled to a carrrier protein, e.g., bovine serum albumim (BSA) or keyhole limpet hemacyanin (KLH), and combined with an adjuvant can be administered in doses as low as 1 µg/injection. In practice, however, the usual dose ranges from 20 to 200 µg per injection. DNA technology has made the isolation, characterization and sequencing of genes possible. From the determined nucleotide sequence it is possible to predict the amino acid sequence of the encoded protein. Such information for a wide variety of proteins is cited in the literature. Synthetic peptides used for the production of anti-*mdr-1* antibody and antibodies to other MDR-related markers in our laboratory were selected after alignment searches of EMBL'S Swiss-Prot (EMBL:http://www.embl-heidleberg.de/) protein sequence database using the Mail-FASTA program (Pearson and Lipman 1982). The entire amino acid sequence of the encoded proteins of interest were obtained from the data bank above using the NETSERV program. By using synthetic peptides it is possible to optimally design an antigen for the purpose intended, i.e., it is feasible to select the region or domain of the protein to which resulting antibodies will react. The major disadvantage with anti-peptide antibodies is that they may not recognize the native protein. Peptides are usually synthesized using the solid-phase techniques introduced by Merrifield (1963), purified (80 % purity for immunological purposes) and coupled to carrier proteins as mentioned above. Peptides used in our laboratory were synthesized by Biosyn Ltd. and Immune Systems Ltd. Free peptide was used for *in vitro* immunization regimes and peptide conjugated to BSA or KLH was used for *in vivo* immunizations and inital screening of hybridomas by ELISA. Peptide conjugation kits with a choice of two carrier proteins and two heterobifunctional linkers were obtained from Immune Systems Ltd. for conjugation of the chosen peptides in our laboratory or alternatively peptides were obtained conjugated to a carrier protein of choice from Biosyn Ltd.

Adjuvants

Adjuvants are essential to induce a strong antibody response to soluble antigens. Most adjuvants incoporate two components, i.e., a substance designed to protect the antigen from rapid degradation and a substance that will nonspecifically stimulate the immune system. Commonly used adjuvants are Freund's complete (FCA) and incomplete adjuvants (FIA) (Sigma, UK). Both consist of nonbiodegradable mineral oils with the addition of heat-killed *Mycobacterium tuberculosis* (which contains muramyl dipeptide in its extracellular matrix which stimulates immune cell function during primary immunization) in FCA. In our laboratory, Freund's is the adjuvant used for most immunization regimes. However these reagents are hazardous to the researcher and can also cause lesions at the injection site in the animal.Another adjuvant that can be used successfully (and reportedly less toxic) is Imject-Alum (Pierce Warringer) which is an easy-to-use preformulated immunogen grade alum (alum is used as a alternative to Freund's adjuvant but traditionally was tedious to prepare for immunization regimes).

Animals for Immunization

The animals chosen for immunization should be healthy and free of all known rodent viruses (e.g., mouse hepatitis virus, Sendai virus) or bacterial infections which may alter immune response. Both mice and rats may be used for monoclonal antibody production but all protocols in this chapter refer to production of antibodies in mice. Balb/c mice are the preferred strain of mice used because derivatives of Balb/c myelomas have become the most commonly used fusion partners and the resulting hybridomas can be grown as ascitic tumours in this mouse strain. Two to six animals should be used for each immunization schedule. Animals used for antibody production in our laboratory are purchased from animal suppliers Harlan UK Ltd.).

Choice of Myeloma Fusion Partner

Myelomas can be induced in some strains of mice by injecting mineral oil into the peritoneum. Myeloma cells provide the genetic material necessary for continued cell division of the fused cells in culture. Many of the first myelomas were induced in Balb/c mice (Potter 1972) and derivatives of these are most commonly used . In our laboratory we use the SP2/O-Ag 14 myeloma cell line (Shulman et al. 1978) which is available from both the American Tissue Culture Collection (ATCC; Rockville, MD, USA) and the European Collection of Animal Cell Cultures (ECACC). Myelomas also have all the necessary cellular machinery for the secretion of antibodies but in order to avoid hybridomas which secrete more than one type of antibody, myelomas used for fusions have been selected for lack of functional antibody production. Since only about 1 % of the total starting cells are fused and approximately 1 in 10^5 form viable hybridomas, many unfused cells remain in culture. The unfused lymphocytes from the immunized animal will not continue to grow in culture due to their finite lifespan and will die off naturally, but the myeloma cells are well adapted to grow in culture and a means of eliminating these unfused cells is necessary. This is achieved by the development of myeloma cell lines that grow in normal culture medium but do not grow in a defined selection medium because they lack a functional gene required for DNA synthesis in this selection medium. If, however, these myelomas are fused with normal immunized spleen cells which provide the necesssary genetic material for DNA synthesis, the hybrids continue to grow in the selection medium. Cells possess two pathways for the synthesis of the nucleotides needed for DNA replication, the *de novo* and salvage pathways. Commonly the mutant myeloma cells possess a defective copy of the enzyme hypoxanthine guanine phosphoribosyl transferase (HGPRT) which is essential for purine synthesis using the salvage pathway. HGPRT cells are selected by culturing mutagenized cells in the presence of 8-azaguanine. Post-fusion medium contains the drug aminopterin (which blocks *de novo* purine synthesis) and therefore mutant unfused myeloma cells die, but myelomas fused with B cells from the immunized animal (in which the *de novo* and salvage pathways are both operational) can revert to the salvage pathway in the presence of aminopterin if the medium is exogenously supplied with hypoxanthine and thymidine (thus post-fusion medium is referred to as HAT medium).

- Myeloma growth medium: Dulbecco's modified Eagle's medium (DMEM) with GlutaMAX 1 (Gibco BRL), supplemented with 10 % heat-inactivated fetal calf serum (Gibco BRL)   **media**

Media requirements for cell fusion are as follows:

- Medium 1 (high glucose DMEM): same as above but with 4500 mg/l glucose, 1 % penicllin (10 000 U/ml / streptomycin (10 000 µg/ml)
- Medium 2: Medium 1 but without serum supplement
- Medium 3: Medium 1 supplemented with 5 % BriClone (BioResearch Ireland, Forbairt, Glasnevin, Dublin 9, Ireland), a cloning medium suitable for use in post-fusion stages of hybridoma production. It replaces the function of feeder cells. Also added to this medium is 1 % HAT supplement (100x) (Gibco BRL).

Procedure

Immunization Regimes

Both *in vivo* and a combination of *in vivo* and *in vitro* immunization regimes are routinely used for the production of monoclonal antibodies in our laboratory. *In vitro* immunizations have many advantages over in vivo methods. Only very small amounts of antigen are required which is especially useful when using synthetic peptides. The use of carrier proteins conjugated to synthetic peptides is unnecessary and eliminates the production of antibodies to the carrier protein. Fusions can take place 3–4 days post-immunization as opposed to several weeks for post *in vivo* immunizations. Substances which may prove toxic upon repeated injection, as required in *in vivo* immunizations, can be successfully used in *in vitro* methods. One of the major disadvantages associated with the *in vitro* methods used to date is that the predominantly primary immune response elicited by this method gave rise to the production of mainly IgM class antibodies. This problem is largely overcome by use of the Cell-Prime *in vitro* immunization kit available from Immune Systems Ltd. which through the use of specialized support cells and medium gives rise to approximately 50 % IgG class antibodies. In order to further increase the possibility of obtaining secondary response IgG class antibodies we have combined both immunizations regimes, the *in vivo* primary immunization injection being administered 28 days prior to *in vitro* immunization. It has been demonstrated (De Boer et al. 1988) that a better secondary response is obtainable if the animal used as a source of lymphocytes is immunized *in vivo* 28 days or more prior to *in vitro* immunization.

in vivo immunization procedure

A typical *in vivo* immunization schedule for both synthetic peptides and cellular antigens is described below.

1. Day 1: Synthetic peptide/conjugate is prepared, 100 µg/250 µl PBS in 250 µl FCA **or** 10^6–10^7 cells removed from culture flask, washed three times in PBS and resuspened in 500 µl PBS for intraperitoneal injection.

2. Day 14: First booster injection is administered as above except that FCA is replaced by FIA for the synthetic peptide.

3. Day 28: Repeat as for day 14.

4. Day 35: Tail bleed immunized animals (100–200 µl blood) and assay serum for specific antibody production.

5. Day 42: Repeat as for day 14 for animals showing the best immune response to the antigen. Synthetic peptide may be administered intravenously (i. v.) at this stage (without the use of adjuvant) but only personnel very skilled in animal handling techniques should attempt this procedure.

6. Day 45: Three days after administration of the final booster injection the animals are sacrificed, spleens removed and cell fusions performed.

A combined *in vivo/in vitro* immunization schedule is as follows:

1. Day 1: As for day 1 of in vivo immunization regime.

2. Day 25: Set up support cells from Cell-Prime kit.

3. Day 26: Prime support cells with antigen either unconjugated synthetic peptide at 4–6 µg/ml support medium (20–30 µg total) or whole cells (10^7) irradiated or fixed with glutaraldehyde (to prevent growth *in vitro*).

4. Day 28: Remove spleen from immunized animal, add splenocytes to support cells in culture, add antigen as for day 26 and leave undisturbed for 3–4 days prior to cell fusion.

Note: All steps in the *in vitro* immunization procedure are comprehensively described in the booklet accompanying the Cell-Prime kit.

Note: If preferred, an *in vitro* immunization alone can be performed, the total time between immunization and cell fusion is then reduced to 6–7 days.

Cell Fusion Protocol

All media to be used prior to and during the fusion process and post-fusion plating out medium should be prepared and sterility-tested in advance. Ensure that an adequate number (at least 1×10^7 cells) of rapidly growing, healthy myeloma cells are available. Myelomas should be tested regularly for mycoplasma contamination and any cultures that test positive should be discarded as mycoplasma are very efficient at converting thymidine in the HAT medium to thymine causing the fusion to fail. On the day before the fusion, cells should be subcultured into fresh myeloma medium at a final concentration of approximately 5×10^5/ml. All incubations are in an incubator with 5 % CO_2 at 37 °C.

1. Pipette 0.5 ml of the HAT feeding medium into each well of 8×48-well plates and transfer plates to the CO_2 incubator until the fusion procedure is completed.

2. Aseptically remove spleen from immunized mouse and place in sterile petri dish to which 10 ml high glucose DMEM has been added.

3. Place a sterile cell strainer in a second petri dish containing 10 ml media high glucose DMEM and transfer the spleen to the strainer

using a sterile forceps. Disrupt the spleen and force through the sieve using the plunger from a sterile 10 ml syringe resulting in a homogeneous cell suspension.

4. Draw the cell suspension into a 10 ml sterile pipette and transfer to a 30 ml universal container and allow to stand for 5–10 min to allow any pieces of spleen capsule/fatty tissue to settle out.

5. Transfer the supernatant to a fresh container and wash 3× in medium 2 (see above) by centrifugation at 1000 rpm for 5 min. Gently resuspend pellet in 5 ml medium 2 and return to CO_2 incubator whilst preparing SP2 myeloma cells for fusion. The pellet may look red due to the presence of red blood cells from the spleen but these will not interfere with the fusion procedure. The spleen from an immunized mouse has approximately $5 \times 10^7 - 2 \times 10^8$ lymphocytes.

6. Remove myeloma cells from culture flasks and wash 3× in medium 2 by centrifugation at 1000 rpm for 5 min. Adjust cell number to at least 1×10^7 and resuspend in 5 ml medium 2. This will give a final ratio of spleen cells:myeloma cells of between 5:1 and 20:1.

7. Pool isolated spleen cells and myeloma cells and pellet at 2000 rpm for 5 min. Remove all the supernatant and centrifuge at 2000 rpm for 2 min and aspirate any residual medium from cell pellet.

8. Take up 1.0 ml polyethylene glycol 1540 (PEG) in a sterile Pasteur pipette and very gently resuspend the pellet in the PEG. Start a stop watch. Gently mix the PEG cell suspension with the pipette tip. After 30 s remove the pipette and gently shake the cell suspension.

9. At 75 s, very gently add 0.5 ml medium 3 into the PEG cell suspension and gently but thoroughly mix.

10. At 1, 3 and 4 min post-addition of PEG add a further 1 ml of medium 3 and mix gently. At 5 min add 5 ml of medium 3. Centrifuge at 500 rpm for 5 min.

11. Gently aspirate the supernatant and resuspend the pellet in 10 ml medium 3, incubate for 15 min at room temperature. After this incubation, add a further 15 ml of medium 3 and gently stir the cell suspension.

12. Remove the prepared 48-well plates from the incubator and add one drop of the cell suspension to each well of the 48-well plates using a 10 ml pipette. If any suspension remains continue to add dropwise back over the wells until none remains.

13. Return the plates to the incubator and leave undisturbed for 8–12 days.

14. Inspect hybrids, aseptically collect and test a portion of the supernatant from each hybrid for specific antibody production, if 100 µl supernatant is removed, replace with 100 µl of fresh plating medium 3 above. This screening is normally done by ELISA; it is **imperative** that the method used for the first screening of newly formed hybridomas should be optimized and perfected **before** the fusion is performed.

Expansion and Cloning of Hybridomas

1. Having identified hybridomas secreting antibody against the specific target antigen, the cells are allowed to grow in the 48 wells until nearly confluent. Hybridomas are then expanded to 6-well plates and then to 25 cm^2 culture flasks (preferably flasks with 0.2 µm filtered caps). After 2 weeks hybridomas should be weaned off aminopterin and fed with medium containing HT only. This is continued for a further 2 weeks and thereafter the cells can be grown in the absence of HT also.

cloning of hybridomas

2. Throughout the period of expanding clones, supernatant should be rescreened to ensure cells are still secreting specific antibody. Stocks of positive clones should be frozen down in liquid nitrogen as soon as possible after expansion.

3. Using the limited amount of hybridoma supernatant available, preliminary characterization of a few chosen cell cultures may be performed, i.e., immunofluorescence studies on live cells (to indicate presence or absence of antigen on cell surface), immunohistochemical studies on fixed cells/tissue (to further define location of antigen) and western blotting studies (to determine relative molecular weight of antigen).

4. At this stage most of the hybridoma cultures have arisen from more than one growing colony of cells and the culture is not homogeneous. By definition monoclonal antibodies are homogeneous cultures which have arisen from a single cell. Therefore it is necessary to clone hybridomas to meet this requirement and it also enables one to isolate stable colonies of specific antibody-producing cells. Cloning is normally done by limiting dilution and repeated after the first cloning. (Details of this and other cloning out procedures can be obtained in the reference books cited at the end of this chapter.)

5. When "monclonal" antibodies have been isolated they should be expanded as above and characterized by the methods which follow.

Note: The scope of this chapter does not allow for a detailed overview of the many aspects of monoclonal antibody production. See referenced text books.

6.2
Immunocytochemical Characterization

Immunocytochemistry involves the use of labeled antibodies as specific reagents for the localization and identification of cellular constituents (antigens) in situ. The technique was first described by Coons and Kaplan (1951), who were first to label an antibody with fluorescein and use this to successfully identify an antigen in tissue sections.

Basic Immunodetection Systems

In the direct method the primary antibody is labeled. This conjugate is then applied to the test material in a single step. This method is rarely used as it results in minimum signal amplification and it does not allow antisera to be used in other methods. To increase the sensitivity and allow for greater economy the indirect method was developed. Cells or tissue sections are incubated with an unlabeled primary antibody, followed by an incubation with a secondary antibody conjugated with a label of choice, which is raised against the species of primary antisera (Coons et al. 1955). Sensitivity may be further increased by employing a third step, i.e., a labeled antibody raised against the species in which the secondary antiserum has been raised; this is known as the three-step method. Soluble enzyme immunocomplex methods are extremely sensitive methods which utilize a preformed soluble enzyme/anti-enzyme immune complex which reacts with secondary antibody.This immune complex consists of an enzyme (the antigen) and the antibody directed against the enzyme. Signal amplification results from the large number of enzyme molecules localized to the antigenic site. PAP utilizes a peroxidase/anti-peroxidase complex and APAAP utilizes an alkaline phosphatase/anti-alkaline phosphatase complex (Mason et al. 1969; Cordell et al. 1984). Avidin-biotin methods are multistep methods which are considered generally to be less complicated to perform than PAP and APAAP and are also extremely sensitive. The basic technique utilizes the high affinity of avidin/strepavidin for biotin; biotin can then be coupled to an antibody (usually the secondary antibody) or labeled; avidin/ strepavidin may also be labeled. The avidin biotin complex (ABC) method is the most widely used (Hsu et al. 1981).

Essential Conditions for Successful Immunostaining

- Preservation of antigen, i.e., correct fixation regime
- Specific staining
- Well characterized antibody of high avidity
- Easily visible and easily detected label

Materials

Labels Used in Immunostaining

The use of fluorochromes provide simple and optically sensitive methods, their major disadvantage being the lack of permanent preparation. Fluorescein isothiocyanate (FITC), rhodamine and Texas red are widely used. Enzymes provide a permanent system of labeling and do not require use of specialized microscopic equipment. The most common label of choice is probably horseradish peroxidase; suitable chromogens include 3', 3'-diamino-benzidine (DAB), 3-ethylcarbazole (AEC) and 4-chloro-1-napthol. (**Note:** Many HRP substrates are suspected carcinogens; due care must be taken particularly when handling powdered forms). Alkaline phosphatase is also widely used; suitable chromogens include fast red/fast blue B. The major disadvantage of these enzyme systems is endogenous enzyme activity in some tissues; however this may be overcome by appropriate blocking steps. Gold labeling systems are mainly used at EM level but are also very sensitive at light microscope level. Reaction product of colloidal gold may be intensified by post-incubation with silver (Beesley 1989).

labels

- Strep AB/horseradish peroxidase (HRP) (Dako K337)

reagents

- Labofuge 400 (Heraeus Instruments)

equipment

Procedure

Preparation of Cells for Immunostaining

preparation

1. Using appropriate control cell lines for antibody of interest, wash cells (3×5 min) in PBS by centrifuging at 1000 rpm.

2. Resuspend pellets in PBS, count number of cells and prepare a stock concentration of 1×10^6 cells/ml.

3. Make cytocentrifuge preparations onto poly-L-lysine (Sigma, P-1274) or 3-aminopropyltriethoxy-silane (APES) (Sigma, A3648) coated slides, air dry overnight, foil wrap and store in −20 °C freezer until required.

fixation

Successful immunostaining requires tissue/cellular antigens to be made insoluble and yet their antigenic sites must be unmasked and made available to applied antibody without any great alteration of their tertiary structure. Acetone is an excellent preservative of immunoreactive sites and is used routinely for all antibodies screened in this laboratory.

In establishing the appropriate fixation schedule, it is important to note that the time required for fixation will vary for each individual antibody. In general an epitope located in the region of plasma membrane will require less fixation than one located within the cytoplasmic region of the cell.

1. Remove slides from −20 °C freezer, leave unopened for 15 min then place slides in acetone (BDH, UK) (which has been kept at −20 °C) for 30 s–10 min. Determine optimum fixation time for each individual antibody screened. If the epitope is located in the nuclear region further treatment may be required:

1a. Directly from acetone, transfer slides into 0.1 % Triton-X-100/1x TBS (0.05 M Tris-HCl/0.15 M NaCl, pH 7.6). Incubate for 30 s–3 min at 4 °C. Determine optimum incubation time for each individual antibody screened.

Note: Initially optimal antibody concentration should be determined by titration; this working dilution can thus be used for all outlined immunostaining procedures.

Confirmation of Epitope Localization

To confirm that the antibody is recognizing an external epitope perform immunofluoresence (indirect method) on live cells; by using fixed cells an intracellular location may be confirmed.

A microcentrifuge kept at 4 °C is required.

immunofluo-
rescence on live
cells (indirect
method)

1. Prepare cell suspension in PBS at a concentration of 1×10^6 cells/ml. Place 100 µl of this and 100 µl of test antibody (PBS for negative control) in Eppendorf tubes.

2. Mix tubes by flicking and incubate for 30 min at 4 °C.

3. Wash cells (3×4 min) in PBS by centrifuging at 2000 rpm. Resuspend pellet in 100 µl PBS.

4. **Note:** From this point ensure that everything is covered in tinfoil and mount slides in dark, as FITC and Vectashield are light sensitive. Add 100 µl of anti-mouse IgG FITC-labeled secondary antibody (diluted 1/50 in PBS) (Boehringer, 814385) and incubate for 30 min at 4 °C.

5. Wash cells as before and remove supernatants from tubes, leaving minimum liquid. Aspirate to get an even suspension and add one drop of Vectashield mounting fluid (Vector Labs Ltd., H-1000) to each tube.

6. Place cell suspension directly onto slide, cover, seal with clear nail varnish. View using fluorescent microscope.

immunofluo-
rescence on
fixed cells
(indirect
method)

1. Fix slides in acetone and treat with 0.1 % Triton-X-100 if required, air dry for 15 min.

2. Apply test antibody appropriately diluted in PBS (use PBS as negative control) to each cytospin. Incubate in humidified chamber for 30 min at room temperature.

3. Wash slides (3×PBS) over 10 min.

4. **Note:** From this point ensure that everything is covered in tinfoil and mount slides in dark, as FITC and Vectashield are light sensitive. Apply anti-mouse IgG FITC-labeled secondary antibody (diluted 1/50 in PBS) (Boehringer, 814385) to each cytospin and incubate for 30 min at room temperature.

5. Wash slides as before, drain off excess PBS, mount using Vectashield (Vector Labs Ltd.), cover, and seal with clear nail varnish. View using fluorescent microscope.

ABC Method Using StreptAB Complex/HRP

immunocyto-
chemical charac-
terization

The next stage in the characterization of each test antibody is to carry out an immunocytochemical ABC method in order to stain cells sensitively, and thus look at staining patterns of each test antibody in detail.

1. Fix slides for appropriate time in acetone (treat with 0.1 % Triton-X-100 if required for appropriate time).

2. Air dry for 15 min.

3. To block endogenous peroxidase activity, incubate for 5 min in 0.6 % hydrogen peroxide in methanol. Rinse in distilled H_2O then place in TBS (0.05 M Tris-HCl, 0.15 M NaCl, pH 7.6) for 5 min.

4. To block nonspecific background, incubate for 20 min with 20 % normal serum from species which donated secondary antibody, i.e., normal rabbit serum (Dako, X902)

5. Tap off excess serum and apply test antibody appropriately diluted in TBS for 2 h at room temperature or overnight at 4 °C.

6. Wash in 3×5 min changes of TBS.

7. Apply biotinylated rabbit anti mouse IgG secondary antibody (diluted 1/300 in TBS) (Dako, E345) for 30 min at room temperature. Wash as before.

8. Apply streptAB complex (Dako, K337) prepared according to instructions. Wash as before.

9. Apply horseradish peroxidase substrate, DAB (liquid DAB solution prepared according to instructions; Dako K3465) for 7–10 min at room temperature. Rinse in water to stop reaction.

10. Counterstain lightly with Harris's hematoxylin (Sigma, HHS-1S) (1 min) if required, rinse.

11. Differentiate in 1 % acid alcohol (10 s), rinse, "blue" in Scotts tap water (15 s), rinse.

12. Dehydrate in graded alcohols (2×3 min in 70, 90 and 100 %) and clear in xylene (BDH) (2×5 min).

13. Mount in synthetic medium such as DPX (BDH). Observe by light microscopy.

Results

The specificity of immunocytochemical results relies totally on the properties of the primary test antibody. The test antibody raised should be able to react uniquely with the specific antigen. Care must be taken in the interpretation of results. Short amino acid sequences of an identical nature may be present in various proteins; thus an antibody directed against such a sequence, despite specifically staining the antigen of interest, may also recognize nonrelated molecules (Bendayan 1995). Nonspecific staining of anti-sera may also be due to the mixed population of antibodies in the donor's serum which may react with cellular components additional to the required specific antigen. It is vital to distinguish between specific staining and false positive staining; correct use of controls and appropriate blocking will enable correct interpretation of the results obtained.

interpretation of results

Appropriate controls must be included in all experiments carried out. A **positive control** must be included, i.e., a cell line which is known to be positive for test antibody. **Negative controls** should omit test primary antibody and replace with control mouse ascites fluid (Sigma, M 8373), used at the same concentration as primary antibody, when screening ascites. When screening supernatants use a control mouse IgG such as that available from Vector Labs Ltd. (I-2000); use at recommended concentration range. Alternatively an irrelevant antibody may be used as first layer; if a known negative cell line is available it should also be included. If time constraints allow, include anti-sera specificity controls by applying test antibody preabsorbed with: (1) specific antigen, (2) inappropriate antigen, and (3) antigen fragments/structurally related antigen.

controls

▪ Troubleshooting

There are a number of steps that may be included to reduce background/ nonspecific staining:

- Use a higher dilution of primary antibody, increase incubation time.

- Dilute primary antibody in buffer of high ionic strength (at least 0.15 M NaCl).

- Include 0.05 % Tween 20/ 0.1 % BSA/ 0.03 % in washing buffers/antibody diluents.

- When working with cells which tend to have very adhesive properties (e.g., mucus or serous secreting cells), include 0.03 % casein in blocking buffer, antibody diluents and washing buffers to reduce stickiness (**Note:** addition of casein or Tween 20 may mask some antigens).

- Block with normal serum (20 %) from species which donated secondary antibody. Inclusion of 1 % normal serum in antibody diluents may also help.

- Ensure adequate blocking steps for the enzyme label employed are included; for peroxidase, blocking is as described; for alkaline phosphatase, add 1 mM levamisole to incubating medium. If required use an avidin-biotin blocking system (Vector Labs Ltd. SP-2001).

6.3
Immunoblotting

Immunoblotting can be a useful tool in the characterization of monoclonal antibodies. This technique combines the specificity of immunochemical detection with the resolution of polyacrylamide gel electrophoresis. A number of important characteristics of protein antigens can be determined using this technique, namely, presence and quantity of the antigen, its relative molecular weight and the efficiency with which it can be extracted from crude protein preparations.

The immunoblotting procedure can be divided into a number of steps: (1) preparation of the sample, (2) resolution of the protein sample by gel electrophoresis, (3) transfer of the separated polypeptides to a support membrane, (4) blocking nonspecific binding sites on the support membrane, (5) probing with a primary antibody and (6) detection of the desired polypeptide via a labeled secondary antibody.

▓ Materials

Sample Preparation

The most commonly used sample is obtained from crude whole cell lysates which are prepared from cells in culture, tissues or organs. Crude samples may be used for electrophoresis without further purification. Alternatively, plasma membrane, cytosolic or nuclear preparations can be isolated and used instead. Total cell lysates may be prepared by various methods.The methods favored in this laboratory are: (a) lysis of cells by sonication (times may vary depending on the cell type) and (b) incubation in a lysis buffer. Buffer constituents will differ depending on the antigen being investigated. NP-40 buffer is a good choice for new antigen investigation. Almost all protein preparations can be mixed in sample buffer which is then loaded onto the gel. Discontinuous Tris/glycine SDS-polyacrylamide gel electrophoresis (SDS-PAGE) is the most common choice of gel system, therefore it is important that the sample buffer be compatible with the buffers in this gel system. Sample buffers usually contain a reducing agent such as mercaptoethanol, a denaturing reagent, SDS and a tracker dye, e.g., bromphenol blue. Some preparations may require boiling which will further denature the protein; if the signal is weak try different boiling times and temperatures.

Gel Electrophoresis

Tris/glycine gels can be made up or purchased precast depending on the apparatus used, e.g., Hoeffer mighty small II or Novex (San Diego, CA, USA) for precast gels. The concentration of acrylamide used in the gel will depend on the size range of the proteins to be separated, e.g., 5–15 % acrylamide can be used to resolve proteins in the range of 200–12 kDa. A "stacking gel" of low acrylamide concentration is layed over the resolving gel to allow the protein sample to concentrate at the interface of the two gels before being separated by the resolving gel.

Transfer of Separated Proteins (Western Blotting)

A number of different types of membranes can be used such as nitrocellulose (Sartorius, Göttingen, Germany) and PVDF (Polyvinylidene difluoride, Boehringer Mannheim, Germany) which is the preferred support membrane used in our laboratory. Actual transfer is achieved by electrotransfer. Electrodes may be in a vertical (wet transfer) or horizontal (semi-dry transfer) configuration. Factors influencing the transfer of polypeptides include percentage of acrylamide used and size of the polypeptides. Thicker gels, although easier to handle ,will give a lower transfer of polypeptides. Our laboratory routinely use 0.75-mm-thick gels in conjunction with a semi-dry transfer system (Bio-Rad, Hercules, CA, USA). It is important to remove any air bubbles as this will reduce the efficiency of transfer.

Blocking the Support Membrane

Unoccupied sites on the membrane must be blocked to prevent nonspecific binding of the primary or secondary antibody. Although 5 % nonfat milk is routinely used in our laboratory, this can mask some antigens. Alternatives include Tween and BSA.

Probing the Immunoblot

Primary antibodies (those recognizing the polypeptide) may be purified or in the form of ascites fluid or culture supernatants. It may be necessary to use BSA or nonfat milk and/or Tween 20 in the antibody diluent to reduce nonspecific background. Incubation times will vary depending on the antigen concentration and antibody affinity for the antigen, e.g., 2 h at room temperature to overnight at 4 °C. The blot should be agitated while being probed with the primary antibody.

Detection

The method used to detect the antigen will depend on the choice of labeled secondary antibody. The most sensitive nonradioactive method currently available is detection by chemiluminescence. A number of companies produce kits (Amersham, Arlington Heights, IL USA; Super Signal and Super Signal Ultra, Pierce, Rockford IL, USA) which are used in conjunction with a HRP labeled secondary antibody. Exposure times will vary depending on the primary antibody and concentration of the antigen.

Materials for Separation of Proteins by SDS-PAGE

– SDS-PAGE (Lammli discontinuous system) recipes for two 0.75 mm **gels** thick gels are as follows:

	Resolving gel (%)				Stacking gel (%)
	7.5	10	12	15	5
Acrylamide stock (ml)[a]	3.8	5.0	5.25	7.5	0.8
Distilled H_2O (ml)	8.0	6.8	6.45	4.3	3.6
1.875 M Tris-HCl (ml)	3.0	3.0	3.0	3.0	–
1.25 M Tris-HCl (ml)	–	–	–	–	0.5
10 % SDS (µl)	150	150	150	150	50
10 % APS (µl)	60	50	50	50	17
TEMED (µl)	9	7.5	7.5	7.5	5

APS, ammonium persulfate; TEMED, N,N,N',N'-Tetramethylethylethylenediamine (Sigma-Aldrich).
[a] For recipe, see below.

– NP-40 lysis buffer: 50 mM Tris-HCl; 150 mM NaCl; 1 % Igepal (replaces **solutions** NP-40). Dissolve Tris and NaCl in 50 ml of distilled water. Adjust to pH 8.0 with concentrated HCl and make up to a final volume of 100 ml. Add 1 ml of Igepal. Store at 4 °C for up to 2 weeks.
– Acrylamide stock solution: 29.1 g acrylamide; 0.9 g N-N-methylene bis-acrylamide. Dissolve in 60 ml distilled water. Make up to a final volume of 100 ml and filter. Store in the dark for up to 2 weeks.
– 1.875 M Tris-HCL buffer, pH 8.8: Dissolve 56.8 g Tris in 200 ml distilled water. Adjust to pH 8.8 with concentrated HCl and make up to a final volume of 250 ml. Store at 4 °C for up to 1 month. If the buffer goes below the desired pH discard the solution and adjust the new buffer slowly with concentrated HCl.
– 1.25 M Tris-HCl buffer, pH 6.8: Dissolve 37.8 g Tris in 200 ml distilled water. Adjust to pH 6.8 with concentrated HCl and make up to a final volume of 250 ml. Store at at 4 °C for up to 1 month. If the buffer goes below the desired pH discard solution and adjust the new buffer slowly with concentrated HCl.
– 10x Electrode buffer (running buffer): 1.9 M glycine; 0.25 M Tris; 0.1 % SDS. Dissolve 144.2 g glycine, 30.3 g Tris and 10.0 g SDS in 1 l distilled water. The pH should be approximately 8.3 without adjusting.

- Sample buffer: 2.5 ml 1.25 M Tris-HCl, pH 6.8 (equal 50 mM in 50 ml); 0.5 g SDS (equal to 1 % in 50 ml); 2.5 ml 2-mercaptoethanol (equal to 5 % in 50 ml); 2.9 ml glycerol (equal to approximately 5 % of 87 % glycerol); 0.1 % bromphenol blue. Make up to 50 ml with distilled water. Aliquots of 5 ml may be stored at −20 °C.
- Equilibration buffer: 25 mM Tris; 192 mM glycine. Dissolve 0.75 g Tris and 3.6 g glycine in 250 ml distilled water. The pH should be approximately 8.3 without adjusting.

Procedure

Immunoblotting Protocol Using a Chemiluminescence Detection System

immunoblotting

1. Prepare the protein samples as desired (crude whole cell lysates, plasma membrane, cytosolic or nuclear preparations) in 2× sample buffer.

2. Load 10–15 μl of protein sample into each well.

3. Mini-gels are run for 55–60 min at 250 V/45 mA or until the tracker dye has reached the end of the gel.

4. Equilibrate the gel in transfer buffer for 15–20 min.

5. Transfer the protein to the selected support, i.e., nitrocellulose or PVDF, by the semi-dry blotting system for a minimum of 20 min with the voltage set at 15 and the limit at 340 mA. Using prestained markers will confirm whether the proteins have transferred. Alternatively the blot may be stained with a suitable reagent such as ponceaus.

6. The blots are then incubated in a blocking solution containing 5 % skimmed milk diluted in TBS. Incubation times may vary from 2–4 h at room temperature.

7. The blots are then incubated with the primary antibody. Incubation times will depend on the antibody being used. Tween 20 at 0.1 % may be added to the diluent to reduce nonspecific backround.

8. Following incubation with the primary antibody the blots are washed three times for 5–10 min in washing buffer (TBS, 0.5 % Tween 20) to remove unbound antibody.

9. Incubate the blot with the HRP labeled secondry antibody (DAKO P447, 1/1500) diluted in TBS, 0.1 % Tween 20.

10. Wash as per step 8.

11. Incubate the blots with the chemiluminescence reagents according to the manufacturer's instructions. Wrap the blots in SaranWrap and gently smoothout any air pockets. Expose to autoradiography film for 15 s to 10 min and develop.

References

Harlow E, Lane D (1988) Antibodies: A Laboratory Manual. Cold Spring Harbor Laboratory, Cold Spring Harbor, NY

Beesley JE (1989) Colloidal gold a new perspective for cytochemical marking. Microscopy Handbooks 17; Oxford University, Royal Microscopy Society, Oxford

Bendayan M (1995) Possibilities of false immunocytochemical results generated by the use of monoclonal antibodies: the example of the anti-proinsulin antibody. J Histochem Cytochem 43:881–886

De Boer M, Ten Voorde GH, Ossendorp FA, Van Duijn G, Tager JM. (1988) Requirements for the generation of memory B cells in vivo for the production of antigen-specific hybridomas. J. Immunol Methods 113 :143–149

Doyle A, Griffiths JB, Newell DG (1996) Cell and tissue culture laboratory procedures. Wiley, Chichester, England

Coons AH, Kaplan MH (1950) Localisation of antigen in tissue cells. J Exp Med 91:1–13

Coons AH, Leduc EH, Connolly JM (1955) Studies on antibody production 1. A method for the histochemical demonstration of specific antibody and its application to a study of the hyper immune rabbit. J Exp Med 102:49–59

Cordell JL, Falini B, Erber WN et al. (1984) Immunoenzymatic labelling of monoclonal antibodies using immuno complexes of alkaline phosphatase and anti alkaline phosphatase (APAAP) complexes. J Histochem Cytochem 32:219–229

Hsu SM, Raine L, Fanger H (1981) Use of avidin biotin peroxidase complex (ABC) in immunoperoxidase techniques. J Histochem Cytochem 29:577–580

Kohler G, Milstein C (1975) Continous cultures of fused cells secreting antibody of predefined specificity. Nature 256:495–497

Laemmli UK (1970) Cleavage of structural proteins during the assembly of the head of bacteriophage T24. Nature 227:680–685

Mason TC, Phifer RF, Spicer SS, Swallow RA, Dreskin RB (1969) An immunoglobulin enzyme bridge method for localising tissue antigens. J Histochem Cytochem 17:562–569

Merrifield RB (1963) Solid phase peptide synthesis 1. The synthesis of a tetrapeptide. Science 85:2149–2154

Pearson WR, Lipman DJ (1988) Improved tools for biological sequence comparison. Proc Natl Acad Sci. USA 85:2444–2448

Polak JM, Van Noorden SV (1992) An introduction to immunocytochemistry: Current techniques and problems. Microscopy handbooks 11. Oxford University, Royal Microscopical Society, Oxford , UK

Potter M (1972) Immunoglobulin-producing tumours and myeloma proteins of mice. Physiol Rev 52:631–719

Shulman M, Wilde CD, Kohler (1978) A better cell line for making hybridomas secreting specific antibodies. Nature 276:269–270

Towbin H et al. (1979) Electrophoretic transfer of protein from polyacrylamide gels to nitrocellulose sheets. Proc Natl Acad Sci USA 76:4350–4354

▪ Suppliers

Dako, 16 Manor Courtyard, Hughenden Avenue, High Wycombe, Bucks HP13 5RE, UK; fax: 01494 452016

Biosyn Ltd. 10, Malone Road, Belfast, BT9 5BN, Northern Ireland

Immune Systems Ltd., PO Box 120, Paignton, TQ4 7XD, UK

Pierce Warringer, 44, Upper Northgate Street, Chester, CH1 4EF, UK

Harlan UK Ltd., Shaw's Farm, Blackthorn, Bicester, Oxon OX6 0TP, UK

Vector Labs Ltd., 16 Wulfric Square, Bretton, Peterborough PE3 8RF, UK, fax: 01733 263048

The Generation of Long-Term Human Antibody-Producing Cell Lines Using the Epstein-Barr Virus

OLIVIA FLYNN AND DERMOT WALLS*

Introduction

Epstein-Barr virus (EBV) is a ubiquitous human herpesvirus. Following primary infection, whether symptomatic or silent, the virus persists in the healthy host for life through mechanisms that are not fully understood. It is estimated that 80 %–90 % of the total world population is infected with this virus. EBV is associated with a broad spectrum of benign and malignant diseases including infectious mononucleosis (IM), oral hairy leukoplakia of the tongue, Hodgkin's disease, African endemic Burkitt's lymphoma (BL), anaplastic nasopharyngeal carcinoma (NPC), acquired immunodeficiency syndrome (AIDS)-related lymphomas and many post-transplantation lymphoproliferative disorders (Henle and Henle 1979; Kieff and Liebowitz 1990; Magrath 1990).

When used (in vitro), the virus infects and transforms human B lymphocytes which will thereafter grow continuously in culture as lymphoblastoid cell lines (Katsuki et al. 1977). EBV infection also causes polyclonal activation of B lymphocytes with the synthesis and secretion of immunoglobulin (Leibold et al. 1975). The combination of these two properties of EBV makes it a useful tool for the immortalization of human antibody-secreting cells.

Outline

This chapter will describe how infectious EBV is prepared and how it may be used to generate long-term human antibody-producing cell lines from the peripheral blood mononuclear cells (PBMC) of immune individuals (Fig. 7.1).

* *Correspondence to* Dermot Walls, School of Biological Sciences, Dublin City University, Glasnevin, Dublin 9, Ireland; phone +353–1–7045600; fax +353–1–7045412; e-mail wallsd@ccmail.dcu.ie

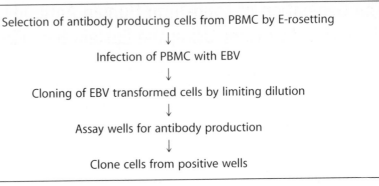

Selection of antibody producing cells from PBMC by E-rosetting
↓
Infection of PBMC with EBV
↓
Cloning of EBV transformed cells by limiting dilution
↓
Assay wells for antibody production
↓
Clone cells from positive wells

Fig. 7.1. Flow chart of the procedure

▪ Materials

precautions   EBV is a pathogen from hazard group 2. These are agents of ordinary potential hazard. This class includes agents that may produce disease of varying degrees of severity from accidental inoculation or injection or other means of cutaneous penetration, but they are contained by ordinary laboratory techniques (Laskin and Lechevalier 1984). The general use of EBV transformed cell lines and of EBV as a tool for immortalizing cells raises concerns as to the general precautions necessary for handling this type of material. Although most adults will have been infected with EBV, precautions should be taken as there is a risk involved for individuals who have not yet come into contact with the virus. The following general rules should be adhered to:

- A dedicated class 2 microbiological safety cabinet and CO_2 incubator should be used for all EBV work.

- Latex gloves, face masks and laboratory coats should be worn during all manipulations of EBV.

- Spillages and waste solutions containing virus should be treated with hypochlorite; benches used for EBV work should be wiped with a hypochlorite solution when work is complete.

- All culture vessels, pipettes, etc., associated with EBV work must be autoclaved before being discarded.

- Laminar flow cabinets should be fumigated before any maintenance work is allowed on both the main and pre-filters.

7.1
Preparation of Infectious EBV

In most laboratories, the cell line B95–8 (ATCC cat. no. CRL 1612) is used as a source of virus (Miller et al. 1972). This is a cell line derived from cotton-top marmoset PBMCs by in vitro infection with wild-type virus from an IM patient. About 2–5 % of cells produce virions which are released into the medium.

Materials

– Tissue culture flasks (Costar 75 cm^3, no. 3375)   equipment
– 0.45 µm filter (Whatman no. 6780–2504))
– CO_2 incubator at 37 °C

Complete RPMI medium is made up of the following: medium

– 1x RPMI-1640 (Gibco BRL no. 041–1870M)
– Fetal calf serum (10 %): heat inactivated at 56 °C for 30 min
– L-Glutamine: 10 mM (Gibco BRL no. 043–5030H)
– Penicillin: 100 IU/ml (Gibco BRL no. 043–5140H)
– Streptomycin: 100 µg/ml (Gibco BRL no. 043–5140H)

Store at 4 °C; fresh L-glutamine must be added every 2 weeks.

Procedure

1. Grow B95–8 cells in suspension using a tissue culture flask at 37 °C   preparation of
   and 5 % CO_2 in complete RPMI medium. Cells should initially be   infectious EBV
   seeded at a density of 5×10^5/ml.

2. Using a healthy culture that is at a density of around 10^6/ml, harvest
   the supernatant by centrifugation at 400 g for 15 min.

3. Pipette the supernatant into a clean tube and filter it through a
   0.45 µm filter.

4. Centrifuge the filtrate at 13 000 g (17 000 rpm) for 2 h at 4 °C.

5. Carefully remove the supernatant and resuspend the pellet in one
   twentieth volume of the original culture medium. Store at −70 °C.

Results

B95–8 cells can be routinely maintained by twice-weekly subculture in this medium. A significant number of cells may firmly adhere to the bottom of the flask and may be loosened with a few gentle taps; the degree of adherence depends on the batch of fetal calf serum used. The filtration step removes cells but allows passage of the enveloped, infectious virus particles. The filtrate prepared in step 3 above can be used to infect human cord blood cells and should have a transforming titer of approximately 10^{-3}. Steps 4 and 5 will significantly increase this titer. Viral infectivity falls off when stored at temperatures above $-70\,°C$, but will be preserved almost indefinitely at or below this temperature.

7.2
Generation of Human Antibody-Producing Cell Lines Using EBV

EBV activates B lymphocytes to produce immunoglobulin (Ig) in a T cell-independent manner (Rosen et al.1977; Henderson et al.1977; Yarchoau et al.1983). All classes of Ig can be detected, but IgM usually predominates. It is preferable to use a donor that has been recently immunized with the antigen concerned. This may be due to a natural primary infection or a seropositive individual who has been given an antigenic boost prior to bleeding.

The selection of specific antibody-producing cells both prior to and following EBV transformation is an essential step in the successful production of long-term monoclonal antibody-producing cell lines. Rosetting with antigen-coated red cells is the method used below to preselect specific antibody-bearing B cells from PBMC (Snow 1995).

Following EBV infection of PBMC, it is common to be able to detect specific antibody in the supernatant medium after 2–3 weeks of culture. Continued growth of the culture invariably leads to a fall in specific antibody levels which become undetectable after 12–20 weeks. Ig-producing cells appear to have a selective disadvantage in the culture, either because they are slower growing or because they have more stringent nutrient requirements. In order to overcome this, virus-transformed cells are cloned early in the culture period by limiting dilution using feeder cells.

Materials

- Peripheral blood mononuclear cells
- 2-Aminoethylisothioronium bromide (AET) (Sigma)
- Sheep red blood cells
- RPMI 1640 (GibcoBRL)
- Cyclosporin A (Sandoz Corp.)
- Microtiter plates, 96-well round bottom (Sigma)
- BSS-FCS-ASP (balanced salt solution-fetal calf serum-aspirin)
- TNP-HRBC (picrylsulfonic acid-haptenated horse red blood cells; TNP from Sigma)
- Percoll stock (Pharmacia)
- 10x Hanks' balanced salt solution (HBSS) (Sigma)
- 1.0 M Hepes, pH 6.8 (Sigma)
- Aspirin solution (40 mg/ml in 95 % EtOH) (Sigma)
- 50 ml centrifuge tubes (Nalgene, no. 3117–0500)

Procedure

Selection of Antibody Producing Cells from PBMC by E-Rosetting

This protocol is adapted from Snow 1995.   E-rosetting

1. Preincubate the rosetting tubes (50 ml round bottomed/pyrex tubes) for 10 min in an ice water bath, then add 6×10^7 cells in 2 ml BSS-FCS-ASP per tube along with 1 ml of the 1.5 % TNP-HRBC solution.

2. Centrifuge the cells at 200 g at 4 °C for 10 min.

3. Gently resuspend the contents with a 6 inch, glass Pasteur pipet and transfer to individual 15 ml centrifuge tubes preincubated in an ice water bath.

4. Centrifuge at 50 g for 40 s at 4 °C; pour off the supernatant and add 200–300 µl of BSS-FCS-ASP to each tube.

5. Gently resuspend the pellets and transfer to a single 50 ml centrifuge tube preincubated in an ice water bath. **Note:** This step significantly reduces the number of non-rosetted cells present in the suspension, thus increasing the efficiency of the Percoll gradients described below.

6. Pellet the rosetted cells at 200 g for 10 min at 4 °C.

Percoll gradient

7. Pour the Percoll gradients into 50 ml centrifuge tubes which have been washed with 95 % ethanol. To maintain the integrity of the gradients they should not be overloaded.

8. The Percoll layers are prepared as follows. **Note:** The recipe is for one tube. For each sample, you need two tubes.

Layer	Percoll stock (ml)	Distilled H$_2$0 (ml)	10× HBSS (ml)	1.0 M Hepes, pH 6.8 (ml)	Aspirin solution (ml)
90 % (1.09 g/ml)	12.85	4.75	2.0	0.2	0.2
100 % (1.10 g/ml)	5.0	3.9	1.0	0.1	–
70 % (1.07 g/ml)	7.5	1.4	1.0	0.1	–

9. The Percoll solutions are returned to neutral pH with 1 N NaOH.

10. Preincubate the Percoll tubes in an ice water bath prior to the addition of the solutions. Place 5 ml of 100 % solution and 0.05 ml of the aspirin solution in each tube.

11. Gently resuspend the pelleted cells (from step 6) in 5 ml of 90 % Percoll solution (2.5 ml from each of the two tubes of 90 % Percoll) and return 2.5 ml into each 90 % tube.

12. Layer the contents of each tube upon the 100 % solutions in each centrifugation tube (a total of 20 ml of the 90 % solution containing the rosetted cells per tube).

13. Layer 5 ml of the 70 % solution onto the gradients. A sharp line of demarcation must exist between each Percoll layer for the gradients to properly separate the free from the rosetted cells.

14. Centrifuge the gradients at 50 g for 20 min at 4 °C.

15. Following centrifugation, suction down each gradient to the bottom 2–3 ml, the rosetted cells are in the pellet. Resuspend the pellets in BSS-FCS-ACP and transfer to a fresh 50 ml centrifugation tube (pool cells from all Percoll gradients in 10 ml of BSS-FCS-ASP).

16. Retrieve 0.1 ml of this and add 0.9 ml of BSS. Evaluate by light microscopy for the percentage of rosette-forming cells present, typically 80 %–90 %.

17. Since T cells do not form stable rosettes under these conditions, virtually all of the rosettes are B cells.

18. Add an additional 30 ml of BSS-FCS-ASP to the centrifuge tube and pellet the cells.

    Note: Preselection of B cells specific for certain haptens or immunoglobulins may also be accomplished by adherence to immobilized antigen.

Infection of PBMC with EBV and Cloning of EBV Transformed Cells by limiting dilution

1. Wash the PBMC twice by centrifugation at 800 rpm for 10 min.

2. Resuspend $1–2\times10^6$ T cell depleted PBMCs in 1 ml of B95–8 virus preparation (see "preparation of infectious EBV", step 5) and incubate for 2 h at 37 °C in a 5 % CO_2 humidified atmosphere. Gently agitate periodically to prevent the cells from sedimenting during the incubation period.

3. Dilute the cells in complete RPMI medium to a concentration of 5×10^5/ml.

4. Place 100 µl of this cell suspension in individual wells of a 96-well round bottom microtiter plates. Two to four plates will be required for this.

5. Dilute residual cells 1:5 with complete RPMI and seed an additional two to four plates.

6. Continue carrying out serial 1:5 dilutions until less than one B cell is reached in an attempt to derive cell lines that are more likely to approximate clones of autoantibody producing cells.

7. To each well add $2–5\times10^4$ irradiated (3000 rads) feeder cells (see "Results," below) suspended in 100 µl complete RPMI medium containing cyclosporin A at 5 µg/ml.

8. Feed the cultures every week or so with complete RPMI medium containing cyclosporin A at 5 µg/ml.

infection with EBV and cloning

9. Screen the wells for specific antibody production as soon as proliferating foci of cells are visible, usually at 10–14 days.

10. Transfer cells from any positive wells to 24-well plates containing 1 ml culture medium **without** cyclosporin A. Assay the supernatant again when the cultures become dense.

11. Clone out cells from positive wells once at ten cells per well and once at one cell per well.

Results

PBMC that have been purified prior to infection with EBV require a cell feeder layer (Pope et al.1974). Many cell types have successfully been used for this purpose: fetal fibroblasts (Rosen et al. 1983); allogenic PBMC, in particular those from individuals who are seronegative for antibodies to EBV antigens (Crawford et al.1983a); and umbilical cord blood mononuclear cells (Stein and Seigal 1983), all of which must be X-irradiated prior to plating. The above method was originally described elsewhere (Crawford 1985; Crawford et al. 1983b; Chiorazzi 1992). Successful cloning has also been carried out in agar (Steinitz et al. 197, Kozbor et al. 1979). In vitro EBV-transformed cell lines show variable growth characteristics, but usually grow well at a cell density of 10^5–10^6/ ml, have a doubling time of 48–72 h, and grow in round clumps of cells.

PBMC from EBV seropositive individuals contain memory T cells that, when stimulated in culture, become specifically cytotoxic for EBV transformed, autologous B cells. This makes it difficult to derive long-term cell lines from PBMC. The above protocol avoids this problem by seeding at low densities (Rickinson et al. 1979) and by the addition of cyclosporin A, an inhibitor of T cell activity, to the medium for 2–3 weeks (Crawford 1981). A T cell depletion step such as density gradient separation of E-rosetted cells prior to infection may also be included for this purpose (Kaplan and Clark 1974).

Reported levels of antibody production of 10 µg/ml and greater have been achieved by this method (Steinitz et al. 1977,1978; Kozbor et al. 1979) and higher (Crawford et al. 1983b). PBMC which have been pretreated with leucine methyl ester(Leu-OMe) or leucyl leucine methyl ester (LeuLeu-OMe) have been shown to exhibit higher cloning efficiencies and improved levels of immunoglobulin production (Ohlin et al. 1989). EBV transformation may be followed by somatic cell hybridization with an appropriate myeloma cell line to rescue or potentiate antibody production (reviewed by Chiorazzi 1992). The use of EBV transformation for the production of human monoclonal antibodies has been reviewed in detail by (Roome and Reading 1984).

References

Chiorazzi N, Wasserman RL, Kunkel HG (1992) Use of Epstein-Barr virus-transformed B-cell lines for the generation of immunoglobulin-producing B-cell hybridomas. J Exp Med 156: 930–935

Crawford DH (1981) Lymphomas after cyclosporin A treatment. In: Touraine JT et al. (eds) Transplantation and clinical immunology XIII. Excerpta Medica, Amsterdam, pp 48–52

Crawford DH, Barlow MJ, Harrison JF, Winger L, Huehns ER (1983a) Production of human monoclonal antibody to rhesus D antigen. Lancet 1: 386–388

Crawford DH, Callard RE, Muggerridge MI, Mitchell DM, Zanders ED, Beverly PCL (1983b) Production of human monoclonal antibody to X31 influenza virus nucleoprotein. J Gen Virol 64:697–700.

Crawford DH (1985) Production of human monoclonal antibodies using Epstein-Barr virus. In: Engleman E et al. (eds) Human hybridomas and monoclonal antibodies . Plenum, New York, pp 37–53.

Henderson E, Miller G, Robinson J, Heston L (1977) Efficiency of transformation of lymphocytes by Epstein-Barr virus. Virology 76:152–163

Henle G, Henle W (1979).The Epstein-Barr virus as the etiologic agent of infectious mononucleosis. In: Epstein MA, Achong BG (eds) The Epstein-Barr virus. Springer, Berlin Heidelberg New York, pp 297–230

Kaplan ME, Clark C (1974) An improved rosetting assay for detection of human T lymphocytes. J Immunol Meth 5: 131–135

Katsuki T, Hinuma Y, Yamamoto N, Abo T, Kumagi K (1977) Identification of the target cells in B-lymphocytes for transformation by EBV. Virology 83: 287–297

Kieff E, Liebowitz D (1989) Epstein-Barr virus and its replication. In: Fields B, Knipe D (eds) Virology. Raven, New York, pp 1889–1920

Kozbor D, Steinitz M, Klein G, Koskimies S, Makela O (1979) Establishment of anti-TNP antibody producing human lymphoid lines by preselection for hapten binding followed by EBV transformation. Scand J Immunol 10: 187–194

Laskin AI, Lechevalier HA (1984). CRC handbook of microbiology (2nd edn) vol.1, Bacteria, CRC, Boca Raton, pp 560–562

Leibold W, Flanagan T.D, Menezes J, Klein G (1975) Induction of EBV-associated NA during in vitro transformation of human lymphoid cells. J Natl Cancer Inst 54:65–68

Magrath I (1990) The pathogenesis of Burkitt's lymphoma. Adv Cancer Res 55:133–269

Miller G, Shope T, Lisco H, Stitt D, Lippman M (1972) Epstein-Barr virus: transformation cytopathic changes and viral antigens in squirrel monkey and marmoset leucocytes. Proc Natl Acad Sci USA 69:383–387

Ohlin M, Danielsson R, Carlsson R, Borrebaeck CAK (1989) The effect of leucyl-leucyl methylester on proliferation and Ig secretion of EBV-transformed human B lymphocytes. Immunology 66:485–490

Pope JH, Scott W, Moss DJ (1974) Cell relationships in transformation of human leucocytes by Epstein-Barr virus. Int J Cancer. 14:122–129

Rickinson AB, Moss DJ, Pope JH (1979) Long-term T-cell mediated immunity to Epstein-Barr virus in man. II Components necessary for regression in virus infected leucocyte culture. Int J Cancer 23:610–617

Roome A.J, Reading CL (1984). The use of Epstein-Barr virus transformation for the production of human monoclonal antibodies. Exp Biol 43: 35–55

Rosen A, Gergely P, Jondal M, Klein G, Britton S (1977) Polyclonal Ig production after Epstein-Barr virus infection of human lymphocytes (in vitro). Nature. 267:52–54

Snow EC (1995) (In vitro) approaches to evaluate deficits in B-lymphocyte function. Methods Immunotoxicol 1: pp 461–463

Steinitz M, Klein G, Koskimies S, Makela O (1977) EB virus-induced B lymphocyte cell lines producing specific antibody. Nature. 269: 420–422

Steinitz M, Koskimies S, Klein G, Makela O (1978) Establishment of specific antibody producing human lines by antigen preselection and EBV-transformation. Curr Top Microbiol Immunol 81:156–163

Yarchau R, Tosato G, Blaese RA, Simon RM, Nelson DL (1983) Limiting dilution analysis of Epstein-Barr virus-induced immunoglobulin production by human B cells. J Exp Med 157: 1–14

Chapter 8

Primary Culture

Finbar O' Sullivan*, Paula Meleady, Shirley McBride, and Martin Clynes

Introduction

Primary culture involves the initial in vitro cultivation of cells obtained directly from animal tissue. While the cultivation of animal cells on serum clots had been performed since the early 1900s, it has really been in the last 20–30 years that significant progress has been made in the culture of specific differentiated cell types. Much information is available in the literature on the cultivation of specific cell types but is beyond the scope of this short chapter.

A number of firms now exist that supply specific primary cell cultures and associated media (Clonetics Corp., Tissue Culture Services Ltd., Skinethic).

Generation of Primary Cultures

The cells of a tissue exist within a matrix of fibrous proteins (e.g., collagen, elastin, and fibronectin), which are interwoven in a hydrated gel of glycosaminoglycan chains. Numerous cellular junctions are also involved in mechanically holding cells together. Thus to generate a primary culture some disruption of the tissue architecture within which the cells exist must occur. An exception to this is when cells exist in a fluid, e.g., blood cells, ascites and pleural effusions. Disruption of tissues can be performed with one of three methods; mechanical disaggregation, chemical disruption, and enzymatic disruption. Mechanical disaggregation of a tissue simply involves finely cutting (to approximately 1 mm or

* *Correspondence to* Finbar O' Sullivan, National Cell and Tissue Culture Center, Dublin City University, Glasnevin, Dublin 9, Ireland; phone +353–1–7045315; fax +353–1–7045484; e-mail 94970271@tolka.dcu.ie

less), mincing, or shearing the tissue through a sieve. The resulting tissue fragments are plated and outgrowth of cells examined. The major limitation of this method is that it is restricted to soft tissues such as spleen. The method of chemical disruption involves removing divalent cations that are essential components of the adhering junctions between cells. This is achieved by the addition of the chelating agents, e.g., EDTA. However this method,when used alone, is limited to soft tissues that are loosely held together by matrix. The third method available is enzymatic digestion, which utilizes different proteolytic enzymes to degrade the extracellular matrix, releasing the cells. Enzymes usually employed include trypsin, pronase, collagenase, dispase, and elastase. These can be used individually or in combination with each other. Two other enzymes often incorporated are DNase I and hyaluronidase. DNase I is added to clear any DNA that has been released from damaged cells. The presence of this DNA causes the reaggregation of cells as well as making the cell suspension viscous and hence difficult to handle. Hyaluronidase attacks the glycosaminoglycan chains of the extracellular matrix (EM) and is thus occasionally used in conjunction with a proteolytic enzyme. Crude preparations of the proteolytic enzymes are usually employed in the digestion of tissue as they contain small impurities of other enzymes, which may aid in the digestion of the tissue. Each enzyme will have associated with it certain advantages and disadvantages. For example, trypsin and pronase will break up a tissue into its component cells but prolonged exposure can be detrimental to cell membranes (loss of vital cell surface receptors and adhesion molecules). In comparison, collagenase and dispase, although less harmful to cells, usually give an incomplete dissaggregation of the tissue. Thus careful consideration has to be given to the selection of an enzyme that will give the optimum yield and viability from the tissue type. Most techniques for disrupting tissues for primary culture employ a combination of these three methods.

Culture Conditions

Following the isolation of cells from the target tissue, the next question to be addressed is under what conditions the cells should be cultivated. This can be of particular importance in the culture of certain normal differentiated cell types, as the expression of specialized cellular functions, in vitro, is often controlled by the composition of the surrounding microenvironment. Hence decisions as to the composition of the basal media, growth supplements, the EM and the presence of other cell types as co-cultures should be made following consideration of the cell type being cultured and the experimental analysis being under taken.

Many commercial media have been developed using various tumor cell lines as models, so they may not always suit the cultivation of normal cells. Unlike tumor cells, which have the potential to divide indefinitely in vitro, normal cells are capable of only a finite number of cell divisions before the onset of senescence and terminal differentiation. This limited number of cell divisions is known as the Hayflick limit and will differ according to the tissue of origin, for example, cells which are derived from constantly regenerating tissue such as skin or bone marrow cells can be maintained in culture for sustained periods of time before the occurrence of terminal differentiation.

Media which have been developed for the primary culture of normal cells include medium 199 and MCDB media, which supports the growth of diploid fibroblast-like cells from various species, e.g., MCDB 104 for human diploid fibroblast-like cells and MCDB 401 for mouse fibroblast-like cells. However the requirements for growth and proliferation of normal cells have yet to be elucidated fully. Supplementation of media with fetal calf serum (FCS) is still required for most primary cell cultures, particularly in the case of normal cells. However as serum supports the proliferation of stromal cells and may contain factors which inhibit the growth of some cells, much research into the development of serum-free systems for primary culture has taken place, e.g., serum-free primary culture of mouse astrocytes (Yoshida and Takeuchi 1991). One successfully developed serum free media is HITES, which allows for the selective culture of small cell lung carcinoma (Carney 1981; Gazdar 1990).

The use of defined culture conditions allows the growth and differentiation of a cell type to be more closely controlled. For example, Rahemtulla et al. (1989) reported on two serum-free media (SFM) suitable for the growth of normal human keratinocytes, in which good growth was achieved in a SFM composed of 10 ng/ml epidermal growth factor (EGF) and 0.3 mM calcium, whereas differentiation was enhanced by a combination of low EGF (0.1 ng/ml) and a high calcium (2 mM) concentration. Many SFM have been described in the literature for the cultivation of progenitor and mature hematopoietic cells (Lebkowski et al. 1995; Sandstrom et al. 1994). Hematopoiesis is a tightly regulated process whereby mature cells in the circulating blood are constantly regenerated. The process is regulated by a cascade of hematopoietic growth factors, especially the interleukins and colony-stimulating factors (CSFs). Many of these cytokines have multiple and overlapping functions.

In some cases the addition of specific growth factors to the culture improves the success of the culture. In the cultivation of neuroblastoma tumor cells, whose growth-regulatory mechanisms have been studied extensively, insulin-like growth factor (IGF)-II has been found to be an important stimulatory ligand for the in vitro growth of NB tumor cell lines and is thought to be implicated in the neoplastic proliferation of tumor cells (El-Badry 1991). Ultimately several different media may need to be tested on the target cells before a suitable one can be found.

The use of various EM can be important to achieve a successful primary culture. For example in the case of nerve cells, which generally do not proliferate in-vitro, the choice of substrate is particularly important for their maintenance. They can demonstrate neurite outgrowth and survive relatively well in collagen (Ebendahl 1979) and poly-D-lysine (Yavin and Yavin 1980) coated plates.

Primary culture systems also often utilize feeder layers of irradiated fibroblasts to supply paracrine factors that the cells normally receive from stromal cells, for example, in the cultivation of human carcinoma cells (Klein et al. 1987).

Safety Considerations

caution! Consideration must be given to both personal safety and safety of those in the environs. When performing primary culture, one should be aware of a potential infection risk, particularly when human tissue is being used. Those performing primary cultures should be immunized against hepatitis B. All operations involving human tissue should be performed in a class 2 downflow recirculating laminar flow cabinet specifically designated for such work. The laminar flow should be validated on a regular basis for containment and user protection. The operator should use good quality surgical-type gloves and wear a lab coat. In some cases safety goggles, face shields and masks should be considered, for example, following tissue homogenization, which generates substantial aerosols, or when working with pleural effusions. All waste should be carefully disposed of, in some cases legislation will dictate the method used, e.g., incineration. At the very least such waste should be placed in clearly marked waste bottles and autoclave bags and thoroughly autoclaved. All sharps should be disposed of carefully and incinerated. Instruments should be cleaned in distilled water and nonionic detergent prior to autoclaving to sterilize them fully. Autoclaving of dirty instruments will accelerate corrosion. Surfaces should be disinfected thoroughly as alcohol is not sufficient on its own (Grizzle and Polt 1998).

8.1
Purification of Cell Populations

Any relatively large tissue mass will be composed of a wide variety of differentiated cell types, e.g., fibroblasts, macrophages, epithelial cells, and smooth muscle cells, all of which are released upon disruption of the tissue. Some of these cell types are capable of rapid proliferation in culture, e.g., fibroblasts, and can cause problems if they overgrow cells of interest. Since the aim of almost all primary cultures is to obtain a highly purified specific cell type, it is often necessary to reduce the numbers of contaminating cell types that exist in the cell population obtained. To this aim there exist a number of strategies that can be used to purify the target cell population.

Materials

– Inverted microscope
– Trypsin EDTA solution (sterile)
– Pipettes (sterile)
– Waste bottle (sterile)
– Fresh growth media (sterile)

Procedure

Differential Trypsinization for the Removal of Fibroblasts   trypsinization

The procedure utilizes the phenomenon that fibroblast cells detach first in a mixed culture treated with trypsin. Thus the presence of contaminating fibroblasts in a culture can be reduced or eliminated by trypsinization of the mixed culture.

Note: All operations must be performed aseptically in a laminar flow cabinet!

1. Remove the culture media from the flask.

2. Pipette in pre-warmed trypsin EDTA solution into the flask and replace the lid.

3. Gently remove the flask from the laminar flow and place on the microscope stage.

4. Select an area of the culture that contains both the cells of interest and the contaminating fibroblasts and observe the trypsinization of this area. As soon as the fibroblasts have detached but before the cells of interest, return the flask to laminar flow.

5. With a pipette, gently remove the trypsin-EDTA solution, which now contains the fibroblast cells.

6. Pipette in fresh culture media and return the flask to the incubator.

Differential Attachment

This is a very common and easy way to reduce the amount of contaminating fibroblasts and macrophages. The cell suspension obtained from the digest is plated on a petri dish with a low-serum media. Both fibroblasts and macrophages should adhere to the surface of a petri dish, preferentially leaving epithelial cells in suspension. The supernatant can then be removed, resulting in purer isolates of adherent and nonadherent cells. Note that the greater the surface area to cell ratio, the more successful this technique is. This purification step is used in the isolation of type II cells described elsewhere in this volume.

Separation by Density

This method can allow the removal of contaminating cell types, or clones from a cell population, and potentially a very pure population of a cell type can be achieved (Viscardi et al. 1992; Resnicoff et al.1987). It relies on different cells possessing different densities, e.g., most mammalian cells have a density range of 1.015–1.120 g/ml. The separation involves layering the digest over a gradient, which is subsequently centrifuged at a specific speed and temperature. A wide range of commercial media are available for cell separations which utilize dextran or silica, e.g., Percoll. Ficoll is commonly used to remove contaminating blood from the digest. The presence of blood is undesirable as it interferes with the attachment of epithelial cells. A limitation is the degree of overlap that may occur between different cell types. In the procedure for type II isolations, a Percoll gradient is employed as the final purification step.

Separation by Monoclonal Antibodies

This method relies upon a monoclonal antibody recognising a specific cell surface marker for that cell. This monoclonal antibody is cross-linked to a magnetic bead, thus allowing those cells with bound antibody to be separated by magnet. This procedure is employed with some success in the separation of lymphocyte subpopulations (Glazer et al. 1990). A modification to this procedure is where the monoclonal antibody is cross-linked to the surface of the petri dish. Thus the target cells remain in the dish following removal of the supernatant. This technique is known as panning. Monoclonal antibodies linked to a fluorescent molecule can be used to separate cells with a flow cytometer, also known as a fluorescent-activated cell sorter (FACS),where cells are separated according to the surface charge.

Selective Media and Addition of Antimitotic Agents

Alterations in the basal media may be employed in cultures in which the selection or elimination of a particular cell type is required. For example the use of D-valine MEM to inhibit fibroblast growth is well characterized. The selection works on the principle that fibroblasts fail to express the enzyme D-amino acid oxidase, thus they are unable to perform the stereoconversion of D-amino acids to L-amino acids. Thus if an essential amino acid is supplied in the culture system only in its D-isoform, fibroblast growth should be inhibited. D-valine MEM is often utilized as a selective basal media. Some fibroblasts, however, show little or no inhibition in growth when cultivated in D-valine MEM (Masson et al. 1993). In primary cultures in which the target cell population shows very slow or no proliferation, there is a risk that rapidly proliferating stromal cells present as a contaminant could overgrow the culture. The addition of antimitotic agents to the culture system may prevent overgrowth in such cultures, such as neuronal primary cultures, which often contain contaminating fibroblasts. These non-neuronal cells are actively proliferating and are capable of overgrowthe nonproliferating neurons in culture. Anti-mitotic agents will kill dividing cells with minimum damage to any long-term culture of neurons. Such an antimitotic agent used in neuronal cultures is cytosine arabinoside (Dichter 1978). The procedure involves the addition of media containing cytosine arabinoside at concentrations of $5-10\,\mu M$ for approximately 24 h and then changing the medium 2–4 days after initial plating of cells.

8.2
Amnion Cell Isolation

This procedure will give a good yield of normal epithelial cells which will last for two passages. However if the cells are required for a physiological or biochemical assay, it is advisable use freshly isolated cells. The procedure also confers the advantages that there are little or no fibroblasts to contaminate the culture and there is a ready supply of this tissue.

▨ Materials

equipment
- Sterile dissection equipment (e.g., scalpel, scissors, forceps)
- Sterile magnetic stirrer
- Sterile PBS A
- Sterile 30 ml universals
- Sterile petri dishes
- Sterile waste bottle

media
- Transport media: DME 1x:Hams-F12 1x (1:1); 10 % FCS; 2 % penicillin/streptomycin; 1 % fungizone; 0.5 % gentamicin; 2 mM L-glutamine. Dispense as 10 ml aliquots into sterile universals and label, i.e., patient name, hospital number, date and time of sample. Arrange for your samples to be collected in this media.
- Digestion media: 42 ml PBS A, 18 ml trypsin (supplied as a 10× stock by Gibco BRL). Final concentration of trypsin is 3x.
- Growth media: Basal media of DME 1x:Hams-F12 1x (1:1) supplemented with 10 % FCS; 2 % penicillin/streptomycin; 0.4 µg/ml hydrocortisone; 2 mM L-glutamine

▨ Procedure

Note: Perform all steps in a class II laminar flow cabinet and take necessary care of sharps and aerosols.

procurement and storage of tissue sample
1. The amnion is collected by the surgical team, placed into sterile transport media and transferred to the lab on ice as soon as possible.
2. Upon receipt of the tissue, wash with sterile PBS and lay out flat on a petri dish with a small amount of sterile PBS. Remove all unwanted tissue such as fat or necrotic tissue.

1. Chop the tissue into 2 mm cubes. Transfer the tissue fragments into a sterile 30 ml universal containing a sterile magnetic stirrer. Add 20 ml digestion media to the tissue fragments.

    dissection and digestion of tissue

2. Stir the tissue sample on a magnetic stirrer so that tissue is gently mixing in the digestion media. This is performed at 37 °C.

3. After 5 min allow the tissue to settle remove and discard the supernatant. This step helps to remove most of the blood and necrotic tissue associated with the tissue.

4. Redigest the tissue fragments with fresh digestion media for 30 min. After the 30 min remove the supernatant and add to it 3 ml of FCS. To the remaining tissue fragments add fresh digestion media and digest for a further 30 min.

5. After the second 30 min of digestion remove the supernatant and add to it 3 ml of FCS. Pool the supernatants and centrifuge at 100 g to sediment the isolated cells.

6. The pelleted cells are resuspended in growth media and plated into the desired culture flask.

8.3
Isolation and Culture of Lung Cancer Cells

Malignant epithelial tumors are the most common lethal lung tumours. These are broadly classified into non-small cell lung cancer (NSCLC) comprising squamous, adeno- and large cell carcinomas, and small cell lung cancer (SCLC). The two classes account for approximately 75 and 25 % of lung cancers respectively. In general, cell lines derived from NSCLCs are adherent while those which originate from SCLCs grow in suspension. However, the morphologic and phenotypic heterogeneity observed in lung tumors in vivo is often mirrored in culture and individual cell lines can grow both attached to the substratum and in suspension (Oie et al. 1996).

▒ Materials

– Sterile dissection equipment (e.g., scalpel, scissors, forceps)   **equipment**
– Sterile magnetic stirrer

- Sterile PBS A
- Sterile 150 ml containers (Bibby Sterilin Ltd.)
- Sterile 30 ml universals
- Sterile petri dishes
- Sterile wire gauze (e.g., tea strainer)
- Sterile waste bottle

media
- Transport media: DME 1x:Hams-F12 1x (1:1); 10 % fetal bovine serum; 2 % penicillin-streptomycin; 1 % fungizone; 2 mM L-glutamine. Dispense this as 30 ml aliquots into sterile universals and label, i.e., patient name, hospital number, date and time of sample. Arrange for your samples to be collected in this media. Transport the sample to the laboratory on ice.
- Digestion media: collagenase 0.4 mg/ml; dispase 0.6 mg/ml; pronase E, 0.6 mg/ml, dissolved in 1x MEM basal media with DNase I 10 U/ml
- Washing media: 1× MEM basal media with DNase I 10 U/ml
- Growth media: basal media of DME 1x:Hams-F12 1x (1:1) supplemented with 5 % fetal bovine serum; 1 µg/ml insulin; 0.5 µg/ml hydrocortisone; 10 ng/ml recombinant human EGF (rhEGF); 1 % fungizone, and 2 % penicillin-streptomycin

Cultivation of Lung Cancer Cells

A wide range of media, both serum-containing and defined serum-free, are currently used to culture the increasing number of lung cell lines which have been established. It is not possible to provide a comprehensive list of those media here. Instead, a selection of the more widely used culture media which are suitable for many lung cell lines will be described.

NSCLC cell lines
- Serum-supplemented media: DMEM: Hams F12 (1:1) or RPMI 1640 supplemented with 5–10 % FCS, 2 mM L-glutamine **or** MEM with same supplements plus 1 mM pyruvate and 1x nonessential amino acids (NEAA)
- Serum-free media (ACL-4, suitable for several NSCLC cell lines): RPMI 1640; insulin 20 µg/ml; transferrin 10 µg/ml; sodium selenite 25 nM; hydrocortisone 50 nM; EGF 1 ng/ml; triiodothyrononine 0.1 nM; ethanolamine 10 µM; phosphorylethanolamine 10 µM; sodium pyruvate 0.5 mM; bovine serum albumin 0.2 %; HEPES 10 mM; glutamine 2 mM (Bower et al. 1986)

For routine culture, NSCLC cell lines with relatively fast doubling times (36–48 h) should be seeded at lower densities than those with slower doubling times (50 h–3 weeks) to avoid excessive passaging manipulations.

- Serum-supplemented medium: The SSM used in NSCLC cell culture can be applied to SCLC cell lines.
- Serum-free media (HITES, a general medium for SCLC lines which can be adapted to suit the requirements of different lines): 400 ml 1× RPMI 1640 supplemented with 2 mM L-glutamine; 10 mM HEPES; 10 nM hydrocortisone; 5 µg/ml insulin; 10 µg/ml transferrin; 10 nM 17-β-estradiol; 30 nM sodium selenite (Carney et al 1981)

SCLC cell lines

Most SCLC cell lines grow in aggregates in suspension. Therefore the trypsinization required when passaging the cells is not involved and all that is necessary is the removal of an aliquot of cells

▓ Procedure

Isolation of Lung Tumor Cells

Note: Perform all steps in a class II laminar flow cabinet and take necessary care of sharps and aerosols.

1. The lung tumor sample is removed by the surgical team, placed into sterile transport media and transferred to the lab on ice as soon as possible.

   procurement and storage of tissue sample

2. Upon receipt of the tissue, wash with sterile PBS and lay out on a petri dish with a small amount of sterile PBS. Dissect the tumor free of the lung mass, removing all unwanted tissue such as fat, bronchioli and blood vessels.

   Note: The center of large tumors tend to become necrotic due to the lack of accessibility to oxygen and nutrients. The outer edge should therefore be the target of the primary culture.

1. 1. Chop the tissue into 2 mm cubes. Transfer to a sterile 50 ml universal containing sterile PBS A. Wash the tissue fragments by inverting three times.

   dissection and digestion of tissue

2. Digest the tissue fragments with 30 ml of digestion media, in a sterile 50 ml tube, which contains a magnetic stirrer.

3. After 30 min remove and retain the supernatant. Redigest the remaining tissue with 30 ml of fresh digestion media for a further 30 min.

4. After a further 30 min remove the supernatant from the digestion and retain. Pool the supernatants and filter through wire gauze to remove large tissue chunks.

5. Pellet the isolated cells by centrifuging at 200 g. Resuspend the cells in washing media and recentrifuge at 200 g; repeat a further two times.

6. Resuspend cells in growth media and plate into desired tissue culture flask.

References

Baron J, Voigt JM (1990) Localisation, distribution and induction of xenobiotic-metabolising enzymes and aryl hydrocarbon hyroxylase activity within lung. Pharmac Ther 47:419–445

Bower M, Carney DH, Oie Hk, Gazdar AF, Minna JD (1986) Growth of cell lines and clinical specimens of human non-small cell cancer in a serum-free defined medium. Cancer Res 46:798–806

Carney DN, Bunn PA, Gazdar, AF, Pagan JF, and Minna, JD (1981) Selective growth in serum free hormone supplemented medium of tumor cells obtained by biopsy from patients with small cell carcinoma of the lung Proc Natl Acad Sci USA 78: 3185–3189

Cunningham AC, Milne DS, Wilkes J, Dark JH, Tetley TD and Kirby JA (1994) Constitutive expression of MHC and adhesion molecules by alveolar epithelial cells (type II pneumonocytes) isolated from human lung and comparison with immunocytochemical findings. J Cell Sci 107:443–449

Devereux TR, Massey TE, Van Scott MR, Yankaskas J and Fouts JR (1986) Xenobiotic metabolism in human alveolar type II cells isolated by centrifugal elutriation and density gradient centrifugation. Can Res 46:5438–5443

Dichter MA (1987) Rat cortical neurons in cell culture: culture methods, cell morphology, electrophysiology, and synapse formation. Brian Research 149: 279–293

Ebendahl T (1979) Stage-dependent stimulation of neurite outgrowth exerted by nerve growth factor and chick heart in cultured embryonic ganglia. Dev Biol 72: 276

El-Badry OM (1991) Insulin-like growth factor II gene expression in human neuroblastoma. Adv Neuroblastoma Res 3: 249–256

Freshney RI (1994) Culture of animal cells: a manual of basic technique, 3rd edition. Alan R Liss, New York

Gazar AF, Linnolia RI, Kurita Y, Oie HK, Mulsine JL, Clark JC, Whistsett JA(1990) Peripheral airway differentiation in human lung cancer cell lines. Cancer Res 50: 5481–5487

Grizzle,WE, Polt SS (1998) Guidelines to avoid personnel contamination by invective agents in research laboratories that use human tissues. J Tissue Culture Methods 11:191–199

Ham RG, McKeehan WL (1979) Media and growth requirements. Methods Enzymol LVII:44–93

Kikkawa Y, Yoneda K (1974) The type II epithelial cells of the lung, 1: methods of isolation. Lab Invest 30:76–84

Klein JC., Zurcher C., van Bekkum DW. (1987) Differential behaviour of human bronchial carcinoma cells in culture Cancer Res 47: 3251–3258

Lebskowski JS, Schair LR, and Okarma TB (1985) Serum free culture of hematopoietic stem cells: a review. Stem Cells 13:607–612

Sandstrom CE, Miller WM, Papoutsakis ET (1994) Review: Serum-free media for cultures of primitive and mature hematopoietic cells. Biotechnol Bioeng 43:706–733

Liu S , Mautone AJ (1996) Whole cell potassium currents in fetal rat alveolar type II epithelial cells cultured on Matrigel matrix. Am J Physiol 270:L577-L586

MassonEA, Atkin SL, White MC (1993) D-valine selective medium does not inhibit human fibroblast growth in vitro. In Vitro Cell Dev Biol 29A: 912–913

Oie HK, Russelle EK, Carney DN, Gazdar AF (1996) Cell culture methods for the establishment of the NCI series of lung cancer cells J Cell Biochem Sup 24: 24–31

Rahemtulla F, Moorer CM and Wille Jr. JJ (1989) Biosynthesis of proteoglycans by proliferating and differentiating normal human keratinocytes cultured in serum free medium. J Cell Physiol 140: 98–106.

Resnicoff M., Medrano EE, Podhajcer OL, Bravo AI, Bouer L, Mordoh J (1987) Subpopulations of MCF-7 cells seperated by percoll gradient centrifugation a model to analyze the heterogeneity of human breast cancer. Proc Natl Acad Sci USA 84:7295–7294

Viscardi RM, Ullsperger S, Resau JH (1992) Reproducible isolation of type II pneumocytes from fetal and adult lung using Nycodenz density gradients. Exp Lung Res18:225–245

Yoshida T, Takeuchi M (1991) Primary culture and cyropreservation of mouse astrocytes under serum-free conditions. Cytotechnology 5: 99–106

Yavin Z,Yavin E (1980) Survival and maturation of cerebral neurons on poly (L-lysine) surfaces in the absence of serum. Dev Biol 75: 454–460.

Abbreviations

CSF Colony stimulating factor

EGF Epidermal growth factor

EM Extracellular matrix

IGF Insulin-like growth factor

NSCLC Non-small cell lung cancer

SCLC Small cell lung cancer

SFM Serum free media

SSM Serum supplemented media

▪ Suppliers

Clonetics Corp, 8830 Briggs Ford Rd., Walkersville, Maryland 21793, USA; fax 001–301–845–1008

Tissue Culture Services Ltd., Boltolph Claydon, Buckingham, MK 18 2LR, UK; fax 0044–1–296–715753

Skinethic, 29 rue Herold, 0600 Nice, France; fax 0033–93–870810

Bibby Sterilin Limited, Tilling Drive, Stone, Staffordshire, ST15 OSA, UK; fax 0785 812121

Gibco BRL Life Technologies Limited, 9 Fountain Drive, Inchinnan Business Park, Paisley, UK; fax 0141 814 6317

Part II

Cell proliferation and Death

Chapter 9

Baculoviral Expression and Partial Purification of Cyclin/CDK Complexes

Mark Jackman*

Introduction

Many different proteins have now been expressed in insect cells using the baculoviral expression system. This system was developed using the *Autographa californica* nuclear polyhedrosis baculoviral strain (AcNPV). Baculovirus protein expression requires infection of insect cells with recombinant virus whose regions encoding for nonessential coat protein(s) have been replaced with a gene encoding for the protein required to be expressed.

Despite protein expression in insect cells being more time-consuming and costly than protein expression in bacteria, the baculoviral expression system offers certain advantages over bacterial protein expression systems. Eukaryotic proteins expressed in insect cells are thought to be more appropriately folded and oligermerized than when expressed in a bacterial environment. Importantly, unlike bacterial expression, expressing a protein in insect cells allows that protein to be posttranslationally modified by phosphorylation, glycosylation, acylation, amidation, signal peptide cleavage, carboxymethylation, prenylation and proteolytic cleavage. Additionally, the baculovirus expression system also permits simultaneous expression of different proteins in the same insect cells allowing the formation of complexes in vivo, making it particularly useful for the study of cell cycle proteins (see, for example, Desai et al. 1992; Parker et al. 1993).

In this chapter we will describe the expression and partial purification of active cyclin/cyclin-dependent kinase (cdk) complexes using the baculoviral expression system. We find levels of cyclin and cdk expression using insect cells to be no greater than those found using bacterial

* Institute of Cancer and Developmental Biology, Wellcome/CRC Institute, Tennis Court Road, Cambridge CB2 1QR, UK; phone +0122–334088; fax +01223–334089; e-mail: mrj@mole.bio.cam.ac.uk

expression systems. Equally, the solubility of cyclins and cdks expressed in insect cell expression systems matches that seen in bacterial expression systems. However, crucially, expression of cyclin/cdk complexes in insect cells allows the production of stable cyclin/cdk complexes owing to the fact that insect cells can provide cdk-activating kinase (CAK) activity, enabling a percentage of cdk co-expressed with its cyclin partner to be phosphorylated on its Thr-160 residue, thus stabilizing the cyclin/cdk complex. It may also be the case that a small percentage of cyclin/cdk complexes expressed in insect cells may be inactive due to the inhibitory Thr14/Tyr15 phosphorylation of cdk; however it seems that some cyclin/cdk is expressed in an active form. It is not clear how toxic the expression of different active cyclin/cdk complexes are to insect cells but cyclin/cdk complexes with a high specific activities (in the range of $1-10\,\mu M$ phosphate/min per mg H1-histone can be obtained using the protocol described below.

Expression of any protein using the baculoviral expression system first requires generation of recombinant baculovirus encoding your desired protein or proteins. An experimental outline for this is given in Fig 9.1. We routinely co-infect insect cells with separate recombinant affinity-tagged cyclin and recombinant cdk virus; however, the emergence of multiple expression vectors for baculoviral expression allows insect cells to be infected with recombinant baculovirus that encodes for both cyclin

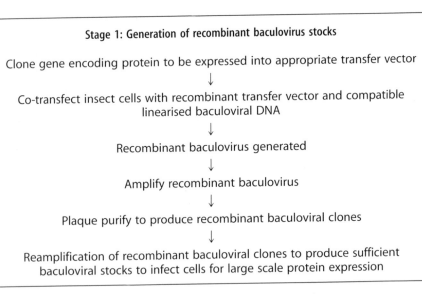

Fig. 9.1. Stage 1: Generation of recombinant baculovirus stocks

---

Stage 2: Coexpression and purification of cyclin/cdk complexes

Co-infect insect cells with a mixture of recombinant baculoviruses encoding
for N-terminally $(His)_6$-tagged cyclin and its cognate cdk

↓

Lyse insect cells and purify $(His)_6$-tagged cyclin and $(His)_6$-tagged cyclin
associated with co-expressed cdk using Ni^{2+}-NTA-Agarose

↓

Separate $(His)_6$-tagged cyclin and $(His)_6$-tagged cyclin associated with
co-expressed cdk by gel filtration or ion exchange chromatography

↓

Monitor purification of cyclin/cdk by assaying all fractions from all purification
steps for H1 histone kinase activity

Fig. 9.2. Stage 2: Coexpression and purification of cyclin/cdk complexes

and cdk. Once recombinant baculovirus has been generated, active cy-
clin/cdk complexes can be co-expressed and partially purified by the
method outlined in Fig 9.2. Briefly, this entails purification of cdk and
cyclin/cdk complexes expressed in insect cells via an affinity tag such as
the $(His)_6$ affinity tag followed by separation of monomeric cdk from
cyclin/cdk complexes. There is no reason why this protocol might not be
adapted for the purification of any multiprotein complex baculovirally
expressed in insect cells. It is beyond the scope of this chapter to go into
any detailed description of protein expression in insect cells using the
baculoviral expression. For this the reader is directed to the excellent
laboratory manual devoted to baculovirus protein expression systems
(O'Reilly et al. 1992). Many different companies now sell products for
baculoviral driven protein expression (for example Pharminogen, San
Diego, CA; Invitrogen, San Diego, CA; Clontech, Palo Alto, CA) whose
product catalogues are a useful source of introduction to the increasing
number of different baculoviral expression systems becoming available.

9.1
Culturing of Insect Cells

▨ Materials

- Sf9 insect cells
- Standard mammalian tissue culture facilities: laminar flow hood, tissue culture plasticware, etc.
- Low temperature cooled incubator: standard incubators have trouble maintaining temperatures below 28 °C
- G10: Graces' insect cell media, 10 % heat inactivated fetal calf serum, supplemented with 50 µg/ml gentamicin sulfate (all from Gibco BRL)

▨ Procedure

There are many different insect cell lines now available for baculoviral expression, but we routinely use the Sf9 insect cell line. Sf9 cells are cultured in G10. Amphetoricin (fungizone; Gibco BRL) can also be used if necessary; however routine use of this may mask low level yeast and fungal infection which may compromise optimal cell growth. Sf9 cells are cultured at 27 °C, do not require carbon dioxide and can be easily switched from monolayer cultures (but we find that cells that have been growing in suspension for longer than 3 months do not immediately readily adapt to monolayer culturing). Both monolayer and suspension insect cell cultures should be split every 3 days. Healthy insect cells should divide roughly every 24 h.

monolayer cultures
For generation and amplification of recombinant baculovirus we use insect cells grown on plastic. Sf9 cells are passaged by gently dislodging cells from plastic by tapping the dish and pipetting a stream of media over the confluent monolayer. It is very difficult to remove all cells using this method (one should only aim to remove as many as necessary, usually roughly 50–70 % of cells). Confluent monolayers should be split three to five times. Insect cells are relatively delicate and should not be vigorously vortexed or subjected to centrifugal forces greater than 1500 g. With a healthy culture greater than 95 % of Sf9 cells should attach to the plastic within 20 min. As monolayers become overgrown some cells become detached. This is normal and these cells are still viable.

For large-scale protein expression we use suspension cultures of Sf9 cells. Sf9 cells should be seeded at an initial concentration of 0.5×10^6 cells/ml in spinner flasks stirring at 50–60 rpm. Although cell densities in suspension can reach 4×10^6 cells/ml they routinely should not be allowed to increase to a density greater than $2–3 \times 10^6$ cells/ml. To subculture remove some of the cell suspension and add fresh media to give a final cell density of not less than 0.5×10^6 cells/ml. Every 14 days it is prudent to give a complete change of media and change the stirring flask. For efficient growth of Sf9 cells it is important to supply good aeration to suspension cultures.

suspension
cultures

9.2
Generation of Recombinant Baculovirus

Cloning of Gene into Appropriate Transfer Vector

There are now many different baculoviral transfer vectors available from many manufacturers. We routinely use the Baculogold system (Pharminogen, San Diego, CA) to obtain recombinant baculovirus. The Baculogold system relies on cotransfection of linearized baculoviral DNA containing a lethal deletion and a *lacZ* gene with a compatible recombinant transfer vector that is capable of rescuing the lethal deletion and replacing the *lacZ* gene. Thus only recombinant baculoviruses are able to replicate within coinfected insect cells. Positive survival selection of recombinant virus makes clonal isolation of recombinant baculovirus unnecessary. However, we recommend that recombinant baculoviruses are cloned using the plaque assay method (see below) to be sure of the identity of your recombinant baculovirus and to optimize protein expression. We express cyclin and cdk using the polyhedrin "late" baculoviral promoter. Our cyclin/cdk baculoviral protocol requires an affinity tag to be placed on the NH_2-terminal of the cyclin and we have used this protocol for the purification of cyclin D, E, A, B1 and B2/cdk complexes. There are many commercially available transfer vectors that enable incorporation of a $(His)_6$ affinity tag.

Cotransfection of Recombinant Transfer Vector with Linearized Baculoviral DNA

Either calcium phosphate precipitation or lipofectin based transfection methods can be used to transfect insect cells. Some constructs appear to be more efficiently transfected into insect cells using the lipofectin based method (GibcoBRL, CellFectin reagent, cat no.; 10362–010). However, as the calcium phosphate precipitation method is cheaper we routinely use this method first. Contaminants in transfer vector DNA may be toxic to insect cells and lead to inefficient cotransfection and thus recombination. Therefore transfer vector DNA should be purified prior to transfection to a degree equivalent to CsCl purity (typically the A_{260}/A_{280} ratio should not be less than 1.7 and ideally 1.8 or above).

Materials

- 2×HBS (Hepes buffered saline): 280 mM NaCl; 10 mM KCl; 1.5 mM $Na_2HPO_4.2H_2O$; 12 mM glucose; 50 mM HEPES. Adjust to pH 7.05 with 0.5 N NaOH and filter through 0.45 μm filter; store in aliquots at −20 °C.
- 2 M $CaCl_2$, filtered through 0.22 μm filter and stored in 1 ml aliquots at −20 °C
- 0.1 X TE:1.0 mM Tris/HCl, pH 8.0; 0.1 mM EDTA, pH 8.0
- Sterile phosphate buffered saline (PBS)
- Transfer vector containing gene encoding for protein to be expressed, precipitated using ethanol and resuspended in sterile water to 1 mg/ml
- Linearized Baculogold DNA (Pharminogen, San Diego, CA, cat. no. 21100D)
- AcNPV C2 wild-type baculovirus DNA (Pharminogen cat. no. 21103D)

Procedure

Transfection of Calcium Phosphate Precipitated DNA

transfection

1. For each transfection seed a 60 mm tissue culture dish with 2×10^6 Sf9 cells, remembering to include one dish as a negative control (non-transfected) and another as a positive control (infected with wild-type AcNPV).

2. While insect cells are attaching to the tissue culture dish precipitate viral and plasmid DNA by mixing 0.2 μg of linearized Baculogold viral DNA with 2 μg of recombinant plasmid DNA in a sterile tube. Very slowly, over a period of approx. 1 min, add 31 μl 2 M $CaCl_2$ to the DNA mixture, shaking the solution often while adding the $CaCl_2$.

3. Allow the DNA to precipitate by leaving the mixture for 30 min at room temperature. A very fine, milky white precipitate may be visible at this stage.

4. Remove media form the attached cells and replace with 0.2 ml G10.

5. Resuspend any precipitated DNA by gently pipetting the solution up and down and add this solution drop-wise to the cells on the dish.

6. Allow cells to take up precipitated DNA by leaving them at 27 °C for 4 h in a humidified environment to stop cells drying out on the dishes. This can be done by placing the dishes on dampened tissues in a plastic container.

7. Wash the DNA from cells by carefully rinsing the cell monolayer 3×with PBS. Add 4 ml G10 and allow cells to grow at 27 °C for 5 days in a humidified environment.

8. After 5 days collect the supernatant from transfected dishes and clarify by centrifugation at 1000 g. Check cells infected with wild-type AcNPV for signs of infection (cells often become enlarged and detached from the plastic when very infected). In some cases signs of infection can be seen in cells that have been co-infected with transfer vector and linearized baculoviral DNA; however infection is not usually obvious until the culture supernatant containing the recombinant baculovirus has been amplified at least once (see below).

Troubleshooting

Transfection efficiency is very dependent upon the size of the DNA precipitate which is determined by the pH of the HBS and speed of $CaCl_2$ addition. To maximize transfection efficiency try making HBS at various pHs (7.00, 7.05, 7.10), freezing aliquots of these and selecting the optimal HBS initially by carrying out trial transfections with any available reporter construct.

9.3
Amplification of Recombinant Baculovirus

To generate further recombinant virus for plaque purification it is necessary to amplify recombinant baculovirus generated in the first cotransfection step. This step described below typically generates 4 ml of baculoviral supernatant containing approximately $1-5\times10^6$ plaque forming units (pfu)/ml supernatant.

■ Procedure

amplification
1. Seed 2×10^6 cells onto a 60 mm tissue culture dish and allow cells to attach.

2. Remove media from cells and add 1 ml of supernatant from the original cotransfection and rock cells gently for 1 h at room temperature.

3. Add 4 ml G10 and incubate in a humidified environment at 27 °C for 5 days.

4. Remove culture media, clarify by centrifugation at 1000 g, and store at 4 °C.

Titering Viral Supernatant by Endpoint Dilution

For maximal protein expression it is necessary to infect insect cells with a multiple of infection (MOI, the average number of viruses that infect a single cell) of 5–10. It is also important when amplifying recombinant baculovirus not to infect insect cells with an MOI>1, as this can lead to selective amplification of defective deleted viruses which reduce protein yield. For these reasons it is useful to be able to titer the amount of virus in supernatants derived from insect cell infections. Titering baculoviral supernatants can be done using plaque assays; however it is much simpler to do this using a limiting dilution method, as described in (O'Reilly et al. 1992), or plaque counting method, as described in (Hannon 1995).

Troubleshooting

If cells show no signs of baculoviral infection compared to a noninfected control dish of insect cells, which is not uncommon, it may be necessary to repeat the above amplification using the supernatant obtained following the first amplification.

9.4
Isolation of Recombinant Baculoviral Clone from Plaque Assay

This stage can be omitted to save time, but we have found that in some cases selection of recombinant clones can produce recombinant baculoviral clones that can express protein to a higher level than others. Clonal selection of recombinant baculovirus should also eliminate any contaminating defective recombinants.

Materials

- 10× stock solution of 5% agarose (SeaKem ME agarose, Flowgen Instruments, Staffs, UK) in double distilled H_2O and autoclaved.
- X-gal (5-bromo-4-chloro-3-indoyl-β-D-galactopyranoside) stock solution (25 mg/ml in DMF)
- Recombinant baculovirus stocks obtained from first or second amplification steps

Procedure

1. Seed Sf9 cells at a density of 2×10^6 cells/60 mm dish being careful to allow cells to settle evenly across dish. Plating out too many cells can cause plaques to become obscured due to an overgrown monolayer. Alternatively cells may be plated at half this density and allowed to grow for 24 h.

2. Dilute virus stocks obtained from first or second amplification steps in G10 medium by 10^{-2}, 10^{-4}, 10^{-5}, 10^{-6}. Aspirate medium from tissue culture dish and infect with 0.5 ml of diluted virus per 60 mm dish, being careful not to let cells dry out. Also seed one dish as a noninfected control and infect another with wild-type virus to act as a positive control.

isolation
of clones

3. Incubate for 1 h at room temperature with gentle rocking.

4. Microwave the 5% agarose solution gently, being careful not to unnecessarily overheat the agarose solution and dilute the melted the agarose to a concentration of 0.5% with complete G10 medium that has been pre-warmed to 60 °C. Mix well by swirling. Cool this agarose solution to 40–42 °C.

5. Remove as much of the virus inoculum from the cells as possible without allowing the cells to dry out. Carefully overlay each 60 mm dish with 4 ml agarose by adding agarose very slowly on to the side of the dish.

6. Allow agarose overlay to harden on a level surface (about 20 min) and add 1 ml of G10 medium to the dish to stop agarose from drying out.

7. Leave cells for at least 5–6 days, until just before cells reach confluency, at 27 °C in a humid environment.

8. Examine for plaques by removing any media that remains on top of the agarose, inverting the dish against a dark background and illuminating the dish from the side. As infected cells become rounded they appear more refractive and thus whiter than noninfected cells. This is difficult and requires some practice. The location of a plaque can be confirmed under an inverted microscope (plaques appear as regions of infected cells that are less dense than the surrounding confluent monolayer). Circle plaques by scratching dish with a needle.

9. Remove circled plaques under sterile conditions using a sterile glass Pasteur pipette. Put each agarose plug into 0.2–0.3 ml G10 and leave overnight at 4 °C to allow virus to elute into media.

10. Reinfect Sf9 cells seeded into 6-well plates (2×10^6 cells/well) with G10 containing virus eluted from each plaque. Incubate for 6 days at 27 °C in a humidified environment before looking for signs of infection.

11. Collect the media and cells from infected cells, centrifuge at 1000 g, 5 min, and store the baculovirus containing supernatant at 4 °C for further amplification. The cell pellet can be kept for confirming that correctly recombined baculovirus have been selected by using the polymerase chain reaction (PCR) as described below.

12. For a quick method to check that a plaque has resulted from infection of insect cells with a correctly recombined virus, take 30 µl G10 containing cloned virus eluted from each plaque picked and add 1/10th vol X-gal stock and incubate overnight. As correct recombination replaces the *lacZ* gene of linearized baculoviral (AcRP23.lacZ) DNA, plaques resulting from infection with correctly recombined virus will not give any color change upon incubation with X-gal. Plaques from incorrectly recombined virus will give a blue color to the G10.

9.5
Confirmation of Identity of Putative Recombinant Baculovirus

Polymerase chain reaction amplification is a quick and convenient method to confirm the identity of recombinant baculovirus after the first amplification following the plaque-pick. One should choose two primers corresponding to opposing regions of the transfer vector that immediately flank the gene that has been inserted into the baculovirus and another which is homologous to sequence in the inserted gene (Fig. 9.3). Using a combination of these primers it should be possible to amplify DNA fragments from recombinant baculoviral DNA of sizes that indicate correct insertion of your gene of interest. PCR amplification of incorrectly recombined baculoviral DNA will amplify DNA fragments of incorrect size.

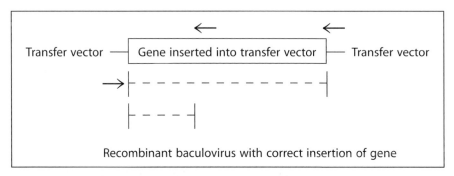

Fig. 9.3. PCR amplification of correctly recombined baculoviral DNA. Using primers (*arrows*) should result in amplification of DNA fragments of sizes indicated by *dashed lines*

To further confirm that the recombinant baculovirus is expressing the correct full length protein we also Western blot insect cell lysates from cell pellets obtained from a first or second amplification of recombinant virus with an antibody raised against the recombinant protein being expressed. Insect cell lysates form cells infected with wild-type baculovirus should also be Western blotted as a control.

▦ Materials

- Lysis buffer: 100 mM NaCl; 0.5 % SDS; 25 mM EDTA; 10 mM Tris-Cl, pH 8.0
- Proteinase K (Boehringer Mannheim, cat no. 161519) added fresh at 0.1 mg/ml
- DNA primers upstream and downstream of inserted gene
- TE buffer: 10 mM Tris/HCl, 1 mM EDTA, pH 7.5
- Insect cell pellet from first amplification following cloning of recombinant virus by picking plaques
- Phenol/chloroform
- Reagents for PCR

▦ Procedure

Preparation of Baculoviral DNA for PCR Analysis

preparation of DNA

1. Add 300–500 μl lysis buffer to cell pellet obtained from amplification of plaqued purified recombinant baculoviral clones. Leave at 60 °C for 6 h or overnight to allow protein digestion.

2. Phenol/chloroform extract DNA from digested insect cells until their is no white interface.

3. Ethanol precipitate DNA and resuspend in 100 μl TE.

4. Carry out PCR; conditions will vary according primers used.

recombinant baculovirus stocks

For generation of large amounts of recombinant baculovirus to enable large scale protein expression baculoviral supernatants can be amplified further by infecting 3×10^7 insect cells plated onto 150 mm tissue culture dishes containing 20 ml G10 and infecting these cells with 50–100 μl baculoviral supernatant obtained from the first amplification following plaque cloning of recombinant baculovirus. Titers of 1×10^8 pfu/ml are commonly obtained following this amplification although it is best to titer supernatants prior to large-scale infections.

Baculoviral supernatants (G10) can be stored in the dark at 4 °C for at least 6 months without significant loss of titer. We keep baculoviral supernatants at -80 °C for longer term storage; however, it is advisable to retiter these supernatants following thawing.

baculoviral supernatants

9.6
Coexpression of Cyclin and CDK in Insect Cells and Preparation of Insect Cell Lysate

Materials

– High titer (0.5×10^7 – 2×10^8 pfu/ml) recombinant cyclin and cognate cdk baculoviral supernatants
– 15 ml glass Dounce homogenizer
– Wash buffer A: 10 mM Hepes/KOH, pH 7.5; 100 mM NaCl; 0.5 mM EDTA
– Hypotonic lysis buffer: 10 mM Hepes/KOH, pH 7.5; 25 mM NaCl; 0.5 mM EDTA; cytochalasin B 10 µg/ml; protease inhibitors; 2 mM PMSF, 2 µg/ml leupeptin, antipain, pepstatin A
– Phosphatase inhibitors: 5 mM NaF, 1 mM sodium orthovanadate
– 5 M NaCl

Procedure

1. Coinfect an 800 ml suspension (1.5×10^6 cells/ml) of insect cells (we use 4×200 ml suspension cultures each being stirred in a 1 liter flask to allow maximum aeration) in log phase growth with recombinant baculoviral (His)$_6$ affinity-tagged cyclin and its cognate recombinant baculoviral cdk at MOI values of 5–10 for each recombinant virus. Typically this requires approximately 30 ml of supernatants each containing recombinant baculoviral (His)$_6$-tagged cyclin or cdk (obtained from second amplifications following plaque picks as described above) to be added to each 200 ml suspension.

2. Harvest infected cells 2–2.5 days post-infection by gently spinning down insect cells (800 g, 5 min; infected cells are very delicate).

3. Carefully wash infected cell pellet in ice cold wash buffer A by gently resuspending cell pellet and spinning as above for 3 min.

4. Resuspend washed cell pellet in ice cold hypotonic buffer using 1 ml hypotonic lysis buffer/ml insect cell pellet. Allow insect cells to swell on ice for 10 min.

5. Dounce homogenize insect cell pellet using tight fitting pestle, check under an inverted microscope for cell lysis.

6. Following cell lysis immediately restore the concentration of NaCl in the cell lysate to 150 mM. This is done to ensure cyclins do not precipitate in low salt.

7. Spin cell lysate 20 000 g for 15 min at 4 °C in HB4 rotor.

8. Remove the supernatant and respin at 100 000 for 30 min, 4 °C. This is important to remove material that would otherwise clog the Ni^{2+} affinity column).

9. Add $MgCl_2$ to a final concentration of 5 mM (this quenches the EDTA in the lysis buffer which would otherwise abrogate the Ni^{2+} affinity purification step).

9.7
Partial Purification of Active Cyclin/cdk Complexes

Materials

- Ni^{2+}-NTA-agarose affinity column(QIAGEN, Chatsworth, CA, cat no. 30210)
- Wash buffer B: 10 mM Hepes/KOH, pH 7.5; 250 mM NaCl; 5 mM $MgCl_2$; 5 % vol/vol glycerol
- Wash buffer B containing 10 mM imidazole
- Wash buffer B containing 100 mM imidazole and protease inhibitors (as used in lysis buffer, see above)
- Gel filtration buffer : 25 mM Hepes/KOH, pH 7.5; 150 mM NaCl, 10 % vol/vol glycerol; 1 mM DTT
- Vivaspin concentrators 10 000 molecular cut off weight (Vivascience, Lincoln UK, cat no.VSO411)
- HPLC/FPLC facilities

Procedure

Note: All purification steps are carried out at 4 °C unless otherwise stated.

1. Pass clarified cell lysate twice through a 0.5 ml Ni^{2+}-NTA-agarose affinity column (approx. 1 ml every 20 min).

2. Wash column with 30 ml wash buffer B.

3. Wash column with 10 ml wash buffer B containing 10 mM imidazole.

4. Elute $(His)_6$-cyclin and $(His)_6$-cyclin-cdk from the Ni^{2+}-NTA affinity column with wash buffer B containing 100 mM imidazole and protease inhibitors (as used in lysis buffer). 4×0.5 ml elutions are usually sufficient to elute bound protein, leaving each elution on the column for 20 min.

5. Immediately add DTT to 2 mM and fresh protease inhibitors to imidazole eluted fractions.

6. Concentrate samples down to 100–200 µl by centrifugation using Vivaspin concentrators (4400 rpm at 4 °C) as described by the manufacturer. Any alternative centrifugation concentrating system could also be used.

7. Separate cyclin/cdk from monomeric cyclin and cdk using a Superose 6 gel filtration column (Pharmacia), 0.4 ml/min flow rate, with gel filtration buffer. Human cyclin B1/cdc2, B2/cdc2, A/cdc2 and cyclin A/cdk2 all elute from this column at approx. 15 ml volume. Collect 0.3 ml fractions, assay for kinase activity using histone H1 as a substrate and phosphocelluose units as described below.

Note: If HPLC/FPLC facilities are not available it is possible to purify cyclin A and B/cdk complexes using Mono Q ion exchange chromatography. We have successfully used the HiTrap series of Mono Q ready made columns (1 ml and 5 ml, Pharmacia). These columns are cheap and can be loaded and washed using a peristaltic pump. Using this method concentration of sample following elution from the imidazole column is not necessary. However, this purification method requires dialysis to remove the high salt prior to loading onto the ion exchange column.

9.8
Assessment of Cyclin/CDK Purification – Measurement of H1-Histone Kinase Activity

To rapidly assess the activity of partially purified cyclin/cdk we measure the amount of P^{33}/P^{32} incorporated onto H1-histone by binding phosphorylated histone to phosphocellulose and using scintillation counting. If this is not possible the H1-histone phosphorylation capability of each fraction can be assessed by adding equal volume of SDS-PAGE sample buffer to each assay and running samples on a 15 % SDS-PAGE gel followed by autoradiography. However, this latter procedure takes considerably longer and one cannot easily calculate the specific activity of the cyclin/cdk.

Materials

- Kinase buffer: 25 mM Tris/HCl, pH 7.5; 150 mM NaCl; 10 mM $MgCl_2$; 1 mM DTT
- H1-histone from calf thymus (Boehringer Mannheim, cat no. 223549)
- 1 mM ATP
- ATP-γ^{33}P (2000 Ci/mmol) or ATP-γ^{32}P (3000 Ci/mmol) (Amersham or NEN DuPont)
- Phosphocellulose units (Pierce, Rockford IL, cat no. 29520)

Procedure

kinase assay

1. Make kinase assay cocktail by mixing 8 µl kinase buffer, 0.4 µl of 1 mM ATP and 0.4 µl ATP$\gamma^{33/32}$P/sample.

2. Add 1 µl of either insect cell homogenate, imidazole eluted, and gel filtration fractions to 8.8 µl kinase assay cocktail, vortex and incubate for samples at 30 °C for 30 min.

3. Add 100 µl 100 mM EDTA to stop kinase reaction and apply samples to phosphocellulose units.

4. Spin and wash samples as described by manufacturer.

5. Add each phosphocellulose insert to 2 ml scintillation fluid and assess H1-histone phosphorylation using a scintillation counter.

6. Snap freeze aliquots in liquid N_2 with BSA or ovalbumin (0.1 mg/ml). We have found that cyclin/cdk activity is stable for at least 6 months in this state; however, cyclin/cdk activity is rapidly lost if aliquots are repeatedly thawed and refrozen.

Note: When assaying for cyclin D/cdk4 activity the retinoblastoma protein Rb is a more suitable substrate than H1-histone.

Troubleshooting

Obviously, purification from lysis of insect cells to freezing of partially purified cyclin/cdks should be carried out as quickly as possible, one should aim to do this within 24–36 h. Poor H1-histone activty is usually a result of poor protein expression in the insect cell culture.

For maximal cyclin/cdk expression it is best to optimize MOIs for cyclin and cdk infections. Different cyclin/cdk complexes, partially purified using this protocol differ in their final purity. This critically depends on the amount of soluble cyclin/cdk expressed. As cdk2 is far more soluble than cdc2 when expressed in insect cells we find that cyclin A/cdk2 and cyclin E/cdk2 complexes are purified to a far greater extent (to approx. 50 % of total protein in final purified fraction) than cyclin A/cdc2. cyclin B1/cdc2 and cyclin B2/cdc2 (approx. 5 % of total protein in final purified fraction). Nevertheless enrichment of at least 10 000-fold H1-histone kinase activity is seen in final fractions compared to insect cell homogenate. Although as the composition and protein concentration of these two fractions are very different it is not possible to directly compare kinase activities. Greater purification of cyclin/cdk complexes may be obtained at the imidazole elution step by eluting Ni^{2+} bound proteins using an imidazole gradient or more gentle stepwise elution profile.

Purifying cdk expressed without cyclin using the above procedure gives no significant H1 histone kinase activity.

References

Desai D, Gu Y, Morgan DO (1992). Activation of human cyclin-dependent kinases in vitro. Mol Biol Cell 3 571–582

Hannon GJ (1995). Expression of cell cycle proteins using recombinant baculovirus. In: M. Pagano (ed) Cell cycle – Material and Methods. Springer, Berlin, Heidelberg, New York, pp 231–242

O'Reilly D, Miller LK, Luckow VA (1992). Baculovirus expression vectors, a laboratory manual. W. H. Freeman New York

Parker LL, Walter, SA, Young PG, Piwnica-Worms, H (1993). Phosphorylation and inactivation of the mitotic inhibitor Wee1 by the nim1/cdr1 kinase. Nature 363:736–738

Suppliers

Vivascience Ltd.
Binbrook Hill, Binbrook,
Lioncoln, LN3 6BL, UK
fax: +44 (0) 1472 398 111

Pharminogen
10975 Torreyana Rd,
San Diego, California 92121, USA
fax: (619) 677–7749

Qiagen Ltd.
Unit 1 Tillingbourne Court,
Dorking Business Park,
Dorking, Surrey RH4 1HJ, UK
fax: 01306–875 885.

Gibco BRL

Chapter 10

Methods for Studying Phosphoinositide Metabolism in Cultured Cells

Conor Duffy*[1], Ailish Hynes[2], and Leo Quinlan[2]

Introduction

In the light of the current focus in scientific research on the role of the phosphoinositides in transmembrane signaling, techniques describing the incorporation of tritiated inositol into the phosphoinositides and inositol phosphates of tissue culture cell lines, and their analysis, are described in this chapter. The phosphoinositide pathway is involved in many aspects of cellular life during the life cycle of the cell including cell growth, differentiation and transformation. It has been implicated in gametogenesis, fertilization, neuromodulation and sensory perception (Michell et al. 1989). In basic terms this signaling route involves the phospholipase C (PLC) catalyzed cleavage of phosphatidylinositol(4,5)bisphosphate (PtdIns(4,5)P_2), yielding two second messengers, inositol(1,4,5)-trisphosphate (Ins(1,4,5)P_3), which acts to release Ca^{2+} from internal Ca^{2+} pools, and diacylglycerol (DAG), which activates protein kinase C (PKC) (Fig. 10.1). These second messengers are responsible for mediating the intracellular consequences of first messengers (e.g., hormones, growth factors, neurotransmitters, agonists; i.e., extracellular molecules) binding to the plasma membrane. These two messenger systems are linked by receptor complexes located in the plasma membrane of the target cell.

The simplest and most effective way to study the phosphoinositdes and inositol phosphates is to label them with [^{3}H]-inositol. Labeling cells with tritiated inositol labels the phosphoinositides (and certain inositol

* *Correspondence to* Conor Duffy: phone +353–1–7045700; fax +353–1–7045484; e-mail CONOR DUFFY@dcu.ie
[1] Conor Duffy, National Cell and Tissue Culture Centre, D.C.U., Glasnevin, Dublin 9, Ireland
[2] Ailish Hynes, Leo Quinlan, Physiology Department, University College, Galway, Ireland

phosphates) so that upon stimulation with a suitable agonist the signaling cascade is switched on and $[^3H]$-inositol(1,4,5)-trisphosphate $(Ins(1,4,5)P_3)$ is released from the plasma membrane. The phosphoinositides and inositol phosphates, $Ins(1,4,5)P_3$ and its metabolites, can be extracted from the cells and separated by thin layer chromatography or anion exchange chromatography, respectively.

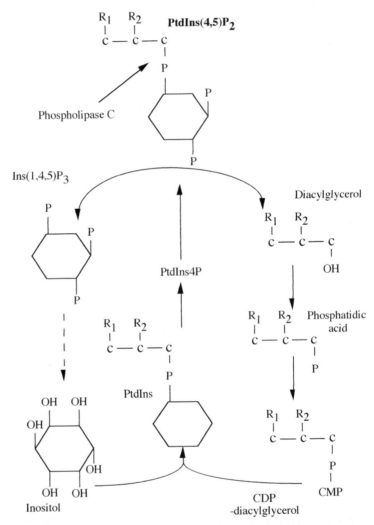

Fig. 10.1. The PtdIns(4,5)P$_2$ signaling system: Phospholipase C cleaves PtdIns(4,5)P$_2$ yielding the two second messengers Ins(1,4,5)P$_3$ and diacylglycerol, which are ultimately metabolized back to PtdIns(4,5)P$_2$

Phospholipids are the major class of membrane lipid, with PtdIns accounting for ~10 % of total phospholipid (Downes et al. 1989). PtdIns is widely distributed among cellular membranes with the polyphosphorylated phosphoinositides, e.g., PtdIns4P and PtdIns(4,5)P$_2$, being more localized in their distribution (Downes et al. 1989), specifically in plasma membranes. PtdIns4P and PtdIns(4,5)P$_2$ account for between 1 and 4 % of the total cellular inositol phospholipid pool (Thomas et al. 1983).

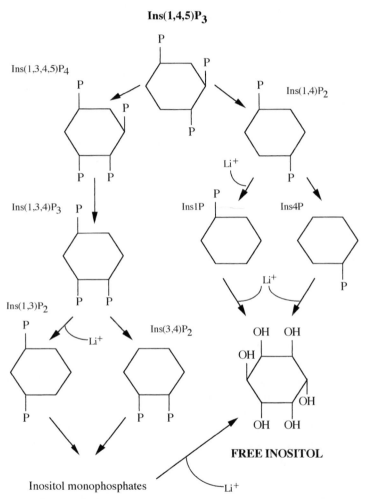

Fig. 10.2. The routes for metabolism of the Ins(1,4,5)P$_3$ signal. The Ins(1,4,5)P$_3$ signal is switched off by 5-dephosphorylation or 3-phosphorylation ultimately yieding free inositol, which is recycled back to PtdIns(4,5)P$_2$. Reactions that are inhibited by Li$^+$ are marked

A recently discovered phosphoinositide, PtdIns(3,4,5)P$_3$ (Stephens et al. 1991), is also labeled and extracted in a similar fashion to the other phosphoinositides using these methods.

Ins(1,4,5)P$_3$ is a short-lived messenger which is rapidly metabolized by two routes, one involving hydrolysis to inositol 1,4-bisphosphate and thence to inositol monophosphates and on to inositol (Fisher et al. 1992), with the other being a 3-phosphorylation yielding Ins(1,3,4,5)P$_4$ (Irvine et al. 1986; Fig. 10.2). It has been shown that lithium can act to inhibit the hydrolysis of the 4-phosphate on inositol 1,4-bisphosphate and also the monophosphatases (Sherman 1989). The phosphatase that acts on inositol 1,3,4-trisphosphate yielding inositol 3,4-bisphosphate can also be inhibited by lithium (Burgess et al. 1985). The routes of dephosphorylation of Ins (3,4)P$_2$ and Ins (1,3)P$_2$ are not known. The end result of both of these pathways is the liberation of free inositol. By pretreating the cells with lithium, before agonist stimulation, it is therefore possible to prevent the completion of the metabolic pathway for InsP$_3$ and augment accumulation of a variety of compounds during stimulation (Fig. 10.2).

The inositol polyphosphates, i.e., inositol pentakis- and inositol hexakisphosphate, comprise the bulk of the inositol phosphate content of mammalian cells (Berridge and Irvine 1989) but their cellular functions are largely unknown. It is these inositol phosphates that are labeled prior to agonist stimulation. Their biosynthesis is independent of the Ins(1,4,5)P$_3$-led metabolic flux that accompanies PLC activation. The intermediates in their biosynthetic pathway are termed agonist insensitive, i.e., agonist binding has no effect. Passive functions such as stores of inositol and phosphate have normally been attributed to the inositol polyphosphates (Berridge and Irvine 1989), primarily because they are seen to function as such in plants, where seeds mobilize substantial quantities of InsP6 upon germination. Recent reports have suggested that the inositol polyphosphates may be more dynamic than had previously been thought (Yang and Shamsuddin 1995). These inositol phosphates are the most prominently labeled prior to agonist stimulation.

For those who are interested, the historical background to the discovery of the signaling pathway is covered in comprehensive detail by Michell 1986. A comprehensive review of techniques for the analysis of DAG is available (Irvine 1990).

Materials

All tissue culture procedures are performed in a laminar flow cabinet; **cell culture and** cells are grown in an incubator maintained at 37 °C in an atmosphere of **labeling** 5 % CO_2 in air. All solutions, unless stated, are prepared using MilliQ water and are filter sterilized using filters with a pore size of 0.22 μm. In the laboratory we have used a variety of techniques to label and study both mammalian cell lines and embryos. As radioactive compounds are used in these analysis, adequate precautions with regard to equipment and personal protection are taken at all times. We use [³H]-inositol (NET-114A) supplied by New England Nuclear in our studies.

- 12-well plates
- Low-speed centrifuge
- Cell scraper
- Conical bottomed glass test-tubes
- Perchloric acid (PCA)
- PBS
- Phytic acid hydrolysate (see below)

The phytic acid hydrolysate is utilized because labeled inositol phospha- **preparation of** tes can be lost during their extraction due to nonspecific binding to glass **the phytic acid** or plastic surfaces. Inclusion of the hydrolysate overcomes this potential **hydrolysate** loss by occupying the potential binding sites (Wregget et al. 1987). The method used is that described by Desjobert and Petek (1956):

1. Sodium phytate (1 g) is dissolved in 5 ml of 0.2 M sodium acetate/ conc. acetic acid (pH 4) in a glass stoppered tube and heated for 8 h in a boiling water bath.

2. The hydrolyzed material is then desalted by passing through a 4 ml column of Amberlite IR-120 [H⁺]-cation exchange resin (BDH). The column is prewashed with three bed volumes of 1 N HCl, followed by a subsequent wash with 12 bed volumes of H_2O, before passing the hydrolyzed material through.

3. The inositol phosphates are washed from the column with 100 ml water and collected in a glass serum bottle. The inorganic phosphorous content of the hydrolysate is determined using a detection kit from Sigma (cat. no. 670-A).

4. The phytic acid hydrolysate sample is freeze-dried and then diluted with water to give a 11.1 mg phosphorus/ml solution, 3 μl of which is

added per sample during the extraction process. The phytic acid hydrolysate is stored at $-20\,°C$.

phosphoinositide extraction
- 100 mM EDTA
- 0.9 % NaCl
- Chloroform
- Methanol
- HCl (concentrated and 1 N)
- N_2

- 50:50 (v/v) freon: tri-*n*-octylamine mixture

phosphoinositide analysis
- Precoated silica-gel 60 F-254 plates (E. Merck, Darmstadt, Germany)
- Oven
- Chromatography tank (9"×10"×3")
- Filter paper (Bio-Rad Laboratories Ltd., Hertfordshire, UK)
- Hamilton syringe with a bevelled needle
- Scintillation vials
- High performance autoradiography film (Hyperfilm-^{3}H, Amersham)
- Potassium oxalate: 1 % (w/v)
- Methanol: chloroform: water: conc. NH_4OH (100:70:25:15)
- Unlabeled standards for PtdIns, PtdIns4P and PtdIns(4,5)P_2 (5 μg of each) (Sigma, cat. nos. P8443; P9736 and P9763, respectively)
- HPLC-pure tritiated inositol phosphate standards for Ins4P, Ins(1,4)P_2, Ins(1,4,5)P_3 and Ins(1,3,4,5)P_4 (NEN Cat. No. NET-970).
- Iodine crystals
- Kodak LX-24 developer
- Kodak FX-40 fixer
- Scintillation cocktail (Beckman, Readysafe)

FPLC
- Pharmacia MonoQ HR5/5 strong anion exchange column
- Eluants A and B are made up as follows:

Eluant A	Eluant B
0.1 mM $ZnSO_4$	0.5 M Na_2SO_4
0.1 mM EDTA	0.1 mM $ZnSO_4$
10 mM HEPES	0.1 mM EDTA
	10 mM HEPES

Both eluants are adjusted to pH 7.4 by the addition of 1 N NaOH (Meek 1986).

- Internal standards: adenosine mono, di, tri and tetra-phosphate and their guanosine equivalents (10 nmol of each)

- Biorad poly-prep chromatographic columns (cat no. 731–1550) or blue (1 ml) pipette tips or Pasteur pipettes plugged with steel wool   **Dowex columns**
- Dowex resin: Dowex 1×8–400 (Sigma cat no. 1X8–400)
- Loading buffer: 5 mM disodium tetraborate/0.5 mM EDTA in water
- Buffers 1–7 are made up as follows:

Buffers (compund eluted)	Components
1 (free inositol)	Water
2 (glycerophosphoinositol)	5 mM Na tetraborate/0.06 M NH_3 formate in water
3 ($InsP_1$)	0.2 M Ammonium formate
4 ($InsP_2$)	0.4 M Ammonium formate
5 ($InsP_3$)	0.8 M Ammonium formate
6 ($InsP_4$)	1.0 M Ammonium formate
7 ($InsP_{>4}$)	2.0 M Ammonium formate

Note: Buffers 3–7 are made up in 0.1 M formic acid.

10.1
Labeling Cells with [³H]-Inositol

▨ Procedure

1. Cells are seeded into 12-well plates and allowed to attach for 24 h   **labeling** before the timed radioactive labeling begins.

2. [³H]-inositol is used at a concentration of 10 μCi/ml in the incorporation investigations described. **Note:** If you are interested in studying the PtdInsP$_3$s it may be necessary to use 20 μCi/ml of [³H]-inositol.

3. After incubating the cells in radioactive medium for the required time period the phospholipids and inositol phosphates can then be extracted. A bifurcating extraction pathway is utilized during which

both elements can be isolated simultaneously (Roldan and Harrison 1989; Fig. 10.3).

Note: All centrifugation steps in this procedure are at $800\,g$ in a low-speed centrifuge for 5 min unless otherwise stated.

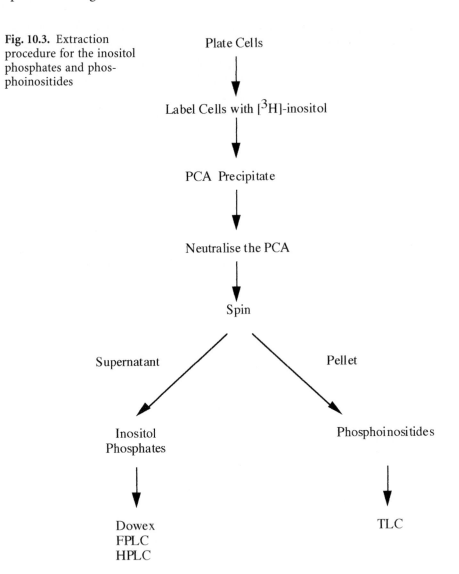

Fig. 10.3. Extraction procedure for the inositol phosphates and phosphoinositides

4. The cell incubations in the radioactive medium are terminated, following removal of the medium, by the addition of: 0.4 ml 10 % PCA (perchloric acid), 0.4 ml PBS and 3 µl phytic acid hydrolysate.

5. The cells are removed from the wells using a cell scraper, precipitated on ice for 5 min in a conical bottomed glass test-tube, and then centrifuged. It is at this stage that the two extraction pathways deviate into the phospholipid and inositol phosphate parts (see "extraction of PtdIns" and "extraction of InsP", respectively).

10.2
Extraction Pathway for the Phosphoinositides and the Inositol Phosphates

▌ Procedure

1. The pellet from the above centrifugation step is washed in 2 ml of 2 % PCA. (An aliquot is taken at this stage and stored at −20 °C to be assayed for protein at a later stage.) This wash is centrifuged and the supernatant discarded.   *extraction of PtdIns*

2. The pellet is resuspended in 3.75 ml of chloroform:methanol:conc. HCl (500:1000:6, by vol.). This is left on ice for 10 min and centrifuged.

3. The supernatant from this spin is transferred to a glass test-tube and the following are added:

H_2O	1 ml
100 mM EDTA	0.25 ml
0.9 % NaCl	1.25 ml
$CHCl_3$	1.25 ml

4. This mixture is vortexed, centrifuged and the upper phase discarded. Then wash the lower phase with 2.5 ml of theoretical upper phase: chloroform/methanol/1N HCl (3: 48: 47), vortex and centrifuge.

5. Discard the upper phase and dry the lipid containing lower phase under a stream of N_2 (30–45 min) and store at −20 °C.

1. Transfer the supernatant from the original centrifugation step ("Labeling Cells with [³H]Inositol) to a plastic test-tube. Add 1 ml of   *extraction of InsP*

a 50:50 (v/v) freon: tri-*n*-octylamine mixture, and 0.8 ml H_2O to the supernatant (the freon: tri- *n*-octylamine mixture acts to neutralize the perchloric acid).

2. Vortex the solution for 30 s and leave on ice for ∼12 min to harden the interphase.

3. Vortex for a further 30 s and centrifuge for 3 min.

4. After the centrifugation step there are three distinct phases: an aqueous phase at the top, a gel phase of tri-*n*-octylamine-perchlorate in the middle, and a freon plus excess tri-*n*-octylamine phase at the bottom.

5. Leave the test-tube on ice for approximately 10 min to solidify the middle phase, which facilitates the removal of the upper inositol phosphate-containing phase. The latter is subsequently stored at −20 °C in a glass test-tube.

10.3
Analysis of the Extracted Phosphoinositides

▣ Procedure

thin layer
chromatography
1. The extracted phosphoinositides are separated by thin layer chromatography (TLC) on precoated silica-gel 60 F-254 plates. Before use, spray the TLC plates with 1 % (w/v) potassium oxalate and "activate" them by heating in an oven at 110 °C for 1 h.

2. While the plate is being activated equilibrate a chromatography tank. This is achieved by lining the sides of the tank with thick blot absorbent filter paper and placing 200 ml of the solvent i.e. methanol:chloroform:water:conc. NH_4OH (100:70:25:15) in the tank (Alter and Wolf 1995).

3. Then air-seal the tank by greasing the lid. The blotting paper absorbs the solvent, thereby equilibrating the tank and leaving a solvent depth of 1 cm in the tank.

4. After activation of the plates, mark an origin 1.5 cm from the bottom of the plate using a soft pencil, with gaps of 2 cm between the sample origins.

5. Redissolve the dried lipids in 20 µl of chloroform and apply to the TLC plate using a 50 µl Hamilton syringe with a bevelled needle. It is extremely advantageous to keep the sample spot as small as possible. This can be achieved by applying the 20 µl of sample onto the origin in 0.5 µl aliquots, allowing the spot to air-dry in between applications.

6. Unlabeled standards for PtdIns, PtdIns4P and PtdIns(4,5)P$_2$ (5 µg of each) are applied to each sample origin as internal standards, to aid running of the sample and detection of the lipids. A standard lane containing 10 µg of each of the three lipid standards is also run on each plate.

7. After loading of the samples place the plate into the equilibrated chromatography tank and allow the solvent to run to within 2 cm of the top of the plate. Mark the solvent front with a pencil for later determination of Rf values. Dry the plate by standing it in a fume hood.

1. The phosphoinositide spots are located by a combination of iodine staining and/or autoradiography. Iodine crystals are allowed to sublimate in a sealed tank, and the plate is placed in the tank until the lipids are visible as discernible yellow spots. As this iodine staining fades with time, circle the spots with a pencil.

staining, autoradiography

2. Autoradiography is performed by placing the TLC plate face down on the emulsion layer of a high performance autoradiography film (Hyperfilm-^{3}H, Amersham) in an autoradiography cassette. The cassette is left for 1–4 weeks at room temperature, in a light sealed location (**Note:** The longer time is necessary if looking for PtdInsP$_3$s).

3. The film is developed in a dark room by immersing in Kodak LX-24 developer for 4–5 min, then fixing in Kodak FX-40 fixer for 1–2 min followed by washing in water, or utilizing an automatic developer. Using the film as a template areas corresponding to the phosphoinositides, and all other areas showing the incorporation of radioactivity, are scraped from the TLC plate and placed into scintillation vials.

4. Add 1 ml of methanol to each vial and leave the samples for 1 h to elute the lipids from the silica.

5. Dispense 5 ml of scintillation cocktail into each vial, vortex and leave overnight to equilibrate before counting in a scintillation counter. All inositol-containing phospholipid compounds are extracted by this procedure and, along with the phosphoinositides, the phosphatidylinositolglycan membrane protein anchors (Low et al. 1986; Saltiel

1991) are also commonly found on the developed autorads, a representative one of which is shown (Fig. 10.4).

Rf values The Rf values for the phosphoinositides, as separated by this method (Fig. 10.4), are shown in Table 10.1.

The lipids can be further analyzed by deacylation whereby the fatty acid components of the lipid are removed and the subsequent glycerophosphates can be identified by anion exchange chromatography on either Dowex or HPLC (Serunion et al. 1991)

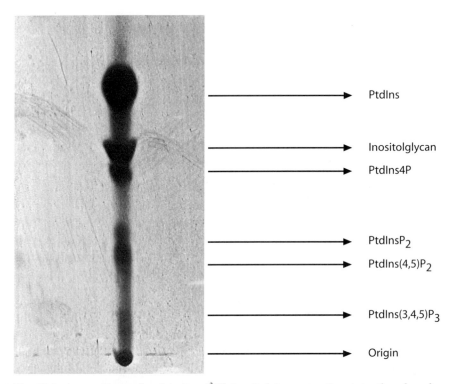

Fig. 10.4. Autoradiography showing [^{3}H]-inositol incorporation into the phosphoinositides of embryonic stem (ES) cells. ES cells were labeled with 20 µCi/ml [^{3}H]inositol in standard ES cell medium for 24 h. The extracted phospholipids were separated by TLC as described in text. The solvent used was MeOH:CH$_3$OH:H$_2$O:conc. NH$_4$OH (100:70:25:15). Lipid identification was based on comigration with lipid standards for PtdIns, PtdIns4P, PtdIns(4,5)P$_2$ and iodine staining

Table 10.1. Rf values

Phosphoinositide	Rf value
Phosphatidylinositol	0.59
Inositolglycan	0.47
Phosphatidylinositol(4)phosphate	0.43
Phosphatidylinositol(?,?)bisphosphate	0.31
Phosphatidylinositol(4,5)bisphosphate	0.23
Phosphatidylinositol(3,4,5)trisphosphate	0.08

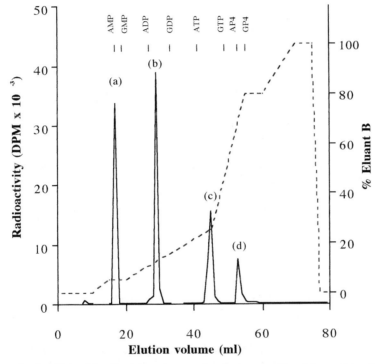

Fig. 10.5. Standardization of the MonoQ column. The elution profile of four tritiated inositol phosphate standards: *a* Ins4P; *b* Ins(1,4)P$_2$; *c* Ins(1,4,5)P$_3$; *d* Ins(1,3,4,5)P$_4$, separated on a MonoQ column using a similar elution gradient to that of Meek (1986). The elution positions of the nucleotide standards are represented as I, with their respective acronym. The eluate was monitored at 254 nm

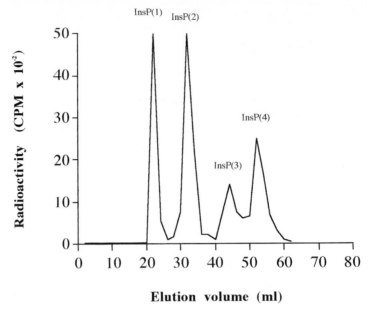

Fig. 10.6. Dowex column standardization. The standards used are the myo-[2-H³ (N)]-forms of: inositol-4-phosphate (Ins4P), inositol-1,4-bisphosphate (Ins(1,4)P₂), inositol -1,4,5-trisphosphate (Ins(1,4,5)P₃) and inositol-1,3,4,5-tetrakisphosphate (Ins(1,3,4,5)P₄). 2 μCi of each standard was added to 10 ml of loading buffer and applied to the column. The elution profile is as described in Materials section

10.4
Analysis of the Extracted Inositol Phosphates

The extracted inositol phosphates can be separated by a variety of means, the most widely used and simple being anion exchange chromatography using Dowex resin and fast protein liquid chromatography (FPLC), both of which will be described in this section, or HPLC (Wong et al. 1992). Irrespective of the method chosen the columns are standardized using HPLC-pure tritiated inositol phosphate standards for Ins4P, $Ins(1,4)P_2$, $Ins(1,4,5)P_3$ and $Ins(1,3,4,5)P_4$. These radioactive standards are run individually and as a cocktail. The samples are collected and scintillation cocktail added. The vials are vortexed, left to stand overnight and then counted in a scintillation counter. Representative graphs showing these standard separations for both FPLC (Fig. 10.5) and Dowex (Fig. 10.6) are shown.

▨ Procedure

FPLC

Separation of the InsPs by FPLC requires a strong anion exchange column (i.e., Pharmacia MonoQ HR5/5). The eluants and elution gradient (Table 10.2) used are similar to that in Meek (1986) (see "Materials"). In order to standardize each sample separation, internal standards, adenosine mono-, di-, tri- and tetra-phosphate and their guanosine equivalents, are used.

1. A flow rate of 1 ml/min is utilized and the eluate is monitored at 254 nm.

2. After injection of the sample 40 ml of eluant A is run through the column to remove any free [^{3}H]-inositol, before the labeled inositol phosphates are eluted.

3. Aliquots that elute at eluant B values greater than 70 % have 1 ml of water added to them, in order to counteract quenching caused by the high salt concentration.

Representative elution profiles are shown demonstrating pre- and post-stimulation profiles (Fig. 10.7A, B). The turnover in the inositol phosphate cascade is manifested in the large increase in the size of the mono, bis-, tris and tetrakis areas of the profiles.

Table 10.2. The elution gradient utilized in the separation of the inositol phosphates on a MonoQ column

Elution volume (ml)	Eluant B (%)
0.0	0.0
10.0	0.0
15.0	5.0
20.0	5.0
45.0	25.0
55.0	80.0
60.0	80.0
70.0	100
75.0	100
77.0	0.0

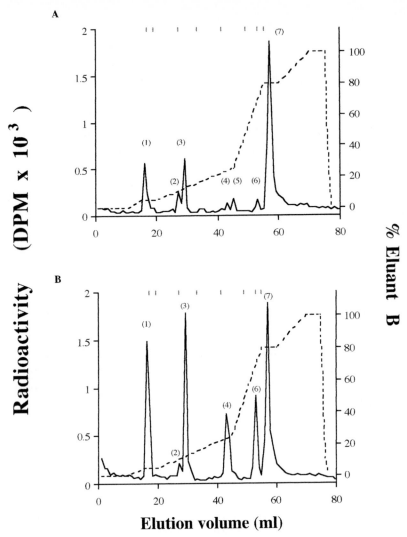

Fig. 10.7A, B. The inositol phosphate profile of embryonic stem (ES) cells cultured in serum-free medium for 24 h (**A**) and following stimulation with dialyzed fetal calf serum (FCS) for 20 min (**B**).Cells were labeled for 24 h with [³H]inositol in serum-free standard ES medium (**A**), and then stimulated with dialyzed FCS (10 % final vol.) for 20 min (**B**). The inositol phosphates were extracted and separated by FPLC on a MonoQ column using the elution gradient of Meek (1986). Identified peaks were (1) Ins4P*; (2) Ins(1,3)P₂; (3) Ins(1,4)P₂*; (4) Ins(1,3,4)P₃; (6) Ins(1,3,4,5)P₄* and (7) inositol pentakisphosphate (*peaks identified by coelution with a tritiated inositol phosphate standard). The elution positions of the nucleotide standards are represented as ı, similar to Fig. 10.5. The profile shown here typify those from several samples, all containing ~ 40 µg protein

Dowex

Dowex 1-x8 is a strong base anion exchanger with benzyl-trimethylammonium functional groups. The resin is supplied in the chloride form and is cross-linked with 8 % divinylbenzene. The effective pH of the resin is 1–14 units.

preparation of the Dowex resin

1. To prepare the working formate form of the Dowex mix it with an equal volume of MilliQ H_2O using a magnetic stirrer for 30 min and then allow to settle under gravity.

2. Remove some of the excess H_2O leaving approximately 20 % by volume of H_2O remaining on top of the resin. Mix this until a homogeneous mixture is obtained and aliquot 1 ml of the slurry per column: Biorad Poly-prep chromatographic columns.

3. Columns can be also be set up in blue (1 ml) pipette tips or Pasteur pipettes plugged with steel wool to prevent the resin falling out. The resin is then converted to the formate form by washing through 15 ml of 5 %Na formate followed by 30 ml MilliQ H_2O. The resin and columns are stored at 4 °C.

All the buffer components essential for this section are listed under "Materials." In our laboratory a 1 ml volume of the extracted inositol phosphates is diluted to 10 ml with loading buffer and applied directly to the column. The column is allowed to run dry before the individual classes of inositol phosphates are eluted with the various buffers.

elution procedure

1. 10 ml volumes of each of the seven buffers are run through the columns. Buffer 1 is applied in 4×2.5 ml aliquots, allowing the column to run dry between each addition.

2. The different inositol phosphate classes are then eluted with increasing ammonium formate concentrations. 5×2 ml fractions of buffers 3–6 are collected into scintillation vials.

3. Add 8 ml of scintillation cocktail to each vial so as to ensure translucent gel formation. Leave the samples to equilibrate overnight before counting.

4. A representative Dowex profile is shown (Fig. 10.6). Following elution of all the inositol phosphate classes, the columns are regenerated by washing with 10 ml buffer 7 followed by 20 ml H_2O. The columns are then stored in approx 2 ml H_2O.

Note: Dowex is primarily utilized for the separation of the inositol mono-, bis-, and trisphosphates. There are some problems with regard to peak resolution when using Dowex, e.g., some remaining [^{3}H]-inositol can be present in the inositol monophosphate fraction. The advantages of using Dowex are that several columns can be run at the same time and that the system is relatively cheap to set up, although it does need to be validated properly.

The FPLC method listed gives much greater resolution of individual isomers and is a much more definitive method for peak resolution, although it is much more expensive to establish and is slower. As the sample size collected is normally >2.5 ml, it is possible to split the sample and use both methods in tandem if possible, i.e., use the simpler Dowex method to evaluate the success of the experiment with regard to incorporation of radioactivity, magnitude of the stimulation response, etc., and then use the FPLC/HPLC method to analyze the specific isomers within each class of inositol phosphate.

Comments

The methods desribed in this chapter have been successfully utilized for the analysis of the inositol phosphate cascade for both cell lines and mammalian embryos (Fahy and Kane 1993, 1994). Irvine (1990) describes a variety of alternative methods for the analysis of inositide related compounds and is a good starting point for further analytical studies in this field.

Acknowledgements. We would like to thank Prof. M.T. Kane, Physiology Department, University College, Galway, for his contribution to the final drafting of this chapter.

References

Alter CA, Wolf BA (1995) Identification of phosphatidylinositol 3,4,5-trisphosphate in pancreatic islets and insulin-secreting β-cells. Biochem Biophys Res Comm 208(1):190–197

Berridge MJ, Irvine RF (1989) Inositol phosphates and cell signalling. Nature 341:197–204

Burgess GM, McKinney JS, Irvine RF, Putney JW Jr(1985) Inositol 1,4,5-trisphosphate and inositol 1,3,4-trisphosphate formation in Ca-mobilizing-hormone-activated cells. Biochem J 232:237–243

Desjobert A, Petek F (1956) Chromatographie sur papier des esters phophoriques de l'inositol. Application a l'etude de la degradation hydrolytique de l'inositolhexaphosphate. Bull Soc Chim Biol 38:871–883

Downes CP, Hawkins PT, Stephens L (1989) Identification of the stimulated reaction in intact cells, its substrate supply and the metabolism of inositol phosphates. In: Michell RH, Drummond AH, Downes CP (eds) Inositol lipids in cell signalling. Academic, London pp 3–38

Fahy MM, Kane MT (1993) Incorporation of [3H]-inositol into phospholipids and inositol phosphates by rabbit blastocysts. Mol Reprod Dev 34:391–395

Fahy MM, Kane MT (1994) The effects of lithium chloride on rabbit blastocyst expansion and accumulation of phophoinositdes and inositol phosphates. J Reprod Fertil 100:347–352

Fisher SK, Heacock AM, Agranoff BW (1992) Inositol lipids and signal transduction in the nervous system: an update. J Neurochem 58:18–38

Irvine RF (1990) Methods in inositide research. Raven, New York

Irvine RF, Letcher AJ, Heslop JP, Berridge MJ (1986) The inositol tris/tetrakisphosphate pathway-demonstration of Ins(1,4,5)P$_3$ 3-kinase activity in animal tissues. Nature 320:631–634

Meek JL (1986) Inositol bis-, tris-, and tetrakis(phosphate)s: analysis in tissues by HPLC. Proc Natl Acad Sci USA 83:4162–4166

Michell RH (1986) Inositol lipids and their role in receptor function: History and general principles. In: Putney JW Jr. (ed) Phosphoinositides and receptor mechanisms. Alan R. Liss, New York, pp 1–24

Michell RH, Drummond AH, Downes CP (1989) Inositol lipids in cell signalling. Academic Press, London, p 534

Roldan ERS, Harrison RAP (1989) Polyphosphoinositide breakdown and subsequent exocytosis in the Ca$^+$/ionophore-induced acrosome reaction of mammalian spermatozoa. Biochem J 259:397–406

Serunian LA, Auger KR, Cantley LC (1991) Identification and quantification of polyphosphoinositides produced in response to platelet-derived growth factor stimulation. Methods Enzymol 198:78–87

Sherman WR (1989) Inositol homeostasis, lithium and diabetes. In: Michell RH, Drummond AH, Downes CP(eds) Inositol lipids in cell signalling. Academic Press, London, pp 39–79

Stephens LR, Hughes KT, Irvine RF (1991) Pathway of phosphatidylinositol-(3,4,5)-trisphosphate synthesis in activated neutrophils. Nature 351:33–39

Thomas AP, Marks JS, Coll KE, Williamson JR (1983) Quantitation and early kinetics of inositol lipid changes induced by vasopressin in isolated and cultured hepatocytes. J Biol Chem 258:5716–5725

Wong NS, Barker CJ, Morris AJ, Craxton A, Kirk CJ, Michell RH (1992) The inositol phosphates in WRK1 rat mammary tumour cells. Biochem J 286:459–468

Wreggett KA, Howe LR, Moore JP, Irvine RF (1987) Extraction and recovery of inositol phosphates from tissues. Biochem J 245:933–934

Yang G-Y, Shamsuddin AM (1995) IP$_6$-Induced growth inhibition and differentiation of HT-29 human colon cancer cells: involvement of intracellular inositol phophates. Anticancer Res 15:2479–2488

Suppliers

Amersham International plc.
Amersham Place
Bucks, HP7 9NA
UK
Phone +44–1494–544000
Fax: +49–1494–542266

BDH Laboratory Supplies
Poole, BH15 1TD
UK
Phone +44–1202–669700

Beckman Instruments (UK) Ltd.
Unit J
Sands Industrial Estate
High Wycombe, Bucks, HP12 4JL
UK
Phone: +44–1494–441181
Fax: +44–1494–537642

Bio-Rad Laboratories Ltd.
Bio-Rad House
Hemel Heptstead
Herts, HP2 7TD
UK
Phone: +44–1442–223552
Fax: +44–1442–259118

Du Pont (UK) Ltd.
Diagnostics and Biotechnology Systems
Wedgwood Way
Stevenage, Herts,SG1 4QN
UK
Phone: +44–1438–734680 (agents for NEN)

E. Merck
Frankfurter Str. 250
D-64293 Darmstadt
Germany
Phone: +49–6151–72–0
Fax: +49–6151–72–3368

Pharmacia Biotech
23 Grosnevor Rd.
St. Albans, Herts, AL1 3AW
UK
Phone: +44–1727–814000
Fax: +44–1727–814001

Sigma Chemical Co.
Poole, Dorset, BH17 7NH
UK
Phone: +44–1202–733114
Fax: +44–1202–418242

Chapter 11

Analysis of Cell Cycle and Cell Death Mechanisms

S. Verhaegen*, Seamus Coyle, Lisa M. Connolly,
Colette O'Loughlin, and Martin Clynes

Introduction

In a multicellular organism tight control of cell number is crucial for tissue homeostasis. Tumor formation is usually thought to occur because of increased proliferation due to disregulated cell cycle control and/or decreased cell death due to suppressed apoptosis (Hartwell and Kastan 1994; Hall and Lane 1994).

The cell cycle is composed of four different phases; G1 (gap 1), which is a phase in which the cell prepares for DNA replication; the S phase, during which DNA is replicated; G2 (gap 2), in which the cell pauses as it prepares to divide; and mitosis (M). Each of the four phases of the cell cycle or entry from one phase into another is controlled by the formation of a complex between a cyclin protein and a cyclin-dependent kinase (CDK). At present there are eight CDKs and more than ten cyclins known. Different combinations of cyclins and CDKs control different phases of the cell cycle, e.g., cyclin D/CDK4 or CDK6 is involved in transition through G1, cyclin E/CDK2 controls entry into the S phase, cyclin A/CDK2 controls transition through the S phase, and cyclin B/CDK1 controls transition through G2/M. During various stages of the cell cycle different cyclins bind to different CDKs, e.g., cyclin A binds to CDK2 during the S phase and to CDK1 during G2. The cyclins can be broadly classed as being G1/S cyclins or G2/M cyclins (Pines 1994).

Numerous techniques are available to investigate the cell cycle. This section will describe the use of flow cytometry to investigate the cell cycle distribution of a population of cells (Sherwood and Schimke 1995), time-lapse videomicroscopy to investigate the kinetics of mitosis and apoptosis (Harrington et al. 1994; Higashikubo et al. 1994) and a kinase

---

* *Correspondence to* S. Verhaegen, National Cell and Tissue Culture Centre/BioResearch Ireland, Dublin City University, Glasnevin, Dublin 9, Ireland; phone +353–1–7045315; fax +353–1–7045484; e-mail steven.verhaegen@dcu.ie

assay to investigate the role of the cyclin/CDK complexes (Eblen et al. 1995).

Apoptosis (also called programmed cell death) is the term used to describe how cells die under a variety of physiological and pathological conditions. Apoptosis is a genetically regulated and active process whereby the dying cell participates in its own demise and disposal. A second mode of cell death, necrosis, is the classically defined form of cell death. Necrosis is passive in nature and produces a dying cell which simply bursts and releases its contents into the surrounding areas. Necrosis is not controlled genetically and is caused by severe or sudden injury (reviewed in Cotter et al. 1990).

Morphological and Biochemical Changes During Apoptosis

Morphological changes during apoptosis include cell shrinkage, membrane blebbing, chromatin condensation and fragmentation. The cells then fragment into membrane-bound apoptotic bodies (for an illustra-

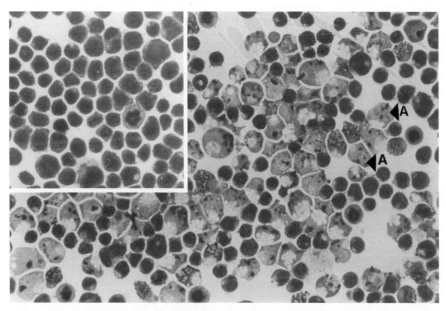

Fig. 11.1. Micrograph of HL-60 cells treated with etoposide (20 µg/ml). After 4 h cells were cytospun onto microscope slides and stained with Rapi-Diff (see text). Typical apoptotic cells with extensive nuclear fragmentation are visible (*arrows*). *Inset* Normal HL-60 cells

tion see Fig. 11.1). The cell's DNA is degraded into internuclesomal fragments (Compton et al. 1992). In vivo apoptotic cells are readily phagocytosed and digested by macrophages or neighboring cells without generating an inflammatory response. These changes distinguish apoptosis from necrosis. Necrosis, which can be caused by severe and sudden injury such as ischemia, sustained hyperthermia or physical or chemical trauma, results in changes in mitochondrial shape and function, and the cell loses its ability to regulate osmotic pressure, swells and ruptures (reviewed in Cotter et al. 1990).

Occurrence of Apoptosis

Apoptosis is important for the development and survival of multicellular organisms, enabling cells to self-destruct when they are no longer needed or become seriously damaged. Multicellular organisms use apoptosis to acquire a balance between generation of new cells and cell death, resulting in little change in the overall number of cells which make up their adult form. Therefore, apoptosis can be seen as a functional counterpart of mitosis. Apoptosis is involved also in normal physiological processes such as developmental limb formation, in which interdigital cells are removed to form the digits of the limb. Strict regulation of apoptosis is required in tissue homeostasis and any imbalance will lead to pathological conditions, such as cancer, and autoimmune and neurodegenerative diseases (Carson and Ribeiro 1993).

Apoptosis-Modulating Genes/Proteins

Apoptosis is controlled by a number of "pro-suicide" and "anti-suicide" genes (Table 11.1)which switch on/off the apoptotic process. Bcl-2 was the first oncogene identified as playing a key role in the control of apoptosis and is a member of the large Bcl family. The expression of the 26 kDa Bcl-2 protein prevents apoptosis and also increases drug resistance to undergo apoptosis due to cytotoxic drugs (Oltvai and Korsmeyer 1994).

Another member of this family is the pro-death gene Bax. Bcl-2 and Bax control apoptosis depending on the level of expression of each. An imbalance towards Bcl-2 and the cell lives, but an imbalance towards Bax and the cell dies. Similarly, the Bcl-2 homologue, Bcl-x_L, and its alternative spliced form, Bcl-x_S, have opposing actions (Nunez et al. 1994)

Table 11.1. Pro- and anti-suicide genes

Anti-suicide genes	Suicide genes
Bcl-2	Bax
Bcl-x_L	Bcl-x_s
bcr-abl	p53
v-abl	c-myc

The pro-death gene p53 stops cells with DNA damage in the G1 phase of the cell cycle to allow the cell to repair the damage. However, if the damage cannot be repaired then p53 will initiate apoptosis. In the majority of human cancers the p53 gene is mutated and the repair processes are impaired resulting in cells with damage to genes that control proliferation leading to tumor development (Hall and Lane 1994).

The c-myc gene is associated with both cell proliferation and apoptosis regulation. This seems to be a contradiction but it has been suggested that c-myc will drive proliferation only under conditions in which apoptosis is suppressed. When apoptosis is not suppressed, such as under insufficient growth factor conditions or expression of apoptosis suppressing genes like Bcl-2, then c-myc drives apoptosis (Harrington et al. 1994).

Execution

Proteases play a major role in the execution of apoptosis. A proteolytic cascade is activated by interleukin 1β converting enzyme (ICE) or ICE-like proteases resulting in the specific cleavage of several substrates. Examples include poly(ADP-ribose) polymerase (PARP) (Kaufmann et al. 1993; Tewari et al. 1995), the 70 kDa protein component of the U1 small nuclear ribonucleoprotein (Casciola-Rosen et al. 1994), lamin B1 (Neamati et al. 1995), topoisomerase 1, α-fodrin and β-actin (reviewed in Martin and Green 1995).

Methods visualizing these morphological and biochemical changes have been used for the detection, quantitation and characterization of apoptotic cell death. Rather than giving a detailed overview of all possible techniques we have opted for a short practical outline of a limited number of techniques that are relatively simple to perform, yet that are able to provide valuable morphological and biochemical information, especially when used in conjunction with each other. An overview of the techniques used is given in Fig. 11.2.

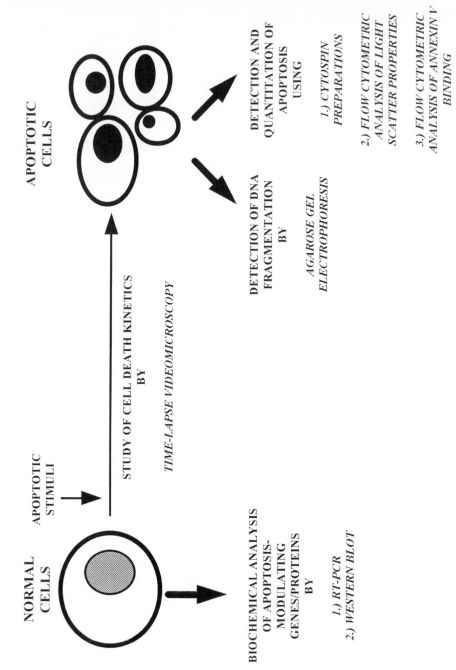

Fig. 11.2. Techniques useful for the study of cellular and biochemical characteristics of apoptosis

11.1
Protein Kinase Assay To Quantitate Activity of Cyclin/CDK Complexes

This section outlines an assay capable of giving quantitative information about the kinase activity of a given cyclin/CDK complex, in this case cyclin E and CDK2, respectively. It is based on immunoprecipitation of the complex using an antibody specific for the cyclin. Histone H1 is then used as a substrate for phosphorylation by the complex, and by using radioactive ATP, levels of phosphorylation can be measured after electrophoretic separation of the histone H1 from the mixture (Eblen et al. 1995).

▨ Materials

- Sonicator (Labsonic U, B. Braun, UK)
- Microfuge
- Gel dryer (GS200, Savent, USA)
- Blotting paper
- Densitometer (GS-670, BioRad, UK)
- Histone H1 (223 549, Boehringer Mannheim, UK): 100 µg/ml stock in H_2O; store at $-20\,°C$.
- 5 mM adenosine-5'-triphosphate (ATP) (127 523, Boehringer Mannheim, UK): 5 mM ATP stock in H_2O, pH 7.0 (using NaOH only); store at $-20\,°C$.
- Trypsin/EDTA solution: 0.25 % trypsin (cat. no. 043–05090, Gibco), 0.01 % (w/v) EDTA in PBS (EDS, Sigma)
- 0.1 mCi/ml [^{32}P]ATP in kinase buffer; store at $-20\,°C$.

Note: Caution should be used in handling all radioisotopes and suitable protection should be employed.

- Mouse anti-cyclin E (sc-248, Santa Cruz Biotechnology, USA)   **antibodies**
- Mouse anti-CDK2 (sc-163, Santa Cruz Biotechnology, USA)
- Normal mouse IgG (I5381, Sigma, UK)

- Radioimmunoprecipitation assay (RIPA) buffer: made up as 20×   **buffers**
  stock; store at 4 °C
  - 50 mM Tris, pH 7.5
  - 150 mM NaCl
  - 1 % Nonidet P-40 (N6507, Sigma, UK)

- 0.1 % Sodium dodecyl sulfate (SDS) (62862,Riedel-deHaen, Germany)
- 1 % Sodium deoxycholate (D5670, Sigma, UK)
  Immediately before use add 10 μl from the following 100x stocks to the lysis buffer:
- 100 mM sodium orthovanadate (store at 4 °C) (S6508, Sigma, UK)
- 100 mM DTT (store at −20 °C) (D5545, Sigma, UK)
- 100× protease inhibitors (all from Sigma, UK): 2.5 mg/ml leupeptin (L2884); 2.5 mg/ml aprotinin (A1153); 100 mM benzamidine (B6506); 1 mg/ml trypsin inhibitor (made up in distilled water and stored at −20 °C) (T9003); 100 mM phenylmethylsulfonyl fluoride (PMSF; 17.4 mg/ml in 100 % ethanol, store at 4 °C) (P7626). **Note:** PMSF has a half life of approximately 30 min in aqueous solution. Add PMSF last and keep it on ice until use.

- Kinase buffer: 20 mM Tris, pH 7.5; 10 mM $MgCl_2$; 1 mM DTT
- 5× loading buffer: 2.5 ml 1.25M Tris-HCl, pH 6.8; 1.0 g SDS; 2.5 ml 2-mercaptoethanol; 5.8 ml glycerol; 0.1 % bromophenol blue. Made up to 20 ml with sterile water

▨ Procedure

sample preparation Samples from adherent cells for all immunoprecipitation experiments are prepared as follows:

1. Set up cells at 5×10^5 cells/25 cm^2 flask 2 days before experiment.

2. Collect samples as follows: pour the media from the flasks into a universal, trypsinize the flask into a single cell suspension and pour the trypsinized cells into the universal (this ensures that any floating cells are collected). Pellet the cells and store at −80 °C.

3. Lyse the cells by sonication (seven pulses at 50 % power) in SDS-RIPA buffer. Check microscopically that the cells have been lysed.

4. Centrifuge the sample at 4000 rpm for 10 min in a microfuge.

5. Collect the supernatant and centrifuge for 1 h at 13 000 rpm.

6. Collect samples, aliquot and store at −80 °C.

7. Assay the sample for protein concentration.

1. Centrifuge 100 µg of total protein at 3000 rpm at 4 °C for 15 min.

2. Transfer the supernatant to another Eppendorf with 1 µg of normal IgG (if a mouse monoclonal is used then normal mouse antibody) and 20 µl of agarose conjugate beads.

3. Centrifuge at 1500 rpm at 4 °C for 5 min.

4. Transfer the supernatant to another Eppendorf with 10 µl of antibody for 1 h at 4 °C.

5. Incubate the solution overnight at 4 °C with 20 µl of agarose conjugate.

6. Collect the agarose beads by centrifugation at 2500 rpm for 5 min at 4 °C.

7. Wash the resulting pellet four times with RIPA buffer.

8. After the final wash discard the supernatant and resuspend the pellet in 50 µl kinase buffer.

immunoprecipitation

1. Collect the agarose beads by centrifugation at 2500 rpm for 5 min at 4 °C.

2. Wash the pellet twice with kinase buffer.

3. Resuspend the pellet in 50 µl of 100 mg/ml histone H1, 5 mM ATP and 0.1 mCi/ml [^{32}P]ATP in kinase buffer for 30 min at 30 °C.

4. Stop the reaction by adding 5x loading buffer and boil for 4 min.

5. Run 20 µl of the reaction mixture on a 15 % SDS gel for 1 h at 200 mA and 1000 V.

6. Place the gel on three sheets of blotting paper and dry on a gel dryer for 3 h at 70 °C.

7. Place the sheet of blotting paper with the dried gel in an autoradiographic cassette with autoradiography film and leave at −80 °C.

8. Develop the film after 1–3 days.

9. The levels of kinase activity can be quantified by using a densitometer to measure the intensity of the labeled histone bands. The histone bands should be just above the dye front of the gel.

protein kinase assay

Note: Negative controls could include: (a) samples treated as above, but adding no cyclin/cdk complex; (b) samples treated as above, but without the addition of radioactive ATP; (c) a sample immunoprecipitated in the absence of anti-cyclin E, for example, immunoprecipitated with normal mouse IgG.

11.2
Analysis of Apoptosis-Modulating Protein Expression by Western Blot

This section describes the application of western blot for analysis of expression of the apoptosis-related proteins Bcl-2, Bax, and Bcl-x_L or Bcl-x_S using commercially available antibodies. For detailed information and guidelines on western blotting techniques see Chapter 6.

We have also analyzed the expression of these genes at mRNA level using a reverse transcriptase-polymerase chain reaction (RT-PCR) approach. A detailed description of this technique is given in the chapter by O' Doherty et al. (this Vol.) and O'Driscoll et al. (1993).

▋ Materials

- Sonicator (Labsonic U, B. Braun, USA)
- BioRad protein assay kit (500–0006, BioRad, UK)
- Semi-dry blotting apparatus: Transblot SD semidry transfer cell (170–3940, BioRad, UK)
- Gel caster: Dual gel caster (SE245, Hoefer, USA)
- Gel electrophoresis apparatus (SE250/260, Hoefer, USA)
- Blotto: 5 % (w/v) nonfat dried milk powder in H_2O
- TBS: For a 10x stock: 0.05 M Tris/HCl; 0.15 M NaCl; pH 7.6
- TBS/0.1 % Tween
- ABC/TBS (Dako): Make up and store as in manufacturer's instructions. From this stock make up a 1/100 dilution of ABC stock in TBS; pH 7.6.
- Enhanced chemiluminesence kit (Amersham)

antibodies
- Mouse anti-Bcl-2 (124) (MCA1279, Serotech)
- Rabbit anti-Bcl-$x_{L/S}$ (S-18) (sc-634, Santa Cruz, USA)
- Rabbit anti-Bax (N-20) (sc-493, Santa Cruz, USA)

buffers
- 2× loading buffer: 50 mM Tris-HCl, pH 6.8; 1 % SDS; 5 % 2-mercaptoethanol; 5 % glycerol (87 %); 0.1 % bromphenol blue

▢ Procedure

Western blotting was performed using a variation of the method of Towbin et al. (1979).

western blotting

1. Collect 2×10^7 cells and wash three times in PBS.

2. Resuspend pellets in 0.6 ml of PBS/100 mM PMSF.

3. Sonicate into crude cell lysates.

4. Centrifuge crude cell lysates at 1000 rpm for 10 min to remove cell debris.

5. Assay the samples for protein concentration using the BioRad protein assay according to the manufacturer's guidelines.

6. Dilute samples in 2× loading buffer.

7. Run 20 µg protein per lane on gels and then transfer to PVDF matrix using a BioRad semi-dry blotting apparatus at 15 V and 0.34 mA for 20 min.

8. Block blots at room temperature for 2 h in 5 % blotto. Wash once in TBS.

9. Add rabbit anti-Bax primary antibody at a concentration of 0.1 µg/ml in TBS/0.1 % Tween at 4 °C overnight **or** add the rabbit anti Bcl-$x_{L/S}$ or mouse anti-human Bcl-2 primary antibodies primary antibodies at a concentration of 0.2 µg/ml in TBS/0.1 % Tween at 4 °C overnight.

10. Wash blots three times in TBS/0.1 % Tween to remove excess primary antibody.

11. Add secondary antibodies (goat anti-rabbit biotinylated or rabbit anti-mouse biotinylated at a 1:2000 dilution and incubated for 2 h at room temperature.

12. Wash blots as above and then incubated in a 1:100 dilution of ABC/TBS (DAKO) for 1 h at room temperature.

13. Wash three times as above.

14. Bands can then be visualized using the enhanced chemiluminescence kit following the manufacturer's guidelines.

Note: Negative controls consist of similar treated samples, except that the primary antibody is left out. Positive controls could include cell lines positive for Bcl-2 and Bax (e.g., HL-60) or Bcl-x$_{L/S}$(e.g., MCF-7).

11.3
Time-Lapse Videomicroscopical Analysis of Cell Death Kinetics

The morphological changes seen during apoptosis can be easily visualized using light microscopy techniques, such as phase contrast microscopy, thus allowing the visualization of apoptotic events in living cultures under normal cell culture conditions.

When combined with video recording, this provides us with a powerful tool to study cell death in a dynamic fashion. Indeed, it allows not only for the visualization of cell death events, but also of cell division events, Thus, it is possible to follow the dynamics of cell number and provides us with information which is difficult to obtain by other cyto-

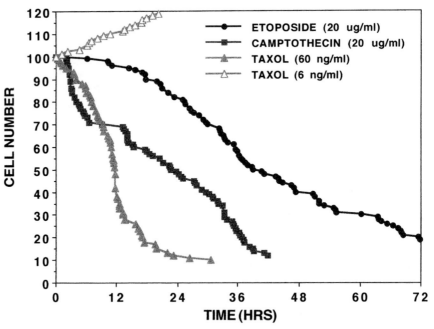

Fig. 11.3. Time-lapse videomicroscopy allows for the detection of both cell division events and cell death events, and thus can be used to follow the cell number of a population in a dynamic fashion

toxicity tests, such as the cellular acid phosphatase assay. Although, these latter have the advantage of being high-through-put assays, they tend to measure endpoints and provide little mechanistic information (Martin and Clynes 1991).

Because the morphological alterations seen during the apoptotic process can be easily visualized using routine phase-contrast microscopy (Desjardins and MacManus 1995), this method lends itself particularly well for studies of the kinetics of apoptosis in *in vitro* cell culture situations. Indeed, the occurrence of apoptotic events can be easily followed by time-lapse videomicroscopy using phase-contrast optics. This allows not only for quantitation of apoptotic events over a period of time, but also for the close study of individual cells (a typical example is shown in Fig. 11.3). It also provides the opportunity to look at cell death in the context of the cell's "history," i.e., did it go through previous divisions after exposure to toxins but prior to apoptosis, did it try to divide before it underwent apoptosis, etc.

The technique has been routinely used for the study of c-myc-induced apoptosis (Harrington et al. 1994).

Materials

- Inverted microscope, equipped with phase-contrast optics (Diaphot, Nikon)
- 10× and 20× phase-contrast objectives with long working distance
- CCD camera (CCD-100, Mitsubishi)
- Video recorder with time-lapse capabilities (HS-S5600E, Mitsubishi)
- Video printer

Procedure

Heating of the microscope stage and sample can be accomplished by either of the following two methods:

- A warm stage heated by a resistive element and equipped with a temperature controller (e.g., Linkam C102)

Note: Although simple and cost efficient this particular heating system has one drawback at higher magnification: to keep the temperature constant the heating element gives little pulses resulting in a small mechanical expansion/contraction of the stage. Although not noticeable at lower

heating of the microscope stage/sample

magnifications, with high power lenses this results in a quick jitter of the phase-halo.

– A warm air blower (e.g., Nikon Diaphot incubation warmer ITC-32) feeding into a closed perspex incubator hood.

Note: The incubator hoods supplied by the microscope manufacturers are generally expensive: alternatively, they can be custom built by a local workshop specializing in plastics, generally at a lower cost.

Scoring of Cell Division and Cell Death Events

The recording speed was set at 3.22 s/field (480 h mode), which at normal playback speed results in an acceleration factor of 160.

Cell division events were scored at the time septa formed between two daughter cells

Cell death was classified as apoptotic on the basis of typical sequence of morphological alterations starting with rounding up of adherent cells, cell blebbing, cell shrinkage and/or break-up into apoptotic bodies. These cell death events were scored at the time when rounding up was first noticed. Typically the fate of about 100 cells in the microscope field was followed.

In some experiments the duration of individual apoptotic events was measured. In these cases the time difference between rounding up and the first occurrence of cellular fragmentation was scored. In these experiments twenty randomly chosen apoptotic events were timed.

Although the technique allows for detailed analysis and quantitation of cell death events, one major drawback is its low throughput, i.e., one can only look at one condition at a time.

11.4
Analysis of Cell Cycle Distribution by Flow Cytometry Based on Propidium Iodide Staining

This method is based on the characteristic of propidium iodide (PI) to intercalate with DNA. Cells are incubated with PI and analyzed by flow cytofluorometric detection. Thus the amount of fluorescence is proportional to the amount of PI intercalated and to the DNA content of the cell (Higashikubo et al. 1994).

Materials

- Vortex
- Flow cytometer (FACScan, Becton Dickinson, UK)
- Trypsin/EDTA solution: 0.25 % trypsin (043–05090, Gibco, UK), 0.01 % (w/v) EDTA in PBS (EDS, Sigma, UK)
- Ice-cold 70 % ethanol, 30 % PBS
- 400 µg/ml propidium iodide (PI) in PBS (P4170, Sigma, UK)
- 1 mg/ml RNase (DNase-free) in PBS (1 119 915, Boehringer Mannheim)

Note: Make up these solutions fresh. PI intercalates between base pairs of double stranded DNA and RNA without base specificity. All DNA-specific compounds should be considered potential carcinogens and therefore handled with some precautions. Solutions should be immediately washed from the skin if contact is encountered.

Procedure

The following is a method for staining cultured cells with PI:

propidium
iodide staining

1. Adherent cells are collected as follows: pour media from a flask into a universal.

2. Trypsinize the flask with a trypsin/EDTA solution (0.25 % trypsin, 0.01 % EDTA solution in PBS) into a single cell suspension and add the trypsinized cells to the universal. **Note:** This ensures that floating cells from the sample are collected.

3. Pellet cells by centrifugation.

4. Resuspend pellet in 200 µl of PBS.

5. To each sample add 2 ml of ice-cold 70 % ethanol, 30 % PBS. **Note:** For consistency of fixation and to avoid clumping of cells it is essential that the ice-cold ethanol is added dropwise over a 1 min period, whilst vortexing the sample at a medium setting.

6. Leave on ice for at least 30 min. **Note:** At this stage samples can be stored for up to 1 month at 4 °C.

7. Harvest cells and resuspend in 800 µl PBS.

8. Check the cell suspension microscopically to ensure that it is composed of single cells.

9. Add 100 μl of PI (400 μg/ml) and 100 μl of RNase (1 mg/ml).

10. Incubate at 37 °C for 30 min.

11. Analyze samples using an argon-ion laser tuned to 488 nm measuring forward and orthogonal light scatter and red fluorescence (measuring both peak and area of the fluorescence). Usually at least 10 000 cells are analyzed.

11.5
Flow Cytometric Detection of Apoptosis Based on Light Scatter Properties

This method has the advantage of being cheap, fast and simple. However, it can only be applied to cell populations with a homogeneous distribution of cell size. In practice, this restricts the technique to cell lines, or situations in which a homogeneous population can be isolated (e.g., granulocytes isolated from blood of patients with chronic myelogenous leukemia). In our hands the technique works very well with suspension cultures, like the myelomonocytic cell line HL-60 (Fig. 11.4) (McGowan et al. 1994; Verhaegen et al. 1995) and the chronic myelogenous leukemia line K562 (McGahon et al. 1994). We recently have it applied to adherent cell lines including the human lung carcinoma line DLKP and the carci-

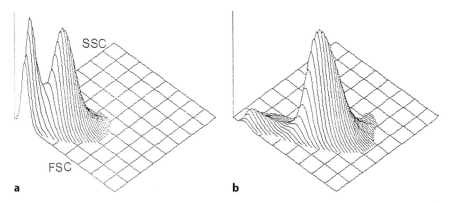

Fig. 11.4a, b. Light scatter properties of HL-60 cells. **A** 3 h after UV irradiation. Note the large populations with reduced forward and side scatter, typical for the presence of shrunken and fragmented apoptotic cells. **B** Distribution typical for normal HL-60 cells. (Verhaegen et al. 1995)

noma lines Hep2 and A549. The technique is very sensitive and allows for the detection of apoptosis during the early stages (cell shrinkage) of apoptosis. In fact the technique can easily be used to routinely monitor the viability of cell populations.

▓ Materials

– Trypsin-EDTA solution
– Ice-cold ethanol (70 % in H_2O; keep stored at $-20\,°C$)
– Flow cytometer

▓ Procedure

1. Collect supernatant (to collect floating death cells).   **flow cytometry**

2. Trypsinize remaining monolayer with a minimum of trypsin solution.

3. Add this suspension to the supernatant.

4. Centrifuge for 5 min at 1000 rpm.

5. Discard supernatant.

6. Gently mix the pellet and remaining liquid on a vortex mixer (medium speed).

7. Over a 1 min period add 1 ml of ice-cold 70 % ethanol.

8. Keep on ice for 1 h. In most cases samples can be stored at 4 °C for prolonged periods

9. Immediately before analysis, pellet cells by centrifuging at 1000 rpm for 5 min and resuspend the cells in PBS.

11.6
Flow Cytometric Detection of Apoptosis Based on Increased Annexin V-Binding

This technique is based on the observation that early during apoptosis phosphatidyl serine is flipped to the outer leaflet of the cell membrane. The detection of this phenomenon makes use of the calcium-dependent binding of annexin V to this phospholipid (Vermes et al. 1995). A FITC-labelled annexin V conjugate is commercially available. The main advantage of the technique lays in its potential to detect the very first signs of apoptosis. Disadvantages are the use of an expensive probe and the fact that the staining has to be performed on unfixed cells which have to be analyzed straight away. Nevertheless the staining procedure is fast and easy to apply.

Materials

– Annexin V probe (BMS306FI, Bendar MedSystems, Austria)
– Staining buffer
– Flow cytometer (FacScan, Beckton Dickinson)

Procedure

flow cytometry

1. Collect sample (about 10^5–10^6 cells).

2. Centrifuge suspension for 5 min at 1000 rpm.

3. Discard supernatant.

4. Add 90 µl staining buffer.

5. Add 10 µl annexin V probe.

6. Mix gently and incubate at room temperature in the dark for 10 min.

7. Resuspend in 1 ml of staining buffer and immediately analyze on flow cytometer.

Note: Do not suspend in PBS or FACS-fluids that do not contain Ca^{2+}. Since annexin V binding is Ca2+ dependent and reversible, doing so will result in loss of bound annexin and hence loss of fluorescent signal.

11.7
Detection of Internucleosomal DNA Fragmentation by Agarose Gel Electrophoresis

This method allows for the detection of the typical internucleosomal DNA degradation seen in apoptotic cells (Compton 1992). A typical DNA ladder is shown in Fig. 11.5.

Fig. 11.5. Agarose gel electrophoresis of DNA isolated from DLKP cells treated with 0.6 µg taxol for 24 h. The typical 200 bp ladder characteristic for internucleosomal cleavage of the DNA during apoptosis is visible

Materials

- 1 mg/ml DNase-free RNase A (1 119 915, Boehringer Mannheim) in 0.1 M sodium acetate/0.3 mM EDTA, pH 4.8   **solutions**
- 20 mg/ml proteinase K in H$_2$O
- 1.5 % agarose in TBE buffer (see below)
- 10 mg/ml ethidium bromide

- Electrophoresis apparatus for mini-gels (Minicell EC370 M, E-C Apparatus Corporation, USA)   **equipment**
- Power source (model 200, BRL-Life Technologies, USA)

- Lysis buffer: 20 mM EDTA; 100 mM Tris, pH 8.0; 0.8 % (w/v) sodium lauryl sarcosinate   **buffers**

– Loading buffer: 10 mM EDTA; 0.25 % (w/v) bromphenol blue; 50 % (w/v) glycerol
– TBE buffer: 2 mM EDTA, pH 8.0; 89 mM Tris; 89 mM boric acid

▪ Procedure

agarose gel electrophoresis

1. Collect cells by centrifuging at 200 g for 5 min at room temperature.

2. Resuspend cells at 2×10^6 cells/ml in lysis buffer (20 µl).

3. Add 10 µl RNase solution.

4. Incubate for 18 h at 37 °C.

5. Add 10 µl of the proteinase K solution.

6. Incubate for a further 1.5–2 h at 50 °C. **Note:** At this stage samples can be stored at 4 °C for up to 72 h.

7. Prepare 1.5 % agarose solution in TBE buffer.

8. Allow the solution to cool down sufficiently, and add 3 µl of the ethidium bromide solution.

9. Cast gel and allow gel to set.

10. After setting, place the gel in the electrophoresis apparatus and fill with TBE buffer.

11. Add 10 µl loading buffer to each sample.

12. Load each sample into the slots using a micropipet with disposable tips. **Note:** Because of the high viscosity of the sample and to avoid shearing of the DNA, it is generally easier to use tips from which about 2 mm from the end is removed.

13. Run gels at 55 V for 4 h in TBE buffer.

14. After electrophoresis bands can be visualized by placing the gel on a UV transilluminator.

11.8
Assessment of Cell Morphology During Apoptosis by Cytocentrifuge Preparation

This must surely rank among the easiest techniques to detect and quantitate apoptosis. It makes use of the morphological characteristics associated with apoptosis, i.e., cellular and nuclear condensation and fragmentation into apoptotic bodies.

Cells are centrifuged onto glass coverslips using a cytospin centrifuge and can be stained using a variety of dyes (Giemsa, hematoxylin-eosin, etc.).

The preparations are then visualized microscopically for the quantitation of normal healthy cells and cells showing characteristics of apoptosis. Thus, a percentage of dying cells can be calculated. Typical results can be found in Verhaegen et al. (1995).

▦ Materials

– Microscope slides
– Rapi-DiffII staining kit (PS128, DiaChem, UK)
– Coverslips
– DPX mounting medium (36029 4H, BDH Laboratory Supplies, UK)

– Cytospin centrifuge   equipment
– Microscope with 40× objective

▦ Procedure

1. Spin cells (5×10^4) onto glass slides using the cytocentrifuge, 500 rpm   cytocentri-
    for 2 min.   fugation

2. Allow slides to air-dry.

3. Cells are then fixed and stained using the Rapi-DiffII kit:

3a. Fix cells by dipping ten times in solution A (methanol-based fixative).

3b. Dip ten times in solution B (eosin-based dye solution).

3c. Dip ten times in solution C (methyl blue-based dye solution).

4. Remove excess stain under running tap water.

5. Allow slides to air-dry.

Note: Slides can be mounted in DPX mounting agent and cover-slipped and stored indefinitely.

This staining procedure enables one to distinguish between cytoplasm and nuclei.

Apoptotic cells can be identified as previously described (Martin and Cotter 1991), generally cells are considered apoptotic when nuclear and/or cellular fragmentation is visible.

At least 300 cells are counted in three different microscope fields on the same slide.

The number of apoptotic cells is expressed as a percentage of the total cell number counted.

A typical microscope field containing apoptotic HL-60 cells is illustrated in Fig. 11.1.

References

Carson DA, Ribeiro JM (1993) Apoptosis and disease. Lancet 341: 1251–1254

Casciola-Rosen LA, Miller D, Anhalt GJ, Rosen A (1994) Specific cleavage of the 70-kDa protein component of the U1 small ribonucleoprotein is a characteristic biochemical feature of apoptotic cell death. J Biol Chem 269: 30757–30760

Compton MM (1992) A biological hallmark of apoptosis: internucleosomal degradation of the genome. Cancer Metastasis Rev 11: 105–119

Cotter TG, Lennon SV, Glynn JG, Martin SJ (1990) Cell death via apoptosis and its relationship to growth, development and differentiation of both tumor and normal cells. AntiCancer Res 10: 1153–1160

Desjardin LM, MacManus JP (1995) An adherent cell model to study different stages of apoptosis. Exp Cell Res 216: 380–387

Eblen ST, Fautsch MP, Burnette RJ, Snyder M, Leof EB (1995) Dissociation of p34cdc2 complex formation from phosphorylation and histone H1 kinase activity. Cancer Res 55: 1994–2000

Hall PA, Lane DP (1994) Genetics of growth arrest and cell death: key determinants of tissue homeostasis. Eur J Cancer 30A: 2001–2012

Harrington EA, Bennett MR, Fanidi A, Evan GI (1994) c-Myc-induced apoptosis in fibroblasts is inhibited by specific cytokines. EMBO J 13: 3286–3295

Hartwell LH, Kastan MB (1994) Cell cycle control and cancer. Science 266: 1821–1828

Higashikubo R, Goswami PC, Roti Roti JL (1994) A comparision of time-lapse cinemicrography and flow cytometry for the study of accelerated cell-cycle transit. Cell Prolif 27: 697–709

Kaufmann SH, Desnoyers S, Ottaviano Y, Davidson NE, Poirier GG (1993) Specific proteolytic cleavage of poly(ADP-ribose) polymerase: an early marker of chemotherapy-induced apoptosis. Cancer Res 53: 3976–3985

Martin A, Clynes M (1991) Acid phosphatase: endpoint for in vitro toxicity tests. In Vitro Cell Dev Biol 27A: 183–184

McGahon A, Bissonette R, Schmitt M, Cotter KM, Green DR, Cotter TG (1994) Bcr-abl maintains resistance of chronic myelogenous leukaemia cells to apoptotic cell death. Blood 83: 1179–1187

McGowan AJ, Fernandes RS, Verhaegen S, Cotter TG (1994) Zinc inhibits UV radiation induced apoptosis but fails to prevent subsequent cell death. Int J Radiat Biol 66: 343–349

Neamati N, Fernandez A, Wright S, Kiefer J, and McConkey DJ (1995) Degaradation of lamin B1 precedes oligonucleosomal DNA fragmentation in apoptotic thymocytes and isolated thymocyte nuclei. J Immunol 154: 3788–3795

Nunez G, Merino R, Grillot D, Gonzalez-Garcia M (1994) Bcl-2 and Bcl-x: regulatory switches for lymphoid death and survival. Immunol Today 15: 582–588

O'Driscoll L, Daly C, Saleh M, Clynes M (1993) The use of reverse transcription-PCR to investigate specific gene expression in MDR cells. Cytotechnology 12: 289–314

Oltvai ZN, Korsmeyer SJ (1994) Checkpoints of dueling dimers foil death wishes. Cell 79: 189–192

Pines J (1994) The cell cycle kinases. Sem Cancer Biol 5: 305–313

Sherwood SW, Schimke RT (1995) Cell cycle analysis of apoptosis using flow cytometry. In: Matsudaira PT, Osborne BA (eds) Methods cell biol, vol 46. Academic Press, London, pp 77–97

Tewari M, Quan LT, O'Rourke K, Desnoyers S, Zeng Z, Beidler DR, Poirier GG, Salvesen GS, and Dixit VM (1995). Yama/CPP32b, a mammalian homologue of CED-3, is a crmA-inhibitable protease that cleaves the death substrate poly(ADP-ribose) polymerase. Cell 81: 801–809

Verhaegen S, McGowan AJ, Brophy AR, Fernandes RS, Cotter TG (1995) Inhibition of apoptosis by antioxidants in the human HL-60 leukemia cell line. Biochem Pharmacol 50: 1021–1029

Vermes I, Haanen C, Steffens-Nakken H, Reutlingsperger C (1995) A novel assay for apoptosis flow cytometric detection of phopshatidylserine expression on early apoptotic cells using fluorescein labelled Annnexin V. J Immunol Methods 184: 39–51

Suppliers

Amersham Int. Plc.
Amersham Place
Little Chalfont
Bucks HP7 9NA
UK

BDH Laboratory Supplies
Broom Road
Poole BH15 1TD
UK
Fax:+44–202–738 299

Becton Dickinson UK

Bendar MedSystem
Dr. Boehringer Gass 5211
P.O. Box 73
A-1121 Vienna
Austria
Fax:+43–1–801 05477

BioRad UK
Bio-Rad House
Maylands Avenue
Hemel Hempstead
Hertfordshire HP2 7TD
UK
Fax: +44–1442–259118

Boehringer Mannheim UK
Bell Lane, Lewes
East Sussex BN& 1LG
UK
Fax:+44–1273–480266

B. Braun Medical Ltd
12–14 Aylesbury Vale Ind. Park
Aylesbury
Bucks HP20 1DQ
UK
Fax:+44–1296–435714

BRL-Life Technologies UK
3 Fountain Drive
Inchinnan Business Park
Paisley PA 9RF
UK
Fax:+44–141–814 6317

DiaChem
Diagnostic Developments,
Southport
UK
Fax: +44–704–664 88

Linkam
8 Epsom Downs Metro Centre
Waterfield, Tadworth, Surrey,
KT20 5HT
UK
Fax:+44–1737–363480

Mitsubishi UK
Travellers Lane
Hatfield
Hertfordshire AL10 8XB
UK
Fax:+44–707–278691

Nikon Europe B.V.
Schipholweg 321
P.O. Box 222
1170 AE Badhoevedorp
The Netherlands
Fax:+31–20 44 96 299

Riedel de Haen
Wunstorfer Str. 40
D-30926 Seelze
Germany
Fax:+49–51–37 999 123

Santa Cruz Biotechnology
2161 Delaware Avenue
Santa Cruz
California 95060
USA
Fax:+1–408 457 3801

Savant Instruments, Inc.
110 Bi-county Boulevard, Suite 103
Farmingdale, New York 11735–3933
USA
Fax:+1–516–249–4639 or 4636

Serotech UK
22 Bankside
Station approach
Kidlington, Oxford OX5 1JE
UK
Fax:+44–1865–373899

Sigma UK
Sigma-Aldrich Company Ltd.
Fancy Road, Poole
Dorset BH12 4QH
UK
Fax:+44–1202–715 460

Part III

In Vitro Models for Cell Differentiation

Chapter 12

Assessing Release of Secretory Products from Individual Cells

Domenico Bosco and Paolo Meda*

▦ Introduction

In a variety of systems, secretion by individual cells, and changes of this function, which depend on cell adhesion/communication, can be studied using modifications of the original hemolytic plaque assay. This assay was developed for evaluating Ig production by lymphocytes (Jerne et al. 1974), and the modifications, referred to as reverse hemolytic plaque assays (RHPAs), are based on the complement-induced lysis of red blood cells (RBCs) bearing immunocomplexes. Practically, the cells of interest are attached to a support, together with RBCs that have been coated with staphylococcal protein A, and tested in the presence of polyclonal Ig raised against the secretory product (most often a protein) to be monitored. Binding of these antibodies to the product released by the cells is followed by the fixation of the immunocomplexes onto the surface of surrounding RBCs. After addition of complement, the immunocomplexes-bearing RBCs lyse, resulting in the formation of a hemolytic plaque around secreting cells (Fig. 12.1).

This approach allows for the identification of secreting cells in a heterogeneous population and provides a quantitative and sensitive method to measure quite small amounts of secreted proteins. Moreover, immunohistochemistry, autoradiography, electrophysiological and other physiological, biophysical and morphological evaluations can be combined with the plaque assay, thus permitting simultaneous monitoring of multiple parameters within the very same cell population.

* *Correspondence to* Paolo Meda, Department of Morphology, 1 rue Michel Servet, University of Geneva Medical School, 1211 Genève 4, Switzerland; phone +41−22−7025210; fax +41−22−705260; e-mail PAOLO.MEDA@MEDECINE.UNIGE.CH

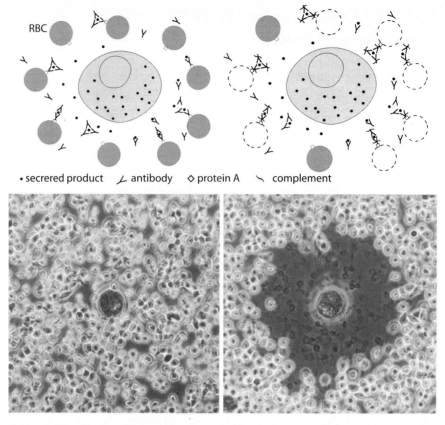

Fig. 12.1. *Upper panel* The mechanism of the plaque assay before (*left*) and after (*right*) addition of complement. *Lower panel*: actual views of a single cell surrounded by red blood cells in a reverse hemolytic plaque assay chamber, before (*left*) and after (*right*) incubation with complement. All halftone illustrations are from a plaque assay detecting insulin release from pancreatic β-cells

Outline

A short protocol is shown in Table 12.1.

Materials

equipment
 – General purpose centrifuge
 – Incubator (37 °C)

Table 12.1. Short protocol

Material to be prepared in advance	Clean slides and coverslips	● Acetone ● Alcohol ● H$_2$0	30 min 30 min 3×30 min
	Trypan blue	● Prepare a 0.4% stock solution in PBS ● At 37 °C ● Filter	30 min
	Protein A-coated red blood cells	● Rinse RBC with 0.9% NaCl, mix 1 volume packed RBCs, 1 volume 0.5 mg/ml protein A and 10 volumes 100 μg/ml CrCl$_3$ ● At 30 °C	3×10 min 60 min
Material to be prepared on the day of the experiment	Coat slides with poly-L-lysine	● Apply on each slide 500 μl poly-L-lysine (100 μg/ml) ● At room temperature ● Rinse with H$_2$0	60 min 15 min
RHPA procedure	Cell preparation	add 4% Protein A-coated "packed RBC" to the cell suspension	
	Cell attachment	● Fill the chambers with 55–60 μl of the mixed suspension of cells ● At 37 °C	45 min
	Secretion test	● Rinse with 200 μl buffer ● Fill with 100 μl buffer supplemented with antibodies ● At 37 °C	Appropriate time
	Complement addition	● Rinse with 200 μl control buffer ● Fill with 100 μl control buffer supplemented with complement (1:40) ● At 37 °C	60 min
	Testing for cell viability	● Fill with 100 μl trypan blue (0.04%) in control buffer ● At room temperature	5 min
	Fixation	● Rinse with 200 μl control buffer ● Fill with 100 μl fixative	Appropriate time

- Microscope equipped for phase-contrast and, if possible, for fluorescence illumination
- Standard microscope slides
- Glass coverslips (22×22 mm)
- Flat stock liner double face tape (410 SCOTCH, 3 M, St. Paul, MN 55144, no. 021200–07137)
- Filter paper drying blocks (Schleicher & Schuell, Postfach 4, 3354 Dassel, Germany, ref. 310992)

reagents   A crucial element for successful RHPAs is the availability of adequate polyclonal antibodies raised against the protein of interest. Thus, when developing this assay, it is wise to test different antibodies, at different dilutions (e.g., 1:10 to 1:200). Before use, the selected antibody must be heat-inactivated (45 min at 56 °C) in order to destroy complement factors which are present in the serum.

It is important to note that monoclonal antibodies are not suitable to perform the RHPA, since the complement-induced lysis of RBCs takes place only after several Ig molecules are bound on a protein.

RHPA can be run using any buffer suitable for the cell type under study. Bovine serum albumin (BSA), at a concentration of 1 mg/ml, must be added to the buffer to avoid undesirable nonspecific adherence of antibodies and secreted proteins to the glass supports.

- $CrCl_3.6H_2O$ (Fluka AG, Buchs, Switzerland, ref. 27096)
- Sheep blood, preserved (Behringwerke AG, Marburg, Germany, ref. ORAW 30/31)
- Guinea pig complement (Behringwerke AG, Marburg, Germany, ref. ORAY 20/21)
- *Staphylococus aureus* protein A (Pharmacia AB, Uppsala, Sweden, ref. No 17–0770–01)
- Poly-L-lysine hydrobromide, MW 150 000–300 000 (Sigma, St. Louis, MO, 63178, ref. P1399)
- Trypan blue (Hermann Gräub, Bern, Switzerland)
- Phosphate-buffered saline (PBS)
- 0.9 % NaCl
- Fixative (e.g., Bouin's or 4 % paraformaldehyde solution if cells will be processed for immunofluorescence; 2.5 % glutaraldehyde solution if cells are to be processed for electron microscopy or analyzed without further processing)
- 66 % glycerin solution in PBS

Preparing for the Experiment

Glassware, the stock solution of trypan blue and coating of RBCs with protein A should be prepared a few days before the experiment.

Glassware

To ensure good adherence of cells and to avoid nonspecific binding of antibodies to Cunningham's chambers, the microscope slides and coverslips should be thoroughly cleaned: Immerse slides and coverslips in acetone (30 min), then in absolute ethanol (30 min) and then in distilled water (3×30 min). Dry at 60 °C for 60 min and store the clean glassware under dust-free conditions.

Trypan Blue Solution

To identify damaged cells, trypan blue staining is recommended at the end of the plaque assay. To prepare a stock solution, dissolve 4 mg/ml trypan blue in 10 ml PBS. Heat 30 min at 37 °C, filter and store at room temperature. This stock solution may be used for months.

Protein A Coating of Red Blood Cells

The aim of this procedure is to prepare RBCs able to fix protein-Ig immunocomplexes using protein A as an attachment site for the Fc region of immunoglobulins. This is achieved as follows:

1. Set up a water bath at 30 °C.

2. In a 50 ml Falcon tube, suspend 10 ml of the commercial RBC suspension in 40 ml 0.9 % NaCl. Centrifuge 10 min at $130\,g$ and room temperature. Discard the supernatant and repeat the centrifugation step twice more. The final pellet is referred to as "packed RBCs."

3. During the centrifugation period, prepare the following solutions:

   – In a 10 ml tube: dissolve 100 μg/ml $CrCl_3.6H_2O$ in 0.9 % NaCl. Protect this solution from light by wrapping the tube in aluminium foil.
   – In a 1 ml tube: dissolve 500 μg /ml protein A in 0.9 % NaCl.

4. After the third centrifugation, in a 10 ml tube mix 500 µl packed RBCs with 500 µl protein A solution.

5. Add 5 ml $CrCl_3$ solution. It is essential that the addition is done drop-wise and under continuous stirring.

6. Close the tube with both a screw cap and Parafilm; place it in the 30 °C water bath for 1 h. Shake gently every 15 min.

7. At the end of this incubation, pour the RBC suspension into another clean 10 ml tube and centrifuge it 10 min at 130 g.

8. Remove the supernatant, add 10 ml 0.9 % NaCl, and centrifuge again 10 min at 130 g.

9. Remove the supernatant, add 10 ml 0.9 % NaCl, and store the protein A-coated RBCs at 4 °C. Coated RBCs should be used within 10 days.

Efficiency of RBC Coating and Antibodies-Protein A Binding

Efficiency of the RBC coating and of the antibody-protein A binding may be assessed using a radiolabeled form of the protein of interest. The test is performed as follows:

1. Mix the following reagents within a microtube (prepare in duplicate):
   - 150 µl stock solution protein A-coated RBCs
   - 100 µl antibodies used in the plaque assay (antibodies should be diluted to about eight times the dilution used for RHPA)
   - 50 µl radiolabeled protein (about 200 ng/ml in PBS)
   - 50 µl PBS supplemented with 0.5 % BSA

2. Prepare two additional microtubes as described above, using a nonimmune rabbit serum instead of the specific antibodies. The source of serum (usually a rabbit) should be the same as that of the antibodies under test.

3. Seal the tubes and incubate them 60 min in a water bath set at 37 °C.

4. Centrifuge 1 min at 12 000 g.

5. Cut the tubes at the interface between pellet and supernatant

6. Count radioactivity of both parts in a counter.

In the tubes containing specific antibodies, the higher the radioactivity in the pellet, the more efficient the protein binding to RBCs. In the tubes containing nonimmune serum, pellet radioactivity reflects nonspecific binding. The usual level of this nonspecific binding should be less than 5 % of the total radioactivity.

Cunningham's Chambers

Some 2–3 h before starting the experiment, prepare the Cunningham's chambers as follows (Fig. 12.2A–E):

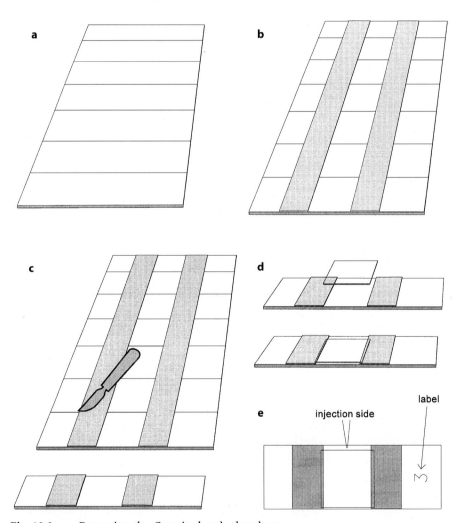

Fig. 12.2a–e. Preparing the Cunningham's chambers

1. Align previously cleaned slides as shown in Fig. 12.2A.

2. Stick two strips of double-face adhesive tape (see "Materials") onto clean glass slides 19 mm apart (Fig. 12.2B).

3. Dissociate the slides by cutting the tape with a scalpel (Fig. 12.2C).

4. Just before starting the secretion test, cover each slide with 500 µl poly-L-lysine (0.1 mg/ml in distilled H_2O) for 1 h.

5. Rinse each slide twice and leave it for 15 min in distilled water. After the slides have dried, apply coverslips onto the tape as shown in Fig. 12.2D.

6. Label each slide on one edge, using a diamond marker (Fig. 12.2E).

Buffers

During the 1 h coating of slides with poly-L-lysine, prepare the control and test buffers required for the secretion test.

12.1
Reverse Hemolytic Plaque Assay

▦ Procedure

RHPA

1. Pour dispersed cells into a 15 ml tube (conical tip) and add the chosen control buffer, up to 15 ml. You will need 5×10^3–10^4 cells per Cunningham's chamber. It is recommended not to prepare more than 10 or 12 chambers per experiment.

2. Centrifuge the tube 5 min at 130 g, room temperature.

3. Remove the supernatant and resuspend the pellet in a volume of control buffer calculated by multiplying 60 µl times the number of chambers to be prepared. Check the adequate concentration of cells by microscope examination of a chamber. Optimal concentration is indicated by a distance of 100–150 µm between individual cells.

4. Add 4 % (v/v) packed RBCs and mix gently.

5. Inject 55–60 µl of this preparation into each Cunningham's chamber using an adjustable-volume pipettor. To this end, apply the pipettor

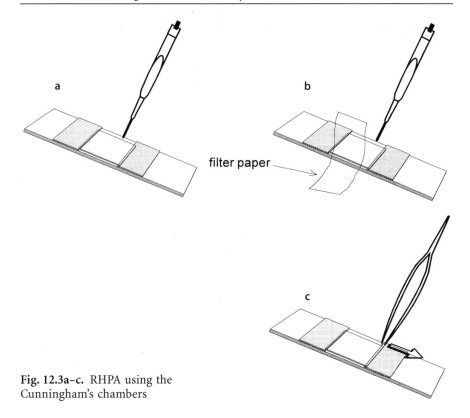

filter paper

Fig. 12.3a–c. RHPA using the
Cunningham's chambers

tip to the wider side of the Cunningham's chamber, as illustrated in
Fig. 12.3A, very close to the coverslip. The cell suspension will easily
enter the chamber by capillary action. A few microliters of the cell
suspension should be left on the injection side in order to avoid the
chamber contents from drying out.

6. Place the Cunningham's chambers in a humidified box (wet filter
   paper on the bottom with water) in order to avoid undesirable drying.

7. Place the humidified boxes at 37 °C for 45 min to permit adhesion of
   the cells to the polylysine-coated glass.

8. Rinse each chamber with 200 µl control (or test) buffer. Then repeat
   the rinsing using 100 µl of the same buffer, now supplemented with
   specific and heat-inactivated anti-protein serum. Solution changes
   are performed by applying a few microliters of the buffer on the

injection side and by simultaneously aspirating the injected volume at the opposite side using filter paper (Fig. 12.3B).

9. Incubate the chambers at 37 °C for the desired time. The appropriate duration of incubation will depend on the type of cells tested and on the amount of protein secreted. In some systems (e.g., amylase secretion from pancreatic acini) a 1 min incubation is sufficient to develop a sizeable hemolytic plaque (Chanson et al. 1989).

10. Rinse each chamber with 200 µl control buffer, and fill it with 100 µl of the same buffer supplemented with guinea-pig complement, diluted 1/40.

11. Incubate the chambers 60 min at 37 °C.

12. Rinse chambers with 100 µl 0.04 % trypan blue (diluted in control buffer) and incubate them for 5 min at room temperature.

13. Rinse the chambers with control buffer and fill them with 100 µl of the desired fixative (e.g., Bouin's or 4 % paraformaldehyde solution if cells will be processed for immunofluorescence; 2.5 % glutaraldehyde solution if cells are to be processed for electron microscopy or ana-lyzed without further processing).

14. If further processing is required, the coverslips forming the top of the chambers may have to be removed. In this case, the adhesive tape on one side of the coverslip is slowly pulled with a forceps which is moved parallel to the slide, and then lifted up to raise the coverslip (Fig. 12.3C).

15. At the end of all the processing, rinse the slides in PBS, cover them with a 66 % glycerin solution in PBS, and mount them with a coverslip which is sealed with nail varnish.

quantitative analysis At the end of the assay, visual inspection of Cunnigham's chambers provides a fast, if crude, way to check for the development of hemolytic plaques and to detect large differences in the number and size of these plaques between the various experimental conditions tested. However, further microscope analysis is needed to accurately assess the percentage of cells forming hemolytic plaques, as well as the individual areas of these plaques. For this analysis, we recommend the following:

● Use a 25× or 40× objective and work with both phase-contrast and bright field illuminations.

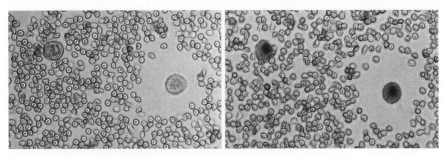

Fig. 12.4. Two cells excluding (*left*) and accumulating (*right*) trypan blue are seen at the end of the RHPA. The occurrence of trypan blue-stained cells surrounded by a hemolytic plaque is not surprising since staining for damaged cells follows the secretion test

- Restrict counts to trypan blue-excluding cells. Trypan blue-stained cells are easily identified under bright field illumination: nuclei of healthy cells appear colorless, whereas those of damaged cells appear dark blue (Fig. 12.4).

- Nonsecreting cells, which are not surrounded by a hemolytic plaque, may not be readily visible, particularly if they have not been specifically stained. Consequently, screen each slide carefully.

- Lymphocytes are present in the commercially available RBC preparation. They may be distinguished from nonsecreting cells by their high nucleus/cytoplasm ratio, their small size and almost consistent trypan blue staining.

- Count about 200 trypan blue-excluding cells throughout each chamber. Avoid the borders of the chambers, where a nonhomogenous development of hemolytic plaques is possible.

- Using phase-contrast illumination, hemolytic plaques are easily visible as dark areas centered around secreting cells. Invariably, these areas comprise nonrefringent ghosts of the complement-lysed erythrocytes.

- Round areas devoid of RBCs, sometimes centered on cells, may be present throughout the preparation. These area are easily distinguished from hemolytic plaques due to the absence of RBC ghosts.

- There are multiple ways (Bosco et al. 1989, 1995) in which to measure the areas of hemolytic plaques. The simplest approach is to use a calibrated grid eyepiece to evaluate the plaque diameter (Fig. 12.5).

Fig. 12.5. This hemolytic plaque was measured using a calibrated grid placed within the microscope eyepiece. It was found to have a diameter of 56 μm, corresponding to a calculated surface of about 2500 μm²

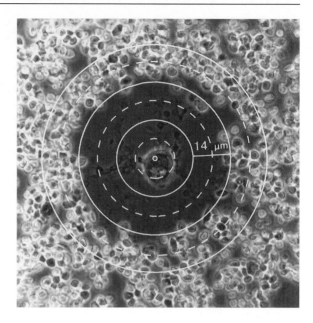

Results

Counting the percentage of cells forming a hemolytic plaque will provide an estimate of the proportion of cells contributing to secretion and will highlight an eventual heterogeneity of the cell population under study. Scoring the mean area of individual hemolytic plaques will provide an estimate of the amount of protein released by each cell. By multiplying the average values of these two parameters, one can further calculate total plaque development, which provides an estimate of the average secretion of the entire cell population (Bosco et al. 1989).

Figure 12.6 shows a typical example of this analysis from a RHPA study of insulin release by single pancreatic β-cells. In this system, increasing the glucose concentration from basal (2.8 mM glucose) to maximal (16.7 mM glucose) stimulatory levels elicits a substantial increase in the proportion of plaque-forming cells, but essentially no change in plaque area. These data indicate that stimulation markedly enhances the recruitment of secreting cells, without affecting in a major way the all-or-none functioning of individual cells.

This feature, however, is not necessarily seen in other systems. For example, RHPA analysis of amylase release from exocrine pancreatic acini (Bosco et al. 1988) reveals marked changes in the area of individual hemolytic plaques after exposure to an increasing carbamylcholine concentration (Fig. 12.7). Hence, in this system, secreting acini were not only recruited in larger numbers during stimulation, but also significantly increased their individual enzyme output with stimulation.

Plotting the frequency distribution of hemolytic plaques as a function of their size (Fig. 12.8) further indicates that stimulation affects, in a strikingly different way, distinct acinar subpopulations (Bosco et al. 1988).

Fig. 12.6. Insulin release was evaluated by RHPA after a 30 min stimulation by different concentrations of glucose. The *upper panel* shows that the percentage of plaque-forming cells is strongly dependent on the glucose concentration. In contrast, the *middle panel* shows that mean plaque area is essentially unaffected by glucose. The *bottom panel* shows that total plaque development increases in a sigmoidal fashion with the glucose stimulation. This increase is similar to that measured by standard biochemical techniques, which however, do not allow differentiation of the variable contribution of individual cells

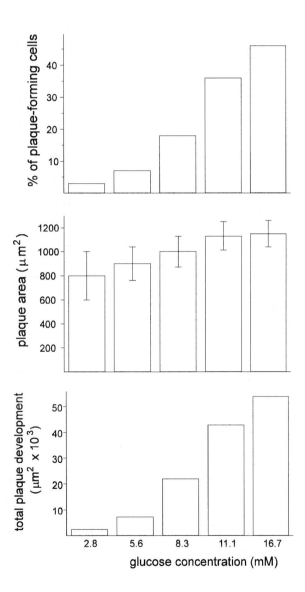

Fig. 12.7. The areas of hemolytic plaques formed around amylase-secreting acini of pancreas increase with carbamylcholine (*CCh*) concentrations, indicating that the enzyme output of individual acini is modulable

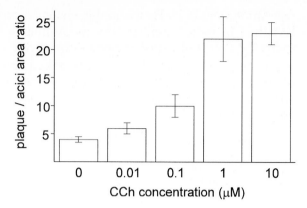

Fig. 12.8. Frequency distribution of hemolytic plaques around pancreatic acini. Under stimulation by carbamylcholine (*bottom panel*) the distribution is much more spread than under control conditions (*top panel*)

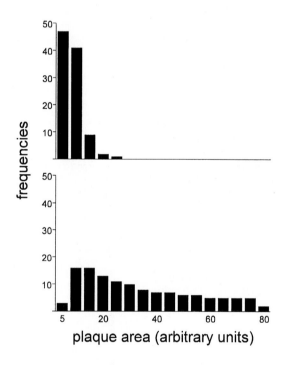

The plaque assay also provides a unique possibility to compare, within the very same preparation, the secretion of single and aggregated cells (Bosco et al. 1989). For example, by analyzing the areas of hemolytic plaques developing around single and paired β-cells, it was possible to identify a major enhancing effect of cell-to-cell contacts on insulin release (Fig. 12.9).

Fig. 12.9. The *columns* show the size distribution of hemolytic plaques measured around single (*top panel*) and paired (*middle panel*) β-cells that were simultaneously stimulated by glucose within the same series of experiments. The *bottom panel* shows curves that can be theoretically modeled to predict the latter distributions, assuming either that only one (*dashed line*) or two (*continuous line*) β-cells are secreting in each pair. Only the latter curve closely matches the experimentally observed distribution with β-cell pairs (*middle panel*), indicating that cell-to-cell contacts promote the recruitment of insulin-secreting cells

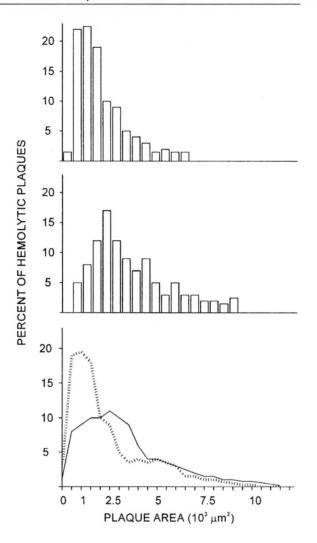

12.2
Reverse Hemolytic Plaque Assay Variants

Combining RHPA with Other Measurements

A major feature of the RHPA approach is the unique possibility to combine the secretion measurement with the evaluation of several other cell parameters, amenable to screening by morphological, electrophysiological or biophysical techniques (Salomon and Meda 1986; Bosco and

Meda 1991; Soria et al. 1991; Bosco et al. 1994) . Most of these approaches require free access to the cells during or after the secretion test. To this end, it is convenient to run the RHPA within petri dishes (Soria et al. 1991; Bosco et al. 1994), rather than in Cunningham's chambers.

Procedure

In this case, the following two modifications should be implemented in the standard RHPA protocol.

1. For attachment, plate 50 µl aliquots of the cell and RBC mixture at the center of a 35 mm plastic culture dish. Spread out the aliquot so that it covers a round area of 15 mm diameter.

2. For all rinsing steps, use 2 ml of the appropriate buffer, which should be applied slowly by maintaining the pipette tip along the edge of the petri dish. Aspirate the buffer, dry the cell-free surface of the petri dish with filter paper, and add 50 µl of the appropriate reagent to continue the assay.

Sequential Reverse Hemolytic Plaque Assay

To identify cells secreting according to a cycling pattern, to demonstrate the response of the same cell to multiple agents, or to identify individual cells secreting different products, it is essential to repeatedly assess the secretion of the very same cell. This can be achieved using a sequential version of the RHPA (Giordano et al. 1991).

Sequential RHPA differs in several respects from the standard plaque assay. The major difference is that RBCs should not be attached to the assay chambers, as they will have to be replaced at each successive incubation. To allow for the attachment of secreting cells, but not of RBC, 10 % serum is added to the medium during the attachment period. Secretagogues, antibodies, complement and RBC are then added simultaneously in a second step. Another difference between the sequential and standard versions of RHPA is that a photoetched grid coverslip (Bellco Glass, Inc. Vineland, NJ; no. 1916–92525) is used to build the top of the Cunningham's chambers. This will permit, throughout the different incubations, the repeated identification of the same cells, which can be located by their position with respect to the photoetched grid.

In practice, sequential RHPA is performed as follows:

1. Prepare Cunningham's chambers using photoetched grid coverslips.

2. Suspend secretory cells in a culture medium containing 10 % serum.

3. Fill the chambers with 55 µl aliquots of the cell suspension.

4. Incubate 45 min at 37 °C.

5. Rinse with 200 µl of the same culture medium.

6. Add 100 µl of reagent buffer containing 4 % protein A-coated RBCs, 0.1 % BSA, a specific and heat-inactivated antiserum, complement (diluted 1:40), and secretagogues (where appropriate).

7. Incubate for the desired time at 37 °C.

8. Transfer the chambers to the stage of an inverted microscope and photograph (or video record) at a magnification of 100x for the subsequent localization of individual cells and measurement of the hemolytic plaques they may have induced (Fig. 12.10).

9. Rinse with 200 µl culture medium supplemented with serum to completely wash out the RBC carpet.

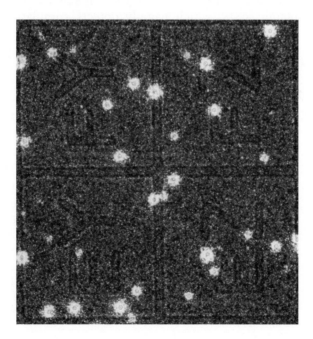

Fig. 12.10. Low magnification view of a plaque assay chamber used in a sequential RHPA. The position of cells is precisely determined with respect to the letters and grid bars photoetched on the coverslip that forms the top of the chamber. Secreting cells are surrounded by a hemolytic plaque which appears as a white round area within the dark carpet of intact RBC

10. Incubate the chambers for a desired period of time at 37 °C, or refill them immediately with a new mixture of reagents and RBCs.

11. The entire washing-refilling-incubating-photographing procedure may be repeated as many time as desired.

12. After the last incubation, fill the chambers with 0.04 % trypan blue, incubate them 5 min at room temperature, rinse with control buffer and fill with fixative.

13. Photographs (or video recording) from successive incubations are compared for the presence of hemolytic plaque around individual cells, which are identified by their position with respect to the numbers and letters photoetched on the top of the Cunningham's chamber.

Comments

limitations/ advantages   Standard biochemical assays, which typically investigate very large numbers of cells at a time, are usually amenable to quantitative assessment of secretion on time scales short enough so as to provide a fair representation of the kinetics of secretion. This may not always be feasible using a plaque assay, which usually detects cumulative amounts of secretory products. In contrast, the plaque assay has several unique advantages over biochemical approaches. Hence it is the approach of choice when only a few cells are available, when cell heterogeneity is suspected, or when direct control of cell arrangement is required. It also permits the combination of multiple approaches on individual cells.

Applications

In principle, and with minor adaptations of the procedure described above, the RHPA approach is amenable to the secretory analysis of a variety of compounds. Existing assays already include those for the following compounds (Table 12.2):

Acknowledgements. We are grateful to G. Andrey, A. Charollais, F. Cogne, P.-A. Ruttimann and E. Sutter for excellent technical assistance.This work was supported by grants from the Swiss National Science Foundation (32–343086.95), the Juvenile Diabetes Foundation International (195077), and the European Union (BMH4-CT96–1427).

Table 12.2. Secretory analysis using reverse hemolytic plaque assays (RHPAs)

Secreted compound	Reference
Adrenocorticotropin	Childs and Burke (1986)
Amylase	Bosco et al. (1988)
Angiotensinogen	Sernia et al. (1992)
Basic fibroblast growth factor	Lewis et al. (1991)
Casein	Frawley et al. (1986)
Chromogranin A	Ritchie et al. (1992)
Epidermal growth factor	Lewis et al. (1991)
Growth hormone	Frawley et al. (1985)
Insulin	Salomon and Meda (1986)
Insulin C-peptide	Lewis et al. (1988)
Interferon	Lewis et al. (1991)
Interleukins (1, 2, 4)	Lewis et al. (1991)
Luteinizing hormone	Smith and Neill (1987)
Oxytocin	Jarry et al. (1992)
Parathyroid hormone	Fitzpatrick and Leong (1990)
Prolactin	Neill and Frawley (1983)
Relaxin	Taylor and Clark (1989)
Testosterone	Pino et al. (1992)
Transferrin	Boockfor et al. (1989)
Transforming growth factor	Lewis et al. (1991)
Tumor necrosis factor	Lewis et al. (1991)

References

Boockfor FR, Schwrz LK, Derrick FC (1989) Sertoli cells in culture are heterogeneous with respect to transferrin release: analysis by reverse hemolytic plaque assay. Endocrinology 125: 1128–1133

Bosco D, Meda P (1991) Actively synthesizing B-cells secrete preferentially after glucose stimulation. Endocrinology 129: 3157–3166.

Bosco D, Meda P, Thorens B, Malaisse WJ (1995) Heterogenous secretion of individual B cells in response to D-glucose and to nonglucidic secretagogues. Am J Physiol 268: C611-C618

Bosco D, Chanson M, Bruzzone R, Meda P (1988) Visualization of amylase secretion from individual pancreatic acini. Am J Physiol 254:G664-G670.

Bosco D, Orci L, Meda P (1989) Homologous but not heterologous contact increases the insulin secretion of individual pancreatic B-cells. Exp Cell Res 184: 72–80

Bosco D, Soriano JV, Chanson M, Meda P (1994) Heterogeneity and contact-dependent regulation of amylase release by individual acinar cells. J Cell Physiol 160. 378–388

Chanson M, Bruzzone R, Bosco D and Meda P (1989) Effects of n-alcohols on junctional coupling and amylase secretion of pancreatic acinar cells. J Cell Physiol 139:147–156

Childs GV, Burke JA (1986) Use of the reverse hemolytic plaque assay to study the regulation of anterior lobe adrenocorticotropin (ACTH) secretion by ACTH-releasing factor, arginine, vasopressin, angiotensin II, and glucocorticoids. Endocrinology 120: 439–444

Fitzpatrick LA, Leong DA (1990) Individual parathyroid cells are more sensitive to calcium than a parathyroid cell population. Endocrinology 126: 1720–1727

Frawley LS, Boockfor FR, Hoeffler JP (1985) Identification by plaque assays of a pituitary cell type that secretes both growth hormone and prolactin. Endocrinology 116: 734–737

Frawley LS, Clark CL, Schoderbek WE Hoeffler JP, Boockfor FR (1986) A novel bioassay for lactogenic activity: demonstration that prolactin cells differ from one another in bio- and immuno-potencies of secreted hormone. Endocrinology 119: 2867–2869

Giordano E, Bosco D, Cirulli V, Meda P (1991) Repeated glucose stimulation reveals distinct and lasting secretion patterns of individual pancreatic B-cells. J Clin Invest 87: 2178–2185

Jarry H, Hornschuh R, Pitzel L, Wuttke W (1992) Demonstration of Oxytocin release by bovine luteal cells utilizing the reverse hemolytic plaque assay. Biol Reprod 46, 408–413

Jerne NK, Henry C, Nordin AA, Fuji H, Koros AMC, Lefkovits I (1974) Plaque forming cells: methodology and theory. Transplant Rev 18: 130–191

Lewis CE, Clark A, Ashcroft SJH, Cooper GJS, Morris JF (1988) Calcitonin gene-related peptide and somatostatin inhibit insulin release from individual rat B cells. Mol Cell Endocrinol 57: 41–49

Lewis CE, McCracken D, Ling R, Richards PS, McCarthy SP, McGee J O'D, (1991) Cytokine release by single, immunophenotyped human cells: use of the reverse hemolytic plaque assay. Immunol Rev 119: 23–39

Neill JD, Frawley LS (1983) Detection of hormone release from individual cells in mixed populations using a reverse hemolytic plaque assay. Endocrinology 112: 1135–1137

Pino AM, Inostroza H, Valladares LE (1992) Detection of testosterone secretion from individual rat Leydig cells. J. Steroid Biochem. Mol Biol 41: 167–170

Ritchie CK, Cohn DV, Fitzpatrick LA (1992) Chromogranin-A secretion from individual parathyroid cells: effects of 1,25-$(OH)_2$vitamin D_3 and calcium. Bone Mineral 18: 31–40.

Salomon D, Meda P (1986) Heterogeneity and contact-dependent regulation of hormone secretion by individual B cells. Exp Cell Res 162: 507–520

Sernia C, Shinkel TA, Thomas WG, Ho KKY, Lincoln D (1992) Angiotensinogen secretion by single rat pituitary cells: detection by a reverse haemolytic plaque assay and cell identification by immunocytochemistry. Neuroendocrinology 55: 308–316

Smith PF, Neill JD (1987) Simultaneous measurement of hormone release and secretagogue binding by individual pituitary cells. Proc Natl Acad Sci USA 84: 5501–5505

Soria B, Chanson M, Giordano E, Bosco D, Meda P (1991) Ion channels of glucose-responsive and unresponsive B-cells. Diabetes 40: 1069–1078

Taylor MJ, Clark CL (1989) Analysis of relaxin release by cultured porcine luteal cells using a reverse hemolytic plaque assay: effects of arachidonic acid, cyclo- and lipo-oxygenase blockers, phospholipase A_2, and melittin. Endocrinology 125: 1389–1397

Chapter 13

The Granulocyte/Macrophage Colony-Forming Unit Assay

Augusto Pessina*

Introduction

The Hematopoietic System

Although the complex dynamics of the interactions among the different components of the hematopoietic system remains poorly understood, much progress has been made in identifying the various cell types and the multitude of growth factors regulating its homeostasis. For example, the existence of pluripotent stem cells (with an infinite capacity of self-renewal), progenitor cells (with a finite capacity for self-renewal), committed stem cells (oriented toward a specific lineage differentiation) and mature cells (fully differentiated that do not divide) has been established (Fig. 13.1).

The existence in bone marrow of stem cells, able to renew themselves and generate progenitors committed to individual hematopoietic lineages, was demonstrated in 1961 by Till and McCulloch. However, many controversies about the origin and the phenotype of the totipotential hematopoietic stem cells persist. Stem cell renewal and differentiation have been widely investigated and the mechanism was found to be consistent with a stochastic model, as suggested by many mathematical studies (Vogel 1969; Nakahata 1982; Suda 1983).

During the last two decades many studies indicated that while the self-renewal or differentiation capacity of stem cells should represent an intrinsic property of this progenitor, the survival of progenitors and their commitment seems to be controlled by the hematopoietic microenvironment (Witlock and Witte 1982; Metcalf 1989). The essential role of the bone marrow microenvironment has been supported by many in

* Augusto Pessina, Laboratory of Cell Culture, Institute of Microbiology, Faculty of Medicine, University of Milan, 36 Via Pascal, Italy; phone +39–2–26601222; fax +39–2–26601218; e-mail pessinaa@imiucca.csi.unimi.it

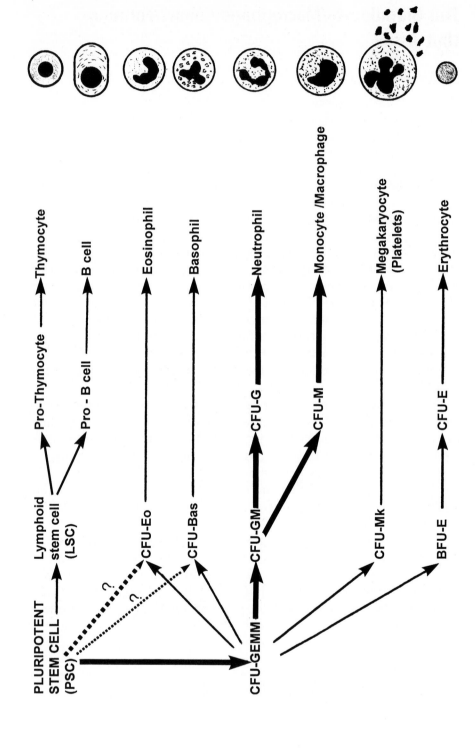

vitro studies on long-term bone marrow culture (LTBM) (Dexter 1979) and it has been demonstrated that it involves a stromal cell compartment and extracellular matrix (ECM). The ECM of bone marrow is in a great part produced by stromal cells and is characterized by important classes of molecules such as collagens, proteoglycans, fibronectin, tenascin, thrombospondin and hemonectin (reviewed by Klein 1995).

The bone marrow stroma is represented by a heterogeneous population of cells which includes fibroblast like-cells, adipocytes, macrophages and, according to some authors, endothelial cells as well. The stromal cells influence hematopoietic stem cell homing and proliferation both by their surface molecules (more than 30 investigated) and by the production of several cytokines and growth factors interacting with cell receptors on the different cell populations of bone marrow (Greenberger 1991)

The cytokines functionally active on the bone marrow compartment support pleiotropic physiological functions and it is impossible summarize their functions in this chapter; however, according to the suggestion of Ogawa et al. (1993), these cytokines or growth factors may be classified into three essential groups:

- Factors affecting very primitive progenitors: interleukin IL-1, IL-2, IL-3, IL-6, IL-7, IL-11, stem cell factor (SCF; Steel factor), leukemia inhibitory factor (LIF) and granulocyte/macrophage colony-stimulating factor (GM-CSF). Among these, IL-3, IL-6 and IL-7 act in combination with SCF as a potent stimulus of early stages of lymphoid development (McNiece et al. 1991). As proposed by Metcalf et al. (1980), the commitment of granulocyte/macrophage progenitors should be regulated by early acting factors such as IL-3 and GM-CSF acting in synergism with SCF. IL-2 may be considered an hematopoietic factor active both on B and T cell proliferation and in synergism with IL-1 is able to promote the natural killer cell production from bone marrow progenitors.

- Intermediate active lineage nonspecific factors: IL-3, GM-CSF, IL4, SCF.

- Late active lineage-specific factors: M-CSF, IL-5, erythropoietin (Epo), G-CSF. For a review on the correlation between the functional and structural diversity of colony-stimulating factors see Konshansky and Karplus (1993).

◀ **Fig. 13.1.** Description of hematopoiesis. *BFU* Burst forming unit; *CFU* colony-forming unit; *E* erythrocyte; *G* granulocyte; *M* macrophage; *Mk* megakaryocyte

In addition, hepatocyte growth factor (HGF), originally isolated as mitogen for adult hepatocytes, has been recently indicated as a factor playing a crucial role in regulating proliferation and differentiation of hematopoietic progenitiors (Nishino et al. 1995).

Besides those factors representing positive regulators a number of inhibitory factors have been described including tumor necrosis factor (TNF)-α, transforming growth factor (TGF)-β, macrophage inflammatory protein (MIP)1-α and, more recently, the inhibitory peptides pEEDCK and AcSDKP (Guigon and Bonnet 1995). Great importance must be given to the nuclear regulatory factors (e.g., GATA transcription factors), able to bind DNA sequences which produce, in hematopoietic cells, the expression of specific genes (Weiss 1995).

In Vitro Techniques For Testing Hematopoiesis

The first clonogenic technique was a mixed "in vivo/in vitro" method suggested by Till and McCulloch (1964) and based on the injection of bone marrow cells obtained from syngeneic donors into X-irradiated recipient mice. The injected cells are able to produce colonies (colony forming unit-spleen, CFU-S) in the spleen or in the bone marrow of the recipients. The development of in vitro cloning assays for hematopoietic stem cells originated with work involving granulocyte/macrophage precursors (Bradley and Metcalf 1966) and mast cells (Pluznick and Sachs 1965). These experiments had a decisive importance because it was observed that when bone marrow cells are seeded in plates in the presence of the appropriate medium and growth factors a certain number of cells is capable of growth and form clusters or colonies. Clusters are groups with a small number of cells ($<$50) whereas colonies contain more than 50 cells and often consist of many hundreds of cells. By subculturing individual clusters or colonies, studies were performed concerning the capacity of cells to proliferate and differentiate until maturation to "end cells"; thus an in vitro standardized system for growing and differentiated bone marrow cells was defined. An extraordinary number of studies has been performed to evaluate the proliferation and differentiation in vitro of the lineages leading to mature cells.

Many reports have been published concerning the various techniques for cloning committed progenitors: erythroids (primitive burst forming units, BFU-E, and erythropoietin responsive CFU,CFU-E), T lymphocytes, B lymphocytes, pure macrophages (M-CFU), pure granulocyte (G-GFU), megakaryocytes (CFU-MK) and multilineage progenitor cells such as GM-CFU or GEMM-CFU. The presence of stromal cells is tested with CFU-F (fibroblast colony forming unit) found in liquid cultures of bone marrow cells that contain fibroblasts, endothelial cells and macrophages (Wang and Wolf 1990).

During the past few years, investigations into hematopoiesis have been greatly aided by the isolation of new cytokines and, more recently, by the availability of factors produced by molecular cloning techniques.

Today many methods are available for culturing in vitro hematopoietic cells both in human and in animal models (dogs, rats, mice). Many of these techniques utilize fresh bone marrow cells as well as peripheral or umbilical cord blood cells. A summary of techniques for testing the clonal growth of various hematopoietic progenitors was reported by Deldar et al. (1995). Other methods applied to LTBM were developed by Dexter and Allen (1977) and Whitlock and Witte (1982). More recently Naughton and Naughton (1989) developed LTBM cultures using, as a support for stromal cells, nylon mesh templates which mimic the three-dimensional structure of the bone marrow.

13.1
Granulocyte/Macrophage Colony-Forming Unit Assay

The following description concerns a standard agar clonal assay which allows scoring of mixed colonies or clusters containing granulocytes and macropahges (GM-CFU). With minor changes and by specific staining techniques this standard assay may be adapted to analyze the growth of mixed (CFU-GEMM) or pure colonies for erythrocytes, macrophages, granulocytes, eosinophils, and megakaryocytes. This basic method may be applied to many types of studies but can be used particularly:

- To investigate the frequencies of different types of progenitors in a bone marrow cell population

- To screen the effect of molecules (drugs, chemicals, etc.) on the proliferation and differentiation of progenitors

- To test the presence of hematopoietic factors in supernatants, conditioned media or biological fluids

- To analyze growth factor requirement and interaction of different hematopoietic progenitors

As suggested by the 14th ECVAM Workshop (Gribaldo et al. 1996) it is very important to initiate a pre-validation study of these techniques (in particular of the GM-CFU assay). In fact, these tests also have the potential to reduce the number of animals required for hematotoxicity studies and could play a key role in bridging the gap between preclinical toxicology studies in animal models and clinical investigations (Parchment et al. 1993). Furthermore, they should help in the risk assessment of food contaminants and other chemicals and products.

Materials

Culture Medium

Prepare complete enriched culture medium as follows (for 100 ml):

- McCoy's 5A culture basal medium (modified by Grace-Iwakata): 72.8 ml
- Glutamine solution (200 mM): 2 ml
- Sodium bicarbonate (7.5 % w/v solution): 0.6 ml
- Fetal calf serum (FCS): 20 ml. **Note:** FCS should be pretested before use to find the batch supporting the most adequate growth. Heat inactivate at 56 °C, 30 min before use. FCS can be stored at −20 °C for up to 1 year.
- Sodium pyruvate: 1 ml
- Minimum essential medium (MEM)-essential amino acids (EAA) (50x): 1.6 ml
- MEM-nonessential amino acids (NEAA) (100x): 0.4 ml
- MEM-vitamins (100×): 0.4 ml
- L-serine (21 mg/ml): 0.04 ml
- L-asparagine (20 mg/ml): 0.16 ml
- Penicillin(10 000 U/ml)/streptomycin (10 000 μg/ml): 1 ml

Note: The medium must be used within 1 month.

Agar Aliquots

1. Prepare 5 % agar solution by mixing 5 g of agar (Difco, USA) in 100 ml of bidistilled water.

2. Heat the mixture with stirring until agar is dissolved and then sterilize by heating to 121 °C, 15 min.

3. Cool to about 80 °C; distribute aliquots of 15–20 ml in 100 ml glass bottles.

4. Store the aliquots at +20 °C until use (within 1 year).

Growth Factors

The more common and cheap source of CSFs, able to stimulate proliferation and differentiation of GM-CFU, is the conditioned medium of WEHI-3B cells, prepared as follows:

1. Culture WEHI-3B cells in McCoy's medium containing 1 % FCS until the medium becomes yellow, indicating the presence of lactic acid (usually until 10–12 days).

2. Harvest the conditioned medium by centrifugation of the cell suspension at 5000 g, 10 min.

3. Heat treat the conditioned medium for 1 h at 60 °C; sterilize by filtration (0.22 μm). Store 1–2 ml aliquots at –20 or –80 °C.

Other good sources of CSFs are the conditioned media obtained from many primary cell cultures (e.g., murine lung cells) or established cell lines such as L-cells, SR-4987 ,Y1 (Pessina et al. 1986, 1992).

For studying the growth of megakaryocyte progenitors in a murine system, a good source of CSF is the conditioned medium of concanavalin-A stimulated spleen cells (mu-ConA-SCCM), which can be prepared as suggested by Deldar et al. 1995.

Erythroid progenitors from mouse, dog, rat and human require the presence of Epo in the culture, as described in the section "GEMM-CFU Assay."

A common source of factors for stimulating human progenitors is leukocyte conditioned medium (hu-PHA-LCM), readily prepared according to the method of Parker and Metcalf (1974).

For particular studies, lineage specific factors (interleukins, cytokines and growth factors) are commercially available as purified or recombinant molecules by many companies (e.g., Genzyme, Sigma, USA).

Procedure

Standard Procedure

cell suspension

Donor: BDF1 female mice, 8–12 weeks old. **Note:** All procedures must be performed at + 4 °C (melting ice) under rigorous sterile conditions in a microbiological safety cabinet using sterile reagents and materials.

1. Anesthetize the animals by ether and kill them by cervical dislocation. Wash them throughly with 70 % ethyl alcohol and leave them for 5 min wrapped in alcohol-soaked gauze.

2. Remove the skin of the legs and isolate the femura and tibiae from the muscles. It is important to operate by cutting the muscle ligaments with a small scissors in order to thoroughly clean the bones.

3. Cut the tibia at the distal end and then disarticulate the knee capsule at the proximal end. The femura, cleaned of muscle tissue, are cut under the head.

4. Put the bones in a petri dish containing culture medium and maintain at 4 °C on ice.

5. Insert a 21-gauge needle mounted on a 1 ml syringe into the end of each bone, held with surgical pincers. The bone marrow cells are flushed out by pressure and the bone cavity is washed by flushing 5–6 ml of the medium up and down.

6. Bone marrow cells are collected in a sterile container by filtering through a 0.1×0.1 mm steel mesh.

7. Wash the cells at 400 g, 10 min . Discard the supernatant and resuspend the pellet in 0.5 ml 0.85 % ammonium chloride; stir gently and hold for 5 min at 4 °C to lyse erythrocytes.

8. Wash the cells and resuspend them in complete medium, count them in a Bürker chamber and adjust the suspension to 10^6 viable nucleated cells/ml (viability must be >95 %).

Note: The suspension must consist of single cells. If cell aggregates are present, they must be dispersed by gentle pipetting!

agar culture

1. On a tray, prepare the petri dishes, numbering them according to the experimental design.

2. Place the glass bottle of agar in a boiling water bath so that it melts.

3. Prepare the plating mixture as follows (for nine petri dishes, 1 ml/dish, prepared in 50 ml plastic tubes; Nunc, USA):
   - Complete culture medium: 7.34 ml
   - Conditioned medium: 1 ml (or 10 ng of recombinant murine (r-mu-) GM-CSF)
   - Cell suspension (10^6 cells/ml): 1 ml

4. Warm the plating mixture to 37 °C and then add to each tube 0.66 ml of 5 % agar suspension.

5. Mix thoroughly and distribute 1 ml/dish of the mixture (usually in triplicate).

6. After cooling the plates to room temperature (or 2 min at 4 °C) incubate the cultures at 37 °C in an air +5.0 % CO_2 incubator for 7 days for CFU-GM scoring. CFU-E are scored after 2 days of incubation.

Note: Adequate humidity during incubation is critical as drying of the culture drastically reduces colony formation!

1. CFU-GM colonies are scored by scanning the whole petri dish using an inverted microscope at 20–25 × magnification.

2. Aggregates containing 50 or more cells are defined as CFU-GM colonies.

3. Aggregates with 20–50 cells are counted as clusters.

4. The count is expressed as number of colonies/dish (equivalent to colonies/10^5 bone marrow cells). The average count of three dishes is generally calculated. In the above-described conditions it is easy to obtain the stimulation of 100–150 colonies/dish.

5. For cytological examination of colonies, the agar cultures can be fixed in situ (or put on a glass slide) and then stained as follows:
   - Dual esterases (α-naphthyl acetate and naphthol-acid sodium-D-chloroacetate) staining for the identification of granulocytic, macrophagic mixed colonies and to discriminate between eosinophil and neutrophil granulocytes.
   - Benzidine staining for identification of hemoglobin containing progenitors(CFU-E)
   - Acetylcholinesterase staining for identification of megakaryocyte progenitors (CFU-Mk)

scoring the colonies

For details of staining techniques, see Deldar (1995).

13.2
Variants

Bone Marrow Cell Source

The source of bone marrow cells can be changed by using different animal species.

- For rats the procedure described above can be followed.

- For dogs and human bone marrow, cells were aseptically obtained from heparinized bone marrow aspirate. The aspirate is layered (4:3 volume ratio) on Ficoll-Hypaque ($d = 1.077$ g/ml) and centrifuged at $400\,g$, 30 min at 15 °C. Mononuclear cells collected at the interface were washed twice and resuspended at 10^6/ml.

- For human studies, mononuclear cells from umbilical cord blood and peripheral blood (very poor in stem cells) may be substituted for bone marrow.

- For human hematopoietic progenitors, it is important to use specific growth factors (see "Growth Factors"); score GM-CFU at 14 days and CFU-E at 7 days of culture.

▪ Procedure

Assay of Chemicals

The effect of chemicals on hematopoietic progenitors can be studied by testing bone marrow cells under three different conditions:

- After mice have been treated in vivo with the chemicals
- After the fresh bone marrow cell suspension has been pretreated in vitro with chemicals
- By adding chemicals directly to the culture dish

In this last condition, the procedure for the agar culture (see "agar culture," steps 1 and 3) must be changed as follows:

1. Distribute in each petri dish 100 µl of the chemical at the 10-fold dilutions to be studied.

2. Prepare adequate control dishes by distributing 100 µl of chemical solvent diluted in culture medium.

3. Prepare the plating mixture by reducing by 1 ml the quantity of culture medium.

Assay of Colony-Stimulating Activity

For testing the capacity of biological fluids (or conditioned media or semi-purified material) to stimulate hematopoietic progenitors the procedure for agar culture (see "agar culture," steps 1 and 3) is changed as follows:

1. Distribute in each petri dish 100 µl of sample at two-fold serial dilutions (from undiluted to 1:64). Controls receive 100 µl of undiluted laboratory standard factor (LSF).

2. Prepare the plating mixture by replacing the addition of growth factors or conditioned medium with an equal amount of complete medium.

The activity of the studied material can be expressed as arbitrary units, **calculation** calculated on the basis of the activity of LSF, as follows:

$$U/ml = D \times A \times B,$$

where D is the dilution factor of the sample at which is counted a number of colonies that is half that counted in controls with LSF (i.e., $D=32$ if this dilution is 1:32); $A=10$ (since 0.1 ml of sample was tested); and $B=50$ (since, arbitrarily, we consider the maximal response of the LSF as giving 100 % of colonies).

GEMM-CFU Assay

To evaluate erythroid and megakaryocyte growth, prepare the plating mixture (see "agar culture," step 3) as follows:
- Complete culture medium: 5.34 ml
- mu-ConA-SCCM or hu-PHA-LCM: 1 ml
- BSA (Grade V, Sigma) (10 %): 1 ml. **Note:** BSA needs to be pretested. It must be deionized and delipidated (Watt and Davis 1990). Stock solutions may be kept frozen in aliquots at −20 °C for 1 year.
- Erythropoietin (10 U/ml): 1 ml
- Cell suspension(10^6 cells/ml): 1 ml

Acknowledgements. The author thanks Dr. Elisabetta Mineo and Dr. Michela Piccirillo for their excellent assistance in preparing the manuscript.

References

Bradley TR, Metcalf D (1966) The growth of mouse bone marrow cells in vitro. Austr J Exp Biol Med 44: 287–289

Deldar A, House RV, Wierda D (1995) Bone marrow colony-forming assays. Methods Immunotoxicol 1:227–250

Dexter TM (1979) Hemopoiesis in long-term bone marrow culture. Acta Hemat 62: 299–304

Dexter TM, Allen TD, Lajtha LG (1977) Conditions controlling the proliferation of hematopoietic cells in vitro. J Cell Physiol 91: 335–344

Greenberger JS (1991) The hematopoietic microenviroment. Crit Rev Oncol Hematol 11: 65–84

Gribaldo L,Bueren J, Deldar A, Hokland P, Meredith C, Moneta P, Mosesso R, Parchment R, Parent-Massin D, Pessina P, San Roman J, Scoeters G (1996) Alternatives to Laboratory Animals 24:211–231

Guigon M, Bonnet O (1995) Inhibitory peptides in hematopoiesis. Exp Hematol 23: 477–481

Klein G (1995) The extracellular matrix of the hematopoietic microenviroment. Experientia 51: 914–926

Konshansky K, Karplus PA (1993) Hematopoietic growth factors: understanding functional diversity in structural terms. Blood 82: 3229–3240

McNiece IK, Langley KE, Zsebo KM (1991) The role of recombinant stem cell factor in early B cell development. J Immunol 146: 3785–3790

Metcalf D (1980) Clonal analysis of proliferation and differentiation of paired daughter cells: Action of granulocyte-macrophage colony stimulating factors on granulo-macrophage precursors. Proc Natl Acad Sci USA 77:5327–5330

Metcalf D (1989) The molecular control of cell division, differentiation commitment and maturation in hematopoietic cells. Nature 274:168–169

Nakahata T, Gross AJ, Ogawa M (1982) A stochastic model of self-renewal and commitment to differentiation of the primitive hematopoietic stem cells in culture. J Cell Physiol 113: 455–458

Naughton BA, Naughton GK (1989) Hematopoiesis on nylon mesh templates. Annals NY Acad Sci 554:125–140

Nishino T, Hisha H, Nishino N, Adach M, Ikehara S (1995) Hepatocyte growth factors as a hematopoietic regulator. Blood 85: 3093–3100

Ogawa M (1993) Differentiation and proliferation of hematopoietic stem cells. Blood 81:2844–2853

Parchment RE, Huang M, Erickson-Miller CL (1993) Roles for in vitro myelotoxicity tests in preclinical drug development and clinical trial planning. Toxicol Pathol 21:241–250

Parker JW, Metcalf D (1974) Production of colony stimulating factor in mitogen stimulated lymphocyte cultures. J Immunol 112:502–510

Pessina A, Neri MG, Muschiato A, Brambilla P, Marocchi A, Mocarelli P (1986) Colony stimulating factor produced by murine adrenocortical tumor cells. J Natl Cancer Inst 76:1095–1099

Pessina A, Mineo E, Neri MG, Gribaldo L, Colombo R, Brambilla P, Zaleskis G (1992) Establishement and characterization of a new murine cell line(SR-4987) derived from marrow stromal cells. Cytotechnology 8:93–102

Pluznick DH, Sachs L (1965) The cloning of normal "mast cells" in tissue culture. J Cell Physiol 66: 319–325

Suda T, Suda J, Ogawa M (1983) Single cell origin of mouse hematopoietic colonies expressing multiple lineages in variable combinations. Proc Natl Acad Sci USA 80: 6689–6693

Till JE, Mc Culloch EA (1961) A direct measurement of the radiation sensitivity of normal mouse bone marrow cells. Radiation Res 14: 213–222

Vogel H, Niewish H, Matioli G (1969) Stochastic development of stem cells. J Theoret Biol 22: 249–270

Wang QR, Wolf NS (1990) Dissecting of hematopoietic microenviroment with clonal isolation and identification of cell types in murine CFU-F colonies by limiting dilution. Exp Hematol 18: 355–361

Watt SM, Davis JM (1989) Flow sorting for isolating CFU-E. In: Pollard JW and Walker JM (eds) Methods in molecular biology, vol.V. Humana, Clifton, New Jersey, pp 347–360

Weiss MJ, Orkin SH (1995) Gata transcription factors: key regulators of hematopoiesis. Exp Hematol 23: 99–107

Whitlock CA, Witte ON (1982) Long term culture of B lymphocytes and their precursor from murine bone marrow. Proc Natl Acad Sci USA 79: 3608–3612

▓ Abbreviations

BSA	Bovine serum albumin
CFU-S	Colony-forming unit-spleen
ConA	Concanavalin A
EAA	Essential amino acids
ECM	Extracellular matrix
ECVAM	European Centre for Validation of Alternative Methods
Epo	Erythropoietin
FCS	Fetal calf serum
G-CSF	Granulocyte colony-stimulating factor
GM-CSF	Granulocyte/macrophage colony-stimulating factor
HGF	Hepatocyte growth factor
hu	human
IL	Interleukin
LCM	Leukocyte conditioned medium
LIF	Leukemia inhibitory factor
LSF	Laboratory standard factor
LTBM	Long-term bone marrow culture
M-CSF	Macrophage colony-stimulating factor

MEM   Minimum essential medium
mu   murine
NEAA   Nonessential amino acids
PHA   Phytohemoagglutinin
r-mu   Recombinant murine
SCCM   Spleen cell conditioned medium
SCF   Stem cell factor (or SF, Steel factor)
TGF   Transforming growth factor
TNF   Tumor necrosis factor

Chapter 14

Investigations into Cell Differentiation Using Cells in Culture

SHIRLEY McBRIDE[*,1,2] and MARTIN CLYNES[1]

Introduction

Investigations into Cell Differentiation Using Cells in Culture

Almost every multicellular organism is a clone of cells descended from a single original cell, the fertilized egg. This cell is the ultimate totipotent stem cell, having the potential to form a complex organism through proliferation, differentiation and organization of the differentiated tissue. Cellular differentiation may be defined as the process by which "a cell acquires or displays a new stable phenotype without changing its genotype" (Ham and Veomett 1980), and embryonic development involves the progressive production of generations of stem cells, each with greater restriction of potential for differentiation. In this way, a single fertilized egg cell gives rise to three germ layers, the ectoderm, mesoderm and endoderm, each of which, in turn, gives rise to the specific tissue types of an adult organism, all the cells of which possess identical genomes. Thus, the ectoderm gives rise to the epidermis and the nervous system, the mesoderm gives rise to the vertebral column, connective tissue, cartilage, bone, fibrous tissue, muscle cells and the vascular system, and the endoderm gives rise to the digestive tract, the trachea and the lungs. Once established, the cells of the germ layers are no longer totipotent but are determined stem cells and so, for example, a mesodermal stem cell will only give rise to mesodermal tissue and generally will not differen-

* *Correspondence to* Shirley McBride: phone +44–131–6502954; fax +44–131–6506528; e-mail Shirley.McBride@ed.ac.uk
[1] Shirley McBride, Martin Clynes, National Cell and Tissue Culture Center, Dublin City University, Glasnevin, Dublin 9, Ireland
[2] Shirley McBride, Current address: The University of Edinburgh, Department of Pathology, The University of Edinburgh Medical School, Teviot Place, Edinburgh EH8 9AG

tiate into ectodermal or endodermal tissue types. Similarly, most adult specific tissue types also contain stem cell populations which have even further restricted potential for differentiation. Differentiation in the adult is principally concerned with maintenance of the differentiated state attained during development, i.e., tissue renewal and the correct replacement of old and damaged cells.

Differentiation processes usually involve the induction of genes which characterize and are responsible for the differentiated state, and the repression of genes for cell proliferation. In some tissues, a master gene has been identified which appears to control phenotypic differentiation, for example the MyoD gene in myoblasts (Weintraub et al. 1991). However, the complex signaling systems responsible for those changes in gene expression which lead to phenotypic alterations as yet remain largely unknown. A number of cell culture models have been established to examine cellular differentiation. These employ a variety of differentiation-inducing agents which allow manipulation of the differentiation pathways open to stem cells and progenitor cells. Differentiating agents are capable of altering gene expression, leading to an induction of phenotypic differentiation. For example, the effects of retinoic acid (RA) and vitamin D_3 on differentiation are mediated by nuclear receptors which act as transcription factors (de The et al. 1989) and bromodeoxyuridine (BrdU) alters the expression of differentiation-specific genes following its incorporation into DNA (Tapscott et al. 1989), although its precise mechanism of action has not yet been elucidated. Many additional differentiating agents with similar effects have been studied including butyric acid (Chen and Breitman 1994), vasoactive intestinal peptide (VIP) (Pence and Shorter 1990), transforming growth factor (TGF)-β (Masui et al. 1986), phorbol esters and diacylglycerol (Jetten et al. 1989).

This chapter will describe several stem cell culture systems in which differentiating agents are utilized in the establishment of in vitro models for cellular differentiation. An example of a stem cell population from each of the germ layers will be given; neuronal cells (ectodermal), HL60 leukemic cells (mesodermal) and lung cells (endodermal). Differentiating agents used are RA, vitamin D_3 and BrdU. It should be noted that these agents are potentially hazardous, as are most such agents, and should be handled with due care and attention as outlined in the individual procedures.

14.1
SK-N-SH Neuronal cells

SK-N-SH, a human cell line established from a thoracic neuroblastoma tumor, is commonly used in the study of neuronal cell differentiation. The cell line contains three subpopulations which are capable of phenotypic interconversion (transdifferentiation): (1) N, which is neuroblastic in appearance, containing many neurite-like processes; (2) S, which is larger, flattened and exhibits Schwannian rather than neuronal properties; and (3) I, with a morphology intermediate between that of N and S cells (Ciccarone et al. 1989). These subpopulations have been isolated and used as models for spontaneous and induced neuronal cell transdifferentiation. Various differentiating agents have been used to modulate interconversion between the SK-N-SH subpopulations (Ross et al. 1994) and described here is a procedure which uses RA to induce S type differentiation in I type SK-N-SH cells (Ross et al. 1991).

▦ Materials

- SK-N-SH cell line: American Type Culture Collection, 12301 Parklawn Drive, Rockville, Maryland 20852–1776, USA (cat. no. ATCC HTB11)
- SK-N-SH culture medium: Minimal essential medium (MEM) supplemented with 5 % fetal calf serum, 2 mM L-glutamine, 1 % (v/v) MEM non-essential amino acids, 1 mM sodium pyruvate and 1 % (v/v) Earls balanced salt soultion
- All-*trans* retinoic acid (RA) (Sigma, cat. no. R2625)
- Ethanol

▦ Procedure

Isolation of SK-N-SH Clonal Subpopulations

SK-N-SH clonal subpopulations may be isolated using a limiting dilution assay, as outlined elsewhere in this manual. The three subtypes can be discriminated morphologically as described above and determination of protein marker expression will confirm their identities – N cells express nerve growth factor (NGF) receptors (Rettig et al. 1987), S cells express type I and II collagens typical of Schwann cells (DeClerck et al. 1987), and I cells differentially express both N and S cell marker proteins (Ciccarone et al. 1989).

Note: Spontaneous interconversion between the clonal subtypes may occur. It is therefore advisable to freeze and store large stocks of the clones in liquid nitrogen, to work within a limited number of passages following isolation and to verify the identity of the clones regularly by checking marker protein expression.

differentiation assay

1. Reconstitute RA in 95 % ethanol to a stock concentration of 1 mM, aliquot into cryovials and store at $-80\,°C$.

   Note: Dimethylsulfoxide (DMSO) is often used to solubilize differentiating agents including RA. However, DMSO is itself a potent inducer of differentiation in several cell types. It may therefore be preferable to use alternative solvents such as ethanol, as described here. RA is light sensitive, therefore all manipulations involving this agent should be carried out in subdued light. RA is also teratogenic: all work with the undissolved powder should be carried out in a fume hood; two pairs of gloves should be worn when handling concentrated solutions; and all waste solutions should be incinerated.

2. Harvest the SK-N-SH N, S and I clones by trypsinization, seed into multiwell tissue culture plates or flasks and allow to adhere by incubating overnight at 37 °C, 5 % CO_2.

3. Dilute stock RA with culture medium and add to the cells to produce a final concentration of 10 µM. Make an equivalent dilution of ethanol to add to control cells.

4. Incubate for 7 days at 37 °C, 5 % CO_2, replacing the medium after the first 3–4 days. Observe the cultures microscopically at varying intervals. The I type clones will change morphology and differentiate into S-type SK-N-SH cells. If multiple plates have been set up, the cells may be harvested at regular intervals during the exposure period to monitor phenotypic changes. Cells may either be trypsinized or fixed in situ with cold acetone as appropriate.

14.2
HL60 Leukemic Cells

HL-60 is a human promyelocytic leukemia cell line commonly used in differentiation studies. These cells proliferate continuously in suspension culture and are bipotent, being capable of either myeloid or monocytic differentiation in response to various differentiating agents (Fontana et al. 1981). Exposure to agents such as RA, DMSO and actinomycin D will cause HL60 cells to differentiate along a granulocytic/myeloid pathway (Breitman et al. 1980; Collins et al. 1978; for details see Mollinedo et al., this Vol.), while vitamin D_3 and 12–0-tetradecanoyl-phorbol 13-acetate (TPA) induce monocytic/macrophage differentiation in HL60 cells (McCarthy et al. 1983; Boyd and Metcalf 1984). Myeloid differentiation can be assessed by assaying superoxide production induced by phorbol myristate acetate (PMA) as determined by nitroblue tetrazolium (NBT) reduction (Yen et al. 1984) or by the expression of cell surface markers such as Mo1 (Yen and Forbes 1990).

Note: PMA is a hazardous agent and should be handled with care. The cell surface markers Mo2 and My4 are expressed during monocytic differentiation (Yen and Forbes 1990). Described here is a method for the induction of monocytic differentiation in HL-60 cells using vitamin D_3.

Materials

- HL-60 cell line: American Type Culture Collection (cat.no. ATCC CCL 240)
- HL-60 culture medium: RPMI 1640 medium supplemented with 10 % fetal calf serum and 2 mM L-glutamine
- Vitamin D_3 (Sigma, cat.no. C9774)
- Ethanol

Procedure

1. Reconstitute vitamin D_3 in 95 % ethanol to a stock concentration of 1 mM, aliquot into cryovials and store at −80 °C. **Note:** Identical precautions to those outlined for RA should be taken when working with this agent.

differentiation assay

2. Harvest the HL-60 cells and seed into multiwell tissue culture plates or flasks.

3. Dilute stock vitamin D_3 with culture medium and add to the cells to produce a final concentration of 1 µM. Make an equivalent dilution of ethanol to add to control cells.

4. Incubate for 7 days, replacing the vitamin D_3-containing medium after the first 3–4 days.

5. Following incubation with vitamin D_3, cells can be harvested and assayed as required or cytocentrifuged onto glass slides and fixed with cold acetone for immunocytochemical analysis.

14.3
Lung Tumor Cells

The adult mammal lung contains over 40 different cell types, at least eight of which are found in the epithelial lining of the tracheobronchial airways. These epithelial cells include ciliated cells, basal cells, brush cells, mucous goblet cells, serous cells, Clara cells, type I and type II cells and neuroendocrine cells. Each of these differentiated cell types carries out specialized functions and expresses specific proteins characteristic of its phenotype. It is generally believed that an undifferentiated stem cell population gives rise to these differentiated cell types and also to the various epithelial tumors which can occur in the lung (Gazdar et al. 1981; Bergh et al. 1989), although these stem cells have not yet been identified. Lung tumors are broadly classed into non-small cell lung cancers (NSCLCs) comprising squamous, adeno- and large cell carcinomas and small cell lung cancer (SCLC) and each can express a range of characteristic marker proteins (World Health Organization 1982).Variant SCLC (SCLC-V) is a subset of SCLC. These cells are poorly differentiated and, while they can express some neuronal marker proteins such as neuron specific enolase and neurofilaments, they often down-regulate expression of epithelial cell specific proteins such as cytokeratins (Broers et al. 1988). The use of such cells in differentiation studies can provide much information on the differentiation pathways open to poorly differentiated lung cells and can also give an insight into the maturation blocks which are believed to result in the tumorigenic phenotype. The following is an outline of procedures for BrdU-induced differentiation in H82 SCLC-V cells.

Materials

- H82 cell line: American Type Culture Collection (cat. no. ATCC HTB 175)
- H82 culture medium: RPMI 1640 medium supplemented with 10 % fetal calf serum and 2 mM L-glutamine.

Note: We have observed variations in the ability of different serum batches to sustain growth of H82 cells. It is therefore necessary, before commencing differentiation experiments, to carry out a serum screen with these cells to identify a serum in which they will grow satisfactorily.

- BrdU (Sigma, cat. no. B5002)
- 0.2 µm filter (Millipore)
- PBS-A (optional)

Procedure

1. Reconstitute BrdU in sterile PBS-A or in culture medium (without serum) to a stock concentration of 10 mM. Sterilize by filtering through a 0.2 µm filter, aliquot into cryovials and store at −80 °C. **Note:** Identical precautions to those outlined for RA should be taken when working with this agent.

    differentiation assay

2. Harvest H82 cells, seed into 6-well plates at a density of 5×10^3 cells per well and incubate at 37 °C, 5 % CO_2 for 48 h.

3. Dilute stock BrdU with culture medium and add to the cells to produce a final concentration of 5 µM.

4. Incubate for 7 days, replacing the medium after the first 3–4 days. Cells may then be harvested and analyzed as required.

    Note: H82 cells normally grow in suspension with only a small percentage of cells (<5 %) attaching to the substrate. BrdU induces adherence and spreading of H82 cells. Both populations (attached and unattached) may be studied following treatment. Attached cells may be harvested by trypsinisation or fixed in situ using cold acetone for immunocytochemical studies, while cells which remain in suspension may be harvested and cytocentrifuged onto glass slides before fixation.

■ References

Bergh J, Arnberg H, Eriksson B, Lundqvist G (1989) The release of chromogranin A and B like activity from human lung cancer cell lines: A potential marker for a subset of small cell lung cancer. Acta Oncol 28:651–654

Boyd AW, Metcalf D (1984) Induction of differentiation in HL-60 leukemia cells: a cell cycle dependent all-or-none event. Leukemia Res 8:27–43

Breitman TR, Selonick SE, Collins SJ (1980) Induction of differentiation of the human promyelocytic leukemia cell line (HL-60) by retinoic acid. Proc Natl Acad Sci USA 77:2936–2940

Broers JLV, Ramaekers FCS, Klein Rot M, Oostendorp T, Huysmans A, van Muijen GNP, Wagenaar SS, Vooijs GP (1988) Cytokeratins in different types of human lung cancer as monitored by chain-specific monoclonal antibodies. Cancer Res 48:3221–3229

Chen Z, Breitman TR (1994) Tributyrin: A prodrug of butyric acid for potential clinical application in differentiation therapy. Cancer Res 54:3494–3499

Ciccarone V, Spengler BA, Meyers MB, Biedler JL, Ross RA (1989) Phenotypic diversification in human neuroblastoma cells: Expression of distinct neural crest lineages. Cancer Res 49:219–225

Collins SJ, Ruscetti FW, Gallagher RE, Gallo RC (1978) Terminal differentiation of human promyelocytic leukemia cells induced by dimethyl sulfoxide and other polar compounds. Proc Natl Acad Sci USA 75:2458–2462

de The H, Marchio A, Tiollais P, Dejean A (1989) Differential expression and ligand regulation of the retinoic acid receptor a and b genes. EMBO J 8:429–433

DeClerck YA, Bomann E, Spengler BA, Biedler JL (1987) Differential collagen biosynthesis by human neuroblastoma cell variants. Can Res 47:6505–6510

Fontana JA, Colbert DA, Deisseroth AB (1981) Identification of a population of bipotent stem cells in the HL-60 human promyelocytic leukemia cell line. Proc Natl Acad Sci USA 78:3863–3866

Gazdar AF, Carney DN, Guccion JG, Baylin SB (1981) Small cell carcinoma of the lung: Cellular origin and relationship to other pulmonary tumours. In: Stratton FA et al. (eds), pp 145–175

Ham RG, Veomett MJ (1980) Mosby CV (ed) Mechanisms of development, St Louis, pp 317

Jetten AM, George MA, Pettit GR, Rearick JI (1989) Effects of bryostatins and retinoic acid on phorbol ester- and diacylglycerol-induced squamous differentiation in human tracheobronchial epithelial cells. Cancer Res 49:3990–3995

Masui T, Wakefield LM, Lechner JF, LaVeck MA, Sporn MB, Harris CC (1986) Type β transforming growth factor is the primary differentiation-inducing serum factor for normal human bronchial epithelial cells. Proc Natl Acad Sci (USA) 83:2438–2442

McCarthy DM, San Miguel JF, Freake HC, Green PM, Zola H, Catovsky D, Goldman JM (1983) 1,25-dihydroxyvitamin D_3 inhibits proliferation of human promyelocytic leukemia (HL-60) cells and induces monocytic-macrophage differentiation in HL-60 and normal human bone marrow cells. Leukemia Res 7:51–55

Pence JC, Shorter NA (1990) In vitro differentiation of human neuroblastoma cells caused by vasoactive intestinal peptide. Cancer Res 50:5177–5183

Rettig WJ, Spengler BA, Chesa PG, Old LJ, Biedler JL (1987) Coordinate changes in neuronal phenotype and surface antigen expression in human neuroblastoma cell variants. Can Res 47:1383–1389

Ross RA, Bossart E, Spengler BA, Biedler JL (1991) Multipotent capacity of morphologically intermediate (I-type) human neuroblastoma cells after treatment with differentiation-inducing drugs. In: Evans Ea (ed) Advances in neuroblastoma research, vol. 3. Wiley-Liss, New York, pp 193–201

Ross RA, Spengler BA, Rettig WJ, Biedler, JL (1994) Differentiation-inducing agents stably convert human neuroblastoma I-type cells to neuroblastic (N) or nonneuronal (S) neural crest cells. In: Evans EA (ed) Advances in neuroblastoma research, vol. 3. Wiley-Liss, New York, pp 254–259

Tapscott SJ, Lassar AB, Davis RL, Weintraub H (1989) 5-Bromo-2-deoxyuridine blocks myogenesis by extinguishing expression of MyoD1. Science 254:532–536

Weintraub H, Davis R, Tapscott S, Thaye M, Krause, M, Benezra R, Blackwell TK, Rupp R, Hollenberg S, Zhuang Y, Lassar A (1991) The *MyoD* gene family: nodal point during specification of the muscle cell lineage. Science 251:761–766

World Health Organization (1982) The World Health Organization Histological typing of lung tumours, 2nd edn. Am J Pathol 77:123–136

Yen A, Forbes ME (1990) c-*myc* downregulation and precommitment in HL-60 cells due to bromodeoxyuridine. Can Res 50:1411–1420

Yen A, Reece SL, Albright KL (1984) Dependence of HL-60 myeloid cell differentiation on continuous and split retinoic acid exposures: Precommitment memory associated with altered nuclear structure. J Cell Physiol 118:277–286

Chapter 15

Cell Culture Models for Mononuclear Phagocytes

Philippe Poindron* and Yves Lombard

Introduction

Mononuclear phagocytes are ubiquitous cells and practically all tissues can be used as sources for obtaining them, most commonly: bone marrow, thymus, spleen, blood, peritoneal cavity, lungs, brain, skin. In this chapter, particular attention will be focused on the primary culture models developed from murine peritoneal macrophages and human peripheral blood monocytes; the use of these systems is recommended by the Society for Leukocyte Biology (Morahan 1980). The description of methods for obtaining other types of mononuclear phagocytes or f studying them can be found in specialized handbooks (Adams 197⁹, Adams et al. 1981; Beelen and Poindron 1994).

Although primary cultures of mononuclear phagocytes are relatively easy to obtain, many attempts have been made to establish macrophage cell lines in order to bypass the primary explantation step (Defendi 1976) and to circumvent the relative inability of these cells to multiply under normal culture conditions. The methods for establishing macrophage cell lines have been reviewed by Walker (1994). Here we will describe the method developed in our laboratory to obtain macrophage cell lines by extended in vitro culture of primary explants (Lombard et al. 1985, 1988).

* *Correspondence to* Philippe Poindron, Université Louis Pasteur, Faculté de Pharmacie, Département d'Immunologie, Immunopharmacologie et Pathologie, BP 24, 67401 Illkirch, France; phone +33–388–676927; fax +33–388–660190; e-mail Philippe.Poindron@pharma.u-strasbg.fr

15.1
Murine Peritoneal Macrophages from Unstimulated Peritoneal Cavity

Advantages of Peritoneal Cavity as a Source of Murine Macrophages

The peritoneal cavity is a very convenient source of macrophages for the following reasons (Padawer 1973; Daems 1980):

- It contains a finite cell population whose constituent cell types are easily recognizable.

- The total population collected can be increased or decreased in a relatively predictable manner by using well-described experimental procedures such as stimulation with irritant substances which either increase the total number of resident and exudate macrophages collected (proteose peptone, thioglycolate) or decrease the number of resident macrophages through a reaction known as the macrophage disappearing reaction (newborn calf serum, melanin).

- The washing fluid contains evenly dispersed cells; aliquots are therefore representative of the whole population.

- The cells of the peritoneal wash fluids are easy to count using routine hematologic methods.

- The cells of the peritoneal cavity are obtained without the aid of proteolytic agents, the effects of which on cell morphology and function are not entirely known

Materials

- 8–12-week-old mice weighing 18–20 g. In order to ensure the reproducibility of results from one experiment to another, choose animals of a given strain, age and sex, and do not change these parameters thereafter.

animals

Note: Some circannual variations in the number and properties of peritoneal macrophages can occur. Take into account this phenomenon in interpreting results obtained over a long period of time.

- Cork plate
- Pins

equipment

- 5 cm forceps
- Scissors for dissection
- Surgical grooved stainless steel sound (rigid surgical instrument used to explore the wounds or to separate skin from underlying tissues)
- Filter paper
- Sterile, single-use, polyethylene Pasteur transfer pipettes (Copan, Poly Labo, Strasbourg, France)
- Guillotine for decapitating small animals
- Polypropylene centrifuge tube (15 ml)
- Plastic flasks (15 cm^2) or 24-multiwell plates
- Glass coverslips (optional)
- Hemocytometer

reagents
- Iodinated 90 % ethanol and ethanol
- Culture medium: RPMI 1640 medium supplemented with 10 % heat-inactivated (30 min, 56 °C) fetal calf serum, and antibiotics (sodium benzylpenicillinate [200 U/ml], streptomycin sulphate [40 μg/ml]).

Note: Since the macrophages are very sensitive to environmental conditions and especially to serum, it is recommended to use the same batch of serum for all experiments in which the results are to be compared with each other. In addition, before use, medium and serum should be checked for the absence of bacterial lipopolysaccharide (LPS).

■ Procedure

dissection of the mouse

1. Exsanguinate the mouse by decapitation.

2. Fix the mouse with pins onto a cork plate wrapped in filter paper.

3. Disinfect the skin over the abdomen by flooding it with iodinated 90 % ethanol.

4. Grasp the lower middle part the abdominal skin with ethanol-flamed forceps and make an incision 2–3 mm just below the teeth of the forceps.

5. Introduce the surgical grooved sound through the incision and pull it cranially up to the basis of the neck.

6. Incise the skin with the dissection scissors by using the groove of the sound as a guide.

7. Lift the skin up with the grooved sound, then incise the skin on the right and left sides, perpendicular to the first incision and towards the fore- and hindlimbs. Fold the skin flaps and fix them with pins onto the cork plate. Do not tear the abdominal wall.

1. Grasp the abdominal wall in its lower middle part with the ethanol-flamed forceps and incise over 3 mm perpendicular to the caudocranial axle of the mouse, just below the teeth of the forceps.

    washing of the peritoneal cavity

2. Introduce the capillary end of a Pasteur polyethylene transfer pipette filled with 2.5 ml of culture medium through the incision, the lips of which are maintained elevated with the forceps, and allow the medium to flow into the peritoneal cavity. Do this once more, taking care not to flood the cavity or to injure the gut.

3. Inclinate the cork plate so that the washing fluid accumulates above the ventral side of the mouse diaphragm. Aspirate and force back the fluid five times and withdraw it with the Pasteur transfer pipette.

1. Place the fluid in polypropylene centrifuge on ice.

    preparing the cell suspension

2. Centrifuge the tubes at $300\,g$ for 10 min at $4\,°C$. Discard the supernatant and resuspend the pellet in a small volume of culture medium. Count the cells in an hemocytometer and adjust the final volume of the suspension to obtain the desired concentration of cells.

3. Seed the suspension in plastic flasks or in 24-multiwell plates, with or without glass coverslips (see below).

1. Incubate the vessels at $37\,°C$ in an air-CO_2 humidified incubator. We have successfully used a mixture of air (95 %) and CO_2 (5 %); however, some authors have observed that a higher partial pressure of CO_2 (10 %), in combination with 5 % of O_2, gives the best results.

    incubation conditions

Note: Peritoneal wash fluid from several animals can be pooled.

Results

Number of Cells Obtained

The total number of cells obtained from an unstimulated peritoneal cavity ranges, depending on the report, between 0.5 and 8.2×10^6, the mean being 3.55×10^6 with a standard deviation of 1.94×10^6 (Daems 1980). The washing fluid mainly contains resident macrophages, eosinophilic granulocytes and lymphocytes.

The absolute number of resident macrophages does not vary to the same extent. It ranges, depending on the report, between 0.9 and 2×10^6, the mean being $1.54 \pm 0.49 \times 10^6$. In our hands, the absolute number of resident macrophages collected varies between 1.2 and 1.5×10^6 per peritoneal cavity (BALB/c or C56BL/6 mice).

Purification

About half the resident macrophages from the unstimulated peritoneal cavity adhere firmly to glass or plastic substrates within 2–3 h. This property allows the purification of these cells, since under these conditions granulocytes and lymphocytes cannot adhere and are easily removed by a thorough washing of the culture.

Use of Macrophage Cultures

Mouse peritoneal resident macrophages can be used either in short-term (48 h) or long-term (7 days and more) cultures. However, environmental conditions (quality and origin of serum, adherence to the substrate) deeply modify the morphology, the functions and the capacities of resident macrophages, which tend to become activated with increasing culture time.

Multiplication of Resident Macrophages

It is generally known that resident macrophages do not multiply in culture except under special conditions (presence of macrophage colony-stimulating factor, M-CSF or macrophage/granulocyte colony-stimulating factor, MG-CSF, in the culture medium, for instance). We have shown (see below) that this finding is not entirely true. Peritoneal macrophages are able to indefinitely multiply without the addition of exogenous growth factors to the culture medium but in the presence of so-called mesothelial cells, which play the role of feeder cells. The nature and origin of these mesothelial cells are unknown.

Transfer of Macrophages

Macrophages cannot be easily detached from plastic or glass substrates by either mechanical or enzymatic procedures without undergoing extensive damage. If transfer from one vessel to another is necessary, macrophages can be cultured on glass coverslips which can be easily handled with the cells remaining attached.

Identification

Macrophages are identified by morphologic, antigenic, biochemical and functional criteria. Usually, at least three criteria should be assessed and at least 90 % of the cells studied should be positive for each criterion (for details, see Vand der Meer 1980; Van Furth 1981). Examples of morphologic, biochemical and functional characterization will be given in Section 15.3 (Fig. 15.3).

15.2
Human Blood Monocytes

Advantages of Human Blood Monocytes as a Model of Human Mononuclear Phagocytes and a Source of Human Macrophages

- Peripheral blood contains enormous amount of leukocytes which can be easily collected using leukapheresis without harm to the donor.

- After purification, blood monocytes can be cultured for several days (7 and more) and mature into macrophages (Andreesen et al. 1986) under standard or special conditions. It is therefore possible to compare the properties of monocytes to those of monocyte-derived macrophages.

- Monocyte-derived macrophages can be used in basic research or reinfused into the donor after being activated to cytotoxicity by human interferon (IFN)-γ (adoptive cellular immunotherapy; Dumont et al. 1988; Andreesen et al. 1990; Faradji et al. 1994).

Problems and Preliminary Precautions

- All patients should provide informed consent prior to leukapheresis.

- Leukapheresis (not obligatorily elutriation) should be carried out in blood transfusion centers under the control of and by authorized persons.

- The immune status of the donor regarding hepatitis B, hepatitis C and human immunodeficiency virus (HIV) should be established before any experiment.

- The procedure requires sophisticated and expensive equipment (see below).

- The procedure is lengthy and should be carried out by experienced persons.

Materials

equipment
- Cobe 2997 or Cobe Spectra apheresis system (Cobe Lab, Lakewood, CO, USA)
- Beckman J6-M/E centrifuge (Beckman Instruments, Palo Alto, CA, USA)
- Large capacity JE-5.0 elutriation rotor (Beckman)
- One or two interchangeable separation chambers of either 40 ml or 5 ml capacity (Beckman)
- Masterflex computerized drive peristaltic pump (Cole-Parmer Instrument Co, Chicago, IL, USA)
- Transmed 40 μm micro-aggregate filters (Sodis, 25, rue des Erables, 78150, Rocquencourt, France)
- Coulter channelizer ZM (Coulter Electronics Inc, Hialeah, MC, USA)
- Laminar air flow hood
- Refrigerated centrifuge

sterile single use disposables
- Polyvinylchloride tubings and collection bags (Maco-Pharma, Tourcoing, France)
- Plastic flasks or multiwell plates, or Teflon culture bags (Stericell, Dupont de Nemours, Wilmington, DE, USA)
- Polyethylene centrifuge tubes

– Calcium- and magnesium-free Hanks balanced salt solution (CMF-HBSS)
– Elutriation medium: solution of human serum albumin (2 % w/v) in CMF-HBSS (final osmolality 290 mOsm ; final pH 7.2)
– Citric acid-citrate-dextrose solution as anticoagulant solution for leukapheresis
– Culture medium: RPMI 1640 medium supplemented with sodium benzylpenicillinate (100 U/ml), streptomycin sulphate (100 µg/ml), 2 mmol/l glutamine, 2 mmol/l sodium pyruvate, 1 % (v/v) nonessential amino acid (NEAA) solution (Gibco, Grand Isalnd NY, USA), 5×10^{-6} mol/l indomethacin, 3×10^{-5} mol/l mercapthoethanol and 2 % (v/v) autologous heat-inactivated (30 min, 56 °C) serum

culture medium and reagents

Procedure

The experimental procedure described here is that of Dumont et al. (1988) (culture conditions), modified by Faradji et al. (1994) (use of counterflow centrifugation elutriation).

1. Prepare a vascular access by double antecubital venipuncture of the donor. Some patients may require the insertion of a silastic catheter into a subclavian vein to obtain an adequate blood flow. Connect the tubings to the apheresis system.

leukapheresis

2. Process 5 l of whole blood at a flow rate of 50 ml/min. Table 15.1 lists the technical data of the operating conditions.

3. Centrifuge the leukapheresis products to remove the platelet-rich plasma and resuspend the cell pellet in elutriation medium at a final concentration of 50×10^6 white blood cells per ml.

1. Mount the semi-closed elutriation sytem as shown in Fig. 15.1. The quick release JE-5.0 rotor should be sterilized at 121 °C for 20 min after having been rinsed with sterile distilled water.

counterflow centrifugation elutriation

2. Allow the cell suspension to flow directly through a Transmed 40 µm micro-aggregate filter into the counterflow centrifugation system, using the following conditions: temperature of centrifugation 18 °C; initial flow rate, 40 ml/min (large chamber) or 9 ml/min (small chamber).

3. After loading, gradually increase the flow rate by 10 ml/min (large chamber) or 1 ml/min incrementally, every 10–15 min (collect the

Table 15.1. Collection of peripheral blood monocytes from normal volunteers and cancer patients using Cobe separators[a]: technical data of the operating conditions. (Faradji et al. 1994, with permission)

Parameters	Cobe 2997[b] (n=170)	Cobe Spectra[b] (n=95)
Centrifuge speed (rpm)	860	815
Volume of blood processed (l)	6.7 (5.8–8.5)[c]	5.5 (5.0–7.5)
Total blood flow rate (ml/min)	50	50
Collection rate (ml/min)	0.8	0.8
Anticoagulant: whole blood ratio	1:10	1:10
Length of the procedure (min)	120 (90–180)	100 (90–120)
White bloods cells ($\times 10^6$/ml)	66.1 ±9.2[d]	57.8±11.5
Red blood cells ($\times 10^6$/ml)	210 ±20	190±30
Platelets ($\times 10^6$/ml)	764 ±281	832±173
Granulocytes ($\times 10^6$/ml)	7.6 ±3.8	4.5±2.1
Lymphocytes ($\times 10^6$/ml)	56.2±5.4	54.7±6.7
Purity of monocytes collected (%)	37.2 ±4.8	34.9±5.2
Monocytes collected per procedure ($\times 10^9$)	1.94 ±0.53	1.55±0.25

[a] Cells were collected by leukapheresis into a volume of 80 ml (range, 60–125 ml).
[b] Data were obtained from leukapheresis performed in 50 normal donors (one leukapheresis/donor) and 50 cancer patients (with a total of 215 procedures).
[c] Parentheses indicate range.
[d] Mean ±SD.

fractions before changing the flow rate ; see below) until a mixed population (ratio: 50/50) of monocytes/lymphocytes is seen to exit the chamber. The final flow rate is 60 ml/min (large chamber) or 11 ml/min (small chamber). The ultimate fraction depends on the clearing of lymphocytes as determined by monitoring the cells emerging from the centrifuge with the Coulter channelizer.

4. Stop the rotor and the medium flow and remove the quick release assembly from the rotor.

5. Position vertically the rotor so as to allow the remaining cells to sediment to the outflow site and to gravitate towards the collection bag. This final step should be carried out without opening the sterile disposable tubings.

preparation of cell suspension

1. Centrifuge the cell suspension obtained above at $300\,g$ for 10 min at 4 °C and wash the pellet in CMF-HBSS.

2. Centrifuge again (same conditions as above); discard the supernatant and resuspend the cell pellet in culture medium at the desired concen-

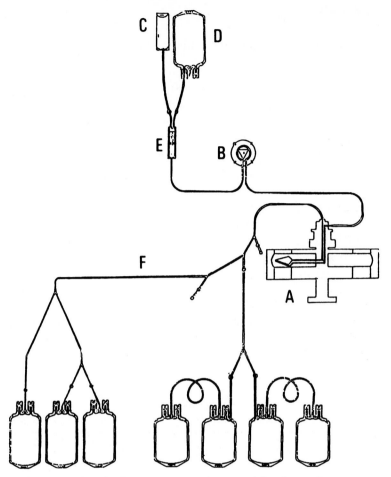

Fig. 15.1A–F. The semi-closed elutriation system. **A** Quick-release JE 5.0 rotor assembly; **B** Masterflex peristaltic pump; **C** leukapheresed blood mononuclear cells; **D** elutriation medium; **E** micro-aggregate filter; **F** single-use set including tubing and collection bag. (Faradji et al. 1994, with permission)

tration. When monocytes are cultured for the purpose of adoptive immunotherapy, they are seeded in Teflon bags at a concentration of 1.5×10^6/ml culture medium.

3. Incubate at 37 °C in an humidified 95 % air/5 % CO_2 incubator.

Note: The whole procedure is carried out under sterile conditions and, when possible, under a laminar flow hood. Media and reagents should be checked for the absence of bacterial LPS before use.

▨ Results

Four fractions are collected by this method. The following data are given for the large chamber:

- Fraction 1 is collected (in 2×500 ml volume) at a 40 ml/min flow rate and contains mostly red blood cells (>95%) and platelets, and approximately 53% of the lymphocytes present in the leukapheresis concentrate.
- Fraction 2 is collected (in 2×500 ml volume) at a 50 ml/min flow rate and consists of approximately 92% lymphocytes and 8% monocytes (38% of the total lymphocytes and 7% of the total monocytes of the leukapheresis concentrate).
- Fraction 3 (one 500 ml volume), collected at a 60 ml/min flow rate, consists of a mixed lymphocyte/monocyte population and represents approximately 2% of the total number of white blood cells present in the leukapheresis concentrate.

Table 15.2. Determination of the counterflow centrifugation elutriation efficiency and monocyte recovery as a function of elutriation chamber used.[a] (Faradji et al. 1994, with permission)

Leukapheresis concentrate

Chamber	White blood cells ($\times 10^9$)	Monocytes (%)	Red blood cells/ white blood cells	Platelets/ white blood cells
40 ml capacity (n=220)	5.236±0.925	37±5	4.51±2.34	32.2±6.5
5 ml capacity (n=45)	4.568±1.237	35±6	3.77±1.56	23.8±7.3

Counterflow centrifugation elutriation concentrate: fractions

Chamber	White blood cells ($\times 10^9$)	Monocyte purity (%)	Monocyte recovery (%)	Red blood cells/ monocytes	Platelets/ monocytes
40 ml capacity (n=220)	1.626±0.389	92.74	85.35	0.2±0.1	0.8±0.2
5 ml capacity (n=45)	1.224±0.321	93.21±0.71	83.21±3.52	0.2±0.1	0.5±0.2

[a] All values are expressed as the mean ±SD.

– The subsequent fraction 4 consists of purified monocytes (89 %–97 %, depending on the experiment) that still remain in the chamber after the rotor has been stopped. It contains approximately 80 % of the total monocytes present in the leukapheresis concentrate.

Table 15.2 shows that the counterflow centrifugation elutriation efficiency and monocyte recovery are similar whatever the size of the elutriation chamber used. Table 15.3 shows that purified human monocytes and monocyte-derived macrophages obtained using the above procedure are able to adhere to plastic substrate, to engulf *Saccharomyces boulardii* and IgG-coated sheep red blood cells, to become cytotoxic for tumor cells after being treated with recombinant human IFN-γ (Dumont et al. 1988), and can secrete interleukin (IL)-1, tumor necrosis factor (TNF) and procoagulant activity (PCA).

Table 15.3. Capacities and functions of human blood monocytes obtained by leukapheresis and counterflow centrifugal elutriation. (Faradji et al. 1994, with permission)

Capacities or functions	Fresh unstimulated monocytes	Recombinant human IFN-γ activated monocytes	Monocyte-derived macrophages	Recombinant human IFN-γ activated monocytes
Secretory products				
IL-1 (ng/10^6 cells)	0.013±0.006[a]	0.38±0.21	13.03±4.36	15.08±6.44
TNF-α (ng/10^6 cells)	0.032±0.017	1.29±0.32	8.14±3.28	9.72±5.73
PCA (U/10^5 cells)	0.1±0.02	63.8±5.3	45.5±5.1	50.1±5.1
Phagocytosis[b]				
Unopsonized S. boulardii	4.1±0.5	4.4±0.6	8.6±1.1	ND
IgG-coated sheep red blood cells	3.8±0.4	5.8±0.9	11±1.5	ND
Cytotoxicity				
Inhibition of [^{3}H]thymidine uptake by U937 cells (%)	22	ND	17.6	95.5
Adherent cells (%)	83±5	88±6	92±3	95±4

IL, interleukin; TNF, tumor necrosis factor; PCA, procoagulant activity; ND, not done.
[a] Mean ±SD.
[b] Mean number±standard deviation of ingested particles par phagocytozing cell after an 8 h incubation.

15.3
Establishment of Murine Macrophage Cell Lines by Long-Term Culture of Primary Explants

Advantages

- The cell lines established using the long-term culture method can be passaged over several years without losing their initial characteristics.

- The macrophages obtained express the essential capacities and functions of normal macrophages (see below).

- It is relatively easy to obtain large amount of macrophages with these cell lines.

  Note: It is possible to establish macrophage cell lines from peritoneal fluid, lungs and spleen (Lombard et al. 1985, 1988). No attempts have been made to establish cell lines from other tissues. Here only the establishment of peritoneal macrophage cell lines is described.

Problems and Disadvantages

- The growth of macrophage cell lines depends on the presence of mesothelial cells or mesothelial-like cells in the culture, so the macrophages cannot be cloned. The term "established coculture" is thus a more appropriate description of these lines (Lombard and Poindron 1985).

- Establishment of a cell line by the long-term culture method is not easy and requires involvement of a laboratory scientist with considerable experience in cell culture methods. The establishment of a line of proliferating peritoneal, splenic and pulmonary macrophages is achieved with only 5, 40 and 20 % success, respectively.

- The mesothelial cells and mesothelial-like cells multiply slowly (Lombard and Poindron 1985) and it takes a while (2–3 weeks) to obtain macrophages and therefore to subculture a flask (see below).

- It is especially difficult to freeze these cell lines. Glycerol should not be added as the cryoprotective agent. Usually, the best medium for freezing these lines consists of culture medium to which is added 8 % dimethylsulfoxide (v/v).

▓ Materials

– Twelve-week-old male or female mice from any strain

Animals

See Section 15.1, "Materials."

reagents and equipment

▓ Procedure

For dissection of the mouse, washing of the peritoneal cavity, and preparation of the cell suspension, see "Procedure" of Section 15.1.

1. Seed 2.5 ml of washing fluid (adjusted to $0.5-1.5\times10^6$ cells). Do not wash the culture. At this time, the total volume of medium is 2.5 ml.

establishment of the line

2. Incubate the flask in an air (94 %)/CO_2 (6 %) humidified incubator. **Note:** The cells are established in 94 % air /6 % CO_2; after a few passages they can be cultured in 95 % air/5 % CO_2.

3. Examine the culture every day using an inverted phase contrast microscope, and change the medium whenever the pH is below 6.8 or whenever microscope examination reveals signs of cell deterioration even in the absence of acidification. This deterioration generally leads to a halt in culture development and possibly to an alteration in cell morphology. The change of medium must be performed as follows: until the mesothelial cells appear (see "Results," below), a known volume of exhausted medium is replaced with an equivalent amount of fresh medium. This volume will vary according to the culture and circumstances but should never exceed half the total volume.

4. When mesothelial cells appear and multiply, increase progressively the amount of fresh medium at each change until the volume attains 7.5 ml/25 cm^2 flask. The volume of medium withdrawn should never exceed half the total volume. At this step, macrophages begin to multiply on the mesothelial cell layer.

1. When mesothelial cells reach confluence and macrophages pass into suspension, the macrophages can be subcultured. Withdraw 3.75 ml of the cell suspension, dilute in an equal volume of fresh medium and change the exhausted medium as described above in step 4.

subculture

Note: It is possible to establish macrophage cell lines from spleen or lung (Lombard et al. 1988). Instead of using cells in suspension as the primary explant, as in the case of peritoneal wash fluid, the whole tissue is cut into fragments of less than $1 \, mm^3$. The fragments are placed in plastic culture flasks ($25 \, cm^2$; 5–6 fragments per flask) and cultured in the same medium as above. Mesothelial-like cells will proliferate around the adhering fragments, macrophages will multiply on the surface of the mesothelial-like cell layer, and pass into suspension.

Results

Evolution of the Long-Term Primary Culture (Fig. 15.2)

After seeding, the primary peritoneal cell suspension consists of two cell classes: one of adherent cells (CAC) and one of nonadherent cells (CNAC).

The CAC contains cells that adhere very rapidly (within 1 or 2 h) to the plastic. Mostly of them are classical macrophages and form an heterogeneous population. A few of these cells are spherical and adhere less strongly to the plastic. The CNAC contains mainly lymphocytes.

Three to 5 days after seeding, the spherical cells in the CAC increase in size and number, and contacts are established between CNAC and CAC cells. These contacts do not result simply from sedimentation of CNAC cells since they can be detached only by gently shaking the flasks.

Eight to 10 days later, cells in the CAC begin to proliferate. They are heterogeneous, with large nuclei, unlike macrophages which initially adhere to the plastic but which retain small nuclei and do not divide. Some of the proliferating cell resemble mesothelial cells (which we find the least incorrect denomination, but which does not mean that they are actuallly identified as true mesothelial cells) and become predominant when the cells of the CAC reach confluence. This confluent layer of adherent, heterogeneous cells we be referred to as the "underlying cells" or "underlying layer." At that time, macrophages spread out on and attach to the underlying cells. Lymphocytes disappear from the CNAC by an unknown mechanism and macrophage-like cells multiply on the surface of the underlying layer. It is very easy to make the cells that are loosely attached to the underlying layer pass into suspension. This suspension is referred to as the "latest CNAC" and usually contains 10^5 cells/ml (mean value; the actual value varies according to the culture).

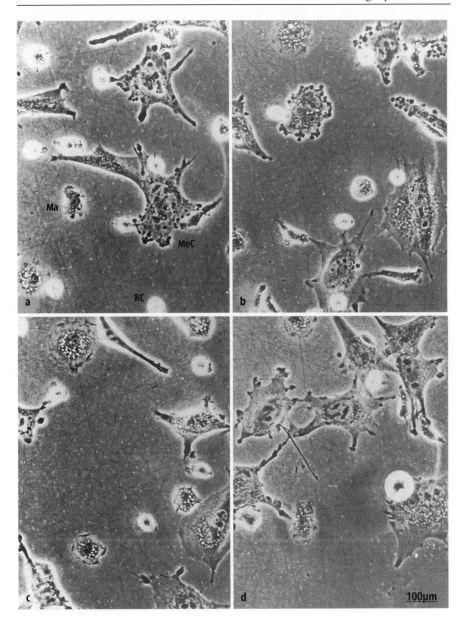

Fig. 15.2a–d.

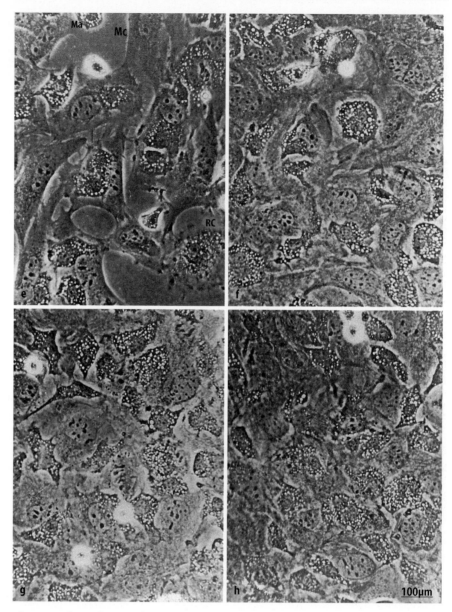

Fig. 15.2a–l. Evolution of a macrophage subculture (35th passage, peritoneal line). The same field was observed during the subculture. **a, b, c, d** After 7, 8, 9 and 10 days, respectively. In **b, c** and **d,** some macrophages (ringed cells) have spread over meso-thelial cells (*MeC*). *RC* Round cell; *Ma* macrophage. **e, f, g, h** After 12, 13, 15 and 16 days in culture, respectively. The mesothelial cells have become confluent. An increasing number of macrophages are spread over the underlying cell layer. The

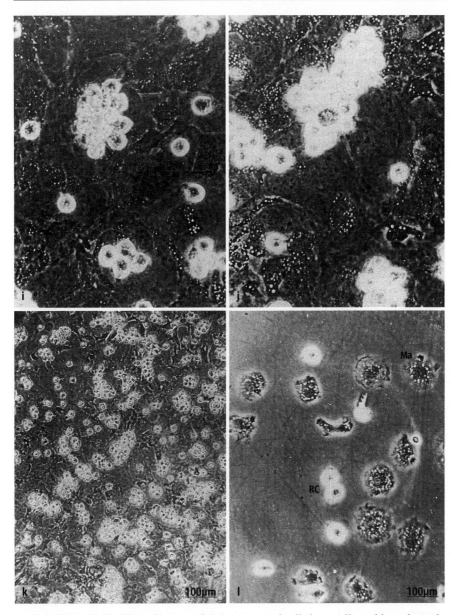

mesothelial cells (*Mc*) have large nuclei. Some round cells have adhered loosely to the underlying cell layer. **i, j** After 18 and 20 days in culture; **k** (lower magnification of **j**) colonies of round cells have emerged at the surface of the mesothelial cell layer and passed into suspension, either spontaneously or after gentle shaking. **l** A 24 h subculture (36th passage) of the preceding culture. (Lombard et al. 1985, with permission)

Care has to be taken not to detach the mesothelial cells or even the mesothelial cell layer from the support by too violent or too lengthy shaking.

Subculture of the Latest Class of Nonadherent Cells

The latest CNAC can be subcultured after dilution, as described in the "Procedure" of Section 15.3. Almost all cells transferred pass very rapidly into the CAC and spread out on the support. Most have the typical morphology of macrophages and in that state cannot divide.

Usually, 3–5 days after transfer, some cells in the CAC undergo major morphological changes. (In some cultures, these changes appear later.) They become spherical and increase in size. Seven to 11 days after transfer, mesothelial cells appear in the CAC, although their relationship with the preceding cells is not established. The cells proliferate actively and form small colonies (1–10 per flask). Some macrophages appear on the surface of these colonies. When they reach a critical size (diameter greater than 1 mm), these macrophages spread out on the central zone; some of them become spherical and less adherent and can be detached by gentle shaking. The cells increase in number and the mesothelial cells continue to proliferate.

Three weeks after beginning subculture, the following features are observed: The underlying cell layer consists almost exclusively of confluent, contact-inhibited mesothelial cells. These cells are large ($40\times50\,\mu$m), have large, oval nuclei and an homogeneous cytoplasm without vacuoles. On the underlying cells, scattered adherent macrophages of medium size are present. They have small nuclei and numerous cytoplasmic vacuoles. Loosely adhering to the underlying cells, are spherical cells which form small colonies of macrophages that pass into suspension, either spontaneously or upon gentle shaking, forming the latest CNAC of the subculture. This latest CNAC can be transferred to new flask after being half-diluted in fresh medium. It contains almost exclusively (99 %) macrophages. After transfer, the culture undergoes the same cycle as above and gives rise to a new latest CNAC which in turn can be transferred.

In these cell lines, the macrophages which multiply are very probably those which have been in contact with the upper side of the mesothelial cells (Lombard and Poindron 1985). In these cultures, indeed, it is very easy to count the macrophages present in the suspension. After having quantitatively eliminated these macrophages and those which loosely adhere to mesothelial cells, one observes that some macrophages remain firmly attached to the mesothelial cells. When these are detached from the plastic substrate by trypsinization, macrophages directly adhering to the plastic substrate remain. The number of macrophages that cannot be detached by trypsinization because they adhere to the plastic substrate does not vary during the culture period, whereas the number of macrophages passed into suspension or adhering to mesothelial cells increases continuously (see Table 15.4).

Table 15.4. Evolution of the number of macrophages and mesothelial cells during the subculture of a peritoneal macrophage cell line obtained by the long-term culture method. (Lombard and Poindron, 1985 with permission)

Days after seeding	Macrophages adhering to glass	Macrophages adhering to mesothelial cells	Macrophages in suspension	Mesothelial cells
0	180 000	0	0	490
1	174 500	0	0	540
2	171 500	0	0	650
3	184 500	0	0	980
4	177 000	0	0	4 250
5	173 500	0	0	7 900
6	177 500	0	0	9 000
7	176 500	0	0	20 700
8	182 500	0	0	21 500[a]
9	182 000	6 500	0	22 500
10	170 500	7 600	0	23 700
11	160 000	8700	0	26 000
12	166 000	10 100	0	31 700
13	161 500	13 600	0	51 000
14	168 000	13 600	0	67 000
15	196 000	14 200	0	108 000
16	197 000	16 600	0	165 000
17	205 000	24 500	4300	265 000
18	200 000	45 300	12 280	436 000[a]
19	210 000	214 000	16 900	650 000
20	224 000	358 000	58 000	942 000
21	216 000	508 000	118 000	1 200 000
22	220 000	663 000	211 000	1 500 000
23	186 000	772 000	350 000	1 750 000[b]

[a] Change of medium because of acidification.
[b] Strong acidification, transfer of the "latest" class of nonadherent cells (CNAC).

Capacities and Functions of Macrophage Cell Lines Obtained by the Long-Term Culture Method (Fig. 15.3)

- Antigenic phenotype: Whatever their origin (peritoneum, lungs, spleen) the macrophage cell lines obtained by the long-term culture method display the antigenic phenotype of cells belonging to the mononuclear phagocyte system and express characteristics indicating a rather high differentiation state. However, there are significant differences in antigen expression between the lines obtained from solid tissues (spleen and lungs), which are very similar in their antigenic pattern, and those originating from peritoneum, which seem to be less well differentiated. The mesothelial cell population is difficult to characterize (Falkenberg et al. 1992).

- Enzymatic markers: Only a peritoneal macrophage cell line has been studied for enzymatic membrane markers. It was shown to express membrane NAD$^+$ glycohydrolase and nucleotide pyrophosphatase (Muller et al. 1988). All the other lines studied expressed cytoplasmic nonspecific esterases (Lombard et al. 1985, 1988).

- Receptors: Usually, the macrophage cell lines obtained by the long-term culture method express complement receptor 3, receptors for crystallizable immunoglobulin G fragment, mannose and β-glucan receptors (Lombard et al. 1985, 1988; Muller et al. 1988; Giaimis et al. 1993).

- Expression of class II major histocompatibility (MHC) antigens: The macrophage cell lines obtained by the long-term culture method spontaneously express class II MHC antigens The degree of spontaneous expression varies depending on the lines. The expression of class II MHC is induced by IFN-γ (Lombard et al. 1988).

- Secretion products: The macrophage cell lines obtained by the long-term culture method can produce superoxide radicals upon stimulation by phorbol myristate acetate (PMA). They spontaneously produce M-CSF and an IFN-γ-like substance in varying amounts depending on the line studied. They produce TNF upon stimulation by IFN-γ. They do not secrete IL-3 but secrete low levels of IL-1. The production of IL-3 and IL-1 cannot be induced by IFN-γ. Finally, the macrophage cell lines produce nitric oxide (NO) spontaneously and after being induced by LPS or LPS and IFN-γ (Lombard and Poindron, unpubl. results) or after being induced by polysaccharides such as pichilan (Vallot et al. 1994).

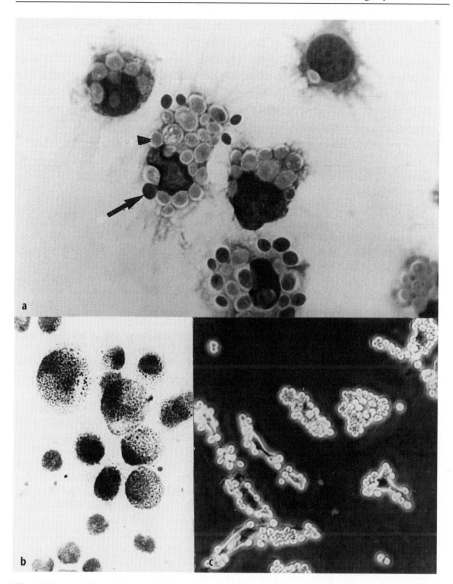

Fig. 15.3a–c. Capacities and functions of macrophage cell lines obtained by the long-term culture method. **a** Phagocytosing cells after 1 h exposure of subculture (pulmonary line) to *Saccharomyces cerevisiae* suspension. Dark particles (*arrow*) and pale particles (*arrowhead*) are extracellularly and intracellularly located, respectively (from Giaimis et al. 1992). **b** Detection of nonspecific esterases in subculture of a peritoneal line. **c** Detection of Fc receptors in subculture of a peritoneal line. Formation of rosettes with antibody-coated sheep erythrocytes. Most of the cells bind more than 10 erythrocytes (**c** from Lombard et al. 1985, with permission)

- Functions: The macrophage cell lines obtained by the long-term culture method are capable of both opsonophagocytosis and lectinophagocytosis (Lombard et al. 1985; Giaimis et al. 1993). They can serve as accessory cells in concanavalin A stimulation of T lymphocytes. The cooperation is more effective between macrophages and lymphocytes belonging to the same haplotype (Lombard et al. 1988). They are both cytostatic and cytolytic towards tumor cells (Lombard et al. 1988).

References

Adams DO (1979) Macrophages. In: Jakoby W, Pastan I (eds) Cell culture, Methods in enzymology, vol LVIII. Academic Press, New York, pp 495–505

Adams DO, Edelson PJ, Koren H (1981) Methods for studying mononuclear phagocytes. Academic Press, New York

Andreesen R, Bross KJ, Emmrich F (1986) Surface antigen analysis of human macrophage maturation and heterogeneity. Leukocyte Host Def 295–300

Beelen RHJ, Poindron P (1994) New methods for studying macrophages and mononuclear phagocytes. J Immunol Methods (Spec Issue) 174:3–320

Daems WT (1980) Peritoneal macrophages. In: Carr I, Daems WT (eds) The reticuloendothelial system. I. Morphology. Plenum, New York, pp 57–127

Defendi V (1986) Macrophage cell lines and their use in immunobiology. In: Nelson DS (ed) Immunobiology of the macrophage. Academic Press, New York, pp 275–290

Dumont S, Hartmann D, Poindron P, Oberling , Faradji A, Bartholeyns J (1988) Control of the antitumoral activity of human macrophages produced in large amounts in view of adoptive transfer. Eur J Cancer Clin Oncol 24: 1691–1698

Faradji A, Bohbot A, Schmitt-Goguel M, Siffert JC, Dumont S, Wiesel ML, Piemont Y, Eischen A, Bergerat JP, Bartholeyns, Poindron P, Witz JP, Oberling F (1980) Large scale isolation of human blood monocytes by continuous flow centrifugation leukapheresis and counterflow centrifugation elutriation for adoptive cellular immunotherapy in cancer patients. J Immunol Methods 174:297–309

Falkenberg U, Lombard Y, Giaimis J, Poindron P, Falkenberg FW (1992) Phenotypic characterization of three long-term cultured resident macrophage lines. Res Immunol 143:25–32

Giaimis J, Lombard Y, Fonteneau P, Muller CD, Levy R, Makaya-Kumba M, Lazdins J, Poindron P (1993) Both mannose and β-glucan receptors are involved in phagocytosis of unopsonized, heat-killed *Saccharomyces cerevisiae* by murine macrophages. J Leukocyte Biol 54:564–571

Lombard Y, Bartholeyns J, Chokri M, Illinger D, Hartmann D, Dumont S, Kaufmann SHE, Landmann R, Loor F, Poindron P (1988) Establishment and characterization of long-term cultured cell lines of murine resident macrophages. J Leukocyte Biol 44:391–441

Lombard Y, Poindron P (1985) Multiplication in vitro et subculture, apparemment indéfinie, de macrophages péritonéaux résidents de Souris par une méthode nouvelle de coculture. In: TerréJ, Pétiard V (eds) Aspects industriels des cultures cellu-

laires d'origine animale et végétale. Comptes rendus du 10éme Colloque organisé-
par la Section de Microbiologie Industrielle de la SFM. SFM, Paris, pp 453–462

Lombard Y, Ulrich B, Poindron P (1985) In vitro multiplication and apparently indefinite subcultures of normal mouse resident peritoneal macrophages. Biol Cell
53:219–230

Morahan PS (1980) Macrophage nomenclature: where are we going ? J Reticuloendothel Soc 27:223–245

Muller CD, Lombard Y, Bartholeyns J, Poindron P, Schuber F (1988) Characterization
of surface markers of continuously growing murine resident macrophages. J Leukocyte Biol 43:165–171

Padawer J (1973) The peritoneal cavity as a site for studying cell-cell and cell-virus
interactions. J Reticuloendothel Soc 14:462–512

Van der Meer JMV (1980) Characterization of mononuclear phagocytes in culture.
In: Carr I, Daems WT (eds) The reticuloendothelial system. I. Morphology.
Plenum, pp 735–771

Vallot N, Boudard F, Bastide M (1994) Effect of a (13)-β-D-glucan, pichilan, extracted
from *Pichia fermentans* on a murine macrophage cell line. Int J Immunopathol
Pharmacol 7:21–35

Van Furth R (1981) Identification of mononuclear phagocytes. Overview and defini-
tions. In: Adams DO, Edelson PJ, Koren H (eds) Methods for studying mononuclear phagocytes. Academic Press, New York, pp 243–251

Walker WS (1994) Establishment of mononuclear phagocyte cell lines. J Immunol
Methods 174:25–31

The Human Leukemia Cell Line HL-60 as a Cell Culture Model To Study Neutrophil Functions and Inflammatory Cell Responses

Faustino Mollinedo*, Antonio M. Santos-Beneit, and Consuelo Gajate

Introduction

Myeloid leukemic cell lines established from patients with either acute myeloid leukemia (AML) or the blastic phase of chronic myeloid leukemia (CML) are arrested at different stages of myeloid development. These lines have been widely studied as models of myeloid differentiation and as a substitute for their more mature, normal myeloid counterparts which sometimes are cumbersome to use. The most broadly studied human myeloid cell line, the human acute myeloid leukemia HL-60 cell line, was derived from a 36-year-old woman diagnosed and treated at M.D. Anderson Hospital in Texas (Collins et al. 1977). This patient was initially considered to have acute promyelocytic leukemia (APL, also named French-American-British classification M3 or FAB-M3), but had atypical clinical features, indicating that the leukemia from which the HL-60 cells were derived was more appropriately classified as an AML with maturation, FAB-M2 (Dalton et al. 1988). APL represents a model for the therapeutic approach of differentiation therapy, as APL cells are able to respond to all-*trans* retinoic acid (RA) treatment by terminal differentiation (Huang et al. 1988). The chromosomal translocation t(15;17) in the blasts is specifically observed in APL, leading to a chimeric gene between the COOH-terminal of the RA receptor (RARα) and the NH_2-terminal of a myeloid gene product initially referred to as Myl (de Thé et al. 1990) and then renamed PML (promyelocytic leukemia). This PML-RARα fusion gene appears to be an instrumental, if not actually the causative, event in the neoplastic process. It has been hypothesized that

* *Correspondence to* Faustino Mollinedo, Laboratory of Signal Transduction and Leukocyte Biology, Instituto de Biología y Genética Molecular, Facultad de Medicina, Universidad de Valladolid-Consejo Superior de Investigaciones Científicas, C/ Ramón y Cajal 7, 47005 Valladolid, Spain; phone +34–83–423062; fax +34–83–423588; e-mail fmollin@med.uva.es

the abnormal PML-RARα gene product represents a dominant-negative gene product able to disrupt the expression of genes involved in the normal development of granulocytes (Kakizuka 1991). So far, three established human cell lines have been reported to be derived from APL patients: HL-60, NB-4 and PL-21 (Drexler et al. 1995). NB-4 carries t(15:17) while HL-60 and PL-21 lack this cytogenetic hallmark for APL, indicating that NB-4 is the only genuine PML cell line. Additional morphologic findings suggest that HL-60 cells represent a discrete stage of differentiation between late myeloblasts and promyelocytes, whereas PL-21 cells show distinct features associated with monocytic cells. Both the HL-60 cell line and the relatively newly reported promyelocytic cell line NB-4, established from a patient with APL (Lanotte et al. 1991), display bilineage potential and can be differentiated either towards the granulocytic or monocytic/macrophage lineages depending on the differentiation inducer used.

HL-60 has proved to be a very useful cell line as it can be induced to differentiate in vitro to a number of different cell types, including granulocytes (mainly neutrophils), monocytes and macrophages, depending on the differentiating agent used (Collins 1987). Neutrophils constitute the main cell type in an inflammatory reaction. In this chapter we will focus on experimental procedures to use HL-60 cells as a model for the study of human neutrophil differentiation and apoptosis. We will also detail some protocols concerning a number of important biochemical processes occurring in differentiating HL-60 cells and mature neutrophils. The HL-60 cell line has proved to be an invaluable tool in different areas of neutrophil biology and has been used for manifold purposes: (a) to provide insight into the regulatory mechanisms of normal neutrophil differentiation; (b) to examine new therapeutic approaches for leukemia treatment; (c) to provide a ready source of cDNA for cloning of important neutrophil genes and for studies of gene regulation; (d) to provide a ready source of mRNA for gene expression studies; (e) to perform prolonged labeling procedures for analysis of metabolic or macromolecule biosynthesis processes that are difficult to carry out in mature granulocytes.

Human neutrophils are short lived and nondividing terminal cells that die rapidly by apoptosis. In fact, once isolated from the peripheral blood, neutrophils must be used immediately for all kinds of assays as they rapidly undergo apoptosis. When cultured in RPMI-1640 containing 10% fetal calf serum (FCS), neutrophils start to undergo apoptosis after 6–12 h (Fig. 16.1), depending on the preparation. This process is even more rapid when cultured in serum-free medium. Due to their short life

Fig. 16.1. Time course of induction of apoptosis in human peripheral blood neutrophils cultured in 10 % FCS-containing RPMI-1640 medium for the times indicated (*lanes 2–6*). A 123-bp DNA ladder used as standard (*STD*) is shown in *lane 1*. DNA loaded in each lane was from 8×10^5 cells

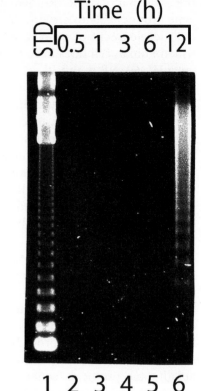

span, the use of HL-60 cells as a substitute for peripheral blood neutrophils is very rewarding. In this regard, it is very convenient to use HL-60 cells in experiments for which a prolonged cell labeling time is required. Two key advantages in using HL-60 cells, compared to their neutrophil counterparts, lie in the fact that HL-60 cells can be cultured and exhibit a very high macromolecule biosynthesis capacity, whereas peripheral blood mature neutrophils do not divide, die quickly by apoptosis and show very poor biosynthesis of macromolecules. Furthermore, HL-60 cells can be induced to differentiate towards neutrophils or to undergo apoptosis; thus, experiment planning is much easier. However, there are also some drawbacks, as the HL-60 cell model is incomplete or defective in some aspects. HL-60 promyelocytes do not contain all the components present in mature neutrophils, and HL-60 cells differentiated towards neutrophils (HL-60 neutrophils) do not behave exactly as mature neutrophils, differing in gene expression and in some biochemical processes (Lübbert et al. 1991; Mollinedo and Naranjo 1991; Mollinedo et al. 1991a). A major characteristic of human peripheral blood mature neutrophils is the presence of three main types of cytoplasmic granules, namely: tertiary or gelatinase-containing granules, which are readily mobilized to the cell surface upon cell activation; secondary or specific granules; and primary or azurophilic granules. HL-60 cells and HL-60 neutrophils contain azurophilic granules, but induction of neutrophil maturation of these cells is not accompanied by the normal acquisition of secondary and tertiary granules or by the expression of some neutrophil proteins. In addition, HL-60 neutrophils are not morphological replicas of normal neutrophils and their respective microscopic appearances differ. Nevertheless, HL-60 cells and HL-60 neutrophils share many functional characteristics with peripheral blood mature neutrophils, and therefore HL-60 neutrophils can be used as a model for the study of neutrophil functions and processes involving this cell type.

HL-60 cells carry an array of cell surface antigens characteristic of immature myeloid cells, are aneuploid showing 44 chromosomes, express *bcl-2* and *bax* (Han et al. 1996), and overexpress c-*myc*. This latter is due to an amplification of DNA sequences encompassing the c-*myc* gene (Collins and Groudine 1982), the expression of c-*myc* mRNA being about nine times higher in HL-60 cells than in a normal cell. Induction of HL-60 cell differentiation with dimethylsulfoxide (DMSO) towards the neutrophil lineage coincides with a decline in both c-*myc* mRNA (Westin et al. 1982) and AP-1 transcription factor activity (Mollinedo et al. 1993a). HL-60 cells do not express p53, as the p53 gene has been largely deleted (Wolf and Rotter 1985); N-*ras* is mutated in codon 61 in these cells (Murray et al. 1983).

HL-60 cells are predominantly promyelocytes, but in some cell cultures about 5 % of the cells exhibit spontaneous differentiation to morphologically more mature cells including myelocytes and metamyelocytes, as well as some, but very scarce, banded and segmented neutrophils. The addition to the HL-60 cell culture of certain compounds markedly increases this spontaneous differentiation, with most cells acquiring several morphological, functional, enzymatic and cell surface antigen characteristics of mature neutrophils. The most widely used neutrophil differentiating agents, as well as the most effective, are DMSO and RA. Many others have also been reported, such as: N,N-dimethylformamide, N-methylformamide, 5-aza-2'-deoxycytidine, tunicamycin, and tiazofurin. However, caution must be taken, as some of the previously reported differentiating agents, e.g. tunicamycin, induce mainly apoptosis in these cells (Perez-Sala and Mollinedo 1995).

In the following sections, we will describe some technical aspects regarding growth conditions, induction of differentiation towards neutrophils, freezing and thawing procedures concerning HL-60 cells, as well as some protocols that allow the analysis of selected processes important to the study of inflammatory cells, and which can be examined using HL-60 cells as a model. These latter protocols include: (a) respiratory burst activity; (b) induction of apoptosis; (c) activation of transcription factors; (d) phospholipase D stimulation. These four activities cover four processes of pivotal importance in inflammation, dealing with crucial biochemical, physiological and signaling processes occurring in human neutrophil biology.

Growth of HL-60 Cells in Serum-Containing and Serum-Free Medium

The usual culture medium for HL-60 cells is RPMI-1640, 10% heat-inactivated FCS, 2 mM glutamine, 100 U/ml penicillin, 24 µg/ml gentamicin. Serum heat inactivation is achieved by incubation of serum at 56 °C for 30–45 min. HL-60 cells continuously proliferate in suspension at 37 °C in a humidified atmosphere of air/CO_2 (19:1), with a doubling time of about 20–30 h, depending on the subline. Cell surface expression of transferrin and insulin receptors is critical to the proliferative capacity of HL-60 cells. Thus, HL-60 cells can be grown in serum-free medium supplemented only with insulin and transferrin (Breitman et al. 1980). The commercially available ITS medium is a convenient serum-free medium that can be prepared easily as follows: an ITS solution (made up of 25 mg insulin, 25 mg transferrin, 25 µg sodium selenite previously dissolved in 10 ml phosphate buffered saline (PBS) and filtered through a 0.22 µm sterile filter) is added in a proportion of 1:500 to a RPMI-1640 culture medium containing 0.4% bovine serum albumin (BSA) (previously sterilized by filtration; for convenience, filter an aliquot of the RPMI-1640 medium containing the total amount of BSA already dissolved) and antibiotics as above. Thus, the final concentrations of the ITS serum-free medium are as follows: 5 µg/ml insulin, 5 µg/ml transferrin, 5 ng/ml sodium selenite in RPMI-1640, containing 0.4% (w/v) BSA, 2 mM glutamine, 100 U/ml penicillin, 24 µg/ml gentamicin. Growth of HL-60 cells in ITS medium is practically identical to that in serum-containing medium (Fig. 16.2A). In both cases, HL-60 cells grow logarithmically to a maximum of about $2.7–3.1 \times 10^6$ cells/ml before leveling off (Fig. 16.2A).

HL-60 cells undergo apoptosis when a cell concentration of over 10^6 cells/ml is reached in serum-containing medium (Fig. 16.2B). Thus, cau-

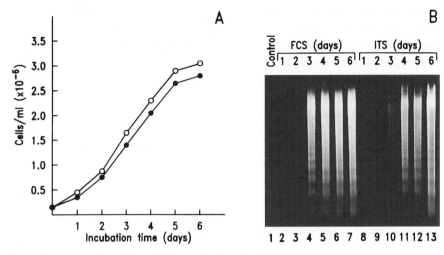

Fig. 16.2.A Growth of HL-60 cells in both 10 % FCS-containing (*closed circles*) and ITS serum-free (*open circles*) RPMI-1640 medium. **B** Induction of apoptosis during growth of HL-60 cells in both 10 % FCS-containing and ITS serum-free RPMI-1640 medium. Analyzed samples correspond to those shown in **A.** DNA loaded in each lane was from 8×10^5 cells

tion must be taken during HL-60 cell culture, and it is important to maintain these cells at a concentration between 1 and 6×10^5 cells/ml in order to eliminate the appearance of deleterious effects, such as apoptosis. If HL-60 cells are overgrown, most of them must be thrown away and the remaining ones highly diluted – at a concentration of about 1×10^5 cells/ml or lower – in order to get rid of the apoptotic cells, which are irreversibly programmed to cell death, and to let the healthy ones grow and initiate a new progeny of HL-60 cells. At very low cell concentrations there is a lag before the onset of growth. When cultured in ITS serum-free medium, HL-60 cells initiate to undergo apoptosis at a somewhat higher cell concentration than cells cultured in the presence of 10 % FCS (Fig. 16.2B). However the onset of apoptosis takes place at a similar cell concentration in HL-60 cell cultures in both ITS serum-free or serum-containing media, and a cell concentration of over 1×10^6 cells/ml must be avoided to maintain healthy cells (Fig. 16.2B). When HL-60 cells are cultured in RPMI-1640 medium in the absence of serum or ITS, the cells undergo apoptosis after about 6 h of culture.

The fact that HL-60 cells cultured in ITS serum-free medium behave similarly to those grown in 10 % FCS medium is important as it allows the study of differentiation and activation of HL-60 cells in the absence of undefined or known serum inhibitors or enhancers for a particular process to be examined. It also allows the effects of certain compounds that have to be added in the absence of serum (to avoid sequestration by serum constituents) to be analyzed.

Neutrophil Differentiation of HL-60 Cells in Both Serum-Containing and Serum-Free Medium

As previously indicated, HL-60 cells can be induced to differentiate to cells showing many of the functions of mature neutrophils and can be used as models for their normal neutrophil counterparts. Thus, HL-60 cells differentiated towards neutrophils (HL-60 neutrophils) are able to display a wide array of the functional characteristics of normal human peripheral blood mature neutrophils, including phagocytosis, lysosomal enzyme release, expression of several cell surface leukocyte antigens, chemotaxis, and generation of superoxide anions. Nevertheless, the HL-60 neutrophils are neither morphologically nor biochemically identical to human peripheral blood mature neutrophils, being deficient in several characteristics of their normal cell counterparts. Thus, HL-60 cells predominantly promyelocytes, and when induced to different, towards the neutrophilic lineage most cells are driven morphologica_/ into the more differentiated myelocyte/metamyelocyte stage with a small proportion of segmented or hypersegmented mature neutrophils. Moreover, as mentioned above, HL-60 neutrophils contain only one of the three main cytoplasmic granules present in human peripheral blood neutrophils. Thus, HL-60 neutrophils lack the rapidly mobilizable secondary and tertiary granules present in human neutrophils and contain only the primary or azurophilic granules. Furthermore, HL-60 neutrophils are deficient in several granule components (e.g., lactoferrin, transcobalamin I), whereas other granule components (e.g., CD11b, gelatinase) are synthesized, but not stored, in the same subcellular particulate as in the normal neutrophil counterpart.

The most efficient agents to induce HL-60 cells to differentiate towards neutrophils are 1.3 % (v/v) DMSO and 1 µM RA. The stock RA solution is prepared at 1 mM in ethanol and stored at −20 °C until use. The time required for maximum HL-60 neutrophil differentiation following RA or DMSO treatment has generally been considered to be 5 or 7 days, respectively. Nevertheless, just as human peripheral blood neutrophils die rapidly by apoptosis, HL-60 cells undergo apoptosis following differentiation towards neutrophils with DMSO and RA. As shown in Fig. 16.3, cells

Fig. 16.3. Time course of induction of apoptosis in HL-60 cells treated with 1.3 % dimethylsulfoxide (DMSO) (*lanes 3–9*) or 1 µM RA (*lanes 10–14*) for the indicated times in 10 % FCS-containing RPMI-1640 medium. *Lane 1* 123-bp DNA ladder used as standard (*STD*); *lane 2* control untreated HL-60 cells. DNA loaded in each lane was from 8×10^5 cells

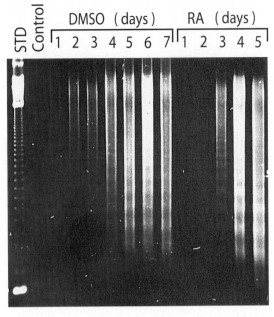

start to undergo apoptosis after 4 and 3 days of treatment with DMSO and RA, respectively. Thus, in prolonged incubations a mixture of apoptotic and differentiating cells is present in the culture of DMSO- or RA-treated HL-60 cells. Overexpression of *Bcl-2* or *Bcl-X$_L$* by gene transfer prevents apoptosis following neutrophil differentiation of HL-60 cells.

A rapid and easy way to monitor neutrophil differentiation depends on the appearance of two typical markers of human neutrophils: (a) cell surface expression of the leukocyte antigen CD11b, and (b) capacity to generate superoxide anion, measured by nitroblue tetrazolium (NBT) reduction (see below). Undifferentiated HL-60 cells lack CD11b and are unable to reduce NBT as these cells fail to generate superoxide anion. The leukocyte antigen CD11b is the α subunit of the CD11b/CD18 antigen, also referred to as Mo1 or Mac1. This antigen represents a major member of a leukocyte family of integrins made up of three glycoproteins sharing the same β subunit (CD18) and differing in the nature of the α subunits (CD11a, CD11b and CD11c). As these proteins always act in dimers, CD11b and Mo1 are used interchangeably. CD11b antigen plays important roles in cell adhesion and phagocytosis and is located in both plasma membrane and tertiary granules in human mature neutrophils,

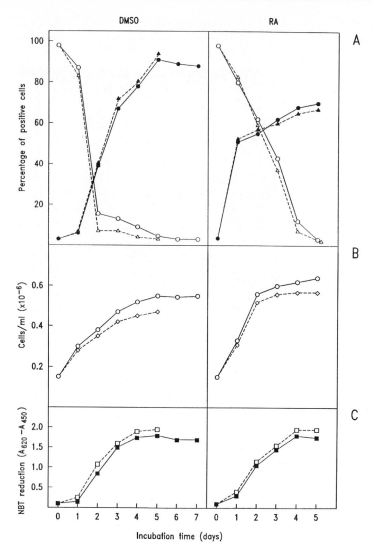

Fig. 16.4A–C. Analysis of HL-60 differentiation towards the neutrophil lineage induced by 1.3% dimethylsulfoxide (DMSO) (*left panel*) or 1 µM all-*trans* retinoic acid (RA) (*right panel*) treatment for the indicated times. **A** Changes in CD11b and CD71 cell surface expression in HL-60 cells treated with DMSO or RA and cultured in 10% FCS-containing (*closed circles* CD11b; *open circles* CD71) or ITS serum-free (*closed triangles* CD11b; *open triangles* CD71) RPMI-1640 medium. **B** Growth of cells cultured in 10% FCS-containing (*open circles*) or ITS serum-free (*open diamonds*) RPMI-1640 medium and treated with DMSO or RA as indicated. **C** Appearance of the capacity to generate superoxide anion assessed by NBT reduction assay during HL-60 neutrophil differentiation induced by DMSO or RA and cultured in 10% FCS-containing (*closed squares*) or ITS serum-free (*open squares*) RPMI-1640 medium

its cell surface expression being up-regulated upon cell activation (Lacal et al. 1988). However, in HL-60 neutrophils, lacking tertiary granules, this leukocyte antigen is present only in the plasma membrane and it is not up-regulated following cell activation. The NBT reduction assay represents a rapid way to measure the capacity of generating superoxide anion.

As shown in Fig. 16.4, DMSO- or RA-induced neutrophil HL-60 cell differentiation is characterized by an increase in CD11b and a decrease in CD71 cell surface expression, as well as by an increase in NBT reduction (Fig. 16.4C). CD71 leukocyte antigen is the transferrin receptor and its expression is linked to the proliferative capacity of the cells. Proliferation of HL-60 cells is greatly diminished by treatment with DMSO or RA (see Figs. 16.2A and 16.4B), and, concomitantly, the cell surface expression of CD71 is almost absent after 2–4 days of treatment (Fig. 16.4A). Maximum CD11b expression and NBT reduction is obtained after 4–5 days of RA or DMSO treatment. However, taking into account the onset of the apoptotic response (Fig. 16.3), it is advisable to work with HL-60 neutrophils obtained following a 3-day treatment with either DMSO or RA, as these treated HL-60 cells exhibit most of the functional characteristics of mature neutrophils and do not undergo extensive apoptosis, as compared to HL-60 cells treated with DMSO for 7 days or with RA for 5 days (Figs. 16.3, 16.4). The neutrophil differentiation pattern is practically identical when cells are grown in 10 % FCS-containing medium or in ITS serum-free medium (Fig. 16.4). Thus, analysis of the effects of certain compounds on HL-60 cell differentiation can be examined in the absence of serum constituents.

DMSO- and RA-induced HL-60 differentiation can be modulated by several agents. Thus, dexamethasone is able to modulate the functional responses of granulocytic differentiating HL-60 cells, increasing and accelerating significantly the functional maturation of RA-induced HL-60 neutrophils (Collado-Escobar and Mollinedo 1994).

Freezing and Thawing of HL-60 Cells

HL-60 cells are frozen and thawn as most cell lines. About $3-5 \times 10^6$ cells are pelleted by centrifugation at 1200 rpm for 8 min and then resuspended in 1 ml of FCS containing 10 % (v/v) DMSO in polypropylene freezing tubes. The tubes are placed first at $-20\,°C$ for 24 h, then at $-80\,°C$ for 24 h, and finally transferred to a liquid-nitrogen container at $-170\,°C$.

Frozen HL-60 cells are reconstituted by incubating the freezing vial in a 37 °C water bath to thaw the cells as quickly as possible. Immediately after thawing, the cells are transferred to a sterile tube containing 10 ml of cell culture medium and centrifuged for 8 min at 1200 rpm to pellet cells. The cells are washed once more with 10 ml of cell culture medium and then the cell pellet is resuspended in 6–8 ml of 10 % FCS-containing RPMI-1640 cell culture medium. The next day, cells are pelleted again to remove cell debris (if necessary, wash again as above), and the cells are resuspended in 6–8 ml of 10 % FCS-containing RPMI-1640 cell culture medium and grown as described above.

The general equipment required for cell culture consists in: laminar flow hood, incubator with a humidified atmosphere of air/CO_2 (19:1), inverted microscope, table centrifuge, water bath and liquid-nitrogen container.

16.1
Detection of Cell Surface Antigens by Immunofluorescence Flow Cytometry

As shown above, neutrophil differentiation of HL-60 cells results in changes in the cell surface expression of some leukocyte antigens, such as an increase in CD11b and a decrease in CD71. Thus, immunofluorescence flow cytometry is required to analyze the cell surface expression of leukocyte antigens (Mollinedo et al. 1992) as a simple and rapid way to monitor HL-60 cell differentiation. This technique is based on the fluorescent staining of cells with antibodies by incubation of the cells with appropriately diluted antibody to allow antigenic binding to take place. The appropriate dilution of an antibody is determined by titration against a fixed number of cells. Saturation of binding sites is obtained when an increase in antibody concentration produces little or no increase in fluorescence intensity measured in the flow cytometer. Two main basic fluorescent staining procedures can be used: (1) direct fluorochrome conjugation to a primary antibody, and (2) indirect staining by a fluorochrome-conjugated second antibody reacting with a nonfluorescent primary antibody. In the first procedure, the sample can be washed and processed directly on the flow cytometer (thus omitting steps 3 and 4 of the protocol given below). The use of the second, indirect, antibody technique allows analysis of a wide array of different primary antibodies with one standard reagent. Commercial fluorochrome conjugated polyclonal antisera to be used as fluorescent secondary antibodies are readily available. Thus, a fluorescein (FITC)conjugated goat (or rabbit) anti-mouse immunoglobulin is routinely used for the detection of mouse monoclonal antibodies, as described below. The secondary antibody must be titrated against a fixed number of cells incubated with a saturating concentration of primary antibody to determine the working solution. The amounts of the primary and secondary antibodies used in the protocol described below usually achieve saturating concentrations for both antibodies.

Materials

- Flow cytometer
- Tabletop centrifuge

equipment

- Phosphate buffered saline (PBS), pH 7.4
- Monoclonal antibody-containing hybridoma culture: Monoclonal antibodies raised against leukocyte antigens expressed at the cell surface during myeloid differentiation, such as CD11b, are used to easily monitor myeloid differentiation.
- P3X63 IgG1 myeloma culture supernatant: negative control for unspecific binding
- Fluorescein (FITC)-conjugated F(ab')$_2$ goat (or rabbit) anti-mouse immunoglobulins. This reagent and its dilutions must be protected from light.
- Formaldehyde solution: 37 %
- Propidium iodide: 1–2 mg/ml in PBS

reagents

Procedure

1. Incubate 100 µl of cell suspension (2–4×10^5 total cells) with 65 µl of monoclonal antibody-containing hybridoma culture supernatant for 30 min at 4 °C. As a negative control use 65 µl of P3X63 myeloma. Bring the final volume to 200 µl with PBS.

flow cytometry

Note: The whole procedure must be performed at 4 °C to avoid endocytosis or other metabolic processes that can affect antigen cell surface expression.

2. Add 1 ml of PBS to each tube and centrifuge for 8 min at 1200 rpm. Remove and discard supernatant by gentle suction with a Pasteur pipette attached to a vacuum line.

3. Resuspend cells in 100 µl of FITC-conjugated F(ab')$_2$ goat anti-mouse immunoglobulin, previously diluted 1/50 in PBS. Bring the final volume to 200 µl with PBS. Protect samples from light with aluminum foil. Incubate for 30 min at 4 °C.

4. Repeat step 2.

5. Resuspend cells in 300 µl of PBS and measure in a flow cytometer in either log (sharper peaks, normally used to calculate percentages of positive cells) or linear (more spread and flattened peaks, used to quantitate differences in the total amount of leukocyte antigens expressed at the cell surface) scales. Specific linear fluoresence is obtained as the mean fluorescence value obtained by subtracting control mean fluorescence in which the first monoclonal antibody is substituted by the myeloma P3X63 immunoglobulin. Cells must be fixed if flow cytometry analysis is postponed for several hours or days. To fix the cells, resuspend them in 300 µl containing 1.5 % formaldehyde. Seal the tubes with parafilm and wrap them in aluminum foil for light protection; store at 4 °C. Cells fixed and stored in this manner can be analyzed by flow cytometry even 2 weeks later.

6. When there is a significant percentage of dead cells in the preparation to be analyzed, it is convenient to discriminate between alive and dead cells. This can be done by scatter measurements in the flow cytometer and by selecting the appropriate gate for living cells, the dead cells and debris appearing near the origin in a forward vs side scatter plot. However, a more accurate and complete exclusion of dead cells can be accomplished by the use of propidium iodide, which stains DNA in dead cells. At step 5, resuspend unfixed cells in 300 µl of PBS and, a few minutes before flow cytometry analysis, add the appropriate amount of a stock solution of propidium iodide (1–2 mg/ml in PBS) so that the final concentration is 20–30 µg/ml. FITC and propidium iodide can be detected simultaneously in the flow cytometer.

Results

A typical result showing the appearance of CD11b-positive cells during neutrophil differentiation is shown in Fig. 16.4. As discussed above, it is sometimes necessary to exclude dead cells through the use of propidium iodide. This fluorochrome is unable to pass through intact cell membranes. However, when the cell dies, the membrane becomes permeable and propidium iodide is able to enter and stain DNA by intercalating with double-stranded nucleic acids. The resulting red fluorescence is proportional to the nucleic acid content. By this method it is possible to discriminate between live nonfluorescent cells and dead fluorescent cells. The 488 nm line of the argon ion laser produces near maximal excitation of FITC (excitation maximum 495 nm) and propidium iodide (excitation maxima 495 and 342 nm). As the emission maxima for FITC (520 nm) and propidium iodide (639 nm) are clearly different, both fluorochromes can be detected simultaneously in the flow cytometer. Thus cells with negative propidium iodide red fluorescence gate (live cells) can be analyzed further for antigen expression (FITC, green fluorescence).

16.2
Respiratory Burst Activity Measured by Nitroblue Tetrazolium Reduction

Nitroblue tetrazolium (NBT) is an electron acceptor that has been widely used to detect indirectly the generation of superoxide anions by granulocytes and macrophages during a respiratory burst (Baehner and Nathan 1968), as outlined in the following equation: $NBT + O_2^- \rightarrow Formazan + O_2$. NBT is a yellow and water soluble dye that is reduced in the presence of superoxide anion to an insoluble black-blue formazan deposit. When cells are incubated in the presence of NBT, the formazan precipitates intracellularly at the sites of superoxide anion production. Thus, formazan can be observed microscopically, thereby providing a qualitative means to identify superoxide anion producing cells, or can be quantitated spectrophotometrically by cell disruption and formazan solubilization using potassium hydroxide (KOH) and DMSO (Rook et al. 1985, Perez-Sala and Mollinedo 1995) as described below.

Materials

equipment
- Spectrophotometer
- Microfuge
- Water bath
- 1.5 ml Eppendorf tubes

reagents
- HEPES-glucose buffer: 10 mM HEPES; 150 mM NaCl; 5 mM KOH; 1.2 mM $MgCl_2$; 1.3 mM $CaCl_2$; 5.5 mM D-glucose, pH 7.5
- 0.1 % (w/v) NBT in HEPES-glucose buffer (freshly prepared for each experiment). As it is somewhat difficult to dissolve the NBT, agitate the mixture for about 1 h in a constant rotatory agitator to dissolve completely. The NBT solution must be covered with aluminum foil, as this product is light sensitive.
- Stimulus, e.g., 4β-phorbol 12-myristate 13-acetate (PMA): 1 mg/ml dissolved in DMSO as stock solution
- Methanol: 70 % in distilled water
- KOH: 2 M in distilled water.
- DMSO

Procedure

NBT reduction
1. Resuspend 8×10^5 cells in 400 µl of HEPES-glucose buffer or serum-free cell culture medium mixed with 400 µl of 0.1 % NBT; activate with PMA as stimulus (0.05–0.2 µg/ml). This reaction must be run in duplicate in 1.5 ml Eppendorf tubes and in a final volume of 800 µl.

2. Incubate at 37 °C for 15 min.

3. Stop the reaction by placing the tubes on ice, and spin for 8 min at 1200 rpm to sediment cells. This step can be also accomplished by centrifugation for 24 s in a microfuge to pellet down the cells. In this latter case, be careful to prevent cell disruption.

4. Remove and discard supernatant by gentle suction (through a Pasteur pipette attached to a vacuum line), and add 250 µl of 70 % methanol to the pellet to wash the cells and to remove unreduced NBT.

5. Spin again for 8 min at 1200 rpm (or for 24 s in a microfuge) and discard the supernatant to remove the unreacted NBT.

6. Add 500 μl of 2 M KOH to lyse the cells; incubate overnight at room temperature. **Note:** The tubes can be stored on the bench for even longer periods of time at the convenience of the researcher. During this step, the formazan is also being dissolved.

7. The next day add 600 μl DMSO and vortex vigorously to dissolve the formazan. An intense blue color develops when superoxide anion has been generated.

8. When all formazan has been dissolved, read on a spectrophotometer at A_{620}–A_{450} (substraction mode, absorbance at 620 nm–absorbance at 450 nm). The developed color is stable over a long period of time, so it is not necessary to read the samples immediately.

16.3
Analysis of Apoptosis Through Visualization of Fragmented DNA in Agarose Gels

Apoptosis is a type of cell death that plays a key role in development and growth of normal adult tissues, being responsible for deletion of cells in normal tissues. Morphologically, apoptosis involves chromatin condensation, nuclear fragmentation and cellular breakdown in the characteristic membrane-enclosed apoptotic bodies, containing well-preserved organelles, which are phagocytosed by nearby resident cells. However, the main biochemical feature of apoptosis is the double-strand cleavage of nuclear DNA at the linker regions between nucleosomes, leading to the generation of the characteristic internucleosomal DNA fragmentation in multiples of 180–200 bp. As fragmented DNA is released into the cytosol from the nucleus, this DNA fragmentation can be easily assessed by isolating and electrophoresing the cytosolic DNA fragments following the method described below (Greenblatt and Elias 1992; Mollinedo et al. 1993b).

Materials

- Tabletop centrifuge   equipment
- Water bath
- Microfuge
- Speed-vacuum concentrator or an oven
- Power supply and agarose gel electrophoresis cuvette
- UV transilluminator and a photographic Polaroid system

reagents
- Sterile Eppendorf tubes: Autoclave
- Phosphate buffered saline (PBS), pH 7.4: Autoclave and store at room temperature or at 4 °C.
- Hypotonic detergent buffer: 10 mM Tris/HCl, pH 7.5–8.0; 1 mM EDTA; 0.2 % Triton X-100. Autoclave and store at room temperature or at 4 °C.
- TE buffer: 10 mM Tris/HCl, 1 mM EDTA, pH 8.0. To prepare 100 ml: 1 ml Tris/HCl 1 M, pH 8.0; 0.5 ml EDTA 0.2 M, pH 8.0. Bring final volume to 100 ml with deionized (or distilled) H_2O. Autoclave and store at room temperature or at 4 °C.
- TBE buffer (5×): 0.45 M Tris-HCl;, 0.45 M boric acid; 10 mM EDTA, pH 8.0. To prepare 1 l: 54 g Tris base; 27.5 g boric acid; 20 ml 0.5 M EDTA, pH 8.0. Bring final volume to 1 l with deionized (or distilled) H_2O. Autoclave and store at room temperature. **Note:** A precipitate usually forms when concentrated TBE solutions are stored at room temperature. Concentrated solutions of 5×TBE are kept in glass bottles at room temperature and any batches that develop a precipitate are discarded. At any case, if this happens, the precipitate can be dissolved by heating at high temperature or autoclaving and used as electrophoresis buffer. For agarose gel electrophoresis dilute the 5×TBE solution 1:10 in distilled water; as a working solution 0.5×TBE provides adequate buffering power for agarose gel electrophoresis.
- RNase A (20 mg/ml, in sterile H_2O). To eliminate any putative DNase activity, the RNase A solution is prepared by heating at 100 °C for 10 min. Then, the RNase A solution is aliquoted and stored at −20 °C.
- Proteinase K: 20 mg/ml in sterile H_2O. Store at −20 °C.
- Sodium dodecyl sulfate (SDS): 20 % (w/v). Wear a mask when weighing SDS and clean the balance as well as nearby area carefully after use, because the SDS is a fine powder and disperses easily. Dissolve in sterile H_2O. There is no need to sterilize SDS solutions. Store at room temperature. SDS solutions precipitate at 4 °C.
- Phenol/Tris: Phenol saturated with 0.1 M Tris/HCl, pH 7.5–8.0. Store at 4 °C.
- Phenol/chloroform/isoamyl alcohol (25:24:1). Store at 4 °C.
- 5 M NaCl: Autoclave and store at room temperature or at 4 °C.
- Ethanol: Store at −20 °C.
- Ethanol: 70 % (diluted in sterile H_2O). Store at −20 °C.
- Gel loading buffer: 0.25 % bromophenol blue; 0.25 % xylene cyanol; 30 % glycerol in sterile water. Store at 4 °C.
- Ethidium bromide: 10 mg/ml in H_2O

▧ Procedure

1. Spin for 8 min at 1200 rpm to pellet the cells ($1.5-2.5 \times 10^6$ cells/sample). Remove the supernatant through gentle suction by a Pasteur pipette attached to a vacuum line. Resuspend the cells in 1 ml PBS and transfer to an Eppendorf tube. Pellet the cells by centrifugation at 1200 rpm for 8 min and discard the supernatant. This centrifugation can be replaced by a rapid centrifugation of 24 s in a microfuge, but caution must be taken in order to prevent cell disruption.

2. Lyse the cells with 200 µl of hypotonic detergent buffer for 30 min at 4 °C. The cells will swell in the hypotonic buffer and be gently disrupted due to the presence of the detergent in the buffer. **Note:** This is a critical step in which only the plasma membrane is disrupted, but the nuclei are preserved. Cells are gently resuspended and allowed to swell in the hypotonic detergent buffer at 4 °C.

3. Centrifuge in a microfuge for 20 min at 4 °C.

Note: Steps 2 and 3 are the most critical, as it is here that separation of cytoplasm from intact nuclei is carried out. The pellet contains nuclei and subcellular organelles whereas the supernatant contains the DNA released into the cytosol due to DNA fragmentation.

4. Collect supernatant (about 200 µl) and transfer to a new sterile Eppendorf tube. At this step, the supernatant can be frozen at -20 °C and step 5 carried out at any time.

Note: An estimation of the success of the ongoing experiment can be roughly obtained at this step by checking the viscosity of the pellet with a yellow micropipet tip. When apoptosis occurs at a high level, the viscosity of the pellet is low, due to the presence of lower amounts of chromosomal DNA. When apoptosis is absent, the viscosity of the pellet is very high, due to the presence of higher amounts of intact DNA.

5. Add 300 µg/ml RNase A (3 µl from a stock solution of 20 mg/ml RNase A) to the supernatant saved in step 4, and incubate for 40–60 min at 37 °C. Then, add 0.5 % SDS (5 µl of 20 % SDS) and 200 µg/ml proteinase K (2 µl of 20 mg/ml proteinase K) and incubate for additional 40–60 min at 37 °C.

Note: These times can be increased at the convenience of the researcher. Protein denaturation by SDS facilitates the proteolytic action of the highly active proteinase K. The RNA and proteins present in the sample are totally degraded in this step.

visualization of DNA in agarose gels

6. Extract twice with 1 volume (200 µl) of phenol/Tris, pH 7.5–8.0. Each extraction consists of the addition of 200 µl of phenol/Tris, followed by vigorous vortexing and centrifugation in a microfuge for 3 min at room temperature. Then, the aqueous upper phase (about 200 µl) is transferred to another Eppendorf tube and the process is repeated once more. An alternative way to proceed is to leave the mixture of fragmented DNA with phenol/Tris overnight (or at least 4 h) in a continuous rotatory mixer at 4 °C. In this case, one phenolization step is enough. **Note:** The nucleic acids (DNA) are isolated in this step in the aqueous phase.

7. Extract once with 1 volume (200 µl) of phenol:chloroform:isoamyl alcohol (25:24:1). Proceed as above. The aqueous upper phase, containing the DNA, is transferred to another sterile Eppendorf tube.

8. Add 300 mM NaCl (final concentration) to the aqueous phase (13 µl of 5 M NaCl) and then add 2 volumes (400 µl) of ice-cold ethanol (or 1 volume -200 µl- of isopropanol). Mix the solution and store overnight at −20 °C to allow the precipitate of DNA to form. An incubation of about 14 h at −20 °C is enough to precipitate completely the DNA. This time can be drastically shortened if precipitation is performed at −80 °C.

9. Microfuge for 20 min at 4 °C. Discard the supernatant by gentle suction. The pellet represents precipitated DNA and salt. Gently add 1 ml of ice-cold 70 % ethanol to the pellet (do not resuspend) in order to remove precipitated salt, and microfuge again for 20 min at 4 °C. Discard the supernatant by gentle suction.

10. Dry the pellet in a speed-vacuum concentrator or in an oven at 40 °C. Resuspend the pellet in TE buffer (usually 15–20 µl of TE buffer for 1.5–2.5×10^6 cells) and store at −20 °C until use.

11. Add gel loading buffer (1:5) and run equal volumes (usually 5–10 µl) of each sample on a 1 % (w/v) agarose gel in 0.5×TBE. The running buffer is 0.5×TBE. The gel is run at 80 V for the first 10 min and then at 110 V. The run takes about 1 h and 45 min for a medium size gel (about 14 cm long). The gel is then stained with 8 µl of a stock solution of ethidium bromide (10 mg/ml in H_2O) in 100 ml H_2O for about 7 min and then destained in H_2O.

12. The stained gel is observed in a UV transilluminator and a Polaroid photograph is taken.

Results

Typical results of apoptosis are shown in Figs. 16.1, 16.2B and 16.3. The characteristic 180–200 bp laddering of internucleosomal DNA breakdown occurring during apoptosis is observed.

16.4
Study of DNA-Protein Interactions: Preparation of Nuclear Extracts and Mobility Shift DNA-Binding Assay Using Gel Electrophoresis

DNA-binding proteins are becoming increasingly important in the study of molecular processes occurring in an inflammatory response. Herein, we describe a simple and rapid protocol to prepare nuclear extracts from a small number of cells, ready to be used for mobility shift assays. Basically, cells are collected and resuspended in a hypotonic buffer in order to allow the cells to swell. Afterwards, cells are gently disrupted with the aid of a detergent, while preserving the integrity of the nuclei. Addition of a high-salt buffer releases soluble proteins from the nuclei (without lysing the nuclei), and the nuclear extract is collected as a supernatant by centrifugation (Schreiber et al. 1989; Hattori et al. 1990b; Mollinedo et al. 1993a). The protocol described here includes the use of buffers containing chelating agents (EDTA, EGTA), nuclei-stabilizing compounds (spermidine, spermine), protease inhibitors (PMSF, aprotinin, leupeptin) and a phosphatase inhibitor (Na_2MoO_4), which increases the yield of phosphorylated DNA-binding proteins. A rapid and easy protocol to examine DNA-protein interactions by mobility shift DNA-binding assay using polyacrylamide gel electrophoresis is also described (Hattori et al. 1990a; Mollinedo et al. 1993a). Mobility shift assays, also named gel retardation assays, are based on the fact that specific protein binding to an end-labeled DNA fragment retards the mobility of the DNA fragment during gel electrophoresis, leading to discrete bands corresponding to individual protein-DNA complexes.

Materials

- Microfuge   equipment
- Power supply and vertical polyacrylamide gel electrophoresis (20×20 cm gel; 1.5 mm thick Teflon spacers) assembly used for protein gel electrophoresis
- Gel dryer

reagents
- Sterile Eppendorf tubes: Autoclave
- TBE buffer (5×): 0.45 M Tris-HCl; 0.45 M boric acid; 10 mM EDTA, pH 8.0. To prepare 1 l: 54 g Tris base; 27.5 g boric acid; 20 ml 0.5 M EDTA, pH 8.0. Complete up to 1 l with deionized (or distilled) H_2O.
- Phosphate buffered saline (PBS), pH 7.4: Autoclave and store at room temperature or at 4 °C.
- Nonidet NP-40: 10 % in sterile H_2O
- Loading buffer: 10 mM HEPES, pH 7.6; 10 % glycerol; 0.01 % bromophenol blue. Stored at 4 °C.
- Buffer A: Prepared as follows (for 50 ml):

Reagent	Amount and preparation
10 mM HEPES, pH 7.6	500 μl 1 M HEPES, pH 7.6: Autoclave and store at room temperature.
10 mM KCl	250 μl 2 M KCl: Autoclave and store at room temperature.
0.1 mM EDTA	25 μl 0.2 M EDTA: Autoclave and store at room temperature.
0.1 mM EGTA	50 μl 0.1 M EGTA: Autoclave and store at room temperature.
0.75 mM spermidine	37.5 μl 1 M spermidine: Filter, store at −20 °C.
0.15 mM spermine	15 μl 0.5 M spermine: Filter, store at −20 °C.
Sterile H_2O	49.12 ml

This solution is stored at 4 °C and the components described below are added immediately before use:

Reagent (to add for a 10 ml solution)	Amount and preparation
1 mM DTT	10 μl 1 M DTT: Filter, store at −20 °C.
0.5 mM PMSF	50 μl 100 mM PMSF: Store at −20 °C.
1×Protease inhibitor (PI)	5 μl 2000×PI (4 mg/ml aprotinin and 4 mg/ml leupeptin in sterile water): Store at −20 °C.
10 mM Na_2MoO_4	100 μl 1 M Na_2MoO_4: Store at room temperature.

– Nuclear lysis buffer: Prepared as follows (for 50 ml):

Reagent	Amount and preparation
20 mM HEPES pH 7.6	1 ml 1 M HEPES, pH 7.6
0.4 M NaCl	4 ml 5 M NaCl (autoclave, store at room temperature)
1 mM EDTA	250 µl 0.2 M EDTA
1 mM EGTA	500 µl 0.1 M EGTA
Sterile H_2O	44.25 ml

This solution is stored at 4 °C and the components described below are added immediately before use:

Reagent (to add for a 1 ml solution)	Amount and preparation
1 mM DTT	1 µl 1 M DTT: Filter, store at −20 °C.
0.5 mM PMSF	5 µl 100 mM PMSF: Store at −20 °C.
2× Protease inhibitor (PI)	1 µl 2000× PI (4 mg/ml aprotinin and 4 mg/ml leupeptin in sterile water): Store at −20 °C.
10 mM Na_2MoO_4	10 µl 1 M Na_2MoO_4: Store at room temperature.

Note: All the indicated solutions are prepared in sterile water, except PMSF. The PMSF solution must be prepared in isopropanol as it is inactivated in aqueous solutions, the rate of inactivation increasing with pH and with temperature. The half-life of an aqueous solution of PMSF is about 15–30 min at pH 8.0. This means that PMSF must be added immediately before use.

Note: The protease inhibitor mixture can be prepared also with additional protease inhibitors (bestatin, pepstatin A, antipain, etc).

▧ Procedure

1. Usually $2-4 \times 10^6$ cells are collected and washed twice with PBS. In the last wash, cells are transfered to 1.5 ml Eppendorf tubes and pelleted by centrifugation at 1200 rpm for 8 min.

2. This step and the following steps must be performed on ice: The cell pellet is resuspended in 400 µl of ice-cold buffer A. First, cells are

preparation of nuclear extracts

resuspended by gentle pipetting with a yellow tip in 200 µl of buffer A; then the volume is completed with additional 200 µl of buffer A.

3. The cells are allowed to swell on ice for 15 min in hypotonic buffer A, after which 25 µl of a 10 % solution of Nonidet NP-40 are added, and the tube is vigorously vortexed for 10 s. **Note:** By this method, cells are disrupted gently, but nuclei remain intact.

4. Centrifuge in microfuge for 30 s. Supernatant containing cytoplasm and RNA is removed and discarded by gentle suction with a Pasteur pipette attached to a vacuum line. Then, 50 µl ice-cold nuclear lysis buffer are added carefully to the nuclear pellet. Do not resuspend the nuclear pellet as the nuclei can be disrupted and DNA can be damaged by pipetting. Obtaining a good nuclear extract preparation could be hampered if nuclei disruption occurs.

5. Shake moderately the tubes in a vortex at about 1/3–1/2 speed. The tubes are ideally shaken continuously at 4 °C for 15 min in a rack using a shaking platform. If this is not possible, the tubes are maintained on ice and are vortexed consecutively, one by one (about 10–20 s each at moderate speed) over 15 min. **Note:** By this process the soluble nuclear proteins are extracted (due to the high salt concentration of the nuclear lysis buffer), but the nuclei are preserved, with the DNA remaining intact.

6. The nuclear extract is centrifuged for 5 min in a microfuge at 4 °C, and the supernatant (approx. 50 µl) is transfered to a new Eppendorf tube. Store immediately at −80 °C. **Note:** Be careful not to take DNA, which is clearly visible by its viscosity and stickiness. If DNA happens to get in the yellow tip, change the tip and continue collecting the clear and nonviscous supernatant.

7. Protein values are calculated in 5 µl of each sample by the Bradford method using a BioRad kit and bovine serum albumin (BSA) as the standard. From $2-4 \times 10^6$ cells, the usual protein concentrations of the nuclear extracts are in the range of 1–2 mg/ml.

preparation of the mobility shift gel

1. Use a 20×20 cm gel and 1.5-mm-thick Teflon spacers. The glass plates must be carefully cleaned and sealed with plastic tape and four big clamps to avoid leakage during casting.

2. Choose a 4 or 5 % polyacrylamide gel depending on the size of the probe. For most cases a 4.5 % polyacrylamide gel is convenient. This is prepared as follows:

Reagent	Amount (ml)
30 % (w/v) acrylamide/bisacrylamide (29/1)	7.8
H_2O	41.7
10×TBE	2.1
10 % (w/v) ammonium persulfate	0.35
TEMED*	0.052
The final volume must be 52 ml.	

* Before addition of TEMED, remove 1 ml of gel solution into a tube. Add 6 µl TEMED to this 1 ml and pour it quickly into the glass plate assembly along the sides to seal it. Allow this sealing solution to move around the assembly until it is polymerized. Proceed quickly as polymerization occurs rapidly. If the sealing with plastic tape is properly placed, this latter step can be omitted.

3. Place the glass plate with the four big clamps on the bench at an angle of about 20°.

4. Pour the gel and insert a comb (12-well white Teflon comb, 0.8 cm broad teeth, 2 cm deep) to create about 1 cm deep lane pockets. Allow the gel to polymerize for about 1 h.

5. Pre-run the gel in a 0.4×TBE buffer at a constant voltage of 200 V until current has reached 22 mA. At 200 V the gel initially draws a current of about 35 mA and drops to 22 mA after about 1 h 30 min.

6. Change buffer, discard used pre-run buffer and use fresh 0.4×TBE buffer for the run.

1. In a 1.5 ml Eppendorf tube, mix the following reagents:

DNA-protein binding reaction

H_2O	Nuclear lysis buffer + nuclear extract	poly(dI-dC) (5 mg/ml)	MgCl$_2$ (120 mM)	Labeled probe
5 µl	10 µl	2 µl	1 µl	2 µl

Note: The usual way to proceed is to prepare a cocktail with H_2O + poly(dI-dC) + MgCl$_2$, and then pour 8 µl into each sample. About 4–6 µg of nuclear extract protein is used. The poly(dI-dC) solution is prepared in sterile water. All the reaction components are kept on ice. The amount of poly(dI-dC) in the reaction mixture can be modified between 4 and

10 µg in order to avoid unspecific DNA-protein binding without affecting specific binding. This must be checked for each particular DNA-protein binding reaction. The nuclear extract (4–6 µg protein) is always added last.

Note: Probes are previously labeled with ^{32}P through labeling the ends of the DNA fragment using Klenow fragment (5' overhang) or polynucleotide kinase (blunt ends or 5' overhang). Chase the Klenow reaction with cold dNTP mix. Probes are prediluted in such a way that the required amount of labeled probe (0.45 pmol of ^{32}P-labeled double-stranded oligonucleotide) is added in 2 µl.

2. Spin briefly in a microfuge to mix all the components. Incubate the mixture for 15 min on ice.

3. Add 1.5–3 µl gel loading buffer and load 12–14 µl of the mixture per lane.

electrophoresis

1. Run the gel at a constant voltage of 200 V and at room temperature until bromophenol blue dye is about 3.5–4.5 cm from the bottom (about 1 h 30 min). This depends on the size of the probe. **Note:** A probe of 22 bp on a 4.5 % gel runs about 3.5 cm ahead of the bromphenol blue dye. From time to time it is convenient to mix buffers from the two compartments by pipetting. This helps to keep the pH constant as the DNA-protein complexes are very dependent on the pH.

2. Carefully remove the side spacers by using a spatula. Slowly pry the glass plates apart, allowing air to enter between the gel and the glass plate. The gel should remain attached to only one plate. Lay the glass plate containing the attached gel on the bench with the gel facing up. Carefully place a piece of dry Whatman 3MM paper on the damp gel. The paper should be large enough to form a border (1–2 cm) around the gel and small enough to fit on the gel dryer. The damp gel should stick immediately and firmly to the Whatman 3MM paper. Do not attempt to move the Whatman 3MM paper once contact has been made with the gel.

3. Place two additional sheets of 3MM paper beneath the 3MM paper with the gel and two additional pieces of filter paper on the drying surface of the gel dryer to avoid radioactive contamination of the gel dryer. Place a sheet of transparent plastic wrap on the top of the gel attached to the 3MM paper. Close the lid of the gel dryer, and apply vacuum so that the lid makes a tight seal around the gel. Heat at 80 °C to speed up the drying process. Under these conditions, the gel should be dried after about 1 h.

4. Remove the gel, which is now firmly attached to the Whatman 3MM paper, from the gel dryer. Remove the transparent plastic wrap, and place a new one on instead. Expose the gel for autoradiography overnight. Sometimes longer exposures are needed to get a good signal.

Results

A typical result is shown in Fig. 16.5. To check the specificity of a particular transcription factor in the DNA-binding reaction, nuclear extracts are preincubated with 1 µl of specific antibody for 12 h at 4 °C; the samples are then subjected to the gel retardation assay as described above.

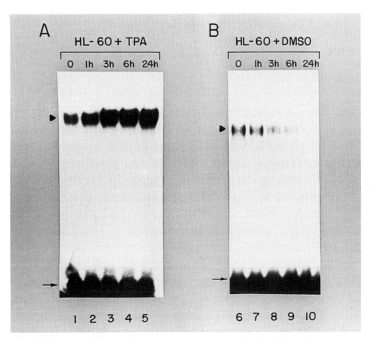

Fig. 16.5. Gel retardation analysis of AP-1 transcription factor binding to the AP-1 sequence of nuclear extracts from untreated HL-60 cells (*lanes 1, 6*) and from HL-60 cells treated with 20 ng/ml 4β-phorbol 12-myristate 13-acetate (PMA or TPA) (inducer of macrophage differentiation) or with 1.3 % DMSO (inducer of neutrophil differentiation) for the indicated times. The *arrowhead* points to the specific binding complex to the AP-1 site. The *arrow* points to the free oligoprobe containing the consensus AP-1 site. (Reproduced from Mollinedo et al. 1993a, by permission)

This should block binding of the specific transcription factor to the DNA sequence, due to binding of the antibody to a site on the transcription factor essential for DNA binding, or affect the mobility of the DNA-protein complex resulting in a supershift of the complex to a slower-migrating position, due to binding of the antibody to a non-essential site for DNA binding on the transcription factor. Controls preincubated with 1 µl of preimmune serum for 12 h at 4 °C must be run in parallel. Further evidences for specific binding of a determined transcription factor to a DNA sequence include: (a) prevention of the DNA-protein binding by addition of a 50-fold molar excess of an unlabeled competitor oligonucleotide containing the sequence of interest (e.g., unlabeled probe); (b) no effect on the DNA-binding response by addition of oligonucleotides containing sequences distinct from the one studied.

16.5
Phospholipase D Activity in Intact Cells

Activation of phospholipase D constitutes a major signaling pathway in mammalian cell stimulation and plays an important role in the signaling processes occurring in human neutrophils and in human myeloid cell lines. Phospholipase D activation results in the production of phosphatidic acid which behaves as a second messenger and can be dephosphorylated by phosphatidate phosphohydrolase to produce the well known second messenger 1,2-diacylglycerol. As a matter of fact, this phosphatidic acid dephosphorylation represents the major pathway for diacylglycerol formation in most cell types. The main substrate for phospholipase D is phosphatidylcholine. Thus, the enzyme phospholipase D hydolyzes physiologically phospholipids generating phosphatidic acid and free choline. Nevertheless, in the presence of a primary alcohol such as ethanol, the enzyme catalyzes a transphosphatidylation reaction that transfers the phosphatidyl moiety to the alcohol, yielding a phosphatidylalcohol (i.e., phosphatidylethanol), which is metabolically stable (Yang et al. 1967). This transphosphatidylation reaction is a unique feature of phospholipase D and constitutes the basis for a sensitive and specific assay to detect phospholipase D activity (Bocckino et al. 1987; Mollinedo et al. 1994).

Materials

- Fume hood
- N$_2$ stream
- Sonicator
- Water bath
- Thin layer chromatography (TLC) tanks
- TLC plates. Silica gel 60 precoated plates 20×20 cm, thickness 0.25 mm

Reagents

- [^{3}H] or [^{14}C]palmitic acid: 1 μCi/ml in toluene/ethanol, 1:1
- HEPES/glucose buffer: 150 mM NaCl; 5 mM KOH; 1.2 mM MgCl$_2$; 1.3 mM CaCl$_2$; 5.5 mM glucose; 10 mM HEPES, pH 7.5
- Chloroform/50 mM HCl: 1:1
- Chloroform/methanol: 1:2
- Chloroform/methanol: 2:1
- Ethyl acetate/iso-octane/acetic acid: 9:5:2
- Scintillation liquid
- β-radioactivity counter.

Procedure

1. About 3–5×10^6 cells/experimental point are required for this assay. Cells are labeled by adding 3–5 μCi/ml [^{3}H] or [^{14}C]palmitic acid overnight. Take the required volume of a solution of [^{3}H] or [^{14}C]palmitic acid (dissolved in toluene/ethanol, 1:1) for a final concentration of 3–5 μCi/ml of cell suspension. Dry under a gentle N$_2$ stream (low pressure) to evaporate the toluene/ethanol. The most convenient procedure to evaporate small volumes of solutions in small tubes is to direct a stream of nitrogen onto the surface of the solution, kept, if possible, in a water bath (approx. 30 °C). The residue is then dissolved in about 0.5 ml of 0.1 N NaOH. After vigorous vortexing and sonication for 10 min, the mixture is neutralized by addition of about 0.5 ml 0.1 N HCl. The substrate suspension is then diluted with fresh culture medium and added to the cell cultures (about 3–5×10^5 cells/ml) to a final concentration of about 5 μCi [^{3}H]palmitic acid/ml of cell culture. Take into account that about 3–5×10^6 cells/experimental point are required.

2. Incubate overnight in a cell culture incubator at 37 °C with a humidified atmosphere of air/CO$_2$ (19:1).

generation of phosphatidylethanol by activated cells

1. Collect cells by centrifugation at 1200 rpm for 8 min. Wash cells once with HEPES/glucose buffer (about 20 ml). Sediment cells by centrifugation at 1200 rpm for 8 min. The cell pellet is resuspended at a concentration of about 15–25×10^6 cells/ml in HEPES/glucose buffer. Take into account that each experimental point requires 200 µl of cell suspension.

2. Aliquot the cells (aliquots of 200 µl for each experimental point) in borosilicate tubes (or Eppendorf tubes). Add 5 µl of ethanol/ml of cell suspension 5 min before adding the stimuli. Proceed in a similar way with control tubes without ethanol.

3. Each tube is incubated with the appropriate stimulus at 37 °C for 30 min or for the required time in a water bath. Unstimulated control cells are always run in parallel. The final reaction volume after addition of the stimulus is 200 µl and contains 0.5 % ethanol.

4. Reactions are stopped by addition of 3.75 volumes (750 µl) of chloroform/methanol (1:2). Vortex vigorously.

lipid extraction

1. Lipids are extracted by subsequent addition of 1.25 volumes (250 µl) of chloroform and 1.25 volumes (250 µl) of 50 mM HCl. Vortex vigorously after each addition. It is advisable to add cold phosphatidylethanol (PEt) as carrier (5 µg/point) before vortexing.

2. Centrifuge at 2000 rpm for 5 min to separate phases (Fig. 16.6A).

3. Carefully collect the organic lower phase into clean tubes. Be careful to not take aqueous phase. **Note:** The experiment can be stopped at this point if the TLC plates are going to be run another day. If so, gas the tubes with N$_2$ stream, and cap and wrap them up well with parafilm in order to avoid condensation. Keep them at −20 °C until use.

4. Dry the above organic phase under a stream of nitrogen (low pressure). If the tubes have been maintained at −20 °C, leave them well capped at room temperature for a while before starting this step. A small amount of methanol can be added to the tube to clear the solution (particularly, if some aqueous phase is present) and to evaporate the solution faster.

Fig. 16.6A, B. Lipid separation. **A** Phase separation following chloroform/HCl extraction. **B** Phosphatidylethanol separation by thin-layer chromatography

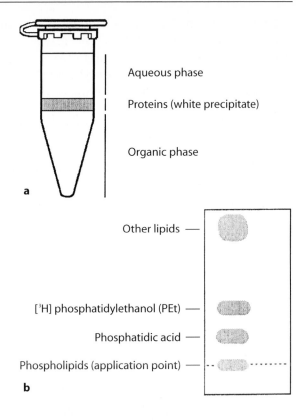

Aqueous phase

Proteins (white precipitate)

Organic phase

a

Other lipids ──

[³H] phosphatidylethanol (PEt) ──

Phosphatidic acid ──

Phospholipids (application point) ──

b

phosphatidyl-ethanol separation by TLC

1. The dried organic samples are dissolved in 20–40 µl of chloroform/methanol (2:1) and samples are applied to a TLC plate to separate the PEt formed. Apply cold PEt (10 µg) as standard in one lane. On each plate, up to ten samples can be applied, including the standards. Samples can be applied as a row (0.5–1 cm long each) of overlapping small spots using P20 or P200 pipette yellow tips. Allow the previous spots to dry before further application of the same sample. This process can be accelerated using a hair dryer. It is advisable to apply samples 1.5 cm apart from each other and 2 cm from bottom of the plate.

2. Plates are developed with a solvent system of ethyl acetate/iso-octane/acetic acid (9:5:2). For 100 ml (usual volume used in a TLC tank): 56.25 ml ethyl acetate, 31.25 ml iso-octane, 12.5 ml acetic acid. The TLC tank is saturated for at least 90 min with this mixture prior to plate development. Tanks can be lined with filter paper to aid saturating the tank with solvent vapor. Place 1 or 2 plates in the tank and

allow the solvent to ascend to the top of the plate. It takes about 60–90 min to run the plate. The plates are then removed and left under a fume hood for about 15–25 min to allow them to dry. The drying period can be shortened using a hair dryer. Stain the chromatogram after solvent evaporation.

3. Lipid spots are visualized by exposure of the TLC plates to iodine vapors and identified by comparison to standards run in parallel in the same plate. Spots are marked with a pencil and, after the plates are completely dried, spots are scraped off and counted for radioactivity. [^{3}H]-labeled compounds can be visualized by autoradiography with the use of a surface autoradiography enhancer (EN^{3}HANCE spray, Du Pont), and compared to the location of standards previously ascertained by iodine staining. Autoradiograms are developed after 6–8 days of exposure. If [^{14}C]palmitic acid is used, plates can be

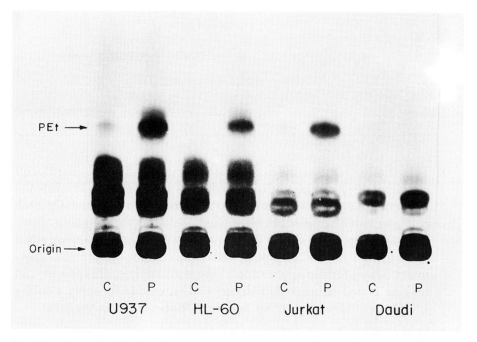

Fig. 16.7. Phosphatidylethanol production in distinct myeloid (U937, HL-60) and lymphoid (Jurkat, Daudi) human cell lines. Cells were incubated with either vehicle (0.5% DMSO) (*C*) or 100 ng/ml 4β-phorbol 12-myristate 13-acetate (PMA) (*P*) for 30 min at 37 °C in the presence of 0.5% ethanol. The *arrows* point to the phosphatidylethanol (PEt) and the origin of application. (Reproduced from Mollinedo et al. 1994, by permission)

exposed for autoradiography without the use of autoradiography enhancers, and autoradiograms are developed in about 2 days. The spots corresponding to PEt and to phospholipids, remaining at the origin of application, are scraped from the plates and radioactivity is determined by liquid scintillation counting.

Note: The developing solvent system used can be reused several times, until significant evaporation is observed. Using fresh developing systems in the TLC tank, the phosphatidic acid usually migrates 4–4.5 cm from the application point; the PEt migrates 5.5–6.5 cm from the application point. In old developing systems, the mobilities are usually higher. Due to this variability, it is necessary to run standards in parallel on the same plate. A typical lipid separation following this method is shown in Fig. 16.6B.

Results

A typical result can be seen in Fig. 16.7, which shows an autoradiogram with [^{3}H]PEt formation.

References

Baehner RL, Nathan DG (1968) Quantitative nitroblue tetrazolium test in chronic granulomatous disease. N Engl J Med 278:971–976

Bocckino SB, Wilson PB, Exton JH (1987) Ca^{2+}-mobilizing hormones elicit phosphatidylethanol accumulation via phospholipase D activation. FEBS Lett 225:201–204

Breitman TR, Collins SJ, Keene BR (1980) Replacement of serum by insulin and transferrin supports growth and differentiation of the human promyelocytic cell line HL-60. Exp Cell Res 126:494–498

Collado-Escobar D, Mollinedo F (1994) Dexamethasone modifies the functional responses of the granulocytic differentiating HL-60 cells. Biochem J 299:553–559

Collins SJ (1987) The HL-60 promyelocytic leukemia cell line: proliferation, differentiation, and cellular oncogene expression. Blood 70:1233–1244

Collins SJ, Gallo RC, Gallagher RE (1977) Continuous growth and differentiation of human myeloid leukaemic cells in suspension culture. Nature 270:347–349

Collins SJ, Groudine M (1982) Amplification of endogenous *myc*-related DNA sequences in a human myeloid leukemia cell line. Nature 298:679–681

Dalton WT, Ahearn MJ, McCredie KB, Freireich EJ, Stass SA, Trujillo JM (1988) HL-60 cell line was derived from a patient with FAB-M2 and not FAB-M3. Blood 71:242–247

de Thé H, Chomienne C, Lanotte M, Degos L, Dejean A (1990) The t(15;17) translocation of acute promyelocytic leukemia fuses the retinoic acid receptor alpha gene to a novel transcribed locus. Nature 347:558–561.

Drexler HG, Quentmeier H, MacLeod RA, Uphoff CC, Hu ZB (1995) Leukemia cell lines: in vitro models for the study of acute promyelocytic leukemia. Leuk Res 19:681–691

Greenblatt M, Elias L (1992) The type B receptor for tumor necrosis factor-alpha mediates DNA fragmentation in HL-60 and U937 cells and differentiation in HL-60 cells. Blood 5:1339–1346

Han Z, Chatterjee D, Early J, Pantazis P, Hendrickson EA, Wyche JH (1996) Isolation and characterization of an apoptosis-resistant variant of human leukemia HL-60 cells that has switched expression from Bcl-2 to Bcl-xL. Cancer Res 56:1621–1628

Hattori M, Abraham LJ, Northemann W, Fey GH (1990a) Acute-phase reaction induces a specific complex between hepatic nuclear proteins and the interleukin 6 response element of the rat alpha$_2$-macroglobulin gene. Proc Natl Acad Sci USA 87:2364–2368

Hattori M, Tugores A, Veloz L, Karin M, Brenner DA (1990b) Laboratory methods. A simplified method for the preparation of transcriptionally active liver nuclear extracts. DNA and Cell Biol 9:777–781

Huang ME, Ye YC, Chen SR, Chai JR, Lu J-X, Zhoa, L., Gu LJ, Wang ZY (1988) Use of all-trans retionic in the treatment of acute promyelocytic leukemia. Blood 72:567–572

Kakizuka A, Miller WH, Umesono K, Warrell RP, Frankel SR, Murty VVVS, Dmitrovsky E, Evans RM (1991) Chromosomal translocation t(15;17) in human acute promyelocytic leukemia fuses RARalpha with a novel putative transcription factor, PML. Cell 66:663–674

Lacal P, Pulido R, Sánchez-Madrid F, Mollinedo F (1988) Intracellular location of T200 and Mo1 glycoproteins in human neutrophils. J Biol Chem 263:9946–9951

Lanotte M, Martin-Thouvenin V, Najman S, Balerini P, Valensi F, Berger R (1991) NB4, a maturation inducible cell line with t(15;17) marker isolated from a human acute promyelocytic leukemia (M3). Blood 77:1080–1086

Lübbert M, Herrmann F, Koeffler HP (1991) Expression and regulation of myeloid-specific genes in normal and leukemic myeloid cells. Blood 77:909–924

Mollinedo F, Naranjo JR (1991) Uncoupled changes in the expression of the *jun* family members during myeloid cell differentiation. Eur J Biochem 200:483–486

Mollinedo F, Vaquerizo MJ, Naranjo JR (1991a) Expression of c-*jun*, *jun* B and *jun* D proto-oncogenes in human peripheral-blood granulocytes. Biochem J 273:477–479

Mollinedo F, Burgaleta C, Velasco G, Arroyo AG, Acevedo A, Barasoain I (1992) Enhancement of human neutrophil functions by a monoclonal antibody directed against a 19-kDa antigen. J Immunol 149:323–330

Mollinedo F, Gajate C, Tugores A, Flores I, Naranjo JR (1993a) Differences in expression of transcription factor AP-1 in human promyelocytic HL-60 cells during differentiation towards macrophages versus granulocytes. Biochem J 294:137–144

Mollinedo F, Martinez-Dalmau R, Modolell M (1993b) Early and selective induction of apoptosis in human leukemic cells by the alkyl-lysophospholipid ET-18-OCH$_3$. Biochem Biophys Res Commun 192:603–609

Mollinedo F, Gajate C, Flores I (1994) Involvement of phospholipase D in the activation of transcription factor AP-1 in human T lymphoid Jurkat cells. J Immunol 153:2457–2469

Murray MJ, Cunningham JM, Parada LF, Dautry F, Lebowitz P, Weinberg RA (1983) The HL-60 transforming sequence: a *ras* oncogene coexisting with altered *myc* genes in hematopoietic tumors. Cell 33:749–757

Perez-Sala D, Mollinedo F (1995) Inhibition of N-linked glycosylation induces early apoptosis in human promyelocytic HL-60 cells. J Cell Physiol 163:523–531

Rook GAW, Steele J, Umar S, Dockrell HM (1985) A simple method for the solubilisation of reduced NBT, and its use as a colorimetric assay for activation of human macrophages by gamma-interferon. J Immun Methods 82:161–167

Schreiber E, Matthias P, Müller MM, Schaffner W (1989) Rapid detection of octamer binding proteins with 'mini-extracts', prepared from a small number of cells. Nucleic Acids Res 17:6419

Westin EH, Wong-Staal F, Gelman EP, Dalla-Favera R, Papas TS, Lautenberger JA, Gallo RC (1982) Expression of cellular homologues of retroviral *onc* genes in human hematopoietic cells. Proc Natl Acad Sci USA 79:2490–2494

Wolf D, Rotter V (1985) Major deletions in the gene encoding p53 tumor antigen cause lack of p53 expression in HL60 cells. Proc Natl Acad Sci USA 82:790–794

Yang SF, Freer S, Benson AA (1967) Transphosphatidylation by phospholipase D. J Biol Chem 242:477–488

Chapter 17

Cell Culture Models for Polarized Epithelial Monolayers

ALAN W. BAIRD* and ENA S. PROSSER

▦ Introduction

Epithelial Culture Conditions

Epithelial cells may be grown on tissue culture plastic or on glass. That cellular polarity is maintained even under these conditions is evinced by, for example, the formation of "domes" (Fig. 17.1) which arise as blister-like structures following transepithelial fluid movement. Indeed, transepithelial fluid and electrolyte transport may be investigated by analyzing the fluid contained within such domes (Lifschitz 1986). However, the normal, physiological environment of epithelial cells is one in which the basolateral domain of each cell is in contact with extracellular matrix which is shared with other nonepithelial cells. It is from this side of the epithelial sheet that nutrition of growing epithelial cells is normally obtained. By using techniques which approach the normal extracellular environment, a wide range of epithelial cells has been cultured with selection of individual phenotype.

Epithelial cell lines (Gstraunthaler 1988) or primary cultures (Evans et al. 1994; Van Scott et al. 1986) are widely used. Applications include studies of cellular polarity, transport and permeability, trafficking, tumor growth and differentiation, motility and chemotaxis, wound healing, angiogenesis, inflammation, cell-cell and cell-matrix interactions. Criteria for evaluating and selecting a relevant cell line (or lines) are governed by the proposed application. This chapter provides a summary review of selected techniques and applications in order to introduce the reader to the field. Specialized texts and references are listed at the end of this chapter.

---

* *Correspondence to* Alan W. Baird, Department of Pharmacology, University College Dublin, Foster Avenue, Blackrock, Dublin, Ireland; phone+353–1–7061557; fax +353–1–2692749; e-mail abaird@macollamh.ucd.ie

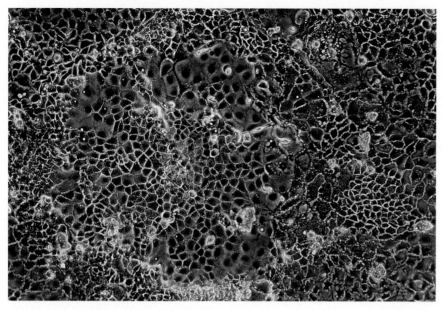

Fig. 17.1. Low power micrographs show dome-like structures are formed when CaCo-2 cells (derived from a human colonic tumor) are cultured to confluence on plastic tissue culture plates. Use of the fine focus control on the microscope clearly indicates the three-dimensional nature of the dome

General Points

Protocols for maintenance and passage of epithelial cells in culture have much in common with other culture methods for adherent cells, whether these cells are primary, passaged lines or transformed lines. Thus general conditions which are covered elsewhere in this book apply here. In addition there are a number of features which are of particular relevance to epithelial cells:

- Make a number of copies of cell lines at as low a passage number as possible. Do not start experiments without back-ups.

- Since epithelial cells may differentiate in culture, work within a set number of passages (ten passages is a useful rule of thumb).

- Maintain the cultures as close to confluence as possible. Subculture at split ratios of 1:2 or 1:3.

- Since cellular junctions between epithelial cells are calcium dependent, incubate monolayers with EDTA (0.02 % in PBS) before splitting with trypsin-EDTA.

- Some epithelial cell lines require vigorous trypsinization during subculture.

Materials

Microporous Supports

A number of porous membranes are commercially available for cell culture applications. Principal suppliers of cell culture porous inserts include Millipore, Nunc, Becton Dickinson and Costar. Sheets of cellulose esters, collagen, polyvinyldifluoride, polycarbonate and polyethylene terephthalate are available in a variety of configurations for different applications. Variable specifications of each of the membrane supports include pore size (0.4–12 µM), insert diameter (compatible with 6-, 12- or 24-well culture plates or with purpose-built Ussing chambers), requirement for matrix coating (which may be cell type-specific) and transparency (transparent membranes permit continuous monitoring of cell growth using phase contrast microscopy). Typical configurations are shown in Fig. 17.2. Growth of epithelial cells on a porous support provides, in addition to a more physiological platform for attachment, the

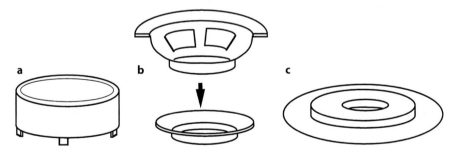

Fig. 17.2a–c. Various geometric arrangements for filter/support assemblies. **a** Porous membranes mounted in a polystyrene housing with projections which permit the assembly to stand in individual wells of multi-well tissue culture plates. **b** Porous membranes suspended from a collar-like structure. This example shows a detachable insert for which custom-built Ussing chambers are available. **c** The cell growth surface is enclosed by a silicone rubber gasket which is firmly glued to the filter

opportunity to deliver media to the basolateral domain (Pitt et al. 1987). Feeder layers or cocultured cells may be grown in the plastic wells in which the inserts are suspended. Nutrients, growth factors and drugs may be applied selectively to the apical or basolateral surfaces in order to determine sidedness of an effect. Finally, by having access to each side of the confluent membrane, the barrier nature of a monolayer may be examined (Lewis et al. 1995).

As with tissue culture plastic, porous supports may be coated with matrix. A large number of natural and synthetic matrices are commercially available including collagen, laminin, fibronectin, gelatin, Matrigel (Becton Dickinson, Bedford, MA)and poly-lysine. Optimal conditions for the growth of an individual cell type, if not published, must be empirically determined.

17.1
Simple Protocol for Preparing Porous Supports

▧ Materials

- Silicon elastomer (Silgard 184; Dow Corning): prepared following the manufacturer's instructions to achieve a cured 2–3 mm thick uniform sheet (20 ml of solution in a standard 90 mm petri dish)
- Silicon adhesive (Silastic 734)
- Membrane filters (HAWP 0500, Millipore)
- γ-Irradiation: used to sterilize the inserts
- Single cell suspensions ($0.5-1\times10^6$ cells/200 µl), obtained by trypsinization of confluent monolayers of epithelial cells grown on tissue culture plastic, are seeded at a 1:1 ratio of areas (e.g., a 25 cm^2 flask will provide enough cells for 24 inserts).
- Rat tail collagen type I (0.25 % in 0.2 % acetic acid)
- Tissue culture media.
- 6-well culture dishes or petri dishes

▧ Procedure

1. Two different diameter cork borers (7 and 20 mm) are used to prepare silicon elastomer O-rings of a suitable size which are attached to the filter, taking care to avoid spreading adhesive in the inner well.

preparing porous supports

2. The inner chamber is coated with collagen and exposed to UV light in a tissue culture hood overnight. The UV exposure cross-links the collagen and can also be used to sterilize the assembly (sterilization can alternatively be carried out by γ-irradiation or autoclaving if the materials are not heat-sensitive).

3. Cell suspensions (10^5–10^6/cm^2) are added to the inner aspect of each insert.

4. Seeded monolayers on the supports are floated in individual wells of 6-well tissue culture dishes or, simply, on culture media in petri dishes. Other configurations of insert stand in the wells on supports or are suspended (Fig. 17.2). Cells are fed by changing both apical and basolateral solutions.

Results

- Cells grow in a polarized fashion with basolateral attachment to the filter or matrix (see below).

- As an alternative, filters can be glued to one open end of a rigid polycarbonate ring to form a cup with the filter as a base and the ring as the side (Handler et al. 1979; Dharmsathaphorn and Madara 1990).

- Virtually any growing area or geometry may be constructed. Typically inserts are made to be compatible with 6- or 12-well culture plates and/or to be used with other laboratory equipment such as, for example, Ussing chambers (see below).

- Growth curves for cells grown on porous supports are obtained by similar procedures as those used for adherent cells grown on tissue culture plastic, for example, tritiated thymidine incorporation or DNA levels.

- Porous supports are ideal for experiments in which the filter separates two populations of cells which share growth media. Examples include coculture or maintenance of cells with feeder layers.

- Many commercially available supports are extremely expensive. It is sensible to consider recycling cell culture inserts (Bell and Quinton 1990). Cells are removed by sonication in several changes of deionized water, examined by phase contrast microscopy to ensure that the filter remains intact and that cells have been removed. Inserts are then ste-

rilized by immersion in 70 % ethanol for 48 h, washed in sterile deionized water and air dried in a tissue culture laminar flow hood. Coating
with substrates, if necessary, may be carried out as described above
before plating the recycled inserts with cells. It is crucial to compare
the characteristics of new and recycled inserts.

Troubleshooting

- The optimal growth conditions of cells on porous supports is dependent upon pore size and density, support material, matrix as well as
  upon media components. It is important to establish these conditions
  if they have not already been described in the literature.

- Although nutrient requirements for cells grown on porous supports
  are the same as those used on nonporous surfaces, growth rates of
  cells on the permeable supports are often slower than those of the
  same cells grown on tissue culture plastic.

- Care must be taken with "home-made" porous supports that the
  adhesive seal is complete. This may be done simply by examining the
  capacity of the cup to contain a fluid such as sterile water.

17.2
Establishing Levels of Confluence of Epithelial Monolayers by
Structural Assessment – Morphological Studies Using Light Microscopy

Epithelia grow naturally as polarized sheets with basolateral attachments
to a matrix and apical membrane in contact with a fluid (or air) environment. In culture, epithelia grow to retain this asymmetry. Examination
of cellular and/or monolayer structure can be used in a number of ways
in order to assess polarity and/or monolayer integrity.

With translucent membranes, a reasonably good estimate of confluence can be obtained by examining the growing cells using phase contrast microscopy. An alternative method (Hughson and Hirt 1996) which
is useful for monolayers cultured on nontranslucent supports is to rinse
the filters with PBS, expose the cells to a 1 % solution of toluidine blue O
(Sigma T3260) for1 h. Wash the monolayer in 70 % ethanol to clear
excess dye. If the monolayer is confluent, a uniformly stained appearance
is observed.

In general, morphological techniques applied to intact tissues may also be applied to cells grown on permeable supports.

Materials

- Glutaraldehyde (3 % in phosphate buffered saline)
- Giemsa stain stock solution (0.4 %) (Sigma G3032)
- Phosphate buffered saline
- Deionized water
- Absolute ethanol
- Xylene
- Mounting medium (Sigma 1000–4)
- Glass slides and coverslips

Procedure

1. Remove growth medium and gently wash cells one time in PBS.

2. Fix cells for 30 min at room temperature in glutaraldehyde, then wash three times with PBS.

3. Stain the monolayer with Giemsa. Incubation times vary for each cell type.

4. Rinse with deionized water to remove excess stain.

5. Remove the permeable support from the insert with a sharp blade or cork borer.

6. Dehydrate the cells and membrane through serial concentrations of ethanol (30, 50, 70, 90 % and absolute) before clearing in xylene.

7. The membrane may be mounted on a slide with the cells uppermost and protected with a coverslip placed upon the monolayer with a drop of mounting medium. (Alternatively the monolayer/filter assembly may be mounted in paraffin wax or resin in order to cut semi-thin transverse sections).

Results

- Transverse section micrographs may be used to determine the gross morphological characteristics of epithelial monlayers. It is interesting to compare the appearance of cells grown on porous supports with that of an otherwise identical batch of cells.

- In order to increase sampling size, monolayers can be rolled up using a rubber policeman to achieve a scroll structure which is then processed normally and examined by transverse section.

Troubleshooting

- Sections must be carefully orientated for cutting sections perpendicular to the plane of the monolayer.

- Some porous supports may be difficult to section cleanly with a microtome. If difficulties are found, individual manufacturers should be consulted for technical information.

17.3
Morphological Studies Using Transmission Electron Microscopy

See also Falcon Technical Bulletin 406 (Becton Dickinson).

Materials

- Cell monolayers grown to confluence on permeable support (organic polymeric supports are easier to section than inorganic supports).
- Phosphate buffer: 0.05 M
- PBS, pH 7.4
- Glutaraldehyde: 2.5 % in PBS, pH 6.8
- Osmium tetroxide: 1 % in 0.05 M potassium phosphate; sonicated for 5 min immediately before use)
- Graded alcohols
- Resin embedding medium (e.g., Epon resin)
- Propylene oxide resin transition fluid
- Lead citrate

Procedure

1. Aspirate medium off confluent monolayers.

2. Wash twice (gently to avoid dislodging cells) with PBS (pH 7.4). Carefully blot tissues to remove excess fluid.

3. Immerse monolayers upon their supports in glutaraldehyde (2.5 %). Leave overnight in a refrigerator.

4. Rinse the monolayer in PBS six times to remove all the glutaraldehyde and fix in osmium tetroxide (2 %) for 2 h at room temperature.

5. Dehydrate by passing through graded alcohols; 30 % for 10 min, 70 % for 15 min and 100 % for 10 min. Finally immerse in fresh 100 % alcohol for 20 min.

6. Transfer the filter to 100 % propylene oxide transitional fluid to fully infiltrate the cells for 15 min. Discard the solution carefully and repeat the infiltration procedure for a further 30 min.

7. Embed in low viscosity resin following the manufacturers' instructions. Align specimens in order to facilitate transverse sectioning.

8. Transfer the monolayer to a 1:1 mix of resin and propylene oxide for 2–3 h and then to 100 % resin solution for a further 2–3 h.

9. Transfer to pans/capsules for embedding. At this stage, orient the sample so that the apical side is uppermost. Place the sample under vacuum to remove air bubbles from the specimen.

10. Mount in grids and stain for 10 min with lead citrate and wash several times in distilled water.

11. Sectioning is carried out using either glass or diamond knives. Diamond knives are more robust for sectioning through inorganic support materials.

Results

- Transmission electron micrographs permit visualization of distinctive epithelial cell characteristics. For example, apical microvilli as well as intercellular junctions can normally be observed.

- Electron microscopy using immunocytochemistry may be used to study surface polarity of epithelial cells.

- Agarose (2 % in water) may be added after the osmium fixation step to prevent dislocation of the monolayer from the filter support.

■ Troubleshooting

- Due to the complex nature of the technology and the large number of variables in each of the steps involved, electron microscopic techniques for any application must be established for each application.

- Some protocols suggest using acetone/alcohol mixtures to dehydrate the samples; however, this may damage the filter supports (particularly true when supports are coated with extracellular matrix).

- Make up osmium tetroxide and propylene oxide immediately before use and always in a fume hood.

- Many chemicals used in this procedure are expensive and toxic. Only make up the amounts which are required and dispose of waste material safely.

17.4
Barrier Function of Epithelial Monolayers

Since epithelial monolayers are formed from sheets of cells joined together by tight junctions, the epithelium can, and does, act as a barrier between the two compartments which it separates. Barrier function of epithelial cells cultured on porous supports can be assessed in a number of ways, as can modification of barrier function (permeability) by macromolecules (Lewis et al. 1995). The degree of confluence of a monolayer of epithelial cells can be determined in terms of resistance to passive, transepithelial movement of marker molecules. A range of markers (with molecular weight) includes lucifer yellow (453), polyethylene glycol, inulin (5000), horseradish peroxidase (40 000) and dextran (70 000). Detection systems for the marker molecules range from colorimetric through enzymatic activity to radioactivity.

Materials

- Epithelial monolayers grown to confluence on Costar Snapwells
- Hank's balanced salt solution (HBSS) with 10 mM glucose (or Krebs-Hensleit solution)
- [^{14}C]mannitol (55 mCi/mmol; New England Nuclear)
- Ecoscint (National Diagnostics, Atlanta, GA) scintillation fluid
- Scintillation counter
- Orbital shaker
- Incubator (at 37 °C)
- Oxygen (95 %) carbon dioxide (5 %) gas mixture

Procedure

Assessment of Cell Monolayer Integrity of Cells Grown on Permeable Supports

using the hydrophilic marker molecule mannitol

1. Remove the culture medium, wash cells gently in prewarmed HBSS to remove any trace amounts of medium.

2. Preincubate monolayers with HBSS, for 30 min, adding 1 ml to the apical and 2 ml to the basolateral side.

3. At time zero replace apical HBSS with 1 ml of HBSS containing 0.5 µCi/ml [^{14}C]mannitol.

Note: Transepithelial electrical resistance may be measured at intervals continuously throughout the experiment using a Millicell ERS (see below).

4. Plates are placed in a 37 °C incubator gassed with 95 % oxygen; 5 % carbon dioxide on an orbital shaker.

5. Take 10 µl samples from stock solution of [^{14}C]mannitol for the determination of initial radioactivity. An apical sample (10 µl) is taken at the end of the experiment to determine loss of radioactivity.

6. At 20 min intervals, for up to 120 min, each insert is transferred to a new well containing 2 ml fresh HBSS. 1 ml samples are taken at 0, 20, 40, etc. min. Radioactivity is counted in each sample by liquid scintillation using a standard scintillation fluid.

7. The average concentration of [^{14}C]mannitol (pmol) is calculated for the apical samples and for the basolateral samples at each time point.

From these values, the rate of appearance or flux of the marker in the recipient chamber (J=mg/s) may be used to determine a simple measure of the apparent permeability coefficient (Papp) from:

Papp$=J/A.Co$

where A is the surface area (cm^2) of the epithelium and Co is the initial concentration of marker at the donor side (mg/cm^3) (Arturrson and Karlsson 1991). Papp is normally expressed in units of cm/s.

Results

- Other perfusion solutions may be used, for example culture medium or Krebs-Hensleit buffer solution.

- Cytochalasin D (1 µg/ml) can be added to the apical side to open intercellular tight junctions and artificially promote flux.

- Experiments should also be carried out with blank filters.

- Papp for passive transport should be independent of direction (absorption/secretion), of concentration of the marker substance (which should not be toxic to the cells) and of active transport processes (Papp determined at 4 °C should be the same as that determined at 37 °C).

- [^{3}H]mannitol may also be used, however, [^{14}C]mannitol is more stable.

- Diffusion chamber systems offer a more symmetrical and controlled environment in which to study cells grown on permeable supports. The Costar Diffusion Chamber System houses cell monolayers cultured on Snapwell permeable supports in an oxygenated, fluid environment which is circulated and maintained at 37 °C . The flow pattern is designed to reduce the unstirred boundary layer at each surface of the membrane which otherwise interferes with the passage of molecules across the monolayer/filter assembly.

Troubleshooting

- Since unstirred layers are formed at the apical and basolateral interfaces with the bathing solutions, it is critical that mixing occurs during the experiment (Karlsson and Artursson 1992).

- All radioactivity (or other marker) must be accounted for to ensure that results are not affected by sequestration of the marker molecule into cells or intercellular spaces.

- Avoid loss of fluid volume by dehydration; for example, cover the assembly with Parafilm between sampling.

- An improved method for calculating Papp (Karlsson and Artursson 1996) takes into account loss of marker from the donor side. Over a number of experimental periods the percent transported/cm^2 is plotted against time (minutes). The slope of the line (the transport rate constant; K/min) is used to calculate an apparent permeability coefficient (Papp):

Papp (cm/s) $= K.V_R/A.60$

where V_R is the volume of the recipient chamber.

- Variables which may influence results from permeability studies include cell seeding density, degree of confluence, differentiation state and composition of the media. In addition, the nature of the porous support, presence or absence of extracellular matrix and effects of hydrodynamic forces may also influence results.

17.5
A Simple, Reproducible and Accurate Electrical Measuring System for Transepithelial Resistances

Epithelial monolayers in vivo and in vitro are capable of separating electrically charged elements (for example by active transport of anions or cations) and sustaining a potential difference between the two compartments separated by the epithelium. The potential difference between the two sides is therefore related to the asymmetric distribution of charge and the resistance or electrical "tightness" of the epithelium.

Simple measures of electrical potential difference are carried out by placing suitable electrode pairs (silver/silver chloride, calomel) in either side of the solutions bathing confluent monolayers cultured on permeable supports. The electrodes are attached via a voltmeter and measurements made. Measurements of transepithelial electrical resistance (or its inverse, conductance) can be made by pulsing direct current across the epithelium through silver/silver chloride electrodes attached to a current source. Transepithelial resistance (R) or conductance ($G=1/R$) is determined by passage of a range of currents (I) with measurements of consequent alterations in the open circuit potential difference (V). The ohmic relationship ($V=IR$) dictates that a linear relationship between V and I is obtained and the slope of this line is related to R (Ωcm^2).

The electrodes are conveniently coupled to the bathing media via salt bridges made from polyethylene tubing (1 mm diameter) containing saturated KCl in 3 % agar.

Materials

- EVOM epithelial voltohmmeter (World Precision Instruments, Sarasota, Florida, USA) or Millicell-ERS (Millipore)
- Endohm tissue resistance measurement chambers (World Precision Instruments) with planar electrodes; 12 mm or 24 mm diameter, matched to culture inserts being used
- Dispersed population of epithelial cells at a high seeding density (e.g., 1:1 split ratio from a stock maintained on tissue culture plastic ware) on 12 mm or 24 mm diameter culture cups (24 mm culture inserts in 6-well plates or 12 mm culture inserts in 24-well plates)

Procedure

1. The meter and electrode chamber are kept in the tissue culture hood. The inner compartment of the electrode assembly is sterilized by filling one time with 70 % methanol which is removed and any residual alcohol washed out with several changes of sterile PBS.

   measuring transepithelial resistances

2. Immediately before use the electrode assembly is filled with an appropriate volume (depending upon the electrode size used) of tissue culture medium at 37 °C.

3. Cells growing on filters are transferred from the incubator into a flow hood.

4. Individual inserts are placed into the tissue resistance measurement chambers and potential difference and tissue resistance are read from the digital display on the voltohmmeter and recorded. Measurements from an individual filter can be made in a few seconds, after which the filters are returned to the multi-well plate.

5. The plates are returned to the incubator.

6. Multiple measurements may be taken over time.

Results

- Measurements of resistance obtained over time may be used as a functional "growth curve" (Fig. 17.3).

- Experiments should be carried out ideally in a 37 °C environment. However, since measurements can be made rapidly and plates restored quickly to the incubator, this is not an absolute requirement.

- Procedures or exposure to drugs which alter resistance can be measured.

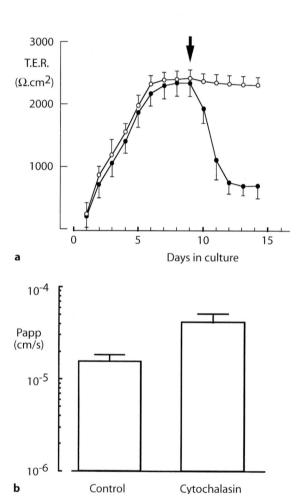

Fig. 17.3.a Transepithelial resistance measurements made using an Endohm electrode chamber with EVOM resistance meter following growth curves of T84 epithelial cells (derived from a colonic adenocarcinoma). Values are given over time on readings obtained from 12 mm Costar inserts which were seeded on day 1 with 10^6 T84 cells/insert. Control values (*open circles*) become stable over time. When cytochalasin D (1 μg/ml) is added (*arrow; closed circles*) transepithelial resistance values fall. **b** [^{14}C]Mannitol flux measurements in T84 monolayers. *P*app values for confluent monolayers are significantly elevated following treatment with the tight junction opener cytochalasin D (1 μg/ml)

- Transepithelial resistance measurements are widely used as an approximate index of passive ion movement through paracellular spaces. However, in "leaky" epithelia the relative contribution of transcellular ion conductance to total ion conductance may be significant. Thus, for low resistance epithelia, permeability is better studied by measurement of fluxes of a passively transported marker molecule (see below).

Troubleshooting

- Measurements must be made of filter blanks and these values subtracted from data obtained with cell monolayers. This is particularly important with leaky epithelia which have low resistances.

- Actual values of resistance may vary between laboratories and even from batch to batch of the same cell line in a single laboratory. Thus control experiments must be included in any investigation of agents which alter resistance.

- If cells are seeded sparsely, it may take a relatively long time to achieve measurable resistances.

- Resistance measurement equipment can be tested using a synthetic membrane such as SynCel (World Precision Instruments).

A purpose-built epithelial voltohmmeter (EVOM, World Precision Instruments; Millicell ERS, Millipore) is a simple and easy to use instrument for measuring membrane potential and resistance of cells cultured on porous supports. The apparatus makes use of alternating current which confers zero net charge on to the monolayer and so eliminates potential adverse effects of direct current on cell membranes. Different electrode systems are available to accompany the EVOM. Simple "chopstick" electrodes are convenient for rapidly screening large numbers of monolayers to estimate whether confluence has been achieved. Much more accurate measurements are made using electrodes made in a planar configuration. Examples of these include Endohm tissue resistance measurement chambers. Concentric electrodes held above and below the monolayer allow a more uniform current density to flow across the membrane. Very low background resistance ($<10\,\Omega$) is also a feature. The Endohm system coupled with an EVOM voltohmmeter is compatible with tissue culture permeable supports manufactured by Costar, Millipore, ICN Biomedicals and Becton Dickinson.

17.6
Electrophysiological Methods for Studying Ion Transport Across Sheets of Epithelia

A number of designs of Ussing-type chambers are available. These are assemblies in which confluent epithelial monolayers on filter supports are mounted between two identical half chambers and bathed on either side by identical physiological solutions. Glass circulation reservoirs (10 ml or 20 ml per side) with jacketed chambers for temperature control have a gas inlet for oxygenation and circulation of the bathing fluids. Potential difference between the two compartments is monitored via matched pairs of electrodes and maintained at zero by passage of a short circuit current through separate electrodes using an automatic voltage clamp device.

Under voltage clamp conditions in Ussing chambers there are no electrical, chemical, osmotic or hydraulic gradients. Thus, net movement of solutes including electrolytes between the two compartments may only occur as a consequence of active transport. Indeed, since net flux of ions causes a change in the short circuit current, this has proved to be a powerful technique for investigating regulation of electrogenic ion transport in absorptive and secretory epithelia (Dharmsathaphorn and Madara 1990; Barrett 1993).

▩ Materials

- Confluent monolayers of epithelial cells on permeable supports which are compatible with the chamber system to be used. Custom-made inserts are available (e.g., Snap Chambers (World Precision Instruments) for Costar Snapwell cups).
- Ussing system: circulation reservoir, two half chambers, two voltage electrodes and two current electrodes, support stand. These systems are available from a number of suppliers in slightly different configurations.
- 37 °C water pump
- Medical gas supply (95 % oxygen/5 % carbon dioxide) with regulated flow
- Voltage/current clamp: e.g., two (DVC-1000) or multi- (EVC4000) channel clamp (World Precision Instruments) with pre-amplifier probes
- Chart recorder or data acquisition system

Procedure

1. Assemble the Ussing chambers with electrodes in place. Fill the chambers with physiological solution (e.g., Krebs-Hensleit solution) which is circulated by the gassing system and maintained at 37 °C by the water-jacketed reservoir.

2. Connect the electrodes to the voltage clamp apparatus via the pre-amplifier probe.

3. Balance the voltage electrodes to zero potential difference and compensate for fluid resistance.

4. Mount the epithelial monolayer and filter in the Ussing chamber, filling each reservoir with equal volumes of physiological solution.

5. Clamp the monolayer to zero potential difference (this is an automatic function of the voltage clamp device). Continuously record the short circuit current. The membrane voltage can be altered over a range of values which, with the consequent changes in short circuit current, can be used to calculate electrical resistance by Ohm's law.

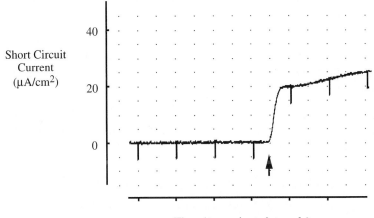

Time (two minute intervals).

Fig. 17.4. T84 monolayers grown on filter supports were voltage clamped at zero potential difference and short circuit current (SCC) continuously recorded. Downward deflections of the current trace indicate perturbations of membrane voltage which were imposed at intervals in order to calculate monolayer resistance by Ohm's law. Forskolin (3 μM) added to the basolateral reservoir stimulated a rapid onset, sustained inward short circuit current which can be accounted for by the drug activating adenylyl cyclase with consequent opening of cyclic AMP-sensitive chloride channels

6. Drugs may be added either to the apical or to the basolateral domain to examine their effect on electrogenic ion transport or monolayer resistance (Fig. 17.4).

Results

Short circuit current is, by convention, termed inward (anion secretion or cation absorption) or outward (anion absorption or cation secretion). The nature of the charge carrying ion(s) can be determined: (a) by radioisotope flux studies; (b) by substitution of selected ions with an impermeant species; or (c) pharmacologically by using selective transport inhibitors.

Alternatives porous supports include the filters prepared on silicon elastomer gaskets (see above) which may be used with traditional Ussing chambers since the mounting pins go through the O-ring support.

Troubleshooting

- It is important to use matched electrode pairs and to compensate for resistance associated with the bathing fluid and agar bridges in the Ussing chamber apparatus.

- Excessive flow rates of the gas circulation system may dislodge monolayers from their support membrane. Similarly, "washing out" drugs by changing the physiological bathing solutions may physically stress the monolayer.

- The gas circulation system will cause vapor formation at the fluid/air interface. Since this represents a possible biohazard, it should be contained by use of condensation chambers above the reservoirs.

- A "dummy" membrane is provided with the DVC-1000. This is useful for training a new operator the principles of the system and, in addition, can be used to check proper function of each clamp.

- Chambers and fluid reservoirs must be carefully cleaned between experiments. **Note:** alcohols, aromatic hydrocarbons and chlorinated solvents may adversely react with chamber materials.

- Agar/KCl -filled bridges may leach potassium into the Ussing chambers. This is minimized if the salt bridge assemblies are stored in PBS or physiological solution.

References

Artursson P, Karlsson J (1991) Passive absorption of drugs in Caco-2 cells. In: Wilson G et al. (ed) Pharmaceutical applications of cell and tissue culture to drug transport. Plenum, New York, pp 93–105

Barrett KE (1993) Positive and negative regulation of chloride secretion in T84 cells. Am J Physiol 265: C859–868

Bell CL, Quinton PM (1990) Recycle those cell culture inserts. In Vitro Cell Dev Biol 26:1123–1124

Borchardt RT, Hidalgo IJ, Hillgreen KM, Hu M (1991) Pharmaceutical applications of cell culture: An overview. In: G. Wilson et al. (ed) Pharmaceutical applications of cell and tissue culture to drug transport. Plenum, New York, pp 1–14

Dharmsathaphorn K, Madara JL (1990) Established intestinal cell lines as model systems for electrolyte transport studies. Methods in Enzymology 192:354–389

Evans GS, Flint N, Potten CS (1994) Primary cultures for studies of cell regulation and physiology in intestinal epithelium. Annu Rev Physiol 56:399–417

Gstraunthaler GJA (1988) Epithelial cells in tissue culture. Renal Physiol Biochem 11:1–42

Handler JS, Stelle RE, Sahib MK, Wade JB, Preton AS, Lawson NL, Johnson JP (1979) Toad urinary bladder epithelial cells in culture: maintenance of epithelial structure, sodium transport and response to hormones. Proc Natl Acad Sci USA 76:4151–4155

Hughson EJ, Hirt RP (1996) Assessment of cell polarity. In: Shaw AJ (ed). Epithelial cell culture. a practical approach. Oxford University, Oxford pp 37–66

Karlsson J, Artursson P (1992) A new diffusion chamber system for the determination of drug permeability coefficients across the human intestinal epithelium that are independent of the unstirred water layer. Biochim Biophys.Acta 1111:204–210

Lewis SA, Berg JR, Kleine TJ (1995) Modulation of epithelial permeability by extracellular macromolecules. Physiol Rev 75:561–589

Lifschitz MD (1986) Prostaglandins may mediate chloride concentrations gradient across domes formed by MDCK1 cells. Am J Physiol 250:F525–531

Madara JL, Colgan S, Nusrat A, Delp C, Parkos C (1992) A simple approach to measurement of electrical parameters of cultured epithelial monolayers: use in assessing neutrophil-epithelial interactions. J Tiss Cult Methods 14:209–216

Pitt AM Gabriels JE, Badminton F, McDowell J, Gonzales L, Waugh ME (1987) Cell culture on a microscopically transparent microporous membrane. BioTechniques 5:162–170

Van Scott MR, Yankaskas JR, Boucher RC (1986) Culture of airway epithelial cells: research techniques. Exp Lung Res 11:75–94

Text Books

Culture of epithelial cells (1992) Freshney RI (ed) Wiley-Liss, New York

Tissue culture of epithelial cells (1985) Taub M (ed) Plenum, New York

Methods in enzymology, vol. 191 (1990) Academic, San Diego 1990

Methods in enzymology, vol. 192 (1990) Academic, San Diego 1990

Epithelial cell culture. A practical approach (1996) Shaw AJ (ed) Oxford University, New York

Cell and tissue culture: laboratory procedures (1995) Doyle A, Griffiths JB, Newell DG (ed) John Wiley, New York

Functional epithelial cells in culture (1989) Matlin KS, Valentich JD (eds) AR Liss, New York

Chapter 18

Ion Channels and Cell Signaling in Cell Cultures

VALÉRIE URBACH*, DEIRDRE WALSH, MARIA HIGGINS,
ISABELLE LEGUEN, CHRISTINA DOOLAN, JOHN CUFFE,
ELIZABETH HORWITZ, CATHERINE HALLIGAN, RUTH GLEESON,
ANTHONY CULLINANE, and BRIAN HARVEY

Introduction

Cell Culture and Ion Transport in Airway Epithelia

Two major diseases, namely cancer and cystic fibrosis (CF), have driven the development of immortalized human airway epithelial cell lines. Primary cultures of airway epithelial cells have also played a central role in the study of other diseases such as airway inflammation and viral infection. However, primary lung culture is limited by the number of viable cells that can be generated from the tissue and by the availability of fresh tissue. Immortalized cell lines of airway epithelium were developed from lung carcinomas or from primary cultures transformed in vitro using viruses (SV40). The development of cell lines has greatly enhanced our understanding of the biochemical and proliferation properties of lung epithelium because these characteristics are not affected by viral transformation. However, many transformed cells lose certain differentiated functions, such as tight junction formation or physiological ionic transport properties. For this reason, most of the knowledge of ionic transport mechanisms and regulation come from studies on primary cell culture.

Here we will describe a primary cell culture technique for human bronchial epithelium and the culture conditions for two cell lines (from human normal bronchus and CF trachea) which maintain Cl⁻ transport properties typical of their genotype. These cell lines are routinely used in our laboratory to study the regulation of ionic transport in normal and CF airway epithelia and some of these properties are discussed.

* *Correspondence to* Valérie Urbach, Wellcome Trust Cellular Physiology Research Unit, Department of Physiology, University College, Cork, Ireland; phone +353–21–903209; fax +353–21–272121; e-mail V.URBACH@ucc.ie

Human Airway Cell Culture Techniques

- Primary cell culture: Primary culture of normal nonmetastatic human airway epithelia is initiated from tissue freshly excised from cancer patients undergoing surgery. Different techniques can be used to cultivate this tissue: explant culture (Lechner and LaVeck 1985; De Jong et al. 1993) or culture from cells dissociated by enzymes (Van Scott et al. 1986; Widdicombe et al. 1985; Wu et al. 1985). Primary cultures have been reported for the proximal part of the respiratory epithelium such as bronchial (Lechner and LaVeck 1985), tracheal (Gruenert et al. 1987, 1988; Widdicombe et al. 1985) or nasal epithelia (Yankaskas et al. 1985).

- Explant culture of human bronchial epithelium: The tissue must be as fresh as possible (not older than 24 h postsurgery) and placed immediately after excision in saline medium containing antibiotics to reduce the potential of microbial contamination. All culture procedures are carried out under aseptic conditions in a laminar flow hood. The lung tissue is placed in a sterile glass petri dish and covered with a few drops of Hank's balanced salt solution (HBSS) (Sigma, cat. no. H2513). Using sterile Pasteur pipettes, the bronchial tissue pieces are repeatedly washed in antibiotic solution for at least 10 min. Using two sterile scalpels, the tissue is chopped into tiny pieces (approx. 1 mm) and washed constantly with HBSS to eliminate excess red blood cells. Using a sterile needle, the base of each flask (25 cm^2 tissue culture flasks, Falcon 3013E) is scored to facilitate cell attachment. Each flask is rinsed with medium Q (see below) and the excess fluid decanted. About nine explants are placed in each flask, i.e., three explants along each scored furrow in the plastic base. The flasks are then incubated at 37°C for 24 h in a humidified 5% CO_2 atmosphere. After the initial 24 h incubation, the explants usually begin to adhere to the flasks.
A Pasteur pipette is used to gently drop medium Q on top of each explant but too much medium at this stage will cause the explants to float off from the base. After 2–5 days, a monolayer of cells is normally observed to grow out from the explants. At this stage, the monolayers contain three different cell types: granular, goblet and ciliated epithelia. Twice to three times weekly the explants should be exposed to fresh culture medium Q. The success rate is highly variable, depending mainly on surgical trauma and health of the biopsy. Explants that are growing well can be used to seed other primary cultures and are removed from the base using a sterile forceps and relocated to another

flask. Following 1–2 weeks maximum of culture, the epithelial monolayers may be isolated by trypsinization for transport studies (see
"Cell trypsinization procedure").

- Transformed cell lines (16HBE14o– and CFTE29o– cells): The transformation and culture techniques of airway epithelial cell lines have
been recently reviewed (Gruenert et al. 1995). The 16HBE14o–, a post-
crisis SV40 transformed cell line, is derived from surface epithelium of
mainstream, second-generation bronchi (Cozens et al. 1994). Ion
transport studies indicate that it retains Cl^- transport properties typical of freshly isolated surface airway epithelial cells, expresses normal
levels of CFTR (cystic fibrosis transmembrane conductance regulator)
mRNA and protein (Cozens et al. 1994), and has tight junctions and
cilia. An immortalized CF cell line (CFTE29o–) derived from tracheal
epithelium homozygous for the ΔF508-CFTR mutation (deletion of
the phenylalanine at position 508) is also used for ionic transport
studies. This post-crisis cell line is defective in cAMP-dependent chloride ion transport, and secretes Cl^- in response to treatment with Ca^{2+}
ionophores (Kunzelmann et al. 1993).

- Cell line culture technique: When not required for experiments, the
cells are frozen down and kept in liquid nitrogen at $-170\,°C$. Cells
from fully confluent flasks ($25\,cm^2$ tissue culture flasks, Falcon 3013E)
are stored in vials containing 2 ml of freezing solution. To initiate
growth from the frozen state, the cells are thawed quickly in a $37\,°C$
incubator. Thawed cells and medium are transferred to centrifuge
tubes and spun down at 1500 rpm for 5 min. The freezing solution is
aspirated from each tube and the pellet resuspended in 6 ml of
medium and split into 3 ml aliquots added to each of two new fibronectin coated flasks (see "Coating procedure"). The cells are fed with
medium twice to three times weekly. Once a confluent monolayer is
formed the cells are trypsinized (see "Cell trypsinization procedure")
and 50 % cells are used for experimental work with the remainder
transferred to a new flask to generate another monolayer.

- Cell trypsinization procedure: The cells are rinsed twice in 3 ml
HEPES buffered saline (HBS) and then bathed in PET solution (1.5 ml)
at $37\,°C$ until the cells are fully detached (5–10 min). It is very important to protect the cells against over-trypsinization. This can be best
achieved by observing the following guidelines. After 5 min incubation the solution should become slightly cloudy in appearance. At this
point, the flasks are agitated firmly by hand and checked under a

microscope. When the monolayer is completely detached and the cells
are unclumped and isolated, the trypsin reaction is inactivated by
addition of fetal bovine serum (FBS) or modified Eagle's medium
(MEM). The cells are then rinsed twice by centrifugation at 1500 rpm
and the pellet resuspended in 2 ml Krebs saline solution.

- Coating procedure: Cell culture is enhanced by coating the flasks with
  fibronectin. The fibronectin coating solution is freeze-stored ($-40\,^\circ$C)
  in 10 ml aliquots. To coat the plastic, 2 ml of this solution are added to
  each 25 cm^2 flask and cured at room temperature for 1 h. Coating of
  the base must be uniform to avoid overgrowth of cells. The excess
  coating solution is then aspirated and the plates carefully rinsed with
  distilled water while avoiding any damage to the treated surface. Plates
  are then ready for use. Fibronectin coated flasks may be stored at 4 °C
  for up to 2 weeks.

Effects of Culture Conditions on Ion Channel Expression

- Hormones and Na$^+$ channel expression: It has been reported that airway epithelial cells may lose their ability to express Na$^+$ channels during proliferation and continuous cell culture (Cott and Rao 1993;
  Benos et al. 1992; Yue et al. 1993). For example, the presence of
  amiloride-sensitive Na$^+$ current was observed in freshly isolated alveolar epithelial cells plated on fibronectin-treated cover slips. However,
  after 48 h of culture, Na$^+$ currents were inhibited and immunocytochemical localization showed a change in the spatial distribution of
  Na$^+$ channels from the plasma membrane to the cytoplasm (Yue et al.
  1993). Hormonal treatment and culturing on permeable filter supports seem to be essential for Na$^+$ channel expression. For example, no
  amiloride sensitive Na$^+$ channel activity was detected in the human
  bronchial epithelial cell line (16HBE14o–) cultured on plastic dishes.
  However, when the cells were cultured on a permeable support or on
  plastic in the presence of aldosterone, the Na$^+$ current could be
  recorded in the whole-cell patch clamp configuration (Kunzelman et
  al. 1996).

- Temperature-sensitivity of ΔF508-mutated CFTR expression: Functional studies failed to detect a cAMP-stimulated Cl$^-$ channel in airway cells normally expressing the CFTR-ΔF508 mutation (Rich et al.
  1990). However when expressed in *Xenopus* oocytes or in SF9 insect
  cells (which are heterologous expression systems typically maintained

at low temperature), the CFTR-ΔF508 mutant is able to generate a cAMP-regulated Cl$^-$ secretion. These observations can be explained by the temperature sensitivity of the processing of the CFTR-ΔF508 protein (Denning et al. 1992). Insertion of the mutant CFTR into the plasma membrane is increased at low temperatures. In our laboratory, we observed that forskolin (a CFTR activator) produced an intracellular Ca^{2+} ($[Ca^{2+}]_i$) increase at temperatures between 14 °C and 37 °C in 16HBE14o– cells or in normal primary lung cultures. However, in ΔF508 CFTE29o– cell line, forskolin failed to produce a $[Ca^{2+}]_i$ change at 32 °C, but did so when the cells were pre-incubated at 14 °C for 2 h before the experiment.

Ion Transport Studies

- Culture supports: Different types of culture supports are better adapted for specific ion transport studies. Usually the culture is started in a plastic flask, and the postconfluent, trypsin-isolated cells can be used in patch-clamp (to study individual characteristics of the channel) or spectrofluorescence (to measure intracellular Ca^{2+} using Fura-2 fluorescent dyes) experiments. During the experiments, the support used is transparent (plastic or glass coverslips for patch-clamp on an inverted microscope, and glass only for spectrofluorescence experiments). The supports are coated with l-polylysine to facilitate the fixation of the cells. For Ussing chamber experiments, which give a measure of transepithelial ionic transport, a monolayer of cells cultured on a permeable support must be used. Monolayers of cells cultured on plastic or transparent permeable filters can also be used for patch-clamp experiments. In this case, the microelectrode access to the apical membrane is possible and epithelial polarity is maintained for sidedness of drug addition. It is also possible to access the basolateral membane of the monolayer by inverting small sections using scalpel blades and a low pressure jet of solution. For fluoresence measurements in intact monolayers, the cells must be grown on glass or on non-autofluorescent filters.

Ion Transport Mechanisms and Regulation in Airway Epithelia

Ion transport by airway epithelia regulates the volume and the composition of liquid at the surface of the epithelium, which plays a central role in lung protection. For example, in CF, a genetic disease associated with an abnormality of NaCl and water transport, more than 90 % of the morbidity and mortality is due to respiratory tract infection.

In 1968, Kilburn predicted that in human lung large volumes of liquid are secreted in distal units and absorbed by the proximal airway epithelia. This conclusion was based upon the disparity between the surface areas of the distal (70 m^2) and proximal (60 cm^2) pulmonary surfaces and the relatively constant height of surface liquid layer (Kilburn 1984; Boucher 1994). The cell model presented in Fig. 18.1 shows the essential mechanisms implicated in the ionic transport through proximal airway epithelial cells. Cl$^-$ secretion is mediated by a two-step process via an Na/K/2Cl cotransporter located in the basolateral membrane and Cl$^-$ channels in the luminal membrane (Willumsen et al. 1989). The Na$^+$ absorption from the lumen to the blood occurs via luminal Na$^+$ channels (Cotton et al. 1987) and basolateral Na/K ATPase pumps (Stutts et al. 1986). The activity of K$^+$ channels contributes both to the charge equilibrium necessary for Cl$^-$ secretion and to the K$^+$ recycling occurring in parallel with Na/K pump activity. Two antiporter systems, Na/H (Paradiso 1992) and Na/Ca (Murphy et al. 1988) exchangers, and a plasma membrane Ca-ATPase (Paradiso et al. 1991) have also been identified in airway epithelia.

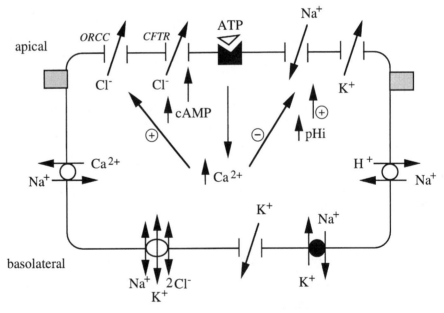

Fig. 18.1. Ion transport and regulation in proximal airway epithelium

Patch-clamp studies have shown that apical Cl^- conductance reflects the activity of several different Cl^- channel types. One of these, the CFTR protein, is an 8–10 pS nonrectifying and cAMP-dependent Cl^- channel that has been identified in immortalized human airway cells (Ward et al. 1991; Haws et al. 1992). In CF, mutation of the CFTR protein, most commonly a deletion of the phenylalanine at position 508 (ΔF508), causes a defect in Cl^- secretion. Another major class of Cl^- channels, an outward rectifier Cl^- channel (ORCC) activated by increasing the intracellular Ca^{2+} concentration (Mason 1991; Anderson et al. 1992; Stutts et al. 1992), has been described in human lung. We have observed a 40 pS outward rectifier Cl^- channnel by increasing the intracellular Ca^{2+} concentration using nucleotides in both normal, primary cultured bronchial epithelium and in ΔF508 CF cells (CFTE29o– cell line) (Urbach et al. 1994; Walsh et al. 1996).

Na^+ transport processes are subject to hormonal or cellular regulation. Hormonal regulation of Na^+ transport in airway epithelia is not clear. In vivo, Na^+ transport in nasal epithelia was described as not regulated by mineralocorticoid hormones (Knowles et al. 1985). However, in vitro, the Na^+ transport rate seems to be regulated by mineralocorticoids in cultured human airway epithelia (Kunzelman et al. 1996). In CF the defect in Cl^- secretion is associated with an hyperabsorption of Na^+ across airway epithelia. This increased Na^+ transport cannot solely be explained by the elevated electrochemical gradient for Na^+ (due to the inhibition of Cl^- transport), but appears also to be associated with a high activity of single Na^+ channels (Chinet et al. 1994). Differents intracellular factors which could regulate Na^+ transport in airway epithelia have been studied. Na^+ transport rates can be regulated by pH. Low intracellular pH inhibits Na^+ transport, whereas raised pH_i accelerates Na^+ transport (Willumsen and Boucher 1992). Intracellular cAMP (Mall et al. 1996; Stutts et al. 1995) and Ca^{2+} (Mason et al. 1991; Graham et al. 1992) are also regulators of Na^+ channel activity in airway epithelia. Intracellular Ca^{2+} seems to act on the Na^+ channel activity by an indirect mechanism which could implicate other intracellular signals such as protein kinase C (Yanase and Handler 1986; Ling and Eaton 1989) or the cytoskeleton (Cantiello et al. 1991).

Although much research has been focused on the Cl^- channel role of the CFTR protein, it is now becoming increasingly apparent that this protein plays an important role in the regulation of transport via other membrane proteins. For example, recent results implicate the CFTR protein in the inhibition of Na^+ channels (Kunzelman et al. 1995; Stutts et al. 1995; Mall et al. 1996) and in the activation of ORCC channels (Schwiebert et al. 1995). An autocrine mechanism involving ATP release via CFTR has been both experimentally supported (Reisen 1995; Schwiebert et al. 1995; Prat et al. 1996) and repudiated (Reddy et al. 1996). In our laboratory, we have demonstrated that the intracellular Ca^{2+} increase produced by external nucleotides can be mimicked by activation of CFTR with forskolin in normal primary lung and in the 16HBE14o– cell line. This Ca^{2+} response to forskolin is prevent by glibenclamide or diphenylamine carboxylate (CFTR channel inhibitors), suramin (a purinnergic receptor inhibitor) and external hexokinase (which cleaves external ATP). In the ΔF508 mutant CFTE29o– cell line we have been able to reproduce the forskolin effect on intracellular Ca^{2+} by pre-incubating the cells at low temperature (14 °C) but not at body temperature. We thus provide evidence for a role of CFTR in the regulation of intracellular Ca^{2+} via an autocrine effect in airway epithelia.

Materials

solutions   – Primary lung medium (medium Q) is prepared as follows:

Medium	Amount	Supplier	Catalogue number
Ham's F12	100 ml	Biowhittaker	12.422.54
3 % Glutamine	4 ml	Gibco	35050.020
Penicillin/strepto-mycin[a]	2 ml	Gibco	15050.022
10 % fetal bovine serum (FBS)	1 ml	Biowhittaker	14.503.A
1 M Hepes	1 ml	Biowhittaker	17.737E
EGF (10 mg/ml)	200 µl	Sigma	E1264
ITS premix	200 µl	Collaborative Research Products	40351
Vitamin A (10 ng/ml)	20 µl	Sigma	R7632
Hydrocortisone (100 µg/ml)	10 µl	Sigma	H0888
EGF, epidermal growth factor			

[a] 10 000 U/ml penicillin G; 10 000 µg/ml streptomycin.

– Eagle's modified medium (MEM) complete: 450 ml Eagle's modified medium; 50 ml FBS; 5 ml L-Glutamine (200 mM); 5 ml penicillin/streptomycin (100×; 10 000 U/ml penicillin G; 10 000 µg/ml streptomycin)

- PET (for 50 ml): 5 ml PVP (Sigma, cat. no P2307); 5 ml 0.2 % EGTA in Hepes buffered saline (HBS; see below); 4 ml 0.25 % trypsin/0.02 % EDTA (Versene) (Gibco, cat. no. 45300–019); 36 ml HBS
- Hepes buffered saline (HBS), pH 7.4 (for 4 l): 19.04 g Hepes; 28.52 g NaCl; 0.8 g KCl; 6.8 g glucose; 15 g $Na_2HPO_4.7H_2O$ (dibasic); 1 ml 0.5 % phenol red. Bring final volume to 4 l with distilled H_2O.
- 0.2 % EGTA (100 ml): 0.2 g EGTA; 100 ml HBS
- Fibronectin coating solution: 100 ml MEM (Biowhittaker); 10 ml bovine serum albumin (BSA; 1 mg/ml; Sigma, cat. no. A7906); 1 ml Vitrogen 100 (Celtrix, cat. no. PC0702); 1 mg human fibronectin (FN) (Becton Dickinson, cat. no. 40008)

- Freezing solution (50 ml): 25 ml fetal bovine serum (50 %); 20 ml MEM complete (40 %); 5 ml DMSO (10 %)

18.1
Ion Transport in Distal Renal Cell Lines

The M-1 Mouse Cortical Collecting Duct Cell Line

The cortical collecting duct (CCD) of the distal nephron plays a role in the fine-tuning of Na^+ reabsorption and K^+ secretion. Tissue culture has become a powerful tool to study renal epithelial function and several cell lines have been established to study ion transport processes (e.g., A6, MDCK, RCCT, LLC-PK$_1$) (Gstraunthaler 1988). A major goal has been to develop a continuous mammalian CCD cell line which, under culture conditions, expresses properties typical for the CCD in vivo and provides a source of cells with minimal variability. M-1 cells retain differentiated transport functions, hormone responsiveness and CCD antigens and thus may serve as a useful model for the study of collecting duct transport (Stoos et al. 1991).

The M-1 cell line was originally derived from CCD microdissected from a mouse transgenic for the early region of SV40. In these mice, the genetic background is modified to enhance cell proliferation without losing the ability for differentiation (McKay et al. 1988). Antibody studies have shown that 75 % of cells express antigens characteristic for CCD principal cells (Stoos et al. 1991). When grown on permeable supports, cells exhibit a high transepithelial resistance and a lumen-negative transepithelial potential difference. The corresponding short-circuit current

(I_{sc}) is amiloride-sensitive and is increased by arginine vasopressin (AVP) (Stoos et al. 1991).

Primary cultures of rat or murine CCD may also be generated from microdissected collecting duct tubules. Each microdissected CCD tubule (150–750 mm) is placed in a 0.32 cm^2 well on top of a collagen gel. Tubules are incubated in RPMI 1640 containing 5 % FBS, 1 nM pyruvate, 2 mM glutamine, 10 mM 2-mercaptoethanol, 50 U/ml penicillin, 50 mg/ml streptomycin, 12 mg/ml tylosin, and 2 % Ultraser G. At confluence, cells are subcultured by washing and incubating in HBSS containing 20 mM Hepes and 0.1 % collagenase (cultures are shaken at 37 °C until the collagen membranes are digested).

Procedure

cell culture M-1 cells were purchased from the American Type Culture Collection, Maryland, USA. Cells were grown in 25 cm^2 sterile polystyrene culture flasks containing a 1:1 mixture of Ham's F-12 and Dulbecco's modified Eagle's medium (DMEM) containing 5 µM dexamethasone and 5 % FBS. This medium also contains 1 % Hepes; 1 % penicillin/streptomycin and was maintained at pH 7.4. Cells were equilibrated with 5 % CO_2/95 % air and kept at 37 °C. Fluid was renewed two to three times weekly. Subconfluent cultures were subcultured by rinsing with 0.25 % trypsin/0.03 % EDTA solution. Most of the trypsin solution was removed and cells were incubated at 37 °C for 15–20 min, until cells detached. Some 3–4 ml of fresh medium were added, aspirated and cells dispensed into new flasks. The subcultivation ratio was 1:3–1:4.

Back stocks were made by growing cells to confluency and trypsinizing as described previously. The cell pellet was resuspended at approximately 2×10^6 cells/ml of freezing medium containing 95 % culture medium and 5 % DMSO; 1 ml of cell suspension was added per freezing vial which was stored and lowered into liquid nitrogen over a 3-h period to achieve a gradual freezing process. To revive cells from prolonged storage, vials were thawed rapidly in a waterbath at 37 °C. Vials were centrifuged at 1500 rpm for 10 min and the supernatant removed. Cell pellets were resuspended with 1 ml culture medium, fresh medium was added and flasks were incubated at 37 °C.

ion transport Characterization of M-1 cells was first performed in Ussing chambers to examine whether monolayers display typical CCD epithelium phenotype. When grown on permeable supports, cells possessed a high transepithelial resistance. Active Na^+ reabsorption was observed and was reduced by 95 % following luminal application of amiloride, confirming that Na^+ is absorbed primarily through a conductive pathway. Arginine vasopressin (AVP) induced an increase in trans-epithelial sodium transport, as measured by short-circuit current (I_{sc}), via an increase in intracellular cAMP. In addition, K^+ secretion was observed by a marked elevation in luminal $[K^+]$. Data indicate that M-1 cells did not undergo major dedifferentiation when compared to native CCD. This may be due to the fact that M-1 cells originate from a transgenic animal carrying SV40 early region genes (Stoos et al. 1991).

Whole-cell patch-clamp studies have been carried out on both isolated and confluent M-1 cells. In confluent cells, $10-100\,\mu M$ amiloride hyperpolarized the membrane potential. In contrast, single cells expressed no significant amiloride-sensitive conductance, and Korbmacher et al. (1993) have suggested that cell-to-cell contact and tight junction formation may be required for Na^+ channel insertion. Neither high K^+ nor Ba^{2+} in the apical bath affected confluent cells, showing that these lack a significant apical K^+ conductance.

The whole-cell conductance (G_{cell}) was 2.6 times higher in confluent cells than in single cells. In single cells it appeared to be dominated by an inwardly rectifying K^+ conductance. Glibenclamide, a known inhibitor of ATP-sensitive K^+ channels, reduced whole-cell currents in both single and confluent cells. Finally, another component of G_{cell} has been detected as a deactivating outward Cl^- current. This is observed during large depolarizing voltages and is abolished by extracellular Cl^-. It is also glibenclamide-sensitive and probably located basolaterally (Korbmacher et al. 1993).

Further studies have been performed to confirm the presence of the classical apical amiloride-sensitive Na^+ channel, similar to that identified in urinary bladder, distal colon, sweat ducts and respiratory epithelial cells (Letz et al. 1995). Both G_{cell} and the single channels are highly amiloride-sensitive. The latter are highly selective for $Na^+>K^+$ with a low single channel conductance (6.8 ± 0.5 picoSiemens, pS). Expression studies demonstrated that poly $(A)^+$ RNA isolated from M-1 cells induced an amiloride-sensitive Na conductance after injection into *Xenopus laevis* oocytes. Transcripts related to the three subunits (α-, β-, and γ) of the epithelial Na^+ channel (ENaC) in rat colon could be detected by northern analysis. The dependence on extracellular Na^+ together with amiloride-sensitivity and $Li^+>Na^+$ selectivity supports the conclusion that the single channels identified are those underlying whole-cell currents previously described. Letz et al. (1995) concluded that the Na^+ channel in

M-1 cells is closely related to ENaC in rat colon and that this cell line provides a useful tool to investigate biophysical and molecular properties of the corresponding channel in native CCD.

Finally, a nonselective cation (NSC) channel (34 ± 2.3 pS) regulated by Ca^{2+} and cGMP has been identified in the apical membrane of M-1 cells (Ahmad et al. 1992). NSC channels show high selectivity for cations over anions but discriminate poorly between Na^+ and K^+. They have been detected in other nephron segments: proximal tubule, cortical thick ascending limb and CCD cells. cGMP inhibits channel activity in the inner medullary collecting duct (IMCD). Northern analysis of poly $(A)^+$ RNA from M-1 cells suggested the expression of more than one gene coding for NSC channel or channel subunits, one of which is identical to the cGMP-gated cation channel of rod photoreceptors. In M-1 cells, Ahmad et al. (1992) found that channel activity was increased by depolarization and abolished by removing cytoplasmic Ca^{2+} and reduced by 100 μM cGMP. Cell signals known to affect ion channel activity in CCD cells are summarized in Fig. 18.2.

intracellular calcium

Recently, much interest has developed in the role of nucleotides and autocoids (prostaglandins and leukotrienes) substances in the kidney. Adenosine and ATP have been shown to regulate electrolyte transport in renal tissue (Middleton et al. 1993). In our laboratory, calcium imaging studies have been performed on M-1 cells to examine the effects of agonists on levels of intracellular calcium, $[Ca^{2+}]_i$. The presence of intracellular stores was demonstrated by addition of the tumor promoter thapsigargin, which mobilizes Ca^{2+} from cellular pools via inhibition of the endoplsmic reticulum Ca-ATPase. Entry of Ca^{2+} via Ca^{2+}-release-activated Ca channels (CRACs) and/or Ca-permeable cation channels was activated using the calcium ionophore 4-bromo A23187 to initially raise $[Ca^{2+}]_i$. Hormones such as aldosterone and AVP also induce transient increases in $[Ca^{2+}]_i$ in M-1 cells

Calcium-dependent signaling mechanisms may be important in regulating salt and water transport in the CCD (Schafer and Troutman 1991). Recent research has shown that ATP may be released alone or coreleased with norepinephrine from sympathetic nerve endings in the kidney suggesting an important modulatory role (Nilius et al. 1995; Rouse et al. 1994). We have shown that nucleotides such as ATP, UTP, cAMP and adenosine produce oscillatory increases in $[Ca^{2+}]_i$ in M-1 cells. The response to ATP was delayed by the purinergic receptor antagonist suramin indicating the presence of P_{2u} purinoceptors. If thapsigargin was added to cells prior to ATP, no $[Ca^{2+}]_i$ transient was observed indicating

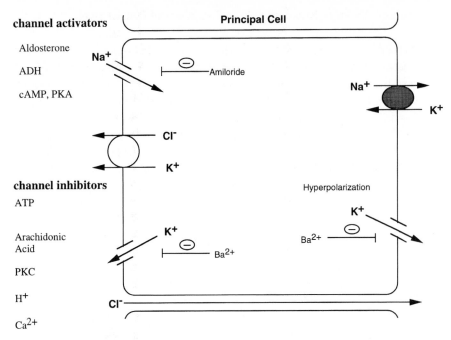

Fig. 18.2. Renal cortical collecting duct principle cell and the factors involved in the regulation of apical membrane sodium and potassium ion channels. Apical Na^+ channels can be inhibited by the diuretic amiloride and K+ channels can be blocked by barium ions

that ATP also raises $[Ca^{2+}]_i$ via IP_3-operated calcium stores. Intracellular calcium responses to dibutyryl cAMP (a hydrolyzable analogue of cAMP) were blocked by protein kinase inhibitor (PKI) suggesting that phosphorylation may be involved in the $[Ca^{2+}]_i$ increase induced by cAMP.

In conclusion, M-1 cells serve as a useful tissue culture model to study renal specific transport processes and the involvement of second messengers. Further research is required to determine the electrophysiology, molecular biology, and regulation of ion channels in the CCD. Little is known about the sequence of biochemical events involved in the action of nucleotides on renal epithelial cells but these pathways may provide an insight into the potential coupling of metabolic demand and electrolyte transport in the kidney.

Ion Transport Characteristics and Culture Conditions of Renal A6 Cells

The A6 cell line derived from the kidney of toad *Xenopus laevis*(Rafferty, 1975) is available from the American Type Culture Collection, USA. A6 cells are morphologically homogeneous and have Na$^+$ transport properties similar to renal distal tubule cells. They have therefore been widely used as a model to study the mechanism of Na$^+$ reabsorption and its regulation by several hormones including insulin, aldosterone and vasopressin (Verrey et al. 1994; Rodriguez-Commes et al. 1994; Candia et al. 1993). As presented in Fig. 18.3, the Na$^+$ absorption is carried out via Na$^+$ channels situated in the apical membrane and a Na/K pump in the baso-

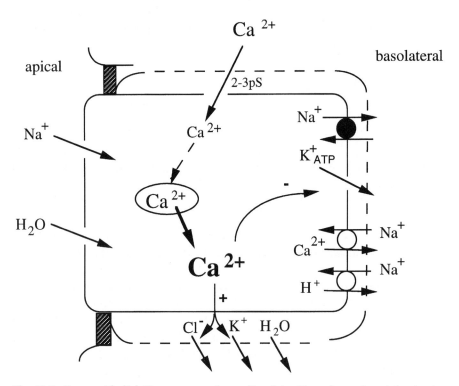

Fig. 18.3. Transepithelial Na transport is mediated by Na+ channel activity in the apical membrane ansd Na/K ATPase in the basolateral membrane of A6 cells. K+ is recycled via ATP-regulated K+ channels (K$_{ATP}$) which are inhibited by a rise in intracellular Ca^{2+}-K channels. Cell swelling activates calcium entry via mechanosensitive channels (SCa) in the basolateral membrane. This Ca^{2+} entry stimulates calcium release from intracellular stores. The swelling-induced increase in intracellular Ca^{2+} causes activation of K+ and Cl$^-$ secretion which drives water efflux and restoration of cell volume (regulatory volume decrease)

lateral membrane. Potassium ions are recycled in parallel to the Na/K pump activity through ATP-dependent K^+ channels situated in the basolateral membranes (Harvey et al. 1991; Harvey and Urbach 1995). Na/H exchange regulating intracellular pH (Casavola et al. 1992; Doi and Marunaka 1995), and Na/Ca exchange (Brochiero et al. 1995) have also been described in the basolateral membrane of A6 cells.

Kidney cells are continually subjected to changes in solute transport load and external osmolarity which can perturb cell volume (Hoffman and Dunham 1995). Several studies have examined the regulatory volume decrease (RVD) of A6 cells after cell swelling. In epithelial cells, swelling produces an increase in cytosolic calcium followed by RVD, which involves loss of KCl through Ca^{2+} activated K^+ channels and Cl^- channels. Using patch-clamp and spectrofluorimetry techniques, we have studied the mechanism of intracellular calcium mobilization after a hypotonic shock in A6 cells.

Cells were maintained in plastic tissue culture flasks at 27 °C in a humidified incubator gassed with 5 % CO_2 in air. Culture medium NCTC 109 or NCTC 105 modified for amphibian cells at 15 % dilution was supplemented with 10 % FBS, 2 mM glutamine, 50 U penicillin and 50 μg/ml streptomycin; final osmolarity was 250 mosmol/l and pH 7.4. This medium was renewed two to three times weekly. Cells were cultured to form confluent monolayers in flasks, and then isolated by trypsinization (0.05 % trypsin/0.02 % EDTA in Mg^{2+} and Ca^{2+}-free saline) for 5 min at 37 °C. The trypsin digestion was stopped and cells rinsed twice with culture medium. After the final centrifugation, the cells were split into two batches. Cells of one batch were suspended in normal Ringer solution for fluorescence and patch-clamp experiments on isolated cells. In the second batch, cells were subcultured in flasks or petri dishes used for patch-clamp experiments on polarized cells when the epithelium reached confluency. On petri dishes, in patch-clamp experiments on confluent monolayers, access to the basolateral membrane with patch pipettes was achieved by inverting small areas (100 μm²) of the monolayer with fire-polished micropipettes.

A6 cell culture

Using patch-clamp techniques, we have identified an inward rectifier Ca^{2+} (S_{Ca-cat}) channel activated by cell swelling or localized membrane stretch (obtained by suction inside the patch-pipette), in the basolateral membrane of A6 cells (Urbach et al. 1993). When pipette [Ca^{2+}] is buffered at less than 1 μmol/l the channel appears to be nonselective for small cations (25–35 pS). However, the stretch-activated channel be-

mechanosensitive calcium entry and release in renal A6 cells

comes selective for Ca^{2+} with a single channel conductance of $2-3\,pS$ when the pipette solution contains physiological concentrations of extracellular Ca^{2+} ($2\,mM$).

Using spectrofluorescence imaging of Fura-2 loaded A6 cells, we have shown that cell swelling or a localized membrane stretch produced a similar increase in intracellular Ca^{2+} due to calcium entry via $S_{Ca\text{-}cat}$ channel and calcium release from thapsigargin-sensitive stores (Urbach et al. 1995). The mobilization of cytosolic stores could be activated by calcium entry via $S_{Ca\text{-}cat}$ channels or/and by a mechano-sensitive transduction pathway. Several studies have demonstrated an important role of the actin cytoskeleton in RVD and in the regulation of some classes of mechanosensitive channels. In A6 cells, disruption of actin filaments with cytochalasin did not change the activation of $S_{Ca\text{-}cat}$ channels by membrane stretch. However, the normal regulation of $[Ca^{2+}]_i$ during RVD is dependent on an intact cytoskeleton (Leguen et al. 1995).

18.2
Human Colonic Epithelium

The mammalian intestine plays a critical role in fluid homeostasis, being a major site of water and electrolyte absorption and secretion. Fluid absorption, both electroneutral and electrogenic, occurs primarily across the epithelial cells lining the villi, whereas fluid secretion, primarily from electrogenic chloride (Cl^-) movement, occurs across the crypt epithelium.

A human intestinal cell line, the T84 cell line, was derived by Murakami and Masui in 1980 from a lung metastasis of a human colonic carcinoma. Morphologically, the T84 cell line retains many of the characteristics of the major population of cells found in intestinal crypts. Functionally, these cells possess many of the receptor mediated regulatory pathways for Cl^- secretion present in normal colon. Since T84 cells carry out electrogenic Cl^- secretion exclusively, and do not possess absorptive properties they are assumed to be a model for colonic crypt epithelium (Fig. 18.4).

The T84 cell line has the capacity to secrete Cl^- in response to regulatory agents (Barret 1993) and identification of transport pathways involved in receptor-mediated Cl^- secretion has been carried out using two complimentary methods: (1) the Ussing chamber, which identifies transepithelial transport (Koefoed-Johnsen and Ussing 1958), and (2) radionucleotide uptake and efflux studies, which identify transport pathways and detect their activation by regulatory agents. Other electrophysiological methods, e.g., microelectrodes and patch-clamp techniques have also been successfuly applied. Using the above methods, the presence of a Na^+-K^+-Cl^- cotransport pathway, a Na^+-K^+-ATPase, a Na^+-H^+

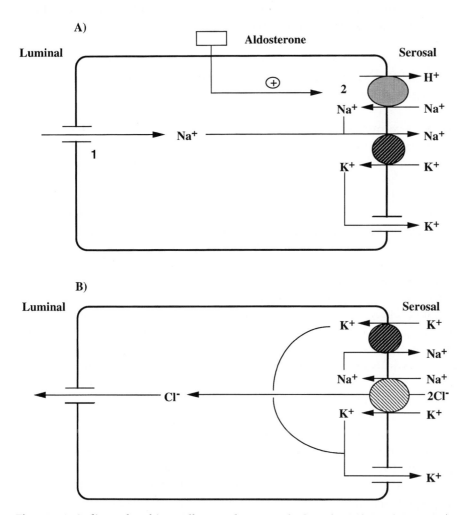

Fig. 18.4.A Sodium absorbing cell: *1* Under normal physiological conditions Na^+ enters the cell through apical Na^+ channels. *2* An early nongenomic effect of aldosterone produces up-regulation of Na^+/K^+ exchange, which in turn causes intracellular alkalinization and activation of basolateral K^+ channels. **B** Chloride secreting cell: Cl^- enters the cell via a basolateral Na^+/K^+/$2Cl^-$ cotransporter and is secreted across the apical membrane via Cl^- channels. Na^+ and K^+ are recycled across the basolateral membrane maintaining electroneutrality

exchanger, two types of K^+ channels on the basolateral membrane (Reenestra et al. 1993) as well as a Cl^- channel on the apical membrane (Halm et al. 1988) have been identified. These transport pathways participate in the Cl^- secretory mechanism and the three "classical" second messengers, cAMP, cGMP (Lin et al. 1992), prostaglandins (Weynier et al. 1985) and intracellular Ca^{2+} (Brayden et al. 1993) have all been implicated in effector mediated Cl^- secretion across crypt cells.

Crypt Isolation and Subculture of Human Colonic Epithelium: Ion Transport Mechanisms

Human colonic epithelial cells can absorb and secrete relatively large volumes of water and electrolytes each day, absorption into the gut lumen is believed to occur in the principal or surface cells (Fig. 18.4), whereas secretion is believed to occur in the crypts of Lieberkühn (Fig. 18.4B). Therefore, together these cells play an important role in the maintenance of salt and water balance within the body. Disease states such as secretory diarrhea (Field and Semrad 1993) or hypertension (Kuchel and Kuchel 1991) may be caused by a shift in salt/water homeostasis at the epithelial level. Absorption and secretion of electrolytes is regulated by endogenous substances, e.g., steroids (Maguire et al. 1994) and amines (Keely et al. 1995), which are released in response to varying whole body osmotic conditions. The degree of interaction between these substances is still not yet fully understood. Steroids, e.g., aldosterone, have been shown to increase Na^+ absorption through a dual action, i.e., they cause classical long-term effects via genomic mechanisms and rapid nongenomic effects (Maguire et al. 1995). These rapid responses may act as a priming effect for the later genomic response and require an upregulation of Na^+/H^+ exchange. Recent studies in rat and human colon implicate a role for intracellular calcium and protein kinase A in the non-genomic effects of aldosterone in target epithelia (Doolan and Harvey 1996a, b). Different human colonic cell lines have been established with the aim of creating a suitable model for the native tissue, examples include T-84, HT-29 and CaCo-2 cell lines. Due to the difficulty in obtaining human colonic specimens, information has only recently become available, using different electrophysiological and pharmacological techniques, concerning the various transporters and hormonal interactions which regulate epithelial function, but there are many questions yet to be answered.

▦ Procedure

Normal samples of human colon were obtained from patients undergoing resection for colonic carcinoma (UCC Hospital Ethics Committee approved). Samples were bathed in a Kreb's solution of the following composition (in mM): NaCl 118, KCl 4.7, $CaCl_2$ 2.5, $MgSO_4$ 1.2, KH_2PO_4 1.2, $NaHCO_3$ 25, glucose 11.1. The pH was maintained at 7.4 by gassing the solution with 95 % O_2/5 % CO_2 and the osmolarity (280–290 mosm) was adjusted by the addition of glucose. The samples were routinely bathed for 1 h in order to remove any circulating hormones present in the tissue. Due to the robust nature of the cells, intact strips of epithelia can be removed from the underlying connective tissue by peeling or blunt dissection.

tissue preparation

Crypts of Lieberkühn can be isolated in the intact state by incubating epithelial cells in a high osmolarity (370–380 mosm), low Ca^{2+} medium. Large strips of epithelium are incubated at room temperature for 15–20 min in 20 ml of dissociation medium of the following composition (mM): NaCl 96, KCl 1.5, Hepes/Tris (35/65 %) 10, Na-EDTA (ethylenediaminetetraacetic acid) 27, sorbitol 55, sucrose 44, and DTT (DL-dithiothreitol) 1; the pH was adjusted to 7.4 by the addition of NaOH. The tissue was shaken gently in order to isolate the crypts from the other cell types and the tissue was removed; at no stage should the tissue be vortexed. The cells are centrifuged for 5 min at 1500 rpm, the supernatant decanted and replaced with a resuspension medium (280–290 mosm) of the following composition (mM): NaCl 140, KCl 5, Hepes/Tris (35/65 %) 10, $MgCl_2$ 1, $CaCl_2$ 2, and glucose 10; the pH was adjusted to 7.4 by the addition of HCl. The cells may be centrifuged or washed again but this is optional since the risk of cell damage increases with the number of centrifugation cycles. Cells remain viable on ice for 6–8 h.

crypt isolation

After isolation, colonic crypt cells are grown as monolayers in a 1:1 mixture of DMEM (Gibco 41966–029) and Hams F-12, supplemented with 10 % FBS, 100 U/ml penicillin, 100 mg/ml streptomycin and 1 % glutamine. Cultures are maintained at 37 °C in an atmosphere of 5 % CO_2/95 % air and stock cultures are fed three times per week. Colonic crypt cells initially exhibit slow growth (doubling time ~3 days) and attach very firmly to the base of the culture flask. Because of this, crypt cells require vigorous trypsinization for subculturing (0.1 % trypsin/0.9 mM EDTA in Mg^{2+} and Ca^{2+}-free phosphate buffered saline). This process

growth and maintenance of colonic crypt cells

usually takes at least 20–30 min (depending on the degree of confluency of the cells) at 37 °C. Colonic crypt cells can be frozen and stored in liquid nitrogen. Freezing medium consists of culture medium supplemented with 5 % DMSO. To break out frozen cells: the cells are shock thawed (37 °C) and washed in ~10 ml of warm culture medium (1500 rpm for 5 min). The pellet is resuspended in 3 ml of fresh culture medium and the cells are plated in a 25 cm^2 flask. Cells are grown to confluence and passaged once before use in experiments.

Investigating Membrane Transport in Human Colon Using the Nystatin Perforation Technique

Nystatin is a polyene antibiotic which inserts pores into biological membranes; these pores are freely permeable to small ions (Lewis et al. 1977). In order to study the serosal membrane in isolation the luminal membrane was bathed in a nystatin bathing solution of the following ionic composition (mM): K-gluconate 110, NaCl 20, KH_2PO_4 1.2, $MgCl_2$ 3, K_2HPO_4 3, EGTA (ethyleneglycol-*bis*- (β-aminoethylether) N,N,N',N'-tetraacetic acid) 5, Hepes-free acid 6, $CaCl_2$ 10–500 nM. The pH was adjusted to 7.2 by the addition of KOH and by continuous gassing of the solution using an air pump. The serosal membrane was bathed in the following solution (mM): NaCl 20, Na-gluconate 101, $NaHCO_3$ 25, KCl 4.7, $CaCl_2$ 2.5, $MgSO_4$ 1.2, KH_2PO_4 1.2, glucose 11.1. The pH was maintained at 7.4 by the addition of H_2SO_4 and gassing with the O_2/CO_2 mix. Following permeabilization of the serosal membrane with nystatin (500 U/ml) a luminal to serosal K^+ gradient was established as observed in the intact cell. Under these conditions, changes in transepithelial current reflect changes in basolateral potassium current (Maguire et al. 1994). Basolateral membrane potassium channels in colon are regulated by endogenous agents and may represent a potential target for therapeutic intervention in disease states.

Fast Steroid Hormone Effects on Ion Transport in Human Colon

In high resistance epithelial cells, mineralocorticoid hormones promote potassium and hydrogen secretion and increase sodium reabsorption. This classical effector mechanism involves the binding of aldosterone to intracellular type I mineralocorticoid receptors initiating genomic events, which have a latency of onset of 2–8 h. Recent studies from our laboratory have demonstrated fast (<1 min) nongenomic activation of K^+ recycling and Na^+-H^+ exchange by mineralocorticoid hormones in human distal colon (Middleton et al. 1993;Nilius et al. 1995). We have also demonatrated rapid stimulation of protein kinase C activity in rat distal colonic epithelium and an increase in free intracellular calcium concentration $[Ca^{2+}]_i$ in isolated rat colonic crypts by mineralocorticoid hormones (Rouse et al. 1994). Here the involvement of Ca^{2+}_i as a possible second messenger in rapid aldosterone effects was investigated in colonic cells by single cell imaging of Fura-2 fluorescence.

Crypt cells were loaded with the Ca^{2+} sensitive dye Fura-2/AM (5 mM) for 30 min at 22 °C. The cells were washed three times (1500 rpm for 5 min) in Krebs solution, pH 7.4 (NaCl 140 mM, KCl 5 mM, $MgCl_2$ 1 mM, $CaCl_2$ 2 mM, Hepes 10 mM, Tris-HCl 10 mM, glucose 10 mM). A micropipette perfusion system was used to apply test solutions to isolated crypt cells plated on a glass coverslip. In all experiments the test solution was identical to the bathing solution except for the dose of drug/hormone used. All experiments were performed at room temperature (20–22 °C) to minimize dye leakage.

spectrofluorescence

Results

Studies from our laboratory have demonstrated rapid (<1 min) nongenomic activation of K^+ recycling and Na^+/H^+ exchange by mineralocorticoids in human colonic epithelium and studies from other laboratories have demonstrated rapid effects of aldosterone on $[Ca^{2+}]_i$ in endothelial cells.

A rapid nongenomic effect of aldosterone on $[Ca^{2+}]_i$ has been demonstrated in the human colonic epithelial cell line T84 (Doolan and Harvey 1996b). Aldosterone induced a rapid increase in $[Ca^{2+}]_i$ within ~2 min. The rise in $[Ca^{2+}]_i$ post-aldosterone appears to result from the activation of a Ca^{2+} influx pathway as: (1) no increase in $[Ca^{2+}]_i$ was observed with aldosterone when cells were bathed in Ca^{2+} free Krebs solution, and (2) emptying of the intracellular Ca^{2+} stores by thapsigargin was not enhanced by aldosterone addition in extracellular Ca^{2+} free solution. In contrast, the Ca^{2+} response to aldosterone, in the presence of 2 mM Ca^{2+} externally, was not decreased following emptying of intracellular Ca^{2+} stores by thapsigargin. Other mineralocorticoid hormones increased $[Ca^{2+}]_i$, whereas the glucocorticoid hydrocortisone failed to increase $[Ca^{2+}]_i$. These results demonstrate the existence of a mineralocorticoid-specific Ca^{2+} signaling pathway in human colonic crypt epithelial cells.

Comments

The involvement of Ca^{2+} as a possible second messenger in rapid aldosterone effects has been reported in human colonic T84 cell culture. Recent studies from our laboratory have demonstrated rapid (within 5 min) activation of PKC activity by aldosterone (0.1 nM) in rat distal colonic epithelium and an aldosterone (0.1 nM) stimulated influx of extracellular Ca^{2+} via a PKC sensitive pathway in isolated rat colonic crypts (Doolan and Harvey 1996a). Other studies from our laboratory have shown aldosterone to produce rapid activation of sulfonylurea-sensitive ATP-regulated K^+ channels (K_{ATP}) in human colonic epithelium (Maguire et al. 1994, 1995). These nongenomic stimulatory effects of aldosterone on K_{ATP} channels can be prevented by inhibition of: (1) the basolateral membrane Na^+/H^+ exchanger or (2) PKC activity. Studies from other laboratories have presented evidence for the activation of K_{ATP} channels in human and rabbit ventricular myocytes and inhibition of Ca^{2+}-sensitive K^+ channels (K_{Ca}) in T84 cells and vascular smooth muscle cells by PKC activation. The K_{ATP} channel is involved in Na^+ reabsorption (a feature of villus cells) and K_{Ca} channel activity is necessary for Cl^- secretion (a feature of crypt cells). We propose that aldosterone enhances net salt reabsorption in the colon by simultaneously activating the pathway for Na^+ reabsorption while down-regulating the Cl^- secretory pathway.

It is clear from the results obtained in this and other studies that the rapid effects of aldosterone are incompatible with classical genomic mechanisms of steroid action and indicate a nongenomic pathway with high affinity for mineralocorticoids and low affinity for glucocorticoids. The rapid aldosterone effects on $[Ca^{2+}]_i$ in colonic cells are the first demonstration of rapid nongenomic steroid effects for these epithelial cells and further support our previous studies that a nongenomic signal transduction mechanism for mineralocorticoid hormone action is operative in epithelia. The rapid in vitro effects of aldosterone are evident within the physiological concentration range of free circulating hormone (0.1 nM) and therefore support its physiological significance.

18.3
Ion Transport in Isolated Subcultured Human and Porcine Chondrocytes

Chondrocytes are the constituent cells of cartilage and are responsible for the maintenance of normal cartilage matrix. The environment of these cells is unusual in comparison to other mammalian cells, as there is no vascular or nervous supply to the tissue. Cartilage matrix contains a high concentration of polyanionic proteoglycans and due to these fixed negative charges there is a high concentration of cations and a low concentration of free anions in comparision to other extracellular fluids, e.g., synovial fluid or plasma (Urban and Hall 1992). Therefore, the tissue osmolality is higher than that found in other extracellular environments. Matrix pH is also approximately 0.5 pH units lower than in synovial fluid. Due to the avascular nature of cartilage there are concentration gradients in metabolites such as protons and in oxygen tension. Obviously, these conditions are important to simulate for chondrocytes isolated outside the joint, particularly in culture conditions.

▓ Materials

– Cartilage medium is prepared as follows: chondrocyte isolation

Reagent (amount)	Supplier (cat. no.)
Dulbecco's modified Eagle's medium (500 ml)	BioWhittaker (12–6147)
Glutamine (5 ml)	Gibco (35050.020)
Penicillin/streptomycin (10 ml)	Gibco (15050.022)
1 M Hepes (5 ml)	BioWhittaker (17.737E)
Fetal bovine serum (50 ml)	BioWhittaker (14.503.A)
Type 1 collagenase	Sigma (C-0130)

▓ Procedure

Only the isolation and culture of articular chondrocytes will be described. Growth-plate chondrocytes are more difficult to obtain due to the presence of the perichrondrium, fibrous connective tissue, around the cartilage.   cell culture

The isolation procedure below is that used in our laboratory, in which chondrocytes obtained by this method are used directly in experiments (Hall et al. 1996; Horwitz et al. 1996).

We obtain pig feet from an abbatoir and these are then transported back to the laboratory as fresh as possible where they are placed in a 70 % alcohol solution before dissection is begun.

1. The metacarpophalangeal joints are opened under aseptic conditions in a flow hood and the cartilage removed. Sterile scalpels and scalpel blades are used for the dissection; in order to ensure sterility scalpels should be autoclaved. It is also essential that a new scalpel and blade be used for each pig foot.

Note: A #22 sterile surgical blade is used to open the joint and #11 sterile surgical blade is used to carefully scrape thin slices of cartilage off. Care must be taken to only take cartilage and not the underlying bone, this can be recognized as pink in color in contrast to the cartilage which is white.

2. Cartilage slices are then collected in a flask containing 30–50 ml Dulbeccos modified Eagle's medium (DMEM) supplemented with fetal calf serum, 1 % penicillin/streptomycin and 1 % glutamine (composition as shown above). These flasks are maintained at 37 °C in 5 % CO_2 atmosphere. The cartilage slices need to be fed every 4–5 days with new medium.

3. Isolated cells are obtained by a 19 h digestion in 20 ml of DMEM with 0.01 mg/ml Sigma crude type I collagenase in a shaking water bath at 37 °C. The solution is then filtered through a metal mesh to remove undigested cartilage and washed three times by centrifugation at 1800 rpm for 10 min. The cells are then resuspended in the experimental solution for use in either patch-clamp or imaging experiments.

Note: Isolated human chondrocytes can be obtained from femoral heads by a modification of the above procedure. In this case the joint is already open so that only a scalpel with a #11 sterile surgical blade is needed to take thin slices of the cartilage.

The cells can then be resuspended and used for experiments or cultured. If the isolated chondrocytes are to be cultured then this should either be as a high density monolayer or in suspension culture, eg. with an agar gel.

We have used such isolated cells in fluorescence experiments to investigate the effect of inflammatory mediators, such as histamine, upon intracellular calcium in chondrocytes (Fig. 18.5). Histamine has been been detected in increased amounts in the joints of rheumatoid arthritis patients (Malone et al. 1986) and histamine receptors, both H_1 and H_2, have been shown to be present in articular chondrocytes (Taylor et al. 1985; Taylor and Woolley 1987). Stimulation of the H_1 receptor was shown to increase prostaglandin E (PGE) production (Taylor et al. 1987), but not to directly affect the breakdown of cartilage.

Cells were loaded with the dye Fura-2 (Molecular Probes, Leiden, The Netherlands) for 45 min at room temperature on poly-L-lysine coated

fluorescence experiments

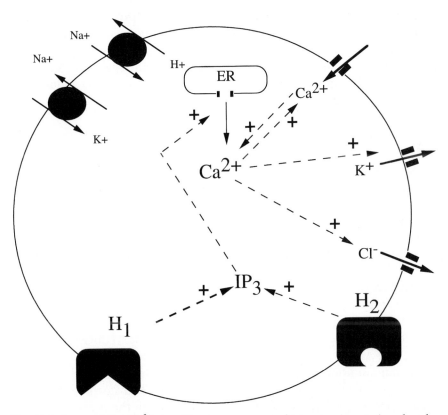

Fig. 18.5. Intracellular Ca^{2+} signaling in response to histamine in porcine chondrocytes. Histamine activation of H1 and H2 receptors stimulate IP3 production and consequent release of Ca^{2+} from the endoplasmic reticulum. The initial rise in Ca^{2+} causes activation of Ca^{2+} influx across the plasma membrane via L-type Ca^{2+} channels. The increase in Ca^{2+} is necessary for the secretory response and activation of other ion channels

glass coverslips. After loading chondrocytes were then washed three times, to remove any excess dye, in the following solution containing (mM):180 NaCl, 10 KCl, 2 $CaCl_2$, 2 $MgCl_2$, 11 glucose, 10 Hepes, pH 7.4 with NaOH and 385 mOsM. A custom-designed puffer pipette system was used to apply test solutions, containing histamine, for a brief period of about 5 s to isolated chondrocytes plated on a glass coverslip. In all experiments the test solution was identical to the bathing solution except for the dose of histamine used. This enabled alteration of the immediate external environment around a single cell without affecting the bulk bathing solution.

We found application of histamine to isolated porcine chondrocytes significantly increased intracellular calcium in a dose-dependent manner and the increase was partially dependent upon the presence of extracellular calcium. This, therefore, implies that there is some role for a plasma membrane calcium transport system in the increase of cytosolic calcium in response to histamine. The increase in intracellular calcium in response to the application of histamine was found to be reduced by both H_1 and H_2 receptor antagonists.

18.4
Cell Culture and Ion Transport in Human Nonpigmented Ocular Ciliary Epithelium

▦ Procedure

tissue culture

The ocular ciliary body cell clone (ODM Cl-2) was established from a primary culture of human nonpigmented ciliary epithelium. Eyes were obtained from the Connecticut Eye Bank and had been enucleated postmortem from a 27 month old male donor. The ciliary processes were cut off and the nonpigmented ciliary epithelial cells were isolated by selected adhesion to surface treated plastic tissue culture dishes (Coca-Prados and Wax 1986). After 1 week of growth in serum-free hormone supplemented media, human nonpigmented epithelial (NPE) cells were transfected with the origin defective mutant 8–4 DNA of SV40. Transformed foci were isolated and transferred into tissue culture flasks in the presence of Dulbeccos modification of Eagle's minimal essential medium containing 10 % fetal calf serum. More than 98 % of the cells were T-antigen positive (Coca-Prados and Chatt 1986). This cell clone (ODM Cl-2) differs from the 8-SVHCE cell clone reported previously, in that in the ODM Cl- 2 no infectious viruses were detected (Coca-Prados and Wax 1986).

The cells are grown in polystyrene culture flasks (25 cm^2, Falcon 3801 Primaria) containing DMEM (Gibco, cat. no. 320–1965) with 10 % FBS and 50 mg/ml gentamicin sulfate (garamycin, Schering). This medium contains 4500 mg/l D-glucose and L-glutamine. NaHCO$_3$ was added to the medium to adjust the pH to between 7.3 and 7.4. The medium was sterilized by membrane filtration with 0.2 μm Corning pore filters. The osmolality of the medium was 328 mOsm. The cells were kept at 37 °C in 5 % CO$_2$ in incubators and passaged every 6–7 days at a split ratio of 1:2 or 1:3; the medium was changed three times a week.

Passage is performed with the following commonly used approach: DMEM was aspirated off and the confluent cells in the flask were rinsed using 10–15 ml PBS, which contained no added Ca^{2+} or Mg^{2+}. The PBS is siphoned off and 3 ml of Ca^{2+} and Mg^{2+}-free HBSS containing 0.05 % trypsin/0.02 % EDTA were added to the flask. The flask is left undisturbed for 3–4 min in an incubator at 37 °C in 5 % C0$_2$/95 % air. The flask is subsequently removed and 5 ml of DMEM and 10 % FCS are added to stop the trypsin reaction. The cells are spun down in a centrifuge tube at 2000 rpm for 2 min. The supernatant is aspirated and 6 ml of medium added to the cell pellet. The cell suspension is then triturated several times to ensure complete dissociation of the cells and then aliquoted out at a ratio of 1:3 into flasks containing 5 ml of warm culture medium. By growing the cells under these conditions, confluency of one flask was achieved between 4 and 6 days.

Back stocks of these cells are made by growing them as above and then freezing the cells as follows: The cells are passaged as described previously. The trypsinization is stopped with DMEM and 10 % FCS and the cell suspension spun down in a centrifuge tube at 2000 rpm for 2 min. The supernatant is aspirated. The cell pellet is resuspended at approximately 2×10^6 cells/ml in freezing solution (70 % DMEM, 20 % FBS and 10 % DMSO). We use 4 ml of freezing solution per confluent 25 cm^2 flask; 1 ml of the suspension is added per freezing vial (Nunc serum vials) which is then tightly closed. The vials are then placed in segregated cardboard boxes and placed on storage canes. The vials are gradually lowered into liquid nitrogen over a 6-h period, being kept at a different level for 1 h. The objective is to maintain a gradual freezing process. To grow the cells from prolonged liquid nitrogen storage, the cells are thawed rapidly in a 37 °C water bath (rapid thawing is necessary to prevent cytoplasmic crystal formation). To each 1 ml of storage suspension 4 ml DMEM +10 % FCS are added and the vials spun at 1500 rpm for 5 min. The supernatant is carefully removed. The pellet of cells is resuspended with 1 ml of warm culture medium and transferred into a 25 cm^2 flask containing 5 ml of culture medium; the flasks are then placed in the CO_2 incubator.

All of the cells studied experimentally were grown on filter cups (Millipore, area $0.6\,cm^2$, Corning, Bedford, MA), with a base composed of collagen (type I, III and IV) which is of similar histological structure to NPE basement membrane. We confirmed that these cultured cells anatomically would function as a secretory epithelial layer. Electron microscopy studies were performed on cells grown on collagen filter cups after 3 days of full confluency (verification of presence of tight junctions and microvilli) . The cells were grown for 3–4 days after full confluency to encourage the full development of all the functional characteristics of this cell type, which may not be fully expressed in the growth phase during cell culture. Experiments were performed only on isolated single cells or on monolayers, between passages 6 and16, cultured according to the conditions described above.

microfluorimetry and imaging Ca^{2+} imaging in living cells requires a suitable incubation chamber with an inverted microscope configuration. The base of a chamber must be made of a material that transmits the required wavelengths of light. With the short working distance and immersion objectives, glass coverslips of $< 0.2\,mm$ thickness were necessary. Open top perfusion chambers were made using plastic rings (0.7 cm deep) which were adhered to the glass coverslips using silicone grease. The isolated cells were placed in these glass bottomed perfusion chambers containing DMEM. The cells were loaded with Fura-2/AM (acetoxymethyl ester form) in DMSO (final concentration $< 0.01\,\%$) at a concentration of 5 µM and incubated in 5% $CO_2/95\%$ air at 37 °C for 35 min. The human NPE (HNPE) cells attached directly to the glass, so polylysine was not used. After 35 min of incubation the cells were washed three times to remove the medium and Fura-2 completely, in modified Krebs solution containing (mM): 138 NaCl, 4.7 KCl, 1.2 MgCl2, 1.2 KH2 PO4, 6 Hepes, 10 glucose and 2.5 CaCl2, pH 7.4 with NaOH, pCa 3 and 297 mOsm. All experiments were performed at room temperature 23°−25 °C to minimize dye leakage. Microfluorimetric measurements were made in the Krebs solution.

imaging system hardware The coverslips were mounted on an inverted epifluorescence microscope (Diaphot 200, Nikon) equipped with 40× and 100× oil immersion lenses. The light from a 100 Watt xenon lamp (Nikon) was filtered through alternating 340 nm and 380 nM interference filters (10 nM bandwidth, Nikon). A custom built chopper circuit controlled the motorized filter wheel and allowed an electronic adjustment of the duration of the exposure at each wavelength. The resultant fluorescence was passed through a 400 nm dichromic mirror, filtered at 510 nm and then collected using an intensified CCD camera system (Darkstar, Photonic Science). Images were digitized and analyzed using the Starwise, Fluo system (Imstar, Paris).

During the aquisition process, at each sampling time, the two source intensity images (at 340 nm and 380 nm) are stored on hard disc. During the aquisition, the information was visualized on-line, that being a pseudocolour calcium image of the individual cells and a curve showing the evolution of the Ca^{2+} changes, since the beginning of the experiment, in a small preselected region of the field, i.e., a single cell or subcellular region. Off-line analysis began with a shading correction for images larger than 256×256 pixels. For background correction, a cell-free region of the field was selected, and the average pixel value, at each wavelength, was subtracted from the corresponding source images. Ca^{2+} values were calculated in circular or polygonal regions (up to four per image) of variable size.

imaging system software

Fura-2 is currently the most widely used of the fluorescent Ca^{2+} indicator dyes and is the indicator of choice for imaging studies. Fura-2 is about 30 times more fluorescent than Quin-2, has a higher K_d for Ca^{2+}, less Mg^{2+} sensitivity, is less susceptible to photobleaching and can be used in the ratio mode. Fura-2 is used at intracellular concentrations in the tens of micromolar as opposed to the millimolar range used for Quin-2, and as a result has much less of a Ca^{2+} buffering effect. With a K_d for Ca^{2+} of about 225 nm, Fura-2 is suitable for $[Ca^{2+}]$; measurements up to several micromolar. The fluorescence excitation maximum of Fura-2 shifts to a lower wavelength on Ca^{2+} binding with a negligible shift in the emission maximum. For optimal separation of fluorescence due to the two forms of the indicator, Ca^{2+}-free Fura-2 is usually monitored at 380 nm and Ca^{2+}-bound dye at 340 nm. Fura-2 has an isofluorescence point at 360 nm which is useful for monitoring Ca^{2+}-independent fluorescence.

calcium dye calibration

For studies of cytosolic free Ca^{2+}, an important consideration is the method by which the indicators are introduced into the cell. Techniques involving microinjection that have been used for Ca^{2+}-sensitive photoproteins and absorbance dyes can be utilized to introduce fluorescent Ca^{2+} indicators. However, these indicators, including Fura-2, also lend themselves to loading as membrane permeant esters, without any requirement for disruption of the cell membrane. This is achieved by incubating intact cells with an esterified form of the indicator (acetoxymethyl esters have been proved to be most useful). Since the esterified indicators are uncharged and hydrophobic, they readily cross the cell membranes. Passive uptake through the membrane and intracellular hydrolysis by endogenous cytosolic esterases releasing the free acid forms results in a final intracellular indicator concentration of 25–100 nM or even higher.

Using Fura-2, which gives shifts in fluorescence spectra, measurements can be carried out at two wavelengths to obtain signals which are proportional to Ca^{2+}-bound and Ca^{2+}-free indicator. The ratio of fluorescence at the two wavelengths is directly related to the ratio of the two forms of the dye and therefore can be used to calculate $[Ca^{2+}]$. The equation which relates the measured Ca^{2+}-bound/Ca^{2+}-free fluorescence ratio (R) to $[Ca^{2+}]_i$ is as follows:

$$[Ca^{2+}]_i = K_d \times (R - R_{min})/(R_{max} - R) \times Sf2/Sb2 \tag{1}$$

where K_d is the product of the dissociation constant of the Ca^{2+}:Fura-2 complex and a constant related to the optical characteristics of the system. R is the ratio F_{340}/F_{380} of the fluorescence signals measured at 340 nm and 380 nm. R_{min} and R_{max} are the limiting values of R in the presence of zero and saturating Ca^{2+} concentrations. Sf2/Sb2 is the ratio of fluorescence values for Ca^{2+}-bound/Ca^{2+}-free indicator measured at the wavelengths used to monitor the Ca^{2+}-free indicator (the denominator wavelength of R). These are all unitless values which depend only on the chemical properties of the indicator and the relative sensitivity of the instrument as a whole for the two wavelengths of measurement. Calibration measurements of zero and saturating levels of Ca^{2+} are required to determine the constants R_{max}, R_{min}, and Sf2/Sb2.

For in vitro calibration, the value of R after background correction was determined in seven calibrating solutions (135 mM KCl; 15 mM NaCl; 10 mM Hepes, pH 7.2; 5 mM EGTA and buffered at various $[Ca^{2+}]$ from 1 nM to 1 mM) containing 10 µM Fura-2 pentapotassium salt. In vivo calibration was done by measuring fluorescence signals in HNPE cells, previously loaded with Fura-2/AM as described, in which the calibration solutions had been introduced by incubation with the ionophone Br-A23187 (10 µM). This was done in Ca^{2+} buffers containing zero $[Ca^{2+}]$, 10 nM, 10 µM, and 1 mM $[Ca^{2+}]$.

bleaching of Fura-2

To reduce dye bleaching, the cells were not exposed to direct light at any stage in preparation and the experiments were performed in a dark room. The intensity UV excitation was set as low as possible whereas the gain of the CCD intensified camera was set high. This allowed us to perform long recordings in the same cells (<45 min).

An estimate of cell volume was obtained as follows: single HNPE cells were viewed with a digital confocal microscope. The cell volume is calculated by analyzing the fluorescent intensity of Fura-2 at the "isosbestic" wavelength. At this wavelength the flourescent intensity is independent of intracellular calcium levels; thus changes in cell volume can be calculated by the corresponding changes in fluorescent intensity at the isosbestic wavelength according to the following equation:

$$\text{Volume/Volume}_{max} = \{(380^* - Bk)^+ (340^* - Bk)\}/2, \tag{2}$$

where 380* represents the fluorescent intensity at 380 nm, 340* is the value of the signal at 340 nm, and Bk is the value of the backround fluoresence signal. Volume changes were found to be associated with equal changes in the size of the cell in all three dimensions.

measurement of cell volume from Fura-2 spectro-fluorescence

Ion Channels and Ca^{2+} Signaling in HNPE Cells

Patch-clamp recording of single ion channel activity and Fura-2 spectrofluorescence have been applied in our laboratory to HNPE cells and the main findings are illustrated in Fig. 18.6. The presence of a voltage-sensitive and calcium-activated large conductance potassium channel (K$_{Ca}$) has been demonstrated and characterized in the HNPE cells. The channel exhibits outward rectification and is of 190 pS conductance.

From single channel and whole-cell patch-clamp analysis it has been confirmed, that this maxi-K$^+$ channel is the predominent ion channel active in this cell in the resting state.

A voltage-sensitive outward rectifying 30 pS Cl$^-$ channel and a voltage-sensitive 17 pS Cl$^-$ channel exhibiting a linear I/V relationship have also been identified (Fig. 18.6).

Maxi-K$^+$ channel activity increases dramatically following agonist binding to angiotensin II, and muscarinic and puringeric receptors, which have been confirmed to be present on this cell type. The K$^+$ channel activity is dependent on an agonist induced increase in free cytosolic calcium levels.

The maxi-K$^+$ channel and intracellular calcium ion activity were increased by adenine and pyramidine nucleotides and the receptor type characterized as P$_{2u}$. Receptor activation increases cytosolic Ca^{2+} levels and is proposed to be mediated predominately by intracellular store release of Ca^{2+}, by an IP$_3$ activated release mechanism (Fig. 18.6). An increase in intracellular Ca^{2+} may in turn be amplified in a cascade mechanism via calcium-induced calcium release (CICR) from other stores.

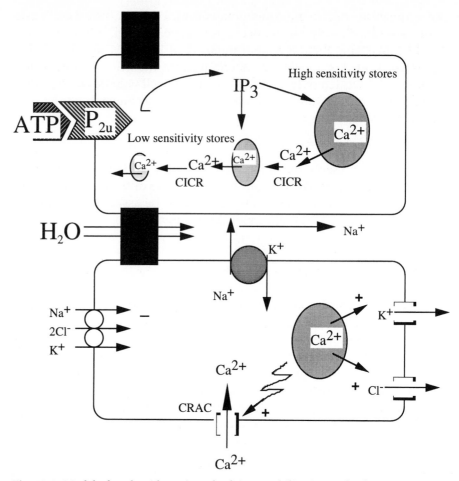

Fig. 18.6. Model of nucleotide-activated calcium mobilization and solute secretion in human nonpigmented epithelium (for details, see text)

The plasma membrane Ca ATPase pump activity is increased by receptor occupancy and provides a regulated extrusion mechanism to remove $[Ca^{2+}]_i$ when the levels become cytotoxic after the agonist induced increase, thus providing a more efficient Ca^{2+} signaling process.

The presence of a Ca^{2+} entry pathway, which is blocked by the L-type Ca^{2+} channel blockers verapamil and nifedipine, has been identified in the plasma membrane of the HNPE cell. This entry pathway is proposed to be involved in the process of refilling of the $[Ca^{2+}]_i$ stores. It is believed that pool depletion triggers activation of these channels (CRAC, calcium release activated calcium entry) via the interaction of a cytoskeletal pathway (Fig. 18.6).

The intracellular calcium signal is found to be heterogeneous in magnitude and displays temperospatial polarity. The signals are initiated from a focal area or "secretory pole" and then travel in an oscillatory-wave like manner throughout the cell via a process involving CICR. Agonist activation of the HNPE cell causes cell secretion and fluid loss with a demonstratable reduction in cell volume.

Based on our experimental data, we have presented a working model for the cellular mechanisms of aqueous humor production (Fig. 18.6). The regulatory mechanisms involved in this fluid secretion process have also been established. Our results may also help to provide an explanation on a molecular level, for the mechanisms by which various pharmacological agents, which are currently in clinical use, induce a reduction in aqueous humour formation. Finally, these studies present us with potential novel pharmacological methods of decreasing aqueous humor production, and provide information leading to the development of possible new alternative adjunct medical treatments for use in the management of glaucoma.

References

Ahmad I, Korbmacher C, Segal AS, Cheung P, Boulpaep EL, Barnstable CJ (1992) Mouse cortical collecting duct cells show nonselective cation channel activity and express a gene related to the cGMP-gated rod photoreceptor channel. Proc Natl Acad Sci USA 89:10262–10266

Anderson MP, Sheppard DN, Berger HA, Welsh MJ (1992) Chloride channels in the apical membrane of normal and cystic fibrosis airway and intestinal epithelia. Am J Physiol 263:L1–14

Barrett KE (1993) Positive and negative regulation of Cl⁻ secretion in colonic cells. Am J Physiol 265:C859-C868

Benos DJ, Cunningham S, Baker RR, Beason KB Oh Y, Smith PR (1992) Molecular caracteristics of amiloride-sensitive sodium channels. Rev Physiol Biochem Pharmacol 120:32–113

Brayden DJ, Krouse ME, Law T, Wine JJ (1993) Stilbenes stimulate colonic Cl⁻ secretion by elevating Ca²⁺. Am J Physiol 264:G325-G333

Brochiero E, Raschi C, Ehrenfeld J (1995) Na/Ca exchange in the basolateral membrane of A6 cell monolayer: role in ca homeostasis. Pflügers Arch 430:105–114

Boucher RC (1994) Human airway ion transport. Am J Resp Crit Care Med 150:271–281

Buchanan JA, Yeger H, Tabcharani JA, Jensen T J, Auerbach W, Hanrahan JW, Riodan JR, Bucwald M (1990) Transformed sweat gland and nasal epithelial cell lines from control and CF individuals. J Cell Sci 109–123

Candia O, Mia AJ, Yorio T (1993) Influence of filter supports on transport characteristics of cultured A6 kidney cells. Am J Physiol 265:C1479–88

Cantiello HF, Stow JL, Prat AG and Ausiello DA (1991) Actin filaments regulate epithelial Na channel activity. Am J Physiol 261:C882-C888

Casavola-V, Guerra-L, Helmle-Kolb-C, Reshkin-SJ, Murer-H (1992) Na+/H(+)-exchange in A6 cells: polarity and vasopressin regulation. J Membr Biol 130:105–114

Coca-Prados M, Chatt G (1986) Growth of nonpigmented ciliary epithelial cells in serum free hormone supplemented media. Exp Eye Res 43:617–620

Doolan CM, Harvey BJ (1996a) Modulation of cytosolic protein kinase C and calcium ion activity by steroid hormones in rat distal colon. J Biol Chem 271:8763–8767

Doi Y, Marunaka Y (1995) Antidiuretic hormone decreases the intracellular pH in distal nephron epithelium (A6) by inhibiting Na+/H+ exchange. Gen Pharmacol 26:723–726

Doolan CM, Harvey BJ (1996b) Rapid effects of steroid hormones on free intracellular calcium in T84 colonic epithelial cells. Am J Physiol 271:(Cell Physiol 40) C1935–C1941

Coca-Prados M, Wax MB (1986) Transformation of Human ciliary epithelial cells by simion virus 40: induction of cell proliferation and retention of b2 – adrenergic receptors. Proc Natl Acad Sci USA 83: 8754–8760

Cott GR and Rao Hydrocortisone promotes the maturation of na dependent ion transport across the fetal pulmonry epithelium. Am J Respir Cell Mol Biol 9:166–171 1993

Cotton CU, Stutts MJ, Knowles MR, Gatzy JT, Boucher RC (1987) Abnormal apical cell mebrane in cystic fibrosis respiratory epithelium. J Clin Invest 79:80–85

Coleman D L, Tuet IK, Widdicombe JH (1984) Electrical properties of dog tracheal epithelial cells grown in monolayer culture. Am J Physiol 245:C355-C359

Cozens AL, Yezzi MJ, Kunzelman K, Ohrui T, Chin L, Eng K, Finbeiner WE, Widdicombe JH, Gruenert DC (1994) CFTR expression and chloride secretion in polarized immortal human bronchial epithelial cells. Am J Respir Cell Mol Biol Vol 10:38–47

De Jong PM, Van Sterkenburg MA, Kempenaar JA, Dijkman JH, Ponec M (1993) Serial culturing of human bronchial epithelial cells derived from biopsies. In Vitro Cell Dev Biol 219A:379–387

Denning G M, Anderson M P, Amara J f, Marshall J, Smith AE, Welsh MJ (1992) Processing of mutant cystic fibrosis transmembrane conductance regulator is temperature-sensitive. Nature 358:761–764

Field M , Semrad CE (1993) Toxigenic diarrheas, congenital diarrheas, and cystic fibrosis: Disorders of intestinal ion transport. Annu Rev Physiol 55:631–655

Graham A, Steel DM, Alton EW, Alton FW, Geddes DM (1992) Second messenger regulation of sodium transport in mammalian airway epithelia. J Physiol 453:475–491

Gruenert DC (1987) Differentiated properties of human epithelial cells transformed in vitro. Biotechniques 5(8):740–749

Gruenert DC, Finkbeiner WE, Widicombe JH (1995) Cultured and transformation of human airway epithelial cells. Am J Physiol 268:L347-L360

Gstraunthaler GJA (1988) Epithelial cells in tissue culture. Renal Physiol Biochem 11:1–42

Hall AC, Horwitz ER, Wilkins RJ (1996) The cellular physiology of articular cartilage. J Exp Physiol 81:535–545

Halm DH, Rechkemmer GR, Schoumacher RA, Frizzell RA (1988) Apical membrane chloride channels in a colonic cell line activated by secretory agonists. Am J Physiol 254:C505-C511

Harvey BJ, Urbach V (1995) Regulation of ion and water transport by hydrogen ions in high resistance epithelia. In: Heisler N (ed) Advances in comparative and environmental physiology, vol 22. Springer, Berlin Heidelberg New York, pp 153–183

Harvey BJ, Urbach V, Van Kerkhove E (1991) Inward rectifier K channels in principal cells of isolated frog skin epithelium and in cultured A6 cells. J Physiol 438:264P

Haws C, Krouse ME, Xia Y, Gruenert DC, Wine JJ (1992) CFTR channels in immortalised human airway cells. Am J Physiol 263:L692-L707

Hoffmann EK, Duham PB (1995) Membrane mechanisms and intracellular signalling in cell volume regulation. Int Rev Cytol 161:173–262

Horwitz ER, Higgins T, Harvey BJ (1996) Histamine induced cytosolic calcium increase in porcine articular chondrocytes. Biochim Biophys Acta 1313:95–100
Keely SJ, Stack WA, O'Donoghue DP, Baird AW (1995) Regulation of ion transport by histamine in human colon. Eur J Pharm 279:203–209

Kilburn KH (1968) A hypothesis for pulmonary clearance and its implications. Am Rev Respir Dis 98:449–463

Knowles MR, Gatzy JT, Boucher RC (1985) Aldosterone metabolism and transepithelial potential difference in normal and cystic fibrosis subjects. Pediatr Res 19:676–679

Koefed-Johnsen V, Ussing HH (1958) The nature of frog skin potential. Acta Physiol Scand 42:293–308

Korbmacher C, Segal AS, Fejes-Toth G, Giebisch G, Boulpaep EL (1993) Whole-cell currents in single and confluent M-1 mouse cortical collecting duct cells. J Gen Physiol 102:761–793

Kuchel OG, Kuchel GA (1991) Peripheral dopamine in pathophysiology of hypertension: Interaction with aging and lifestyle. Hypertension 18:709–721

Kunzelmann K, Kathofer S, Greger R (1995) Na and Cl conductances in airway epithelial cells: increased Na conductance in cystic fibrosis. Pflugers Arch 431:1–9

Kunzelmann K, Kathofer S, Hipper A, Gruenert DC, Greger R (1996) Culture -dependent expression of Na^+ conductances in airway epithelial cells. Pflugers Arch 431:578–586

Kunzelmann K, Schwiebert EM, Zeitlin PL, Kuo WL, Stanton BA, Gruenert DC (1993) An immortalized CF tracheal cell line homozygous for the ΔF 508 CFTR mutation. Am J Respir Cell Mol Biol 8:522–529

Lechner JF, LaVeck MA (1985) A serum-free method for culturing normal human bronchial epithelial cells at clonal density. J Tissue Culture Methods 9:43–48

Leguen I, Urbach V, Harvey BJ (1995) Cytochalasin effect on mechanosensitive calcium chgannel activity and cell swelling-induced calcium mobilisation in renal A6 cells. 489:92P

Letz B, Ackermann A, Canessa CM, Rossier BC, Korbmacher C (1995) Amiloride-sensitive sodium channels in confluent M-1 mouse cortical collecting duct cells. J Membr Biol 148:127–141

Lewis SA, Eaton, DC, Clausen C, Diamond JM (1977) Nystatin as a probe for investigating the electrical properties of tight epithelium. J Gen Physiol 70:427–440

Lin M , Nairn AC, Guggino SE (1992) cGMP-dependent protein kinase regulation of a chloride channel in colonic cells. Am J Physiol 262:C1304-C1312

Ling BN and Eaton DC (1989) Effect of luminal Na+ on single Na channels, a regulatory role for protein kinase C. Am J Physiol 256:F1094-F1103

Maguire D, O' Sullivan G, Harvey BJ (1994) Potassium ion channels and sodium absorption in human colon. Surgical Forum 80:195–197

Maguire D , O' Sullivan G, Harvey BJ (1995) Membrane and genomic mechanisms for aldosterone effect in human colon. Surgical Forum 81:203–205

Mall M, Hipper A, Greger R, Kunzelman K (1996) Wild type but not ΔF 508 inhibits Na conductance when coexpressed in *Xenopus* oocytes. FEBS Letters 381:47–52

Malone DG, Irani AM, Schwartz B, Barrett KE, Metcalfe DD (1986) Mast cell numbers and histamine levels in synovial fluids from patients with diverse arthritides. Arthritis Rheum 29:944–955

Mason SJ, Paradiso AM, Boucher RC (1991) Regulation of transepitehlial ion transport and intracellular calcium by extracellular adenosine triphosphate in human normal and cystic fibrosis airway epithelium Br J Pharmacol 103:1649–1656

Middleton JP, Mangel AW, Basavappa S, Fitz JG (1993) Nucleotide receptors regulate ion transport in renal epithelial cells. Am J Physiol 264:F867-F873

Murphy E, Cheng E, Yankaskas J, Stutts MJ, Boucher RC (1988) Cell calcium levels of normal and cystic fibrosis nasal epithelium. Pediatr Res 24:79–84

Nilius B, Sehrer J, Heinke S, Droogmans G (1995) Ca^{2+} release and activation of K^{+} and Cl^{-} currents by extracellular ATP in distal nephron epithelial cells. Am J Physiol 269:C376-C384

Paradiso AM (1992) Identification of Na/H exchange in human normal and cystic fibrosis ciliated airway epithelium. Am J Physiol 262:L757-L764

Paradiso AM, Cheng EHC, Boucher RC (1991) Effects of bradykinin on intracellular calcium regulation in human ciliated airway epithelium. Am J Physiol 261:L63-L69

Rafferty KA (1975) Epithelial cells grown in culture of normal and neoplastic forms. Adv Cancer Res 21:249–272

Reenestra WW (1993) Inhibition of cAMP- and Ca^{2+}-dependent Cl^{-} secretion by phorbol esters: Inhibition of basolateral K^{+} channels. Am J Physiol 264:C161-C168

Rich DP, Anderson MP, Gregory RJ, Cheng SH, Paul S, Jefferson DM, McCann JD, Klinger KW, Smith AE and Welsh MJ (1990) Expression of cystic fibrosis transmembranaire conductance regulator corrects defective chloride channel regulation in cystic fibrosis airway epithelial cells. Nature 347:358–363

Rodriguez-Commes J, Isales C, Kalghati L, Gasalla-Herraiz J, Hayslett JP (1994) Mechanism of insulin-stimulated electrogenic sodium transport. Kidney Int 46:666–674

Rouse D, Leite M, Suki WN (1994) ATP inhibits the hydrosmotic effect of AVP in rabbit CCT: evidence for a nucleotide P_{2u} receptor. Am J Physiol 267:F289-F295

Schafer JA, Troutman SL (1990) cAMP mediates the increase in apical membrane Na^{+} conductance produced in rat CCD by vasopressin. Am J Physiol 259:F823-F831

Schwiebert EM, Egan ME, Hwang TH, Fulmer SB, Allen SS, Cutting GR, Guggino WB (1995) CFTR regulates ORCC thriugh an autorine mechanism inolving ATP. Cell:1063–1073

Scolte BJ, Kansen M, Hoogeveen AT, Willemse R, Rhim J S, Vander Kamp AWM, Bijman J (1991) Immortalization of nasal polyp epithelial cells from cystic fibrosis patients. Exp Cell Res 182:559–571

Stoos BA, Naray-Fejes-Toth A, Carretero OA, Ito S, Fejes-Toth G (1991) Characterisation of a mouse cortical collecting duct cell line. Kidney Int 39:1168–1175

Stutts MJ, Knowles MR, Gatzy JT, Boucher RC (1986) Oxygen consumption and oua-
   bain binding sites in cystic fibrosis nasal epithelium. Pediatr Res 20:1316–1320

Stutts MJ, Chinet TC, Mason SJ, Fulton JM, Clarke LL, Boucher RC (1992) Regulation
   of chloride channels in normal and cystic fibrosis airway epithelial cells by extracellular ATP. Proc Natl Acad Sci USA 89:1621–1625

Stutts MJ, Canessa Cecilia, Olsen JC, Hamrick M, Cohn JA, Rossier BC, Boucher RC
   (1995) CFTR as a cAMP-dependent regulator of Na channels. Sciences 269:847–851

Taylor DJ, Yoffe JR, Brown DM, Woolley DE (1985) Histamine H2 receptors on
   chondrocytes derived from human, canine and bovine articular cartilage. Biochem J:315–319

Taylor DJ, Yoffe JR, Brown DM, Woolley DE (1986) Evidence for both histamine H1
   and H2 receptors on human articular chondrocytes. Arthritis Rheum 29:160–165

Taylor DJ, Woolley DE (1987) Histamine stimulates prostaglandin E production by
   rheumatoid synovial cells and human articular chondrocytes in culture. Ann
   Rheum Dis 46:431–435

Urbach V Prosser E, Raffin JP, Thomas S, Harvey B (1994) Activation of Cl channel
   by extracellular UTP in human lung epithelium. J Gen Physiol 104:84

Van Scott MR, Yankaskas, Boucher RC (1986) Culture of airway epithelial cells:
   research techniques. Exp Lung Res 11:75–94

Walsh D, Urbach V, Harvey B (1996) Nucleotide regulation of intracellular calcium
   and ion channel activity in human normal and cystic fibrosis lung. The respiratory
   system in health and disease. Welcome Trust Meeting. March 1, 1996

Ward CL, Krouse ME, Gruenert DC, Kopito RR, Wine JJ (1991) Cystic fibrosis gene
   expression is not correlated with rectifying Cl-channels. PNAS 88:5277–5281

Widdicombe JH, Welsh MJ, Finkbeiner WE (1985) Cystic fibrosis decreases the apical
   membrane chloride permeability of monolayers cultured from cells tracheal epithelium. Proc Natl Acad Sci USA 82:6167–6171

Willumsen NJ, Boucher RC (1992) Intracellular pH and its relationschip to regulation
   of ion transport in normal and cystic fibrosis human nasal epithelia. J Physiol
   455:247–269

Willumsen NJ, Davis CW, Boucher RC (1989) Cellular Cl transport in cultured cystic
   fibrosis airway epithelium Am J Physiol 256:C1045-C1053

Wu R, Nolan E, Turner C (1985) Expression of trchelal differenciation in serum-free,
   hormone-supplemented medium. J Cell Physiol 127:167–181

Yanase M, Handler JS (1986) Activators of protein kinase C inhibit sodium transport
   in A6 epithelia. Am J Physiol 250:C517-C523

Yankaskas JR, Cotton CU Knowles MR, Gatzy JT, Boucher RC (1985) Culture of
   human nasal epithelial cells on collagen matrix supports. Am Rev Respir Dis
   132:1281–1287

Yue G, Hu P, Oh Y, Jilling T, Shoemaker R L, Benos D J, Cragoe E J, Matalon S (1993)
   Culture-induced alterations in alveolar type II cell Na conductance. Am J Physiol
   265:C630-C 640

Urbach V, Andersen H, Hall J, Harvey BJ (1993) A calcium permeable cation channel
   activated by membrane stretch and cell swelling in renal A6 and frog skin epithelia. J Physiol 467:273P

Urbach V, Leguen I, Harvey BJ (1995) Volume and mechano-sensitive calcium entry
   and release in renal A6 cells. J Physiol 489:87P

Urban JPG, Hall AC (1992) In: Kuettner K et al. (ed) Articular cartilage and osteo-
   arthritis. Raven, New York, pp 393–406

Verrey F (1994) Antidiuretic hormone action in A6 cells: effect on apical Cl and Na conductances and synergism with aldosterone for NaCl reabsorption. J Membr Biol 138:65–76

Weynier A, Huott PA, Liu W, McRoberts JA Dharmasthaphorn K (1985) Chloride secretory mechanism induced by prostaglandin E_1 in a colonic epithelial cell line. J Clin Invest 76:1828–1836

Chapter 19

Isolation, Cultivation and Differentiation of Lung Type II Epithelial Cells

Paula Meleady*, Finbar O' Sullivan, Shirley McBride, and Martin Clynes

Introduction

Type II alveolar epithelial cells account for approximately 5 % of the epithelial cell population of the alveoli and play many roles in the physiology and biology of the lung. They function as stem cells of the alveolar epithelium and proliferate and differentiate into type I cells during lung growth or after injury. These thin, squamous type I cells account for the other 95 % of the alveolar lining and form the barrier over which gas exchange takes place in the lung. Following damage to these fragile cells, type II cells proliferate and differentiate into type I cells to restore the integrity of the alveolar surface. Type II cells also function in the metabolism of xenobiotics (Baron and Voight 1990) and the regulation of ion transport (Liu and Mantone 1996) and in the immune responses of the lung (Cunningham et al. 1994). They also function in the synthesis and secretion of surfactant (Bates et al. 1994). The ability to obtain pure isolates in vitro of type II cells from animal lungs is of considerable benefit to investigate the activities of this extremely specialized cell type. Once placed into culture, type II cells begin to differentiate and to acquire morphological and biological characteristics of type I cells (Borok et al. 1994). Type II to type I differentiation, which can be controlled to some extent in culture, serves as an extremely useful model for examining the mechanisms involved in both spontaneous and chemically induced epithelial differentiation.

* *Correspondence to* Paula Meleady, National Cell and Tissue Culture Centre, Dublin City University, Glasnevin, Dublin 9, Ireland; phone +353–1–7045726; fax +353–1–7045484; e-mail pmeleady@ccmail.dcu.ie

▓ Outline

A flow chart illustrating the purification procedure of human type II cells is shown in Fig. 19.1.

19.1
Isolation of Human Type II Cells

Various methods for the isolation of relatively pure type II cells from lung tissue have been described previously. These include differential attachment (Kikkawa and Yoneda 1974), use of monoclonal antibodies (Funkhouser et al. 1987), IgG panning (Dobbs et al. 1986) and centrifugal elutriation (Devereux et al. 1986). However, problems with purity of isolates, specificity of antibody markers and expensive equipment mean that these methods are not always practical. Described here is a method for the isolation of highly purified type II cells from human tissue which employs a Percoll gradient cell density separation technique. Following centrifugation, type II cells locate at a specific point on a Percoll gradient which is controlled by the lipid rich, surfactant-containing lamellar bodies present in each cell.

▓ Materials

- Approximately 10 g "normal" human lung tissue. **Note:** Samples are usually obtained during surgical resection of lung tumors. The samples used for type II isolations should be, at least visibly, nontumoral.
- Filter paper
- Beaker
- Sterile 500 ml bottles
- Sterile surgical gauze
- 50 ml centrifuge tubes
- Non-tissue culture grade sterile petri dishes
- Sharp dissection instruments (pointed scissors, scalpel, forceps)
- McIlwain tissue chopper. **Note:** If not available, a sharp scissors and/or scalpel blade may be used.

Fig. 19.1. Flow chart illustrating the purification of human type II cells using a Percoll density gradient ▶

lung sample

Dissection and chopping

Wash - shake for 2-3 mins in PBSA   (removes blood cells and macrophages)

filter through guaze

Digestion   (sterile trypsin - 1x5mins, 2x20mins)

Filtration

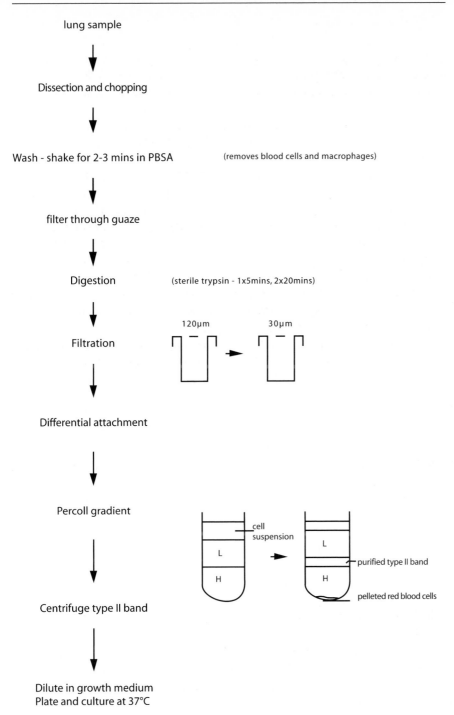

Differential attachment

Percoll gradient

Centrifuge type II band

Dilute in growth medium
Plate and culture at 37°C

- Shaking water bath
- CO_2 incubator
- Refrigerated centrifuge
- Krebs-Ringer phosphate buffer: 130 mM NaCl; 5.4 mM KCl; 1.9 mM $CaCl_2$; 1.29 mM $MgSO_4$; 10 mM Na_2HPO_4; 11 mM glucose. Adjust to pH 7.4.
- Ca/Mg-free phosphate-buffered saline (PBS–): 130 mM NaCl; 5.4 mM KCl; 11 mM glucose; 10.6 mM Hepes; 0.1 mM EGTA; 2.6 mM Na_2HPO_4. Adjust to pH 7.4.
- PBS+: PBS with 1.9 mM $CaCl_2$; 1.29 mM $MgSO_4$; without EGTA
- +Ca/Mg solution: 500 ml 0.9 % NaCl; 10 ml 0.11 M $CaCl_2$; 20 ml 0.15 M KCl; 5 ml 0.15 M $MgSO_4$; 15 ml 0.10 M phosphate buffer, pH 7.4; 30 ml 0.20 M Hepes; 630 mg glucose. Adjust to pH 7.4, filter sterilize.
- –Ca/Mg solution: As for +Ca/Mg solution but without Ca and Mg. Do not filter sterilize.
- Trypsin solution: 500 mg trypsin (Sigma T8003; **Note:** This is a relatively pure and gentle crystalline form of trypsin which results in high yields of good condition cells.); 200 ml +Ca/Mg solution. Filter sterilize. **Note:** This may take a while to dissolve.
- Differential attachment medium: 21 ml Waymouths medium (Gibco 31220–023); 8 mg DNase 1 (Boehringer Mannheim 1284 932); 1 ml fetal bovine serum; 0.5 ml fungizone; 0.5 ml penicillin/streptomycin; 2 mM L-glutamine
- Stock 10× –Ca/Mg solution: 200 ml 0.9 % NaCl; 25 ml 1.5 M KCl; 25 ml 1.0 M Na_2HPO_4; 2.0 M Hepes; 1.26 g glucose. Filter sterilize.
- Percoll heavy gradient: 5 ml stock 10× –Ca/Mg solution; 0.25 ml fetal bovine serum; 12.5 ml distilled H_2O; 32.5 ml Percoll (Pharmacia 17–0891–01)
- Percoll light gradient: 5 ml stock 10× –Ca/Mg solution; 0.25 ml fetal bovine serum; 31.5 ml distilled H_2O; 13.5 ml Percoll; 1 drop phenol red
- Growth medium: DCCM1 medium (Biological Industries 05010A), 1 % Ultroser G (Gibco 15950–017), 2 mM L-glutamine

▦ Procedure

procurement and storage of tissue sample

1. Tissue samples are generally obtained during lobectomy or pneumectomy procedures for lung cancer. A large portion (at least 10 g) of macroscopically normal tissue is required; this is obtained during surgical resection.

2. Place the sample in Krebs-Ringer phosphate buffer for transport and begin the isolation procedure as quickly as possible. If the transportation of the sample is longer then 15 min then the sample should be carried on ice.

Note: Work is carried out in a class 2 laminar flow hood from this point.

dissection of tissue

1. Dissect away as much of the pleura (a thin membrane which covers the lungs) as possible from the lung section using a sharp scissors. Remove also any visible airways and blood vessels.

2. Cut the tissue into smaller sections (approximately 250 mg).

3. Place filter paper on the cutting base of the McIlwain tissue chopper and spread the lung pieces out on this.

4. Cut the tissue into 0.7 mm slices. **Note:** If a McIlwan chopper is not available, the tissue may be chopped into 1 mm cubes using a curved scissors.

washing tissue

1. Place the sliced tissue into a 500 ml sterile bottle containing 200 ml sterile PBS– and shake vigorously for 2–3 min to remove blood cells and macrophages.

2. Remove the solution from the tissue by filtering through sterile gauze (if unavailable, a sterile stainless steel tea strainer is just as useful) over a beaker and retrieve the tissue from the gauze.

3. Repeat the washing procedure (steps 1 and 2) a further three times.

digestion of tissue

1. Following the final wash, replace the tissue into the 500 ml bottle, add 50 ml sterile trypsin solution and incubate at 37 °C for 5 min in a shaking water bath (60 rpm).

2. Remove the trypsin by filtration through gauze as before. **Note:** This is merely a washing step, therefore the trypsin solution is not retained. The subsequent trypsin steps are the actual digestions and the trypsin solutions removed from the tissue are retained.

3. Incubate with 3×50 ml sterile trypsin for 1×5 min and 2×20 min, respectively. The trypsin solutions are removed each time by filtering through sterile surgical gauze as before and are retained in a sterile 500 ml bottle.

4. To this trypsin solution, add 10 ml fetal bovine serum/200 ml trypsin.

5. Following this, to retrieve the maximum number of cells possible, the tissue may be further chopped, washed with PBS−, passed through the gauze and pooled with the previous solution. This step may not be necessary if the starting material is relatively large or if the digestion appears to have gone well.

filtration

1. The trypsin:cell solution is then passed through a 25–30 μm sterile filter (Falcon, cat no. 2340). It may be necessary to pre-filter the solution through a larger gauze size first, e.g., 120 μm (Falcon, cat no. 2360).

2. Centrifuge the final filtrate in 50 ml centrifuge tubes for 10 min at 250 g and resuspend the resulting pellets in 3–5 ml differential attachment medium.

differential attachment

1. Plate the cell suspension in non-tissue culture grade sterile petri dishes with the largest possible surface area to volume ratio to allow as many cells as possible to attach to the plastic and incubate at 37 °C, 5 %CO_2 for 1 h. For example, a 90 mm^2 petri dish could be used with 5 ml of cell suspension.

2. Following differential attachment, remove the supernatant from the petri dish and wash the dish very gently with 10 ml PBS−. Pool this with the supernatant.

Percoll gradient and centrifugation

Note: A properly prepared Percoll gradient is crucial to achieve a good type II cell isolation. The gradient is prepared as follows:

1. Pipette 10 ml Percoll heavy gradient solution into a 50 ml centrifuge tube.

2. Very gently, pipette 10 ml Percoll light gradient solution on top of the heavy gradient. The aim is to obtain a distinct interface between the two layers, with as little mixing as possible. This can be achieved by touching the tip of the pipette against the inside of the tube and allowing the light gradient solution to flow down onto the heavy layer. It may be necessary to prepare an excess of the Percoll solutions and pour several gradients, selecting only those with the sharpest interfaces.

3. The supernatant from the differential attachment is then layered onto the Percoll: Pipette the cell supernatant onto the Percoll. It is important to obtain a sharp gradient at this stage also and the solution is poured as in step 2 above.

4. Centrifuge the gradients for 20 min, 250 g at 10 °C. **Note:** Centrifugation must be carried out at 10 °C as the Percoll densities vary with temperature.

5. Following centrifugation, the type II cells should be visible as a white, creamy band between the two Percoll layers. Surfactant lipids and cell debris will remain above the light Percoll layer while red blood cells will have formed a pellet at the bottom of the tube. Very carefully, using a sterile pipette, remove the upper layers of solution down to approximately 1 cm from the white type II layer. Using a new pipette, remove the remaining solution above the type II cells, along with the type II band itself, down to approximately 1 cm below the band. It is convenient to mark the approximate volume to be removed (usually 8–10 ml) on the outside of the tube beforehand. Transfer this type II suspension to a new 50 ml centrifuge tube. If at this stage a large number of red blood cells appear to remain, centrifuge the type II suspension at 250 g for 5 min, 10 °C. The red blood cells will pellet while the type II cells should remain in suspension.

6. Dilute the cell suspension with an equal volume of PBS– and centrifuge at 250 g for 10 min, 10 °C.

7. Resuspend the pellet in 50 ml PBS+ containing 1.4 mg DNase. Mix well by inverting the tube and then centrifuge at 250 g for 10 min, 10 °C.

8. Resuspend the cells in the appropriate growth medium.

19.2
Culture of Lung Type II Cells

Type II cells generally differentiate quite rapidly (hours/days) into type I cells in vitro. This has resulted in much debate as to whether the current culture procedures are successful in maintaining true type II cultures or are actually failing to do so and resulting in type II-like, type I-like or type I cultures (Paine and Simon 1996). While identification of type I and type II cells is reasonably straightforward the less well-differentiated phenotypes are not well characterized and so are more difficult to identify. It is important to verify the differentiation status of so-called type II cultures before interpreting data obtained from experiments invoving these cells. The conditions described here are those in which have been used for short-term (days) culture of type II cells for use in, for example, toxicity or ion transport experiments, rather than conditions for longer-term cultures.

A procedure for the isolation of highly purified human type II cells can be found in Sect. 19.1 of this chapter. Here, serum supplemented and serum-free media are described for the culture of type II cells.

▓ Materials

- Freshly isolated type II cells
- Culture medium: Waymouth's supplemented with 10 % FCS, 1 % pen/ strep solution, 1 % fungizone, 2 mM L-glutamine
- Extracellular matrix (ECM)-coated plates (Biological Industries E-TCMT-F)

▓ Procedure

serum-supplemented medium

The following procedure was established by Hoet et al. (1994) for the culture of type II cells for toxicity experiments.

1. Rinse each well of the ECM-coated plate three times with sterile PBS prior to use. **Note:** ECM plates are used as type II cells retain their cuboidal morphology better on this surface than on uncoated plastic which promotes spreading and retention of morphology. This appears to be important in preventing type II to type I differentiation.

2. Dilute the type II cells with culture medium to obtain a concentration of 5×10^5 cells/ml and inoculate the ECM plates with 200 μl cell suspension per well. The cells are incubated at 37 °C, 5 % CO_2 and allowed to attach overnight before being assayed.

3. The following day, remove the medium from the cells and rinse the wells once gently with PBS to remove any unattached cells which may be either dead or damaged type II cells or other contaminating cells such as alveolar macrophages, which are the main contaminant of type II cell isolates. The type II cells may then be assayed as required.

Paine et al. (1995) described a medium for the culture of type II cells isolated from rats on tissue culture treated plastic. This medium consists of the following components: Dulbecco's modified Eagle's medium (DMEM); 10 % heat-inactivated fetal calf serum; 1 % pen/strep; 1 % sodium pyruvate; 1 % L-glutamine; 1 % nonessential amino acids and vitamins.

Kumar et al. (1995) have reported on a serum-free medium which they have found capable of maintaining the differentiated phenotype of type II cells for at least 6 days in vitro. This medium is as follows: MCDB 201 medium supplemented with 5 mg/ml fatty acid-free (FAF)-BSA; 30 µg/ml transferrin; 6 µg/ml soybean lipids; 3 µg/ml cholesterol; 1 µg/ml sphingo-myelin; 0.2 µg/ml vitamin E acetate.

defined serum-free medium

1. Dilute the cells to a concentration of 3×10^6 cells/ml and inoculate the 96-well plates with 100 µl/well. **Note:** While uncoated plastic promotes cell spreading, the absence of serum, which also induces spreading, from the cultures appears conducive to the retention of the type II cell morphology which is important to the maintenance of the differentiated phenotype.

2. Incubate at 37 °C, 5 % CO_2 for 48 h with no medium change. If cultures are to be maintained for longer, the medium may be changed every 3 days.

Borok et al. (1994) described a serum-free medium for adult rat alveolar type II cells grown on tissue-culture treated polycarbonate filters. This medium is as follows: DMEM/Hams F12 (1:1) supplemented with insulin (5 µg/ml); transferrin (5 µg/ml); selenous acid (5 ng/ml); linoleic acid (5.35 µg/ml); Hepes (10 mM); nonessential amino acids (0.1 mM); L-glutamine (2 mM); dexamethasone (0.1 µM); penecillin (100 U/ml); streptomycin (100 µg/ml).

Other serum-free formulations have been described in the literature. For example, Fraslon et al. (1991) developed a medium for rat fetal alveolar type II cells. Since fetal cells divide more readily in culture than their adult counterparts, serum-free medium is more easily attainable for such cell types.

19.3
Differentiation of Lung Type II Cells

Once placed into culture, type II cells begin to differentiate and to acquire morphological and biological characteristics of type I cells (Borok et al. 1994). This in vitro type II to type I cell differentiation serves as an extremely useful model for examining the mechanisms involved in both spontaneous and chemically induced epithelial cell differentiation. Type II to type I cell differentiation can be manipulated in vitro in the absence of differentiating agents such as retinoic acid and vitamin D₃ by manipulating culture conditions such as seeding densities, culture medium used and the substratum to which the cells attach. When attempting to culture these cells initially, it is advisable to assess a range of each such variable on one's own cultures to determine which is the most suitable for the type II characteristic or system one wishes to study. For example, serum contains a number of growth factors and soluble matrix components that may influence alveolar functions and differentiation. Bovine serum has been shown to accelerate loss of type II differentiated phenotype and inhibit DNA synthesis whereas exposure to rat serum promotes preservation of type II cells phenotype (Borok et al. 1994). Several culture media, both serum supplemented and defined serum-free media, have been developed for culturing type II cells as outlined above in Section 19.2.

Several types of matrices have been reported for the culture of type II cells. These include tissue culture treated plastic (Kumar et al. 1995), Matrigel matrix (Sigma, cat no. E1270) (Liu et al. 1996) and ECM-coated plates (Hoet et al. 1994).

Seeding densities and especially matrix used seem to influence the 3-dimensional shape of type II cells. Conditions which appear to allow retention of type II cuboidal morphology, such as high seeding densities (Paine et al. 1995) and certain ECM proteins (Kumar et al. 1995), appear to facilitate the retention of other type II cell characteristics, while those which induced spreading of cells, such as low seeding densities, induce type I differentiation.

The following protocols outline ways to manipulate type II cell differentiation in vitro and ways in which to identify type I and type II cells in culture.

Materials

- Freshly isolated type II cells
- Culture medium
- Sterile glass coverslips
- 35 mm petri dishes
- Methanol and acetone
- Extracellular matrix proteins (fibronectin, collagen, laminin)
- Multiwell tissue culture plates

Procedure

1. Plate cells at high (3.5×10^5 cells/cm^2) and low (0.5×10^5 cells/cm^2) densities onto sterile glass coverslips and incubate in 35 mm petri dishes and incubate at 37 °C, 5 % CO$_2$ for 2 days.
   seeding densities

2. Fix in methanol (−20 °C) for 5 min followed by acetone (−20 °C) for 30 s.

3. Determine the presence of type I and type II specific markers (see below) by immunocytochemical analysis.

1. Precoat multiwell plates as follows: Reconstitute ECM in PBS or in medium which contains no serum (fibronectin at 10 µg/cm^2, collagen at 5 µg/cm^2 and laminin at 6 µg/cm^2) and place each (individually or in combination) into culture wells in sufficient volumes to cover the floors of the wells. Incubate at 4 °C overnight, remove solutions from wells and rinse with PBS ×3.
   culture in extracellular matrix-coated plates

2. Inoculate coated wells with type II cells and incubate at 37 °C, 5 % CO$_2$ for 2–3 days.

3. Cells may then be harvested or fixed (as above) for analysis.

Results

Identification of Type I and Type II Cells

Electron microscopic (EM) analysis of type I and/or type II cell preparations is one of the most definitive methods of cell identification. This technique allows detection of surfactant-containing lamellar bodies which are present in mature, healthy type II cells but absent from type I cells (Ten Have-Obroek et al. 1991). However, EM is not always available and is often not practical in terms of both costs and experimental endpoints.

Several proteins have been identified as specific for type II cells and these may be used as markers in immunological procedures (Guzman et al. 1994; Makker et al. 1989). Stevens et al. (1995) identified a surfactant protein A (SP-A) binding protein specific to rat type II cells.

Many type II identification methods employ stains which are specific for lamellar bodies unique to type II cells of the lung such as the tannic acid stain (Mason et al. 1985) and the Papanicolaou stain (Dobbs 1990).

However alveolar macrophages from human isolates of injured lungs have many inclusions and sometimes show positive staining thus hindering identification of the type II cells (Edelson et al. 1988).

The lectin of *Maclura pomifera* (MPA) has been shown to bind specifically to a glycoprotein on the surface of type II cells (Marshall et al. 1988; Dobbs 1990). This is useful in following the differentiation of cultured type II cells toward type I cells because the latter do not show MPA binding. *Bauhinia purpurea* lectin (BPA) has been shown to bind specifically to type I cells in vivo (Kasper et al. 1994). Freshly isolated type II cells were shown by Adamson and Young (1996) not to bind this lectin but as the cells were cultured for a number of days, the number of positive cells increased as the cells differentiated into type I cells, with greater than 70 % of the cells positive for BPA at day 4.

Alkaline phosphatase (AP) is often used as a marker in type II cell populations. This enzyme, however, is common in epithelial cells but alveolar macrophages, which represent the main contaminant of type II cell isolations, are AP negative (Edelson et al. 1988).

Acknowledgements. We would like to thank the members of the Biomed I/EURespIn-Vitro programs especially Dr. Roy Richards, Dr. Samantha Murphy (both University of Cardiff), Dr. Ben Nemery and Dr. Peter Hoet (both Katholieke Universiteit, Leuven) for their help in setting up these procedures in our laboratory.

References

Adamson IYR, Young L (1996) Alveolar type II cell growth on a pulmonary endothelial extracellular matrix. Lung Cell Mol Physiol 14:L1017-L1022

Baron J, Voight JM (1990) Localisation of distribution, and induction of xenobiotic-metabolising enzymes and aryl hydrocarbon hydroxylase activity within the lung. Pharmac Ther 47:419–445

Bates SR, Dodia C, Fisher AB (1994) Surfactant protein A regulates uptake of pulmonary surfactant by lung type II cells on microporous membranes. Am J Physiol 267:L753–760

Borok Z, Danto SI, Zabski SM, Crandall ED (1994) Defined medium for primary culture de novo of adult rat alveolar epithelial cells. In Vitro Cell Dev Biol 30A:99–104

Cunningham AC, Milne DS, Wilkes J, Dark JH, Tetley TD, Kirby JA (1994) Constitutive expression of MHC and adhesion molecules by alveolar epithelial cells (type II pneumocytes) isolated from human lung and comparison with immunocytochemical findings. J Cell Science 107: 443–449

Devereux TR, Massey TE, Van Scott, MR, Yankaskas J, Fouts JR (1986) Xenobiotic metabolism in human alveolar type II cells isolated by centrifugal elutriation and density gradient centrifugation. Can Res 46: 5438–5443

Dobbs LG (1990) Isolation and culture of alveolar type II cells. Am J Physiol 258:L138-L147

Dobbs LG, Gonzalez R and Williams MC (1986) An improved method for isolating type II cells in high yield and purity. Am Rev Respir Dis 134:141–145

Edelson JD, Shannon JM, Mason RJ (1988) Alkaline phosphatase: a marker of alveolar type II cell differentiation. Am Rev Respir Dis 138:1268–1275

Fraslon C, Rolland, G, Bourbon JR, Rieutort M, Valenza C (1991) Culture of fetal alveolar epithelial type II cells in serum-free medium. In Vitro Cell Dev Biol 27A: 843–852

Funkhouser JD, Cheshire LB, Ferrara TB, Peterson RDA (1987) Monoclonal antibody identification of a type II alveolar epithelial cell antigen and expression of the antigen during lung development. Dev Biol 119:190–198

Guzman J, Izumi T, Nagai S, Costabel U (1994) Immunocytochemical characterization of isolated human type II pneumocytes. Acta Cytol 38:539–542

Hoet PHM, Lewis CPL, Demedts M, Nemery B (1994) Putrescine and paraquat uptake in human lung slices and isolated type II pneumonocytes. Biochem Pharmacol 48:517–524

Kasper M, Schuh D, Muller M (1994) *Bauhinia purpurea* lectin (BPA) binding of rat type I pneumocytes: alveolar epithelium alterations after radiation-induced lung injury. Exp Toxicol Pathol 46:361–367

Kikkawa Y, Yoneda, K (1974) The type II epithelial cells of the lung.1: Method of isolation. Lab Invest 30: 76–84

Kumar RK, Li W, O'Grady R (1995) Maintenance of differentiated phenotype by mouse type 2 pneumocytes in serum-free primary culture. Exp Lung Res 21:79–94

Liu S, Mautone AJ (1996) Whole cell potassium currents in fetal rat alveolar type II epithelial cells cultured on Matrigel matrix. Am J Physiol 270:L577-L586

Makker SP, Kanalas JJ, Tio FO, Kotas RV (1989) Definition of an immunologic marker for type II pneumonocytes. J Immunol 142:2264–2269

Marshall BC, Joyce-Brady MF, Brody JS (1988) Identification and characterisation of pulmonary alveolar type II cells *Maclura pomifera* agglutin-binding membrane glycoprotein. Biochim Biophys Acta 966:403–413

Mason RJ, Walker SR, Shield BA, Henson JE, Williams MC (1985) Identification of rat alveolar type II epithelial cells with tannic acid and polychrome stain. Am Rev Respir Dis 131:786–788

Paine R, Gaposchkin D, Kelly C, Wilcoxen SE (1995) Regulation of cytokeratin expression in rat lung alveolar epithelial cells in vitro. Am J Physiol 269:L536-L544

Paine R, Simon H (1996) Expanding the frontiers of lung biology through the creative use of alveolar type II cells in culture. Am J Physiol 270(14) L484–486

Stevens PA, Wissel H, Sieger D, Meienreis-Sudau V, Rüstow B (1995) Identification of a new surfactant protein A binding protein at the cell membrane of rat type II pneumocytes. Biochem J 308:77–81

Ten Have-Opbroek AAW, Otto-Verberne CJM, Dubbeldam JA, Dÿkman JH (1991) The proximal border of the human respiratory unit, as shown by scanning and transmission electron microscopy and light microscopical cytochemistry. Anat Rec 229:339–354

Suppliers

Biological Industries Limited
Kibbutz Beit
Haemek 25115
Israel

Boehringer Mannheim UK Ltd.
Bell Lane
Lewes, East Sussex BN7 ILG
UK

Life Technologies Ltd. (GIBCO)
Trident House
P.O. Box 35
Renfrew Rd.
Paisley PA3 4EF, Renfrewshire
UK

Pharmacia Biotech Ltd.
Davy Avenue
Knowlhill
Milton Keynes
Bucks. MK5 8PH
UK

Sigma Chemical Company
Fancy Road
Poole
Dorset BH17 7BR
UK

Culture Systems for Hepatocytes for Use in Toxicology and Differentiation Studies

Gennady Ilyin, Anne Corlu, Pascal Loyer, and Christiane Guguen-Guillouzo*

Introduction

Owing to the high level of biotransformation enzymes in the liver, hepatocyte cultures are potentially valuable in vitro test systems for determining the metabolic profiles of drugs as well as for measuring the activities and substrate specificities of the various hepatic enzyme systems involved in their biotransformation. Hepatocyte cultures are also of major interest as they can be used to determine the potential toxicity of newly formed metabolites. In this chapter, our aim is to describe both in vitro assays that can readily be adopted by several laboratories and prediction models that can easily be interpreted. To understand the benefits and, more importantly, the limitations of the different hepatocyte models used in toxicology and drug metabolism, it is necessary to be acquainted with the main features of those systems currently in use. This chapter updates the discussion of different model systems of hepatocyte primary cultures and examines one of the main functions of the cells, i.e., drug metabolism, in relation to proliferation, differentiation status or apoptotic response.

Hepatotoxicity

The liver is the key processor of nutrients and xenobiotics coming from the intestinal tract into the body. It plays a central role in providing sources of glucose, in synthesizing plasma proteins and lipoproteins and in regulating the metabolic biotransformation and elimination of most drugs and xenobiotics. In general, the liver can recover from acute injury

* *Correspondence to* Christiane Guguen-Guillouzo, INSERM U 49, Hôpital Pontchaillou, 35033 Rennes, France; phone+33–2–99543737; fax +33–2–99540137; e-mail christiane.guillouzo@univ-rennes1.fr

or weight loss by hepatocellular regeneration, the production of new cells, which restores liver functions and normal tissue architecture (Mehendale 1994). In contrast, chronic injury, often leads to fibrogenesis and abnormal architecture. Presently, multiple episods of chronic and acute injury are difficult to reproduce in vitro.

Specific Enzymatic Patterns

Manifestation of hepatotoxicity is often evidenced by a decrease in different metabolic hepatic functions which can be easily measured. Hepatotoxicity can also be due to active metabolites produced during hepatic biotransformation. Drug metabolizing enzymes are divided into two groups: (1) phase I reactions are generally oxidative and also responsible for hydrolytic processes. The phase I enzymes are represented by cytochrome-P450 (CYP) monooxygenase system. CYPs form a supergene family composed of at least 74 distinct identified genes, 14 of which are common to all mammals (Bock 1995). They are mainly localized in the endoplasmic reticulum but CYP has been reported to be associated with the plasma membrane too, a fact that could explain the presence of neoantigens at the surface of hepatocytes (Loeper 1993). (2) The phase II reactions are mainly glucuronidation, sulfatation and glutathione conjugations. UDP-glucoronyl-transferases are the most abundant phase II enzymes. Localized in the endoplasmic reticulum, they form a multigenic family which catalyzes the transfer of UDP-glucuronic acid to a substrate. At least seven isozymes have been identified in rat; they are grouped in three classes according to their ability to be induced either by 3-methylcholanthrene, clofibrate or phenobarbital (Guengerich 1985). In human liver, six isozymes have been well defined (Miners 1991). Glutathione-S-transferases (GST) are a family of abundant drug metabolizing enzymes which catalyze the conjugation of reduced glutathione to a variety of electrophilic compounds. Four common classes of GST have been identified in several mammalian species (Hayes 1995). The induction of these phase II enzymes represents an important mechanism of protection against toxic and neoplastic effects of chemical carcinogens.

Interspecies and Interindividual Variations

To be predictive of in vivo hepatotoxicity, the ideal in vitro systems should reproduce the in vivo biotransformation of compounds. The difficulty is in maintaining the enzymatic reaction patterns along with the general functional stability of hepatocytes in culture. It is also important to note that there are interspecies variations in the pathways of drug metabolism as well as in the levels and catalytic activities of detoxication enzymes, making direct extrapolation from animal cell models to the human body difficult (Sandker 1994; de Sousa 1995). Moreover interindividual differences exist as well; in humans there are two individual classes of metabolism, low and rapid. Clinical evidence correlates with interindividual variations in the levels of phase I and phase II metabolic enzymes. Thus, oxidation by cytochrome P-450 enzymes and acetylation are metabolic pathways which are low or deficient in some individuals, making them more susceptible to the toxic effects of various drugs (Larrey 1988). In addition, the polymorphism of GSTs may also contribute to variability in the intracellular transport of drugs and activity of detoxication pathways. Ideally, valuable conclusions with human cells should arise from different assays and from adequate cell samples.

Involvement of Different Liver Cell Types

The biotransformation enzymes such as CYP, are mainly expressed in hepatocytes. Compared to the other cell types present in the liver, hepatocytes have higher levels of glutathione and bile acid uptake and secretion. Therefore, primary cultures of hepatocytes represent valuable models for toxicity and drug metabolism testing. However, endothelial and Kupffer cells can also biotransform compounds because of the presence of some CYP proteins and peroxidases (Steinberg 1989). These cells can also release a variety of mediators which regulate the function of hepatocytes and nonparenchymal cells. Furthermore, some drugs or xenobiotics may also injure liver cells other than hepatocytes, i.e, bile duct, endothelial or Kupffer cells. Thus, primary cultures or cell lines derived from these nonparenchymal cell types also represent useful models for certain toxicological screenings.

Role of Cell Interactions in Liver

Analysis of the biological mechanisms which control hepatocyte function is necessary to better understand the phenotypical modifications that occur in the different in vitro systems. The participation of proximal and/or contacting cells in the control of development and differentiation of various tissues, including liver, is now well-established (Edelman

1987). Regarding liver differentiation, interaction of presumptive hepatic endoderm with the mesenchymal area is necessary for activation and maintenance of hepatic differentiation, reflecting cooperation among different cell types (Houssaint 1980). Even at the adult stage, the role of hepatocyte communication with surrounding cells remains crucial for maintaining hepatic tissue specificity. Therefore, it is not surprising to observe that liver dissociation, needed to purify the hepatocyte population, induces major alterations of the functional activity of these cells, such as changes in the transcription of liver-specific genes (Clayton 1983) and entry into the cell cycle. The latter is correlated with the activation of several functions related to growth, such as expression of the immediate early proto-oncogenes (Etienne 1988), but also with the induction of different isoenzymes involved in drug transport and biotransformation which are poorly detectable in normal adult tissue (Guguen-Guillouzo 1981; Vandenberghe 1990; Fardel 1992; Padgham 1993). These events demonstrate the existence of a balance between proliferation and differentiation in normal hepatocytes: depending on the microenvironmental culture conditions and on the presence of growth factors, the cells will undergo differentiation, proliferation or cell death (Fig. 20.1).

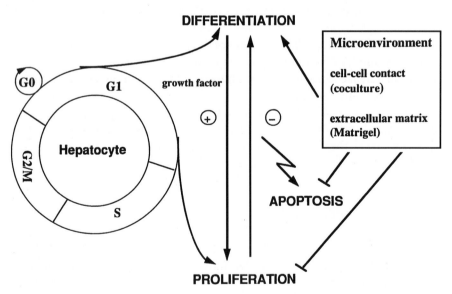

Fig. 20.1. Cross-talk between proliferation, differentiation and apoptosis in liver cells

Hepatocyte Growth Characteristics

It is clearly established that the G0/G1 transition takes place during collagenase perfusion (Etienne 1988; Wright 1992). However, we may emphasize that this G0/G1 transition during cell isolation is necessary but not sufficient for hepatocyte progression through late G1, S phases and mitosis.

Completion of the cell cycle depends in vitro, as in vivo, on both growth factor stimulation and cell-cell interaction signals (Loyer 1996). Evidence exists that hepatocytes in pure culture progress through G1 regardless of growth factor stimulation, until a restriction point located in mid-late G1 phase. In the absence of mitogen, the cells remain arrested at this point, do not replicate and finally die. In contrast, in the presence of growth factors, i.e., hepatocyte growth factor (HGF), transforming growth factor-α (TGFα) or epidermal growth factor (EGF), hepatocytes can cross over the restriction point and progress to late G1 and S phases (Loyer 1996). An important observation is that hepatocytes acquire the ability to respond to growth factor during the G1 phase. This fact reflects a step by step progression through G1. Interestingly, this

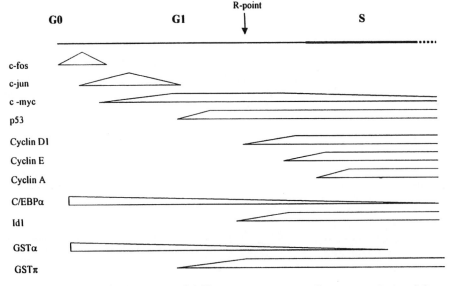

Fig. 20.2. Sequential expression of different oncogenes, cyclins, transcriptional factors and phase II drug metabolism enzymes, during G0-G1-S phase progression of rat hepatocytes. Data for schematic presentation were obtained from densitometric scanning of northern blots

progression is characterized by sequential changes in the expression of genes related to the cell cycle and to transcription factors as well as those belonging to several liver-specific gene families (Fig. 20.2).

Although variations from one species to another might exist, hepatocytes can undergo a limited number of cell cycles in vitro. Thus, adult rat hepatocytes divide once or twice in the presence of EGF alone (Loyer 1996), while several rounds of division can be observed in the presence of both EGF and HGF (Block 1996).

Differentiation vs Proliferation

Progression of hepatocytes through the cell cycle is also regulated by cell density and cell-cell interactions. There is an inverse relationship between cell density and the DNA replication rate, involving unidentified plasma membrane proteins (Ichihara 1986). Also, when heterotypic cell-cell interactions are re-established, e.g., by coculture, hepatocytes are arrested in G1 and are unable to respond to growth factor signal and enter into S phase (S. Cariou and A Corlu, pers. comm.). In parallel, they acquire a higher stability of their specific functions (Fraslin 1985; Corl 1991). These observations lead to the conclusion that, in normal cells, high level of tissue-specific differentiation status and proliferation are mutually exclusive and that cell microenvironment plays a major role in this equilibrium.

Two groups of factors can be used in vitro to modulate the functional activity of hepatocytes in culture: (1) those involved in cell-cell contacts or extracellular matrix components; (2) soluble factors such as growth factors. Consequently, as shown in Fig. 20.3, there are two types of in

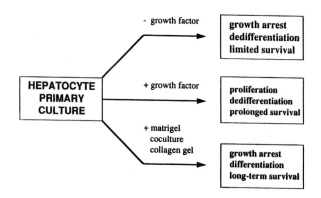

Fig. 20.3. Hepatocyte primary culture models. Growth factors and microenvironment cause significant changes in hepatocyte proliferation activity, differentiation phenotype and survival in culture

vitro model systems: (1) one that preserves the original cell-cell contacts and architecture of the tissue or restores part of these contacts with other cell types or components of the extracellular matrix. Under these conditions the cells are growth-arrested and remain differentiated. (2) Alternatively, the cells are exposed to conditions under which they may respond to growth factors and divide. In the absence of growth factors or the appropriate microenvironment the cells rapidly dedifferentiate and progress to cell death.

The Isolated Perfused Liver

The closest in vitro model of the in vivo situation is the isolated perfused liver. The major advantages are that cell-cell cooperation and the three-dimensional architecture are preserved, that capillary and vessel distribution is left undisturbed and that bile flow can be collected and analyzed separately. In addition, numerous liver diseases can be induced: alcool and galactosamine toxicity, acute hepatitis by xenobiotics or endotoxins or inflammatory reactions. The isolated perfused liver has been used for investigating drug- and chemical induced toxicity (Ballet 1988) and is particularly appropriate for studying alterations induced by ischemia, or hypoxia and hypothermic preservation (Bradford 1986). However, this system is difficult to handle and its functional integrity is not maintained beyond a few hours. Moreover, analysis of various experimental conditions is not possible from a single organ and human organs are not available.

The Tissue Slice Model

Tissue slices retain the normal liver architecture, acinar localization and cell-cell communications. In addition, because various cell types can participate in the biotransformation of compounds, the slice methodology provides a means to investigate interactive toxicity. This in vitro system has recently gained more interest since a mechanical slicer capable of rapidly producing minimally damaged and uniform slices has been developed (Vickers 1994). Slice thickness can be determined and controlled with high reproducibility and the histological findings indicate that slices of 250 µm thickness cultured for 20 h are optimum regarding functional stability while slices thinner than 100 µm have a large number of damaged cells.

To evaluate the effect of the conditions used during the slicing procedure on cell viability, a rapid version of the reduction of the tetrazolium salt (MTT test) is often useful. The leakage of GSTα into the medium can also be used as control of tissue preservation. Interestingly, slices have been found more viable in roller cultures even up to 48 h of culture. Markers of liver cell function including protein synthesis and secretion are linear for 20 h. The slices also synthesize glycogen in the presence of glucose and insulin and respond to glucagon during this culture time. However, although cell microenvironment is preserved, this is not sufficient to maintain CYP at the in vivo levels (Vickers 1994; Wright 1992). An explanation of this poor stability may be that perhaps tissue slicing induces a signal which induces cell entry in G1 but not mitosis, so that the cells remain arrested in G1 and rapidly die because of the lack of an appropriate growth factor signal. The decline can, in part, be prevented by supplementation of the culture medium with hormones and substances known to be efficient for maintaining CYP levels and other specific functions in hepatocyte cultures (Waxman 1989).

20.1
Hepatocyte Isolation

The two-step in situ collagenase perfusion technique is widely used for disaggregation of the adult rat liver in order to select hepatocytes. It can be used as a basic procedure for different species. The method used in our laboratory is based on that of Seglen (1975) with some modifications.

▥ Materials

- Calcium-free Hepes buffer: 137 mM NaCl; 2.7 mM KCl; 0.28 mM $Na_2HPO_4.12H_2O$; 10 mM Hepes, pH 7.65 at 37 °C
- Collagenase solution: Difficulties reside in that collagenase preparations greatly vary in their enzyme specific activity and contain variable amounts of other proteolytic enzymes which may alter the cells. It appears that the lots with higher specific activity produce higher yelds of viable cells. Liberase RH (Boehringer Mannheim) represents a significant advance in that it corresponds to a purified collagenase preparation allowing a high efficiency and reproducibility.

– Medium: Basal media such as Ham's F12, Williams E or a mixure of 75 % MEM and 25 % 199 medium (MEM/199) are commonly used. They are generally added with bovine serum albumin (BSA, 1 mg/ml) and bovine insulin (5 µg/ml).

▨ Procedure

1. Male rats weighing 150–200 g are anesthetized with intraperitoneal injection of Nembutal (1.5 ml/kg), and heparin (1000 IU) is injected into the femoral vein. The liver is cannulated and flushed via the portal vein with calcium-free Hepes buffer. The flow rate should be 30 ml/min for 15 min.

2. This step is performed with the same buffer but containing 0.025 % collagenase and 0.075 % $CaCl_2$ at a flow rate of 20 ml/min for 15 min. The concentration of collagenase can be modulated according to the specific activity of the enzyme batch.

3. At the end of the perfusion the cells are dispersed in Leibovitz L15 medium enriched with 0.2 % bovine serum albumin.

4. After filtration through a gauze the cells are washed twice with Hepes buffer by centrifugation at 50 g for 45 s in order to remove cell debris, damaged cells and nonparenchymal cells, and once with the culture medium used for cell seeding. Cell yelds range from 4×10^8 to 6×10^8 hepatocytes with a viability often greater than 95 %. Very few nonparenchymal cells remain in the final suspension.

This basic procedure can be used for preparing hepatocytes from various other species, including mouse, rabbit, chick, dog, baboon and human. For small rodent species such as mouse or 2-week-old rats, the flow rate is reduced to 3 ml/min for 10 min for the washing step and to 2 ml/min for collagenase perfusion. Cell yields range from 1×10^8 to 1.5×10^8. Organs can be hypothermically preserved for several hours before their disaggregation without marked loss of the viability of isolated cells (Guyomard 1990).

Perfusion of Human Liver Biopsies

The biopsies range in size between 1.5 and 5 g. Silicon catheters (0.75 mm in diameter) are preferentially used. They are inserted into the arterial orifices of the cut surface. It is essential to place several of these microcatheters to increase the yield of isolated cells from one biopsy. If the second step of collagenase perfusion uses recirculated solution, this solution is continuously oxygenated with carbogen and warmed to reach 37 °C in the liver. The flow rate is close to 10 ml/min for 30–45 min until sufficient softening of the tissue is obtained.

■ Comments

Survival of isolated hepatocytes in suspension is short, not exceeding 4 h. However, they retain responsiveness to signals to which they reacted prior to isolation and can be useful for short-term assays or acute toxicological tests. Freshly isolated cells show high transcriptional activity which drastically decreases thereafter with time of incubation. Associated with the transition from G0 to G1 during perfusion, are early phenotypic changes (Fig. 20.2) including induction of the proto-oncogenes c-*fos*, c-*myc* and c-*jun* (Etienne 1988; Loyer 1996), a decrease in transcription factors such as C/EBPα (Rana 1994), and changes in abundance of the fetal isoforms of glycolytic enzymes (Guguen-Guillouzo 1981) but also of CYP and GST isozymes (Vandenberghe 1990).

Note: As far as human liver is concerned, the conditions of organ or biopsy preservation before dissociation are critical. In addition, individual variations and premedication may greatly influence the metabolic activities of isolated human hepatocytes.

20.2
Hepatocytes in Coculture

The aim of coculture is to induce long-term stable functional activity in hepatocytes isolated from the liver tissue placed in culture by restoring cell-cell contacts with another hepatic cell type. In this cell system, increased cell survival from 1 week to several weeks with maintenance of liver specific functions is observed (Guguen-Guillouzo 1983; Clement 1984; Corlu 1991). Interestingly, cells of either hepatic or nonhepatic origin can induce a coculture effect on hepatocytes. Cells that may be used for coculture include endothelial cells from liver sinusoids (Morin 1986), epithelial cells from primitive biliary cells, and fat storing cells before their activation into myofibroblasts (Loreal 1993). Other cell types from nonhepatic origin can also influence hepatocyte activity, for example, mouse embryonic 3T3 cells (Kuri-Harcuch 1989). These observations demonstrate that cells either from epithelial or mesenchymal origins can induce a coculture effect. Moreover, no species specificity of the cell effects seems to exist, since, e.g., it is possible to associate rat hepatocytes with mouse 3T3 cells. However, not all the cell types have the ability to induce an effect onto hepatocytes that is indicative of a defined regulating process. In addition, homotypic interactions, such as those between hepatocytes and hepatocytes, have no significant influence for a prolonged period. All these observations lead us to postulate that the coculture effect is a highly controlled signal.

The model associating hepatocytes with rat liver epithelial cells (RLEC) has been preferentially developed because it allows a simple and selective detachment of pure hepatocytes after several days or weeks of coculture followed by molecular analyses at a high quality level (Fraslin 1985).

Procedure

RLEC Isolation and Cell Line Maintenance

The RLEC are isolated by trypsinization of 10-day-old rat livers according to Williams (1971) and Morel-Chany (1978):

1. The cell suspension obtained after two 15 min incubations of liver fragments in a Hepes-buffered trypsin solution (0.25 %) at 37 °C is washed twice before seeding (see Hepes buffer solution above).

2. The cells are seeded in serum supplemented Williams E medium.

3. Contaminant fibroblastic cells of the suspension are eliminated by taking advantage of their faster attachment to the plastic dish: 20 min after seeding, the supernatant with unattached cells is collected and seeded in another dish. This step is repeated three times so that the third and fourth dishes mainly contain epithelial cells.

Williams E medium is essential for active epithelial cell growth; the cells are maintained by serial subcultures in this medium supplemented with 5–10 % fetal calf serum. RLEC are similar to those described by several authors (Williams 1971; Fausto 1993). Their epithelial origin is ascertained by expression of the characteristic cytokeratins CK8 and CK18. They are positive for the γ-glutamyl transpeptidase but have no or very low levels of albumin or other liver specific markers. Interestingly, RLEC are characterized by their ability to express one isoform of the CYP family, the 2E1 form, mainly involved in alcohol metabolism, and a specific epoxy-hydrolase (Lerche 1997). RLEC also express some growth factors, those well characterized being TGFα, Steel factor and interleukin (IL)-6 (Hampton 1990; M. Rialland, pers. comm.).

When carefully cultivated RLEC are stable for 100 passages. Otherwise, they transform very rapidly and gradually lose contact inhibition at confluency and adhesion properties, both of which hamper their further use in coculture.

Hepatocyte Coculture

1. Plate 8×10^5 hepatocytes in 2 ml of medium in 3.5 cm petri dishes. The medium is made up as follows: Ham F12, Williams E or MEM/199 supplemented with 5 µg/ml bovine insulin; 1 mg/ml BSA; and 10 % fetal calf serum.

2. Three hours after seeding, unattached cells are discarded and 1×10^6 RLEC in fresh medium are seeded per dish in order to reach confluency with hepatocyte colonies within 24 h.

3. Thereafter, the same medium but supplemented with 3.5×10^{-6} M hydrocortisone hemisuccinate is renewed every day. After cell attachment and spreading, a serum-free medium can be used.

Note: Separation of hepatocytes from RLEC after several days of coculture is possible. This property is essential when biochemical or molecular biology studies are performed. For this purpose, cocultures are washed with a calcium-free Hepes buffer and incubated with a 0.075 % collagenase solution, prepared in the same calcium-free buffer (pH 7.6), for 10 min. Hepatocyte colonies specifically detach by vigorous repeated pipetting.

Comments

When cocultured with these RLEC, hepatocytes from various species survive for several weeks and retain high functional capacities. Among the preserved functions are the production of plasma proteins including albumin, transferrin, the group of proteins involved in the inflammatory reaction, drug metabolism enzymes involved in phase I and phase II reactions, and specific enzymes of hepatic glycogen metabolism (Bégué 1984; Guguen-Guillouzo 1983; Morin 1986; Corlu 1991). As generally observed in in vitro systems the high rate of protein synthesis results in part from post-transcriptional regulation and mostly from increased stabilization of mRNAs. However, and in contrast to the functions of hepatocytes maintained in a serum-free hormonally defined medium, hepatocytes cocultured with RLEC retain the ability to transcribe specific genes at a rate identical and even higher to that found in DMSO-treated cells and in cells seeded on Matrigel (Fraslin 1985; Isom 1987). The enhanced survival and function of hepatocytes in coculture with RLEC have been confirmed by many investigators and the induction of CYP2B1/2 by phenobarbital has been demonstrated (Nieman 1991; Rogiers 1990; Akrawi 1993).

In addition, a spontaneous early production and deposition of extracellular matrix components is observed, mimicking the composition of the hepatic extracellular matrix (Corlu 1991). Of interest is the fact that maximal deposition and organization of this matrix correlates well with the maximal functional capacities reached by the cells. It is also important to note that remodeling of this extracellular matrix involves activation of enzymes, mainly metalloproteinase-2, the process of which is dependent on cell-cell interactions (Théret 1997). It also correlates with an important rearrangement of the cytoskeleton, particularly the organization of cytokeratins (Corlu 1991).

Unexpectedly, the reappearance of "fetal-like" functions observed early after seeding is regularly detected throughout the coculture period. It includes the appearance of fetal isoforms of enzymes such as pyruvate kinase M2 and GSTπ (Vandenberghe 1990). In addition, overexpression of oncogenes including c-*myc* or of some components of the extracellular matrix is clearly evidenced. These fetal markers can presumably be explained by the cell arrest in G1 as described above. The mechanism by which cell-cell communication takes place is not completely understood, it could involve a plasma protein molecule, the liver regulating protein (LRP) (Corlu 1991). Paralleling this arrest in G1, cocultured hepatocytes are unable to respond to a mitogenic signal (C. Guguen-Guillouzo, pers. comm.). This observation can be correlated to that described by Khost and Michalopoulos (1991) using DMSO and by Gardner et al. (1996) using Matrigel, both systems which preserve hepatocyte differentiation and block proliferation.

20.3
Hepatocytes on Biomatrix Gels

The role of the extracellular matrix in the regulation of liver specific genes was shown in rat hepatocytes both in vitro and in vivo. Two approaches, described below, were made to reproduce these biomatrix interactions in culture.

Complex Biomatrices

One approach consists of seeding the cells onto a complex biomatrix derived either from liver (Enat 1984; Saad 1993), or from the mouse Engelbreth-Holm-sarcoma (EHS) and referred to as Matrigel (Bissell 1987; Vukicevic 1992). Liver-derived biomatrix is poorly defined and hardly reproducible and has yet to be used for human hepatocyte cultures. In contrast, Matrigel has been extensively used. Matrigel is very rich in laminin and contains other extracellular components such as collagen IV and heparan sulfate proteoglycan. However, growth factors might also be present, particularly TGF-β (Vukicevic 1992). We therefore, strongly recommend the use of cytokine-free preparations such as the GFR-Matrigel from Collaborative Biomedical Products (Becton Dickinson).

▊ Procedure

Protocol for Coating with Matrigel

1. Hepatocytes are seeded in serum-free defined medium (see composition above), in untreated bacterial grade plastic dishes. In general, 400 or 500 μg of Matrigel are used to coat 3.5 cm dishes. Coating is performed at 4 °C in order to delay gel formation.

2. A 30 min incubation at 37 °C produces a dry film ready for cell seeding. It is very important to note that regular addition of Matrigel in suspension into the culture medium is needed to keep a high level of cell functional activity after 1 week of culture. The concentration of 2 μg/ml is generally used and added daily at each medium renewal.

Matrigel has been reported to greatly improve the maintenance of specific functions, similar to coculture. As in coculture, hepatocytes do not respond to growth factors and presumably remain blocked in G1, while their differentiation state increases (Gardner 1996). In parallel, hepatocytes fail to spread and form aggregates.

Single Matrix Component-Made Gel

The other strategy consists of using a single component of the biomatrix as coating gel. Of major interest is that these different components may have opposite effects: collagen IV and laminin induce differentiation, slow and moderate spreading while collagen type I and fibronectin only favor cell attachment and spreading but not long-term differentiation and survival (Caron 1990). Furthermore, collagen type I and fibronectin allow DNA synthesis in response to growth factor stimulation while collagen type IV does not.

Sandwich Collagen Gel

Sandwich collagen gel represents an attempt to reproduce the in vivo three-dimensional microenvironment of hepatocytes (Lee 1993). The method consists of setting hepatocytes in a two layers of collagen, generally type I collagen.

1. A solution of collagen is prepared by mixing nine volumes of 1.5 mg/ ml collagen in 1 mM HCl.

2. The pH of the solution (about 4.0) is adjusted to 7.4 with NaOH mixed with one volume of 10× concentrated culture medium. For example, for Petri dishes of 10 cm in diameter, 5 ml of collagen solution are mixed with 0.6 ml of 10× concentrated medium solution containing 0.34 M NaOH, pH 8.5, and transferred to the dish. Formation of the gel occurs at room temperature but it is faster at 37 °C.

3. After 4–6 h the gel is washed three times with culture medium before use.

4. Hepatocytes are seeded on top of the gel, and after cell attachment and medium removal, a second collagen layer is gently poured on the cell layer. Once this second layer has gelled, it is covered with fresh culture medium.

Hepatocytes survive more than 2 weeks and preserve cuboidal morphology as well as expression of specific functions including drug biotransformation activities. However, it is difficult to evaluate the capacity of the system to transport drugs and metabolites, to determine cell viability and to recover all the cells for measurement of cell count, DNA or protein content.

20.4
Hepatocyte Pure Cultures

Hepatocyte culture on plastic is the simplest model system for studies of acute toxicity. Different observations show that, whatever the density and the composition of the medium used, cell viability and functional stability of hepatocyte monolayers do not exceed a few days. Growth activity is possible but limited to a few rounds, while maintenance of differentiation needs specific environmental conditions which clearly differ from those allowing hepatocyte proliferation.

▓ Procedure

Proliferating Hepatocyte Primary Cultures

1. Plate 6×10^5 hepatocytes in 2 ml of basal medium in 3.5 cm petri dishes coated or not with collagen type I or fibronectin. The recipe for the basal medium is given above.

2. Three hours after seeding, unattached cells are discarded and fresh medium is added. According to experiments, this medium contains EGF alone (25 ng/ml) or EGF+pyruvate (20 mM) or HGF alone (10 ng/ml) or TGFα (20 ng/ml). Growth factor can be added just after attachment or 1 day later. When added later than day 2 the peak of DNA synthesis is delayed.

3. Thereafter, the medium is renewed daily.

   Note: Insulin, which is a potent comitogen, can be added at the concentration of 5 μg/ml.

Characteristics

In rat hepatocytes plated on plastic, the peak of DNA synthesis occurs about 72 h after plating and a maximum of mitosis at 96 h. This result indicates that the G1 phase lasts more than 2 days under these conditions. It is greatly reduced when the cells are seeded on collagen I. The hepatocytes undergo one or two cell cycles with the mitogens cited above. However, HGF is more efficient especially for human hepatocytes (Fig. 20.4). Interestingly, HGF and EGF together have the strongest proliferative effect. Several rounds of division can be performed over a period of 15 days (Block 1996).

In parallel to the growth activity, a gradual drop of different specific functions such as albumin and transferrin secretion is observed. However, the proliferating hepatocytes can return to a mature hepatocyte phenotype in the presence of Matrigel.

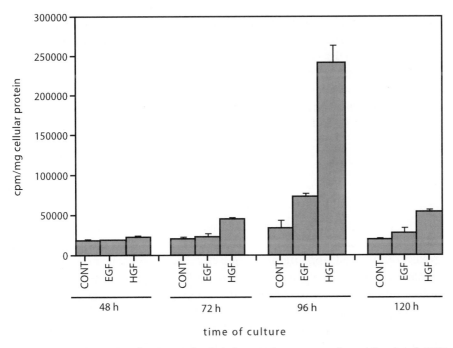

Fig. 20.4. Time course of DNA synthesis in human hepatocytes in unstimulated, EGF – and HGF- stimulated primary cultures. Human hepatocytes were stimulated with mitogens from 20 h after seeding and throughout the culture times. Cultures were pulsed with 2 μCi/ml [³H]thymidine for 24 h

Nonproliferating Hepatocyte Pure Cultures

Basal medium is added with 3.5×10^{-5} M hydrocortisone hemisuccinate at the first medium renewal 3 h after seeding and every day thereafter. In the absence of mitogen, hepatocytes survive for 5–6 days and then die. No proliferation is observed and they gradually lose specific functions. Thus, CYP levels decrease approximately 50 % during the first 48 h in rodent hepatocytes. Induction of specific CYP enzymes by phenobarbital is difficult to achieve while other CYP isozymes, although decreased, remain capable of responding to inducers. Concerning the phase II enzymes, for instance UDP-glucuronyl-transferases and GST isoforms, differential changes are observed. Phenotypic changes with time in culture also include overexpression of P-glycoprotein, the product of multidrug resistance genes, after 2–3 days of culture.

Some soluble factors, e.g., isonicotinamide (7.5 mM) or anionic compounds can also transiently stabilize drug metabolism enzymes (Decad 1977; Miyazaki 1985; Mitaka 1991). DMSO, a dipolar solvent, is probably the most powerful factor able to stabilize liver specific functions for a prolonged period in hepatocyte primary cultures when added to serum-free medium at a final concentration of 1.5–2 % (Isom 1987). Under these conditions most liver-specific functions are well preserved while the cells remain cuboidal, form numerous bile canaliculi and do not divide. Although very powerful for analyzing gene regulation, the usefulness of this model may be questioned for several studies, since DMSO is used at quite a high concentration.

Comments

Proliferation and differentiation are essential but alternative processes in liver cells. In addition, although the different model systems have been greatly improved, proliferative activity and survival of hepatocytes remain limited. New strategies are currently being developed in order to define new cell systems in which the genes controlling the mechanisms that couple these mutually exclusive phenomena, i.e., proliferation, differentiation and cell death, could be modulated in a time-dependent manner. Thus, "controlled" hepatocytes would enter into proliferation by addition into the medium of a regulator and reverse to differentiation at an appropriate time-schedule. Such model systems would be very useful in pharmacology and toxicology, particularly in the development of new in vitro tests for use in with human cells.

References

Akrawi M, Rogiers V, Vandenberghe Y, Palmer CNA, Vercruysse A, Shephard EA, Phillips IR (1993) Maintenance and induction in co-cultured rat hepatocytes of components of the cytochrome P-450-mediated mono-oxygenase. Biochem Pharmacol 45:1583–1591

Ballet F, Chretien Y, Rey C, Poupon R (1988) Differential response of normal and cirrhotic liver to vasoactive agents. A study in the isolated perfused rat liver. J Pharmacol Exp Ther 244:233–235

Bégué JM, Guguen-Guillouzo C, Pasdeloup N, Guillouzo A (1984) Prolonged maintenance of active cytochrome P-450 in adult rat hepatocytes co-cultured with another liver cell type. Hepatology 4:839–842

Bissell DM, Arenson DM, Maher JJ, Roll FJ (1987) Support of cultured hepatocytes by a laminin-rich gel. J Clin Invest 79:801–812

Block GD, Locker J, Bowen WC, Petersen BE, Katyal S, Strom SC, Riley T, Howard TA, Michalopoulos GK (1996) Population expansion, clonal growth, and specific differentiation patterns in primary cultures of hepatocytes induced by HGF/SF, EGF and TGFα in a chemically defined (HGM) medium. J Cell Biol 132:1133–1149

Bock KW (1995) Human UDP-glucuronosyl transferases: classification and properties of isozymes. In: Paifici GM, Francchia GN (eds) Advances in drug metabolism in man. European Commission, Bruxelles, pp 289–309

Bradford BU, Marotto M, Lemasters JJ, Thurman RG (1986) New, simple models to evaluate zone-specific damage due to hypoxia in the perfused rat liver: time course and effect of nutritional state. J Pharm Exp Ther 236:263–268

Caron JM (1990) Induction of albumin gene transcription in hepatocytes by extracellular matrix proteins. Mol Cell Biol 10:1239–1243

Clayton DF, Darnell JE (1983) Changes in liver-specific compared to common gene transcription during primary culture of mouse hepatocytes. Mol Cell Biol 3:1552–1561

Clement B, Guguen-Guillouzo C, Campion JP, Glaise D, Bourel M, Guillouzo A (1984) Long-term co-cultures of adult human hepatocytes with rat liver epithelial cells: modulation of active albumin secretion and accumulation of extracellular material. Hepatology 4:373–380

Corlu A, Kneip B, Lhadi C, Leray G, Glaise D, Baffet G, Bourel D, Guguen-Guillouzo C (1991) A plasma membrane protein is involved in cell contact-mediated regulation of tissue-specific genes in adult hepatocytes. J Cell Biol 115:505–515

de Sousa G, Florence N, Valles B, Coassolo P, Rahmani R (1995) Relationship between in vitro and in vivo biotransformation of drugs in humans and animals: pharmacotoxicological consequences. Cell Biol Toxicol 11:147–153

Decad GM, Hsieh DPH, Byard JL (1977) Maintenance of cytochrome P-450 and metabolism of aflatoxin B_1 in primary hepatocyte cultures. Biochem. Biophys Res Commun 78:279–287

Edelman GM (1987) CAMs and Igs: cell adhesion and the evolutionary origins of imunity. Immunol Rev 100: 11–45

Enat R, Jefferson DM, Ruiz-Opazo N, Gatmaitan Z, Leinwand L, Reid LM (1984) Hepatocyte proliferation in vitro: its dependence on the use of serum-free hormonally defined medium and substrata of extracellular matrix. Proc Natl Acad Sci USA 81:1411–1415.

Etienne PL, Baffet G, Desvergne B, Boisnard-Rissel M, Glaise D, Guguen-Guillouzo C (1988) Transient expression of c-fos and constant expression of c-myc in freshly isolated and cultured normal adult rat hepatocytes. Oncogene Res 3:255–262

Fardel O, Ratanasavanh D, Loyer P, Ketterer B, Guillouzo A (1992) Overexpression of the multidrug resistance gene product in adult rat hepatocytes during primary culture. Eur J Biochem 205:847–852

Fausto N, Lemire JM, Shiojiri N (1993) Cell lineages in hepatic development and the identification of progenitor cells in normal and injured liver. Proc Soc Exp Biol Med 204:237–241

Fraslin JM, Kneip B, Vaulont S, Glaise D, Munnich A, Guguen-Guillouzo C (1985) Dependence of hepatocyte specific gene expression on cell-cell interactions in primary culture. EMBO J 4:2487–2491

Gardner MJ, Fletcher K, Pogson CI, Strain AJ (1996) The mitogenic response to EGF of rat hepatocytes cultured on laminin-rich gels (EHS) is blocked downstream of receptor tyrosine-phosphorylation. Biochem Biophys Res Comm 228:238–245

Guengerich FP, Liebler DC (1985) Enzymatic activation of chemicals to toxic metabolites. CRC Crit Rev Toxicol 14:259–307

Guguen-Guillouzo C, Clément B, Baffet G, Beaumont C, Morel-Chany E, Glaise D, Guillouzo A (1983) Maintenance and reversibility of active albumin secretion by adult rat hepatocytes co-cultured with another liver epithelial cell type. Exp Cell Res 143:47–54

Guguen-Guillouzo C, Szajnert MF, Glaise D, Gregory C, Schapira F (1981) Isozyme differentiation of aldolase and pyruvate kinase in fetal, regenerating, preneoplastic and malignant rat hepatocytes during culture. In Vitro 17:369–377

Guyomard C, Chesné C, Meunier B, Fautrel A, Clerc C, Morel F, Rissel M, Campion JP, Guillouzo A (1990) Primary culture of adult rat hepatocytes after 48 hour preservation of the liver with cold UW solution. Hepatology 12:1329–1336

Hampton LL, Worland PJ, Yu B, Thorgeirsson SS, Huggett AC (1990) Expression of growth-related genes during tumor progression in v-raf-transformed rat liver epithelial cells. Cancer Res 50:7460–7467

Hayes JD, Pulford DJ (1995) The glutathione S-transferase supergene family: Regulation of GST and the contribution of the isozymes to cancer chemoprotection and drug resistance. Crit Rev Biochem Mol Biol 30:445–600

Houssaint E (1980) Differentiation of the mouse hepatic primordium. I. An analysis of tissue interactions in hepatocyte differentiation. Cell Diff 9:269–279

Ichihara A, Nakamura T, Noda C, Tanaka K (1986) Control of enzyme expression deduced from studies on primary cultures of hepatocytes. In: Guillouzo A, Guguen-Guillouzo C (eds) Isolated and cultured hepatocytes. John Libbey Eurotext Ltd/INSERM, London, pp 187–208

Isom H, Georgoff I, Salditt-Georgieff M, Darnell JE (1987) Persistence of liver-specific messenger RNA in cultured hepatocytes: different regulatory events for different genes. J Cell Biol 105:2877–2885

Kost DP, Michalopoulos GK (1991) Effect of 2% dimethyl sulfoxide on the mitogenic properties of epidermal growth factor and hepatocyte growth factor in primary hepatocyte culture. J Cell Physiol 147:274–280

Kuri-Harcuch W, Mendoza-Figueroa T (1989) Cultivation of adult rat hepatocytes on 3T3 cells: expression of various liver differentiated functions. Differentiation 41:148–157

Larrey D, Pessayre D (1988) Genetic factors in hepatotoxicity. In: Guillouzo A (ed) Liver cells and drugs. Les Editions INSERM and John Libbey Eurotext, Paris, pp 143–152

Lee J, Morgan JR, Tompkins RG, Yarmush ML (1993) Proline-mediated enhancement of hepatocyte function in a collagen gel sandwich culture configuration. FASEB J 7:586–591

Lerche C, Fautrel A, Shaw PM, Glaise D, Ballet F, Guillouzo A, Corcos L (1997) Regulation of the major detoxication functions by phenobarbital and 3-methylcholanthrene in co-cultures of rat hepatocytes and liver epithelial cells. Eur J Biochem 244:98–106

Loeper J, Descatoire V, Maurice M et al. (1993) Cytochromes P-450 in human hepatocyte plasma membrane: recognition by several autoantibodies. Gastroenterology 104:203–216

Loreal O, Levavasseur F, Fromaget C, Gros D, Guillouzo A, Clément B (1993) Cooperation of Ito cells and hepatocytes in the deposition of an extracellular matrix in vitro. Am J Pathol 143:538–544

Loyer P, Cariou S, Glaise D, Bilodeau M, Baffet G, Guguen-Guillouzo C (1996) Growth factor-dependence of entry and progression through G1 and S phases of adult rat hepatocytes in vitro. J Biol Chem 271:11484–11492

Loyer P, Ilyin G, Cariou S, Glaise D, Corlu A, Guguen-Guillouzo C (1996) Progression through G1 and S phases of adult rat hepatocytes. In: Meijer L, Guidet S, Vogel L (eds) Progress in cell cycle research, vol 2. Plenum Press, New York, pp 37–47

Mehendale HM, Roth RA, Gandolfi AE, Klaunig JE, Lemasters JJ, Curtis LR (1994) Novel mechanisms in chemically induced hepatotoxicity. FASEB J 8:1285–1295

Miners JO, MacKenzie PI (1991) Drug glucuronidation in humans. Pharmacol Ther 51: 347–369

Mitaka T, Sattler CA, Sattler GL, Sargent LM, Pitot HC (1991) Miltiple cell cycles occur in rat hepatocytes cultured in the presence of nicotinamide and epidermal growth factor. Hepatology 13:21–30

Miyazaki M, Handa Y, Oda M, Yabe T, Miyano K, Sato J (1985) Long-term survival of functional hepatocytes from adult rat in the presence of phenobarbital in primary culture. Exp Cell Res 159:176–190

Morel-Chany E, Guillouzo C, Trincal G, Szajnert MF (1978) "Spontaneous" neoplastic transformation in vitro of epithelial cell strains of rat liver: cytology, growth and enzymatic activities. Eur J Cancer 14:1341–1352

Morin O, Norman C (1986) Long term maintenance of hepatocyte functional activity in co-culture: requirements for sinusoidal endothelial cells and dexamethasone. J Cell Phys 129:103–110

Padgham CRW, Paine AJ (1993) Altered expression of cytochrome P-450 mRNAs, and potentially of other transcripts encoding key hepatic functions, are triggered during the isolation of rat hepatocytes. Biochem J, 28: 621–624

Rana A, Mischoulon D, Xie Y, Bucher NL, Farmer SR (1994) Cell-extracellular matrix interactions can regulate the switch between growth and differentiation in rat hepatocytes: reciprocal expression of C/EBP α and immediate-early growth response transcription factors. Mol Cell Biol 14:5858–5869

Rogiers V, Vandenberghe Y, Callaerts A, Verleye G, Cornet M, Mertens K, Sonk W, Vercruysse A (1990) Phase I and phase II xenobiotic biotransformation in cultures and cocultures of adult rat hepatocytes. Biochem Pharmacol 40:1701–1706

Saad B, Scholl FA, Thomas H, Schwalder H, Streit V, Waechter F, Maier P (1993) Crude liver membrane fractions and extracellular matrix components as substrata regulate differentially the preservation and inducibility of cytochrome P-450 isoenzymes in cultured rat hepatocytes. Eur J Biochem 46:805–814

Sandker GW, Vos RM, Delbressine LP, Slooff MJ, Meijer DK, Groothuis GM (1994) Metabolism of three pharmacologically active drugs in isolated human and rat hepatocytes: analysis of interspecies variability and comparison with metabolism in vivo. Xenobiotica 24:143–155

Seglen PO (1975) Preparation of isolated rat liver cells. Meth Cell Biol 13:29–83

Steinberg P, Schramm H, Schladt L, Robertson LW, Thomas H, Oesch F (1989) The distribution, induction and isoenzyme profile of glutathione -transferase and glutathione peroxidase in isolated rat liver parenchymal, Kupffer and endothelial cells. Biochem J 264:737–744

Theret N, Musso O, L'Helgoualc'h A, Clément B (1997) Activation of matrix metaloproteinase-2 from hepatic stellate cells requires interactions with hepatocytes. Am J Pathol 150:51–58

Vandenberghe Y, Morel F, Pemble S, Taylor JB, Rogiers V, Ratanasavanh D, Vercruysse A, Ketterer B, Guillouzo A (1990) Changes in expression of mRNA coding for glutathione S-transferase subunits 1–2 and 7 in cultured rat hepatocytes. Mol Pharmacol 37:372–376

Vickers AEM (1994) Use of human organ slices to evaluate the biotransformation and drug-induced side-effects of pharmaceuticals. Cell Biol Toxicol 10:407–414

Vukicevic S, Kleinman HK, Luyten FP, Roberts AB, Roche NS, Reddi AH (1992) Identification of multiple active growth factors in basement membrane matrigel suggests caution in interpretation of cellular activity related to extracellular matrix components. Exp Cell Res 202:1–8

Waxman DJ, Morrissey JJ, Leblanc GA (1989) Hypophysectomy differentially alters P-450 protein levels and enzyme activities in rat liver: pituitary control of hepatic NADPH cytochrome P-450 reductase. Mol Pharmacol 35:519–525

Williams GME, Weisburger EK, Weisburger JH (1971) Isolation and long-term cell culture of epithelial like cells from rat liver. Exp Cell Res 69:106–112

Wright MC, Paine AJ (1992) Evidence that the loss of rat liver cytochrome P-450 in vitro is not solely associated with the use of collagenase, the loss of cell-cell contacts and/or the absence of an extracellular matrix. Biochem Pharmacol 43:2337–2343

Part IV

In Vitro Models
for Toxicology and Pharmacology

Chapter 21

Cellular Models for In Vitro Toxicity Testing

Flavia Zucco*[1], Isabella De Angelis[2],
and Annalaura Stammati[2]

Introduction

Compared to other biomedical disciplines, cell culture systems have only recently been systematically adopted for use in toxicology (Paganuzzi Stammati et al. 1981), due, in particular, to the difficulty in extrapolating in vitro data to the in vivo situation for regulatory purposes. However, in the last decade, technological advancements using cell culture systems, together with the important results obtained in biomedical research, have impressively promoted its use in toxicology investigations. Growth factors and substrates able to stimulate cell proliferation and/or expression of specific highly specialized functions, and sophisticated analytical techniques for mechanistic studies, together with automated procedures, have played a pivotal role.

Advantages and Disadvantages in Using Cell Cultures in Toxicology

The interest in cellular in vitro models is due to the fact that they represent the elementary living unit and, at least theoretically, cells from any species and tissues can be cultivated, including those of human origin, bypassing most (if not all) the ethical problems connected with human experimentation. However, cells in vitro are a very simplified system with respect to the complexity of the entire organism so that they can be

* *Correspondence to* Flavia Zucco: phone +39–6–44230122; fax +39–6–44230229;
  e-mail istituto@itbm.rm.cnr.it
[1] Flavia Zucco, Istituto Tecnologie Biomediche, C.N.R., Via G. B. Morgagni 30/E, 00161 Rome, Italy
[2] Isabella De Angelis, Annalaura Stammati, Laboratorio di Tossicologia Comparata ed Ecotossicologia, Istituto Superiore di Sanità, Viale Regina Elena 299, 00161 Rome, Italy

used only in limited steps of certain studies or to investigate very specific and single mechanisms (Nazdone 1977). While there has been significant improvement, e.g., by coculturing cells or using batteries of different cell types, the lack of complexity will remain the most important shortcoming of using cell culture in toxicology.

The advantages of the various cellular models of in vitro toxicology can be summarized as follows:

- Cellular and molecular mechanisms are easily explored.

- Detailed and early identification of cellular damage is possible.

- Human cells may be used.

- Reversibility of the effect is easily verifiable.

- The system is simple and highly reproducible.

- No difficulties due to the small quantities of substances are presented.

- A statistically significant number of trials is possible.

- The technology is relatively easy (possibility of automatization).

- In vitro systems are less costly and time consuming than in vivo experiments.

- The number of animal experiments can be reduced.

The disadvantages of the various cellular models of in vitro toxicology are:

- The system is overly simplified compared to the complexity of the organism.

- Systemic toxic effects cannot be studied.

- Very few toxic mechanisms can be studied in each test.

- Acute more than chronic toxicity testing may be performed.

- Substance concentrations are difficult to choose.

- The physicochemical properties of test substances are problematic for the exposure conditions.

- Difficulties in correlating the in vitro situation to the in vivo one both in terms of experimental design and results.

Cellular Models

Various in vitro cellular models are available with characteristics which make them of particular interest for specific purposes.

Freshly isolated cells and primary cultures are the best choice when the aspects of the tissue/organ of origin must be preserved and the karyotype must be normal. However this type of preparation is neither long-lasting (from a few hours to weeks) nor can it be standardized.

Finite and continuous cell lines usually do not preserve all their in vivo characteristics, the former can be still diploid but the latter are aneuploid. The finite cell lines have a limited life span: they may last some months or even years but, after a certain number of passages, they become senescent and die. The continuous cell lines are immortal: most of them transformed or of tumoral origin. However, the system can be optimally standardized thereby guaranteeing a high reproducibility (Freshney 1994).

Cell culture systems have different areas of applications in toxicology according to the aim of the study, which can be either to screen chemicals (toxiciy testing) or to investigate a particular mechanism, such as is done in biotransformation studies (Table 21.1).

Cytotoxicity Testing

In vitro toxicity testing using cultured cells is based on a few main assumptions: (a) a wide range of pathologies is related to a small number of crucial events at a cellular level, especially where acute toxicity is concerned; (b) several elements of the organism's complexity may (or will) be mimicked in vitro; and (c) the exposure conditions may be reproduced in vitro.

The broad aim of in vitro toxicity testing is to screen compounds in order to get an indication of their general cytotoxicity (see below). In this case the cellular experimental model should be easy to handle and highly standardizable. Generally the cells used are continuous cell lines, since the cell injury taken into account is related to basic cellular structure and/or functions. It is possible, however, to perform screening on specialized cells in culture when the mechanism of action of the compound under study is already known or suspected. For example, neurotoxic molecules may be screened on in vitro neuronal cell culture models and effects on specific targets can be monitored.

Table 21.1. Different toxicological areas in which cellular models are used

Cytotoxicity testing of chemicals (intrinsic cellular toxicity)
    Aims
        Identify the range of doses
        Select priorities
        Assess the toxic potency against a reference list
        Acquire information on mechanisms
    Endpoints
        Mainly related to cell death
        Related to specific basal cellular functions and/or structures
    Cells
        Mainly cell lines
Mechanistic studies
    Aims
        Mechanism related to basic cellular functions
        Organo-specific mechanisms on differentiated functions
    Endpoints
        Very specific according to the mechanism to be investigated
    Cells
        Various cell types, mainly established cell lines
        Organospecific primary cultures and cell lines
Biotransformation studies:
    Aims
        Phase I and II enzymes investigations
        Metabolites or secondary products detection/measurements
    Endpoints
        Mainly enzymatic reactions
    Cells
        Freshly prepared hepatocytes (rat, chicken, human)
        Cells derived from other tissue or organs (fresh preparations and cell lines)
        Genetically engineered cells

Mechanistic Studies

Compared to toxicity testing, mechanistic studies are more narrowly targeted, being devoted to determining specific toxic mechanisms or elucidating organospecific toxic effects. Therefore a great variety of in vitro cellular models are used, according to the aspect dealt with (Jolles and Cordiez 1992). For mechanistic studies related to basic cellular functions and/or structure continuous cell lines, even if usually dedifferentiated, may be used. By contrast, when analyzing the expression of specific cellular functions or the maintenance of specialized cellular structures in culture, then the freshly prepared cells or primary cultures are preferred.

Table 21.2. Specialized cellular models in toxicology studies

System	Primary cultures or isolated cells	Cell lines
Nervous	Chick embryo ganglia and brain; mouse and rat cerebellum cells; rat dorsal ganglia; mouse and rat glia; human mouse and rat astrocytes	C 1300 (mouse neuroblastoma); SR CDF.DBT (mouse glioma); B 103 (rat neuroblastoma); C 6 (rat glioma); PC 12 (rat pheochromocytoma); U-251 (human glioma); IMR 32 (human neuroblastoma)
Digestive	Rat, guinea pig, mouse, hamster, chick isolated intestinal cells	HT 29 (human colon adenocarcinoma); $CaCo_2$ (human colon adenocarcinoma); T_{84} (metastasis of human colon carcinoma); IEC 17 (rat duodenum diploid cell line)
Respiratory	Human, rabbit, rat, bovine, guinea pig alveolar macrophages; canine, rabbit, rat airway epithelial cells	A 549 (human macrophage like cells); W138 (human lung fibroblast); HTE (hamster tracheal epithelial cells); 2C5 (rat hamster tracheal epithelial cells); L2 (rat lung type II pneumocytes)
Genito-urinary	Rat, rabbit kidney cells	Vero (monkey kidney fibroblast cells); MDCK (dog kidney epithelial cells); MDBK (bovine kidney epithelial cells); $LLPCK_1$ (porcine kidney epithelial cells)
Cardio-vascular	Rat, mouse, chick heart cells	
Reticulo-endothelial	Human, mouse lymphocytes and erytrocytes, leucocytes and neutrophils; rat and mouse peritoneal macrophages; mouse spleen, bone marrow, mast cells	EB 3 (human lymphoblastoid cells); V937 (human lymphoblastoid cells); TK6 (human lymphoblast); K562 (human leukemia cells); L1210 (mouse leukemia cells); S-49, L 5178YS (mouse lymphoma); G M86 (745 A) Friend mouse leukemia cells
Exocrine: Liver	Human, rat, chick mouse, hamster hepatocytes; rat Kupffer cells (liver macrophages)	ARL, BRL-3A, MHICI, Reuber H35, HTC (rat hepatoma cells); Hep G2 (human hepatoma cells); Hepa 1 (mouse hepatoma)
Pancreas	Rat cells	
Mammary		MCF 7 (human mammary carcinoma)

Table 21.2. (Continue)

System	Primary cultures or isolated cells	Cell lines
Endocrine:		
Leydig	Pig, mouse cells	
Sertoli	Rat	
Luteal	Rat	
Testicular	Rat	
Thymic	Human, rat	
Paratoid	Human, bovine	
Adrenal	Rat, bovine	
Pituitary	Rat	GH3/BS (rat pituitary tumor)

However, more recently, due to the increased insight into cellular properties in vitro and to the capability of inducing and maintaining the expression of specific cell functions, many cell lines, especially of human origin, can be used. According to a recent review of the literature, cells derived from any organ or tissue, either as fresh preparations or as cell lines, have been used in toxicology, to investigate organospecific toxicity (Table 21.2).

Biotransformation Studies

The main organ for biotransformation is the liver. Several substances are (de)toxified by specific enzymatic activities, which are expressed at different levels and in different forms in the various animal species. The first approach to the in vitro study of metabolism was performed using freshly isolated rat hepatocytes, and these became the most widely used experimental model. While hepatocytes have been prepared from other species, such as chicken embryo and mouse, their use has been restricted to only a few studies. Efforts to obtain human hepatocytes able to survive and retain their specific functions in vitro are still underway, and have not yet been fully successful (Rodrigues 1994; Skett et al. 1995; Wrighton et al. 1993). In recent years attention has also been paid to the expression of biotransforming activities in vitro by cells derived from other organs or tissues than liver: lung, intestine, dermis, kidney, etc. Several cell lines have been recently characterized for the presence of phase I and phase II enzymes (Hornardt and Wiebel 1996). In order to investigate specific steps of (de)activation of certain compounds, genetically engineered cells are available, in which the gene encoding a defined activity has been inserted.

Toxicity testing by cellular models is nowadays a wide and complex area. Many books have been published in the last decade on this topic (Frazier 1992; Watson 1992; Barile 1994; Gad 1994; Tyson and Frazier 1994; O'Hare and Atterwill 1995). In this chapter we want to deal strictly with the use of cellular in vitro models for toxicity testing, i.e., mainly for screening.

Animal Cell Cultures for Toxicity Testing

Some theoretical and practical aspects, already mentioned above, need to be further elaborated on in relation to the use of cultured cells in vitro toxicology. Toxicity testing, for screening purposes, is mainly based on general toxicological principles, defined as including "all the studies where mainly undifferentiated cells are used to evaluate the effects of tested substances on common basic biological functions and biochemical processes essential for life" (Paganuzzi Stammati et al. 1981). These studies represent the first approach to be dealt with in in vitro toxicology and they can help in obtaining different kinds of information:

- Development of tests for the screening of synthetic and natural compounds with unknown mechanisms of action, in order to determine possible injurious effects, before these compounds come in contact with the human organism through the skin, or gastrointestinal or respiratory tract

- Establishment of priorities and selection of appropriate concentrations for further testing

- Classification of different molecules on the basis of their toxicity in a particular field of toxicology (irritation, acute toxicity, etc.), against a reference list

The aim of developing in vitro methods for screening chemicals is, among others, to reduce the number of animals utilized every year for the fulfillment of toxicological tests, as required by EU Directive 79/831, also in agreement with EU Directive 86/609, which encourages the use of alternative techniques.

cells   Screening studies mainly involve the use of established cell lines either derived directly from tumors, or from primary cultures or diploid cell lines following a transformation process. The most used tumoral cell lines are: cells of human origin (Hela, HEp-2), or fibroblasts deriving from human tissues (HEL-299, WI-38), mouse (L 929, Balb/3T3), or hamster (V 79, BHK 21). The wider use of cell lines rather than primary cultures is based on their easy handling, which allows good standardization of experimental conditions and protocols, with consequently better reproducibility of the results (Zucco 1992). Moreover, the cell lines can be easily purchased from cell and tissue banks, which are able to provide a complete characterization of the cells together with an overall check, including possible contaminations. Also used in some cases are primary cultures of fibroblasts derived from embryonic tissues of different species or, more seldom, primary cultures derived from target organs for the screening of compounds with specific effects (e.g., cells derived from nervous system for the screening of neurotoxic compounds). Transfection of several different cell types by a variety of methods may allow immortalization of cells such that they are able to retain their normal karyotype and functions in vitro.

Since the main emphasis of toxicology is directed toward humans, it is worthwhile to underline that the use of human cells is consequently highly desirable, in order to reduce the difficulties of extrapolating data from one species to another. Indeed, beside tumoral cell lines of human origin, cells derived from different target organs do exist and are already employed in pharmacotoxicological research. However, real advances using these systems will be made only after removal of the difficulties in obtaining human tissues for research purposes; this matter is still not regulated in many countries. It is also interesting to note that there is a trend towards the use of genetically modified human cell lines (Rogiers et al. 1993).

endpoints   The main problem in assessing cytotoxicity is the choice of indexes suitable for early detection of the damage (Cook and Mitchell 1989). This is further complicated by the fact that there are different cellular targets which can be compromised by a toxic compound, in contrast to what happens, for example, in mutagenesis, in which the only target is DNA. As shown in Table 21.3, it is possible to single out structural (the different cellular components) and functional cellular targets. Obviously, an alteration of the latter is a direct consequence of an impairment of the former, but in any case the two targets are approached differently. Therefore, great effort has been made in the last 20 years to develop tests

Table 21.3. Most commonly used targets in basal cytotoxicity

Structural targets	Parameters
Membrane	Exclusion or uptake of dyes
	Leakage of cellular constituents:
	– Enzymes (LDH, alkaline phosphatase, etc.);
	– Nucleic acids/radiolabeled precursors
	– Nucleotides
	– Ions or cofactors (Ca^{2+}, K^+, NADPH)
	Cell detachment from substrate
	Morphological observation (blebbing, tight junctions)
	Lipid peroxidation measured as:
	– malondialdehyde formation
	– Chemiluminescence and fluorescence of peroxides
	– Ethanol production
	– Diene conjugates
Lysosomes	Uptake of neutral red
	Acid phosphatase content
Mitochondria	Mitochondrial integrity (MTT assay)
Ribosomes	Synthesis of macromolecules; induction or inhibition of enzymes
Macromolecules	DNA, RNA and protein content/synthesis
Cytoskeleton	Observation of cytoskeletal elements (actin, tubulin, etc.)
Morphology	Observation of
	– Live cells (phase contrast)
	– Fixed/coloured cells (light microscopy)
	– Ultrastructural changes (organelles, tight junctions, microvilli, etc.)

Functional targets	
Cell viability	Cell number measured by:
	– Cell count
	– DNA, RNA and protein content
	– Total macromolecules
	– Uptake of vital dyes
Cell growth	Cell number measured as above
	DNA, RNA and protein synthesis
	Colony-forming ability
	Interference with cell cycle:
	– Mitotic index
	– Cycling fraction
Respiration	Oxygen consumption
	Oxidative metabolism
	Anaerobic glycolysis
Energy metabolism	Intracellular ATP, ADP and AMP
	Succinate dehydrogenase activity (MTT assay)
	Protein synthesis as incorporation of C^{14} leucine or S^{35} methionine
Others	Electrophiles formation
	Covalent binding

related to the different functions and/or structures of the cells; these tests must be sensitive enough to detect the possibly injurious effects at early stages. Clearly, the membrane, which protects the cell from the outside environment, is the first structure damaged by xenobiotics and for this reason there are a number of assays evaluating the membrane. However, it must be taken into account that the toxicity could be underestimated in the presence of slight effects on the membrane if the target of the test compound is an intracellular one. Morphologic alterations, either of the monolayer or of single cells, is another immediate and easily detectable endpoint, but for screening purposes quantitative tests, able to give a clear dose-effect answer, are preferred.

validation A necessary condition for the acceptance of alternative methods at regulatory level for screening purposes is their validation, that is, recognition of their scientific soundness. The theoretical basis of this process has been discussed by a group of experts, who defined it as "the process by which the reliability and relevance of a procedure are established for a specific purpose" (Balls et al. 1990). In fact, the proposed in vitro method must be reliable regarding its reproducibility in different laboratories and at different times. Moreover, it must be relevant, i.e., predictive with respect to a possible risk to human health and useful in achieving the aim of the test itself. Different parameters are adopted during the evaluation of a new assay:

- Sensitivity refers to identifying "true positive" results and represents the probability that a new test correctly identify compounds classified as positive by the reference test.

- Specificity indicates the probability that a new test correctly identifies compounds classified as negative by the reference test.

- Accuracy refers to the number of compounds correctly identified (positive and negative) by the new test with respect to the reference one.

- Predictive value is the ratio between true positive results and total positive results (both true and false) (Purchase 1986).

The validation procedure is a very complex one and requires different steps:

- Development of the assay

- Intra- and interlaboratory validation

- Evaluation of cost/benefit with respect to already existing tests

- Evaluation of its applicability to practical problems

- Acceptance by regulators

Recently, the importance of "protocol refinement and transfer" before the start of the formal validation study has been emphasized: this phase is called prevalidation (Curren et al. 1995). A number of problems must be overcome before the method under study can be accepted. One of the major difficulties is the choice of the reference chemicals and of reference results, not always easily available. Another aspect which must be taken into account is that in in vitro toxicology the considered toxicity index is often a reductive one compared to the more complex in vivo situation, or it can be only correlated to the effect under examination.

In the last years a series of national (CTFA in USA, ZEBET/BGA in Germany, OPAL in France) and international programs have been undertaken with different aims: standardization of protocols and detection of a suitable test battery (Balls et al. 1987), validation of in vitro tests towards available human (Bondesson et al. 1989) or animal data (DOC CEC 1988, 1991). Some of these studies are still in the evaluation stage, while others have been concluded with enough encouraging results, even if a series of difficulties must be overcome, in order to achieve the ultimate goal of the validation process. For this purpose, a second workshop on practical validation aspects was organized in Amden in 1994 by the European Research Group for Alternatives in Toxicity Testing (ERGATT), on behalf of the European Centre for the Validation of Alternative Methods (ECVAM). During the meeting, and in light of the results obtained in the previous validation studies, "recommendations have been made concerning the practical and logistical aspects of validating alternative toxicity testing procedures" (Balls et al. 1995).

international validation studies

Due to EEC directive 97/18, which prohibits the marketing of cosmetic products containing ingredients or combinations of ingredients tested on animals after June 30, 2000, most efforts have been directed toward the development of in vitro toxicity assays to be used in the field of eye or skin irritation. The alternative tests proposed for ocular and for skin irritation and the principles on which they are based have been thoroughly described in a number of excellent books and reviews (Loprieno 1995; Herzinger et al. 1995; Sina and Gautheron 1994; Gordon et al. 1994; Gad 1993; Green et al. 1993; Bruner 1992; De Leo 1992) and are summarized in Tables 21.4 and 21.5. It is important to underline that, in spite of the great efforts spent in the development of alternatives to the traditional Draize test, none of the reported assays has been fully validated,

Table 21.4. Most widely used alternative tests proposed for ocular irritation

Tests	Parameters
Cytotoxicity tests	Cell count
	Trypan blue uptake
	Neutral red uptake/release
	Cell adhesion/detachment
	Total protein, RNA, DNA content
	Incorporation of radiolabeled precursors
	Colony-forming ability
	Red blood cells hemolysis
	Enzyme release
	Fluorescein release
	Fluorescein diacetate/ethidium bromide
	Silicon microphysiometer
	MTT reduction
	Corneal repair
	Synthesis and release of plasminogen activator
Ex vivo	Chorioallantoic membrane (CAM) and modifications (HET-CAM,CAMVA, BECAM)
	Chick yolk sac assay
	Bovine corneal opacity/permeability (BCOP)
	Isolated eye
Others	EYTEX
	Motility of *Tetrahymena thermophila*
	Luminescence of *Photobacterium phosphoreum*
	Tube wall material produced by the *Nicotiana sylvestris* pollen

Table 21.5. Most widely used alternative tests proposed for skin irritation

Tests	Parameters
Physicochemical methods	pH
	Partition coefficient
	Absorption spectra
Cytotoxicity	Neutral red uptake/release
	Total protein, RNA, DNA content
	Incorporation of radiolabeled precursors
	Enzymes release
	Keratin staining with rhodamine B and Nile blue
	Cell number and colony area measurements
Others	TESTSKIN (human skin equivalents)
	SKINTEX
Human volunteers	Cutaneous blood flow
	Transepidermal water loss by evaporimetry

which means that they are still in the development or in the evaluation step, but they have not yet been accepted by regulators. In any case most of them gave encouraging results and thus they should be included in the legislation, at least in the prescreening step. Indeed, prescreening tests can have less requirements from a legislative point of view than tests proposed for a complete replacement, provided that both are based on a strong rationale, provide reliability and relevance and can be validated.

Choice of Experimental Model

In setting up a toxicity testing protocol, great attention must be paid to every aspect, as several problems may arise from only an approximation of its design (Bridges et al. 1983; Stark et al. 1986; Garle et al. 1994). For general cytotoxicity testing, continuous established cell lines are usually preferred, due to their relative stability, generally good characterization and easy handling. The culture techniques are standardized, guaranteeing a higher reproducibility of the results.

The physiological condition of the cells is also relevant: toxic response may differ according to the phase of growth (stationary or replicative), especially in relation to exposure time. By using cells in a stationary growth phase, toxicants with effects on the cell cycle can be missed. Moreover, metabolism is usually slowed down and the high cell density may influence presentation (availability) of the target sites. In growing cultures, the cells are more dispersed, intercellular cooperation is lower and the cells may, in general, be more sensitive. Freshly trypsinized and subcultured cells may also be more sensitive due to possible membrane receptor alterations, so that one should wait about 12 h before beginning testing, in order to let the cells fully recover.

Toxic Agent Preparation and Exposure Conditions

The compatibility of the compound under study with the culture conditions should be investigated in advance in order to avoid unwanted alterations of the culture medium, or interference with it, which may alter the exposure conditions. Among the various aspects to be checked are: variations in pH and osmolarity (especially with compounds to be tested at high concentrations) and possible denaturation of medium proteins or lipid oxidations following addition of the compound.

chemicophysical properties, solubility of agents

Several compounds are insoluble in water and in the medium. The use of a solvent (usually ethanol, methanol, DMSO) presents the previously mentioned problems of interaction with the medium. Moreover a solvent may have its own intrinsic toxicity, which may complicate interpretation of the results. For all these reasons, the maximum final solvent concentration into the medium considered to be acceptable is 1 %, and the control of reference should have the same solvent concentration as the treated samples.

Other compounds may be volatile or be in the form of particles or micelles. In all these cases special precautions must be taken in order to be sure of the real exposure conditions: to avoid volatility the culture system must be sealed and to guarantee the dispersion of the particulate, sonication procedures may be adopted.

concentrations   As a first approach the compound should be tested over a wide range of concentrations, each differing from the next one by an order of magnitude, for example, from 10^{-3} to 10^{-8}. Once the dose-response curve is established, the effective concentrations range may be identified and further detailed, starting from the maximum concentration with no effect to the first concentration producing the maximum effect (generally 4–5 dilutions). For unknown compounds the concentration is reported in mg/ml, but since the toxic effect is related to the number of molecules, the best way is to report the concentration, when possible, as mM or µM/ml. It is also useful to refer to the number of cells used. When in vivo data are available, the in vitro concentrations should be in the range of the in vivo plasma concentration.

exposure times   Different exposure times can be chosen according to the general strategy of the experiment but also to the endpoint taken into account. If the in vivo kinetics of the compound are known, it should be referred to for setting the exposure times. This may vary from minutes and hours to several days. Short exposure times are generally adopted when functional alterations must be monitored, i.e., metabolic changes, permeability mechanisms, etc. Longer exposure times are used when effects on cell reproduction or viability are investigated. In the latter case, the stability of the culture conditions should be taken into account, because indirect or secondary effects on the culture medium may interfere with the overall toxicity observed.

In some cases it is also advisable to verify the reversibility of the effect by removing the medium with the toxic compound and adding new fresh medium.

Endpoints as Toxicity Markers

The ideal endpoint in toxicity screening should be representative of a specific mechanism; in other words in order to be biologically significant, the endpoint must be relevant to an in vivo physiological effect. This is not the case for many compounds whose mechanism of action has yet to be fully identified. Many toxicities are investigated in blind trials, in which the active principle is not known: that is the case for contaminations or for new products. Moreover, with few exceptions, there is more than one intracellular target.

For all these reasons, attention should be focused on endpoints which may be indicative of a great variety of insults and which may lead to a sort of common pathway to cell death. Many tests indeed are devoted to monitoring this all-or-none phenomenon, by measuring different parameters. Nonetheless, while cell death-related endpoints are sensitive, they are not without complications: They can give false negative results, as there are very specific mechanisms that are not related to cell death but which may still impair the life of an organism in vivo. By contrast, with respect to the in vivo situation, the use of very specific endpoints may also give false positive outcomes in vitro. Finally, cell death-related endpoints do not allow verification of the possible reversibility of the effect.

For all the reasons mentioned above, a battery of tests is usually suggested, designed in order to cover different aspects of the toxicity. If the results obtained by various endpoints are in agreement in terms of ranking or of LC_{50} determination, it is possible that a common general mechanism is involved. If, however, in some of the tests a discrepancy is recorded, the type of test in which the compound appears more toxic may suggest a specific mechanism of action and further tests should be made in that direction.

It must be pointed out that morphological monitoring of the cultures by the optical microscope is a crucial step in toxicity studies; cell injury can be appreciated already at early stage of exposure and it is advisable to record the observations made, either in writing or by taking pictures. Such observations are often very meaningful in the final overall evaluation of toxicity.

Materials

For the Lowry assay:

- Solution A: sodium carbonate 4 %
- Solution B: sodium potassium tartrate 4 %
- Solution C: cupric sulphate 2 % in water
- Phosphate buffered saline (PBS): 20 mM sodium phosphate, pH 7.4; 140 mM NaCl
- 0.5 N NaOH
- Standards: bovine serum albumin (BSA) serial dilutions in 0.5 N NaOH (0–50 µg protein/well)
- Folin-Ciocalteu's phenol reagent (1:1 in water)

For the Kenacid blue assay:

- Kenacid blue stock solution: 0.4 g in 250 ml ethanol and 630 ml water. Immediately before use add 12 ml of glacial acetic acid to 88 ml of stock solution and filter.
- Glacial acetic acid/ethanol/water: 1:50:49 (v/v/v)
- Glacial acetic acid/ethanol/water: 10:5:85 (v/v/v)
- Solvent solution: 1 M potassium acetate in 70 % ethanol

For the neutral red assay:

- Neutral red stock solution: 5 mg/ml in distilled water; can be stored in the dark at 4 °C for several months. The day before its use, dilute the neutral red stock solution in the culture medium to a final concentration of 50 µg/ml. Keep overnight at 37 °C and centrifuge at 1500 g for 5 min immediately before use to remove undissolved crystals.
- Wash-fix solution: 4 % formaldehyde containing 1 % $CaCl_2$ in distilled water
- Solvent solution: 1 % glacial acetic acid in 50 % ethanol/water

For the MTT assay:

- MTT stock solution: 5 mg/ml in PBS. Pass through a 0.22 µm filter in order to sterilize and remove small amounts of insoluble residues.
- DMSO
- Sorensen's glycine buffer: 0.1 M glycine; 0.1 M NaCl. Adjust to pH 10.5 with 1 M NaOH.

For LDH determination

- 0.1 M phosphate buffer (PBS)

- NADH solution (0.17 mg/ml): Dissolve 3.4 mg NADH in 20 ml buffer. Keep at 0°−4 °C and prepare fresh daily.
- Pyruvate solution (9.76 mmol/l): Dissolve 0.107 g monosodium pyruvate in 90 ml buffer. Adjust the final volume to 100 ml with buffer. Store in aliquots at −20 °C; the solution is stable for about 2 months.

For colony formation:

- Gentian violet: 1 g gentian violet in 5 % glacial acetic acid, 15 % ethanol, 80 % distilled water

For oxygen consumption

- Respiration buffer: Chance buffer; 6.2 mM KCl; 145 mM NaCl; 11 mM sodium phosphate, pH 7.4
- Sodium dithionite crystals

Procedure

Total Protein Content

The determination of total protein content is one of the simplest and most commonly used parameters as an endpoint of cytotoxicity. A toxic substance that interferes with growth or which causes the death of all or part of the exposed cell population results in a reduction in cell number. The total number of cells may be determined either directly (by cell counting) or indirectly, by measuring an endogenous component present in uniform quantity in the cells, such as DNA, RNA, proteins.

Several colorimetric methods are available for measuring the protein content of a cell population, one of the most widely used being those proposed by Lowry (1951), Bradford (1976), Skehan (1990; sulforodamine B method). All of these have been optimized for use with an automatic microplate reader, which allows rapid and economic quantitation, particularly suitable for screening purposes (Doyle et al. 1993; Martin and Clynes 1993).

All these methods require extraction of the proteins from the monolayer; once in solution the proteins react with specific dyes that shift the absorption spectrum and may be spectrophotometrically recorded. The samples can be preserved at 4 °C for about 15 days without significant alteration.

An alternative method, proposed and used by the FRAME (Knox 1986, Clothier 1995), is the Kenacid blue method; the cells are fixed and stained with a solution of Coumassie blue 250. In this case the cell layer is not solubilized and the whole process may be repeated or, even better, the same layer may be used for other tests, such as neutral red or MTT.

The main disadvantage of these tests is the possibility of overestimating cell number, either when dead cells remain strongly attached to the substrate or when the cells are treated with drugs inhibiting cell replication without inhibiting protein synthesis, e.g., BUDR or methotrexate.

total protein determination with Lowry methods

1. Seed an appropriate number of cells in 96-well tissue culture plates so that about 60–70 % confluency is reached at the time of treatment (approx. 1×10^4 cells/well/250 µl medium). Incubate at 37 °C for at least 24 h.

2. Remove the medium and add fresh medium containing graded dilutions of the toxic compound to be tested. At least eight wells must be used as controls. Incubate for the chosen intervals of exposure times (generally 24, 48, or 72 h).

3. Remove the medium with the test agent, wash the layer 3× with warm PBS.

4. Lyse the cells adding 50 µl 0.5 N NaOH to each well and incubate the plate for 2 h at 37 °C, or overnight.

5. Mix 100 parts solution A with 1 part solution B and 1 part solution C.

6. Add 125 µl of the ABC mixture (step 5) to each well. Incubate for 10 min at room temperature.

7. Add 25 µl of a of Folin Ciocalteu's reagent (1:1 in water) to each well, mix for 10 s and keep at room temperature for 30 min.

8. Measure the absorbance at 630 nm against reference filter of 405 nm.

9. Use BSA serial dilutions in NaOH as standards.

protein determination using Kenacid blue assay

1. Treat the 96-well plates as described for the Lowry assay.

2. Remove the medium, wash 3× with PBS and fix in 150 µl glacial acetic acid/ethanol/water (1:50:49 v/v/v) for 20 min at room temperature.

3. Remove the fix solution and stain with 150 µl Kenacid blue solution for 20 min at room temperature with gentle shaking.

4. Wash 3× with ethanol/glacial acetic acid/water (10:5:85 v/v/v). The final wash should be done by shaking for 20 min.

5. Add 150 µl of solvent solution to each well (1 M potassium acetate in 70 % ethanol) and mix rapidly for 20 min in order to obtain an homogeneous distribution of the dye.

6. Read the absorbance at 577 nm against a reference filter of 404 nm.

Neutral Red Uptake

The neutral red (3-amino-7-dimethyl-2-methylphenazine-hydrochloride; NR) assay is based on incorporation of the dye into lysosomes of viable cells by passive transport across the plasma membrane. Damaged or dead cells are not able to retain the NR. The use of a vital dye avoids underestimation of the toxicity due to the presence of dead cells still adhering to the substrate. The performance of the NR assay has been extensively assessed and as a result is widely used for various in vitro cytotoxicity validation studies. The numerous advantages of this test, such as its simplicity, economy, sensitivity and reproducibility, have been demonstrated in many laboratories (Babich and Borenfreund 1990, Borenfreund and Luemez 1984, Doyle et al. 1993, Watson 1992).

The uptake of NR may vary according in different cell types since the number of lysosomes per cell differs in different tissues. The incubation time should also be optimized for each cell line: NR uptake is frequently preceded by a variable lag phase. In general, 3 h are sufficient for the majority of cells.

Often the cells in culture, once close to confluency, show a progressive decrease in NR uptake, probably due to reduced lysosomal activity or a slowing down of cellular metabolism. For this reason it is better to use exponentially growing cells. **Note:** This assay is not useful when testing substance that act on lysosomes, such as chloroquine sulfate, which inhibits NR uptake by altering lysosomal pH.

1. Seed and treat the cells as described for the Lowry assay.

2. Prepare the NR solution as described in "Materials."

3. Remove the medium from the treated plates and replace with 100 µl of NR solution. Incubate for 3 h at 37 °C.

4. Remove the NR solution and rinse the well with 200 µl wash-fix solution for 2 min.

5. Add 100 µl of solvent solution to each well.

neutral red
uptake

6. Extract the dye for 20 min on microtiter plate shaker.

7. Read the absorbance at 540 nm.

MTT Cell Viability Assay

The MTT test is a rapid and sensitive colorimetric assay based on the formation of a colored insoluble formazan salt. The amount of formazan produced is directly proportional to the cell number in numerous cell lines and, therefore, it can be used to measure cell viability and proliferation. This technique is particularly useful for cells which are still metabolically active but not proliferating and for cell suspension cultures. The assay is based on the capacity of the mitochondrial dehydrogenase enzymes to convert a yellow water-soluble tetrazolium salt, 3-(4,5-dimethylthiazol-2-yl)-2,5-diphenyltetrazolium bromide (MTT), into a purple insoluble formazan product by a reduction reaction. The insoluble crystals are dissolved in DMSO and the absorbance is read with a spectrophotometer (565 nm) (Mosmann 1983, Carmichael et al. 1987, Supino 1995).

Nonetheless, the MTT assay presents a number of serious problems:

- The intensity of the reduction varies considerably from one cell line to another and frequently decreases with culture age

- The reaction tends not to be linear with cell number and its rate varies with pH and with cellular and medium glucose levels. All these difficulties give a notable variability, for example IC_{50} values can vary 20-fold according to MTT concentration, culture age, population density and length of assay. For this reason, when using this test as viability parameter much attention must be paid to its optimization for each cell line and to its use within the limits of documented validity.

MTT assay
1. Seed and treat the cells in 96-well microtiter plates as described for the Lowry assay.

2. Prepare the MTT stock solution (5 mg/ml PBS) as described in "Materials."

3. Add 50 µl MTT stock solution to 200 µl of medium in each well.

4. Incubate in a humidified incubator at 37 °C for 4 h.

5. Remove the medium. For nonadherent cells: centrifuge the plates at 200 g for 5 min.

6. Add 200 µl DMSO and incubate for 10 min on a microplate shaker to dissolve the formazan crystals.

7. Add 25 µl of Sorensen's glycine buffer.

8. Read the plate at 570 nm against a reference of 620 nm. Plates should be read within 30 min after addition of DMSO.

Release of Lactate Dehydrogenase

Analysis of the release of intracellular enzymes can be used to monitor changes in culture viability. Intracellular enzymes are only released after damage to the cytoplasmic membrane, which is generally involved in the early phase of the toxic action. Among the various enzymes released into the culture medium, lactate dehydrogenase (LDH) is the most commonly used in cytotoxicity studies. It provides an estimate of instantaneous damage, for example, by cell freezing and thawing, or progressive loss of cell viability over a few hours.

The activity of LDH can be measured as a reduction of pyruvate to lactate coupled to the oxidation of NADH to NAD^+. Its disappearance is followed spectrophotometrically at 340 nm. Both LDH released from the cells into the culture medium and the amount remaining in the cells can be quantitated (Vassault 1983, Welder and Acosta 1994, Doyle et al. 1993). It is also possible to determine LDH activity by a colorimetric assay (Decker and Lohmann-Matthes 1988) based on the reduction of tetrazolium salts; this technique permits the processing of a higher number of samples by using microplate readers, but as it is an indirect measurement of enzyme activity, it is not very specific.

The stability of LDH can vary considerably between different cell lines and this assay is not suitable for all cell types, in particular fibroblasts, which have lower endogenous levels of this enzyme. Moreover, some xenobiotics can interfere with this assay.

1. Expose the culture to the appropriate concentration of the test agent. At the end of the treatment period collect the medium from controls and treated cultures, centrifuge (2000 g×5 min, 4 °C), collect the supernatant and place on ice.

LDH determination

2. Wash the monolayers with 1 ml PBS and scrape the cells. The cellular suspension is transferred to a test tube and sonicated on ice three times for 10 s.

3. Add to 0.1 ml of supernatant and to 0.1 ml of cellular sample: 1.2 ml of NADH solution (0.17 mg/ml in PBS, pH 7.4), and 0.2 ml of sodium pyruvate solution (1 mg/ml in PBS). Vortex for 5 s.

4. Transfer the solution to a cuvette and place it into a spectrophotometer with controlled temperature (25–30 °C).

5. Monitor the absorbance variation (ΔA) at 1 min of intervals, for 4 min at least, at a wavelength of 340 nm.

6. Calculate the mean ΔA/min for the 4 min monitored. The LDH/l is obtained with the following formula:

Units LDH/min $= \Delta A \times D \times F$

where D is the dilution factor (10 000) and F is a temperature factor (1.16 at 30 °C). The results are expressed as percentage of the total cellular LDH, in which: total LDH = released LDH + cellular LDH.

Colony-Forming Ability

The colony-forming assay (CFA) determines the proportion of single cells in a cellular population able to give rise to growing colonies. While cytotoxicity tests (LDH, neutral red, MTT) measure alterations in structural integrity or metabolic pathways and are only indirectly correlated with cell death, testing cell survival in terms of reproductive integrity actually measures cell recovery or cell death after exposure to a toxicant.

This test is usually performed in the presence of the toxic agent throughout colony formation, but in this case cell death vs impairement of cell reproduction cannot be distinguished. Therefore the following approach, which gives clearer results, is recommended: Cells in a subconfluent culture should be treated for 24 h with the test chemical, then dispersed and replated in fresh growth medium at very low density. The seeding density depends on different parameters, such as growth rate and plating efficiency of the cells used, and should be selected so that the final number of colonies expected in the control are between 50 and 200. After about a week the dishes are fixed, stained and the colonies with more than ten cells are counted (Wilson 1992).

The CFA can be measured by plating cells either on a solid substratum, such as tissue culture plastic, or by seeding them in nonadherent medium such as soft agar. The latter is particularly suitable for cells growing in suspensions, e.g., hematopoietic cells. In this case the dish contains two separate layers with different percentages of agar: (1) a bottom layer, more concentrated (0.5 % agar), which prevents cell adhesion to the bottom of the dish and (2) a top layer consisting of 0.3 % agar and containing the cells (Skehan 1995).

This test is not suitable for those cell lines showing a limited proliferative capacity and low plating efficiency; generally, the plating efficiency is inversely correlated to the level of differentiation and specialization. Moreover the duration of the test is longer, about 1–2 weeks, thus the risk of contamination is increased.

1. Treat semiconfluent cultures with graduated concentrations of the test substance for 24 h.   **colony formation**

2. Collect cells, centrifuge and count very carefully.

3. Dilute the cell suspension to a final concentration in the range of 100–400 cells/dish, depending on the cell line. The final volume of the cell sample to be distributed into the plates for the assay should be in the range of 50–200 μl in order to avoid errors in sampling (too small volumes may give rise to volume errors, while large volumes may give rise to errors in cell number, the cells being too dispersed).

4. Seed the cells in a 60 mm dish already prepared with 4 ml medium. Mix gently to uniformly distribute the cells. Prepare at least four dishes for each dilution and for the control.

5. Incubate the plates for about a week, changing the medium during this period, if necessary.

6. At the end of the incubation, wash the plates with PBS and stain them with gentian violet. Gently agitate for 20 min at room temperature.

7. Take off the dye and wash the plates extensively with distilled water.

8. Using a colony counter or under the microscope, count all the colonies with more than ten cells. Express the results as (number of colonies counted/number of cells plated) ×100.

Oxygen Consumption

The measure of cellular oxygen consumption is frequently applied to studying the mechanism of toxic action, particularly for substances that cause alterations in the mitochondria. Mitochondrial alterations can play a critical role in the processes of cellular death and this technique is one of the simplest, fastest and most reliable. Measurement can be done using a variety of preparations including isolated cells, subcellular fractions, mitochondria, or proximal tubules tissue slices (Cain and Skilleter 1987, Schnellmann 1994).

Generally, oxygen consumption is measured with a Clark oxygen electrode (Clark 1953); this apparatus consists of a water-jacketed unit into which is inserted the Clark electrode (Ag/AgCl platinum electrode), isolated by a thin Teflon membrane, which is freely permeable to oxygen. A magnetic stirrer is necessary to maintain a steady state of oxygen tension. The chamber is maintained at the desired temperature (usually 37 °C) by a heat circulator and oxygen consumption is registered by a recorder.

The rate of oxygen consumption (nmol O_2) can be determined from the traces obtained on the chart recorder using the following equations:

$$\frac{\text{chart speed (mm/1 min)} \times \text{nmol} O_2/\text{mm}}{\text{number of cells or cellular protein (mg/ml)}}$$

where nmolO_2/mm is the total oxygen concentration in the buffer divided by the chart distance between 0 % O_2 consumption and 100 % O_2 consumption (mm).

measurement by Clark electrode

1. About 30 min before the start of the experiment, switch on the electrode, magnetic stirrer, chart recorder and water circulator with thermostat. Bring the chamber to the appropriate temperature (usually 37 °C).

2. Assemble the Clark oxygen electrode at each experiment, according to the manufacturer's instructions. Set the electrode voltage to 0.8.

3. Pipette 1.8 ml respiration buffer (previously equilibrated at the experimental temperature and fully saturated with air) into the oxygraph chamber (make sure there are no air bubbles), start the stirrer and the recorder (speed 0.1 mm/s).

4. After 5–10 min of stabilization, adjust the recorder to the 100 % position, that is the line in which the buffer is fully saturated with air. Control that this line is stable over the next 5 min.

5. Add a few crystals of sodium dithionite which rapidly consumes the O_2 dissolved. At this stage the buffer is anaerobic and the pen position on the recorder is referred to as being at the 0 % position and must correspond to the zero voltage of the electrode.

6. Collect the cells, wash with respiration buffer, count and resuspend at a final concentration of about $2.5-5\times10^6$ cells/200 µl buffer.

7. Add the cell suspension to the oxygraph chamber. This step should not appreciably alter the oxygen concentration of the buffer: microliter syringes are ideal for this purpose.

8. Record the basal cellular respiration for 5 min, then add the test compound (no more than 20 µl each time) into the chamber and record for a further 5 min.

References

Babich H, Borenfreund E (1990) Application of the neutral red cytotoxicity assay to in vitro toxicology. Altern Lab Anim 18: 129–144

Balls M, Riddell R, Horner R, Clothier R (1987) The Frame approach to the development, validation and evaluation of in vitro alternative methods. In: Goldberg AM (ed) Alternative methods in toxicology, Mary Ann Liebert, New York, pp 45–58

Balls M, Blaauboer B, Brusick D, Frazier J, Lamb D, Pemberton M, Reinhardt C, Roberfroid M, Rosenkranz H, Schmid B, Spielmann H, Stammati AL, Walum E (1990) Report and recommendations of the CAAT/ERGATT Workshop on the validation of toxicity test procedures. ATLA 18: 313–337

Balls M, Blaauboer BJ, Fentem JH, Bruner L, Combes R, Ekwall B, Fielder RJ, Guillouzo A, Lewis RW, Lovell DP, Reinhardt CA, Repetto G, Sladowski D, Spielmann H, Zucco F (1995) Practical aspects of the validation of toxicity test procedures. ATLA 23:129–147

Barile F A (1994) Introduction to in vitro cytotoxicology – Mechanisms and methods. CRC, Boca Raton

Bondesson I, Ekwall B, Hellberg S, Romert L, Stenberg K, Walum E (1989) MEIC: a new international multicentre project to evaluate the relevance to human toxicity of in vitro cytotoxicity test. Cell Biol Toxicol 5: 331–348

Borenfreund E, Puerner JA (1984) A simple quantitative procedure using monolayer cultures for cytotoxicity assays (HTD/NR90). J Tissue Cult Methods 8: 7–9

Bradford M (1976) A rapid sensitive method for the quantitation of microgram quantities of protein utilizing the principle of protein dye binding. Anal Biochem 72: 248–254

Bridges JW, Benford DJ, Hubbard SA (1983) The design and use of in vitro toxicity test. In: Turner P (ed)Animal in scientific research: an effective substitute for man? MacMillan, Basingstoke, pp 47–68

Bruner LH (1992) Ocular irritation. In: Frazier JM (ed)In vitro toxicity testing, Marcel Dekker, New York, pp 149–189

Cain K, Skilleter DN (1987) Preparation and use of mitochondria in toxicological research. In: Snell K, Mullock B (eds) Biochemical toxicology – a practical approach, IRL, Oxford, pp 217–254

Carmichael J, De Graff WG, Gazdar AF, Minna JD, Mitchell JB (1987) Evaluation of tetrazolium based semiautomated colorimetric assay: assessment of chemiosensitivity testing. Cancer Res 47: 936–942

Clark L C, Wolf R, Granger D, Taylor Z (1953) Continuous recording of blood oxygen tension by polarography. J Appl Physiol 6: 189–193

Clothier RH (1995) The Frame cytotoxicity test (Kenacid blue). In: O'Hare S, Atterwill CK (eds) Methods in molecular biology, vol. 43. In vitro toxicity methods. Humana, Totowa, New Jersey, pp 109–118

Cook JA, Mitchell JB (1989) Viability measurements in mammalian cell systems. Anal Biochem 179:1–7

Curren RD, Southee JA, Spielmann H, Liebsch M, Fentem JH, Balls M (1995) The role of prevalidation in the development, validation and acceptance of alternative methods. ATLA 23: 211–217

Decker T, Lohmann-Matthes ML (1988) A quick and simple method for the quantitation of lactate dehydrogenase release in measurements of cellular cytotoxicity and tumor necrosis factor (TNF) activity. J Immunol Meth 115: 61–69

De Leo WA (1992) Cutaneous irritancy. In: Frazier JM (ed) In vitro toxicity testing. Marcel Dekker, New York, pp 191–203

CEC (1988) Collaborative study on relationship between in vivo primary irritation and in vitro experimental models. EC Document V/E/LUX/157/88, Luxembourg

CEC (1991) Collaborative study on the evaluation of alternative methods to the eye irritation test. EC Document XI/632/91-V/E/1/131/91, Luxembourg

Doyle A, Griffiths JB, Newell DG (1993) Cell and tissue culture: laboratory procedure. John Wiley, New York

Freshney RI (1994) Culture of animal cells. A manual of basic technique, 3rd edn. Wiley-Liss, New York

Frazier JM (1992) In vitro toxicity testing – application to safety evaluation. Marcel Dekker, New York

Gad SC (1994) In vitro toxicology. Raven, New York

Gad SC (1993) Alternatives to in vivo studies in toxicology. In: Ballantyne B, Marrs T, Turner P (eds) General and applied toxicology, vol. I. Stockon, New York, pp 179–206

Garle MJ, Fentem JH, Fry JR (1994) In vitro cytotoxicity tests for the prediction of acute toxicity in vivo. Toxicol In Vitro 8: 1303–1312
  Goldberg AM (1983–1995) Alternative methods in toxicology, vol. 1–11. Mary Ann Liebert, New York

Gordon WC, Harvell J, Bason M, Maibach H (1994) In vitro methods to predict dermal toxicity. In: Gad SC (ed) In vitro toxicology, Raven, New York, pp 47–55

Green S, Chambers WA, Gupta KC, Hill RN, Hurley PM, Lambert LA, Lee CC, Lee JK, Liu PT, Lowther DK, Roberts CD, Seabaugh WM, Spronger JA, Wilcox NL (1993) Criteria for in vitro alternatives for the eye irritation test. Food Chem Toxicol 31: 81–85

Herzinger T, Korting HT, Maibach HI (1995) Assessment of cutaneous and ocular irritancy: a decade of research on alternatives to animal experimentation. Fund Appl Toxicol 24: 29–41

Hornhardt S, Wiebel FJ (1996) Catalogue of cell lines in toxicology and pharmacology. GFS-Forschungszentrum für umwelt und gesundheit, Bericht 3/96. ISSN 0721–1694

Jolles G, Cordier A (1992) In vitro methods in toxicology. Rhone-Poulenc Rorer Foundation, Academic Press, New York

Knox P, Uphill PF, Fry JR, Benford J, Balls M (1986) The Frame multicentre project on in vitro cytotoxicology. Food Chem Toxicol 24: 457–463

Loprieno N (1995) Alternative methodologies for the safety evaluation of chemicals in the cosmetic industry. CRC, Boca Raton

Lowry OH, Rosenbrough NJ, Farr AL, Randall R L (1951) Protein measurement with the folin phenol reagent. J Biol Chem 193: 265–275

Martin A, Clynes M (1993) Comparison of 5 microplate colorimetric assays for in vitro cytotoxicity testing and cell proliferation assays. Cytotechnology 11: 49–58

Mosmann T (1983) Rapid colorimetric assay for cellular growth and survival: application to proliferation and cytotoxicity assay. J Immunol Meth 65: 55–63

Nardone RM (1977) Toxicity testing in vitro. In: Rothblat GH, Cristofalo VJ (eds) Growth, nutrition and metabolism of cells in culture. Academic Press, New York, pp 471–491

O' Hare S, Atterwill CK (1995) Methods in molecular biology, vol. 43: In vitro toxicity testing protocols. Humana, Totowa, New Jersey

Paganuzzi Stammati A, Silano V, Zucco F (1981) Toxicology investigations with cell culture systems. Toxicology 20: 91–153

Purchase IFH (1986) Unpredictability: an assay in toxicology. Food Chem Toxicol 24: 343–349

Rodrigues AD (1994) Use of in vitro human metabolism studies in drug development. An industrial perspective. Biochem Pharmacol 48: 2147–2156

Rogiers V, Sonck W, Shephard E, Vercruysse A (1993) Human cells in in vitro pharmacotoxicology. VUB, Brussels

Schnellmann RG (1994) Measurement of oxygen consumption. In: Tyson CA, Frazier JM (eds) Methods in toxicology, part B: in vitro toxicity indicators, Academic Press, New York, pp 129–139

Skehan P, Storeng R, Scudiero D, Monks A, McMahon J, Visitica D T, Warren J, Bokesch H, Kenney S, Boyd MR (1990) New colorimetric assay of anticancer drug screening. J Nat Cancer Inst 82: 1107–1113

Skehan P (1995) Assay of cell growth and cytotoxicity. In: Studzinski GP (ed) Cell growth and apoptosis – a practical approach series, IRL, Oxford, pp 169–191

Sina JF, Gautheron PD (1994) Ocular toxicity assessment in vitro, In: Gad SC (ed) In vitro toxicology. Raven, New York, pp 21–45

Skett P, Tyson C, Guillouzo A, Maier P (1995) Report on the international workshop on the use of human in vitro liver preparations to study drug metabolism in drug development. Biochem Pharmacol 50: 280–285

Stark DM, Shopsis C, Borenfreund E, Babich H (1986) Progress and problems in evaluating and validating alternative assays in toxicology. Food Chem Toxicol 24: 449–455

Supino R (1995) MTT assay. In: O'Hare S and Atterwill CK (eds) Methods in molecular biology, vol 43. In vitro toxicity testing protocols. Humana, Totowa, New Jersey, pp 137–149

Tyson CA, Frazier JM (1994) Methods in toxicology, vols. 1A, 1B. Academic Press, San Diego

Vassault A (1983) Lactate dehydrogenase: UV method with piruvate and NADH. In: Bergmeyer HU, Bergmeyer J, Grassl M (eds) Methods of enzymatic analysis. III. Enzymes 1: oxidoreductases, transferases, 3rd edn. Verlag-Chemie, Weinheim, pp 118–126

Watson RR (1992) In vitro methods of toxicology. CRC, Boca Raton

Welder AA, Acosta D (1994) Enzyme leakage as an indicator of cytotoxicity in cultured cells. In: Tyson CA, Frazier JM (eds) Methods in toxicology, part B. In vitro toxicity indicators. Academic Press, New York, pp 46–49

Wilson AP (1992) Cytotoxicity and viability assay. In: Freshney RI (ed) Animal cell cultures – a practical approach series, 2nd edn. IRL, Oxford, pp 263–303

Wrighton SA, Vandenbranden M, Stevens JC et al.: Shipley LA, Ring BJ, Rettie AE, Cashman JR (1993) In vitro methods for assessing human hepatic drug metabolism: their use in drug development. Drug Met Rev 25: 453–484

Zucco F (1992) Use of continuous cell lines for toxicological studies. In: Castell JV, Gomez-Léchon M (eds) In vitro alternatives to animal pharmaco-toxicology. Farmindustria, Serie Cientifica, Madrid, pp 43–68

Chapter 22

Miniaturized In Vitro Methods in Toxicity Testing

Robert O' Connor*, Mary Heenan, Conor Duffy, and Martin Clynes

Introduction

In vitro miniaturized colorimeteric assays are extensively used in the determination of a substance's ability to enhance cell growth or promote cell death. The widespread use of colorimeteric endpoint assays in the determination of cell number, the most common measure of cell growth, is due to their simplicity and sensitivity and also to their ability to be scaled up using a semi-automated high-throughput system.

The most commonly used measure of toxicity is impairment of cell growth, generally determined by reduced cell numbers, as compared to untreated controls. However, the colorimetric endpoints used in determining cell number will not distinguish between cytotoxic and cytostatic effects, when used 6 or 7 days after exposure of cells to the test chemical. Cytostatic effects may only effect the growth rate of the cells in a temporary and reversible manner (once the cytostatic agent is removed). Cytotoxic agents cause irreversible cell damage, resulting in cell death, either by an apoptotic or necrotic pathway. As both agents cause impaired cell growth, both will exhibit similar effects in colorimetric endpoint toxicity assays.

Some agents may be toxic to cells at all stages of the cell cycle, while others are cell cycle specific. Therefore, in the course of the toxicity assay, it is essential to expose the cells to the toxin for a sufficient time for the agent to exert its toxic effect and to incubate the cells for a sufficient length of time following toxic insult. This is often necessary for the consequences of the exposure to become apparent in the assay system.

There are both advantages and disadvantages to all the methods of endpoint determination of cell number and some are more applicable to

* *Correspondence to* Robert O' Connor, National Cell and Tissue Culture Centre, Dublin City University; Glasnevin, Dublin 9, Ireland; phone +353–1–7045703; fax +353–1–7045484; e-mail oconnorr@dcu.ie

certain types of experiments. Consideration must be given when setting up a miniaturized colorimetric assay system to choose the best system for the analysis of cell number under the particular experimental condition. Different assays exhibit varying sensitivity levels and ranges of linearity with cell number, and as colorimetric endpoints measure the levels of particular cellular targets, different systems may have a bias in certain circumstances. Therefore, in the initial calibration of the assay system, it is important to check, by an independent method, that the assay system is giving a true and proportional reflection of the endpoint. For example, if the colorimetric assay is a general assay to determine cell number, then cell number may be checked by performing a cell count, and the range of cell numbers encountered within the assay system investigated to show proportional increases in color intensity, when analyzed by the chosen colorimetric endpoint. Otherwise, this endpoint is unsuitable as a reflection of cell number and an alternative endpoint should be employed or assay conditions altered, as appropriate.

It is critical when using colorimetric assay endpoint determinations to consider which cellular parameter to study, as the endpoint chosen may be biased. Certain cellular functions may be effected by a particular toxin or chemical which, while not causing cell death, would give a misleading result in the assay. For example, Horakova et al. (1978) showed that although total cell number in control cultures was directly proportional to the total cell protein or nucleic acids, in cultures exposed to 6-thioguanine and vermiculine the relationship between the parameters was perturbed. Therefore, it is important to consider if cell number is an accurate reflection of the effect of the compound on cells or if its effect on an alternative target is more informative, or whether more than one endpoint should be analyzed.

It is vital to prescreen the cell line which is being used in the assay to determine that a suitable level of sensitivity and a sufficiently extensive range of cell number can be analyzed, as the sensitivity and linear range for the colorimetric assays may vary, depending on both the cell line being analyzed and the colorimetric endpoint being employed. Some of these factors are discussed below. The sensitivity of the assay endpoint is offset against the range of linearity of optical density (OD) to cell number. Generally, endpoints with high sensitivity, give a low range of linearity of OD with cell number. The levels of specific enzyme activity may also change during cell growth and it is necessary to complete a standard curve of each cell line under the precise conditions of the experiment, if the exact relationship between cell number and activity is required. Overall, the acid phosphatase (AP) assay, the protein staining assays and the neutral red assay (NR) are more sensitive than MTT in determining cell number, although the linearity of the highly sensitive assays is substantially decreased (Givens et al. 1990; Martin and Clynes 1993).

The MTT endpoint determination of cell number exhibits a good linear relationship of OD to cell number for a large range of cell types and over a wide range of cell numbers (Alley et al. 1988). However, the sensitivity of the assay is quite poor. Cells actively growing will reduce MTT to a greater extent than cells which are not (Mossman 1983), and this, in combination with inherent differences in the activity of MTT reducing enzymes among cell lines (Ford et al. 1989), may lead to differences in assay sensitivity in various cell lines. There are also differences of opinion as to the stability of MTT stock solution, with the recommended storage time ranging from fresh preparation to at least 6 weeks (Green et al. 1984; Twentyman and Luscombe 1987; Ford et al. 1989; Leprince et al. 1989) which may result in differences in assay sensitivity. MTT is also a mutagenic agent, and extreme care must be taken in its handling and disposal.

The NR assay is based on the incorporation of a dye by lysosomes. However, dead cells have been stained by NR (Allison and Young 1973), so an exact relationship with cellular viability should not be assumed. This assay is generally more sensitive than MTT, but less so than crystal violet dye elution (CVDE), sulforhodamine B (SRB) and AP.

The SRB assay is very sensitive, but exhibits a pronounced loss of linearity of OD vs cell number at higher cell densities. This effect is more notable than for NR and CVDE.

The AP assay gives high sensitivity but, as a consequence, has a low range of linearity between OD and cell number. The AP activity has also been found to vary when cells are grown under different growth conditions.

22.1
Setting Up Miniaturized in Vitro Assays

- Always wear disposable gloves.   safety note!

- Wear a clean lab coat.

- Perform all chemical additions in a class II laminar air flow cabinet which operates to BS 5726 specifications.

- Dispose of waste according to local regulations.

Two days prior to setting up the assay the cell line to be tested is passaged and seeded at sufficient density so as to be 70–80 % confluent on the day of the assay. All the assays described below are set up in 96-well plates and cell number is evaluated colorimetrically using a microplate reader (Titertek Multiskan PLUS, Mk II).

Procedure

1. Day 1: Trypsinize one, almost confluent, flask of cells. Resuspend the pellet in 5 ml of medium and count using a hemocytometer. Dilute the cells with medium to give a final concentration of 10 000 cells/ml. Dispense 100 µl of cell suspension into each well of a 96-well plate using a multichannel pipetter, i.e., 1000 cells/well. The assays are devised to last for 7 days so the initial amount of cells seeded can be altered depending on the growth characteristics of the cell line being used. Do not pipette cells into the first column of wells, as medium will later be added to this column and it will serve as a blank. When the cells have been added to the wells, shake the plates vigorously to ensure a homogeneous spread of cells and check using a microscope. Incubate the plates overnight at 37 °C.

2. Day 2: Remove plates from the incubator and examine to ensure that cells have adhered to the base of the wells. If cells appear viable commence adding the toxin/s to be tested, beginning at the lowest concentration and working up to the highest. The final volume in each well on the plate is 200 µl. If adding one toxin only the chemical solutions are made up to a concentration twice that of the final desired solution, to take account of the diluting effect of adding 100 µl of chemical to 100 µl of cell suspension. When all the solutions have been added, gently shake the plates to ensure adequate mixing and incubate at 37 °C.

3. Day 7: Remove plates from the incubator and process as indicated for the relevant assay. All of these assays are read at 570 nm, except for the acid phosphatase assay which is read at 405 nm, on a multiplate reader. It is important to remove any bubbles from the wells before reading the absorbances.

22.2
Acid Phosphatase Assay

The acid phosphatase (AP) assay is based on the ability of the AP enzyme in the lysosomes of cells to hydrolyze the *p*-nitrophenyl phosphate yielding the *p*-nitrophenyl chromphore (Martin and Clynes 1991).

▨ Materials

– PBS
– Substrate-containing buffer: 10 mM *p*-nitrophenyl phosphate (Sigma, cat. no. 104) in 0.1 M sodium acetate, pH 5.5, containing 0.1 % Triton X-100. The sodium acetate can be prepared in advance and stored at 4 °C in the dark. The *p*-nitrophenyl phosphate is added immediately prior to performing the assay.
– 1 M NaOH

▨ Procedure

1. Remove the medium from the cells and rinse the wells once with 100 µl of PBS.

2. Add 100 µl of freshly prepared substrate-containing buffer to each well and incubate plates at 37 °C for 2 h.

3. After incubation remove the plates from the incubator and add 50 µl of 1.0N NaOH to each well. This causes an electrophilic shift in the *p*-nitrophenol chromophore and thus develops the yellow color. The plates may then be read at 405 nm on the ELISA reader.

22.3
Neutral Red Assay

This assay is based on the accumulation of the neutral red (NR) dye in the lysosomes of viable cells. In damaged or dead cells the NR dye is not held in the lysosomes and the plasma membrane does not act as a barrier to hold the dye in the cells (Borenfreund and Puerner 1983; Fiennes et al. 1987; Triglia et al. 1991).

Materials

– Neutral red dye: Prepare a 1:80 dilution of 0.4% neutral red (3-amino-7-dimethyl-2-methylphenazine hydrochloride) (Sigma, cat. no. N-7005) in culture medium to give a final concentration of 50 µg/ml. Centrifuge at 1500γ for 10 min and retain the supernatant. The 0.4% NR stock can be kept at 4°C in the dark for several weeks.
– Formol-calcium mixture: 10 ml 40% formaldehyde and 10 ml 10% anhydrous calcium chloride in 80 ml of ultrapure water
– Acetic acid-ethanol mixture: 1% glacial acetic acid in 50% ethanol

Procedure

1. At the end of the incubation period remove the medium from the plates and rinse each well with 200 µl PBS.

2. Add 200 µl freshly prepared NR solution and incubate for a further 3 h.

3. At the end of this period the NR solution is removed and the wells are washed with 50 µl of the formol-calcium mixture.

4. Then add 200 µl of the acetic acid-ethanol mixture to elute the dye, agitate and read at 570 nm.

22.4
Sulforhodamine B Assay

This assay is a protein binding dye assay in which cells are first fixed in TCA before staining with the pink dye (Skehan et al. 1990).

Materials

– Trichloroacetic acid (TCA): 50%
– Sulforhodamine B (SRB): 0.4% solution dissolved in 1% acetic acid (Sigma, cat. no. S-9012)
– Glacial acetic acid: 1%
– 10 mM Tris buffer, pH 10.5

▦ Procedure

1. At the end of the incubation period cells are fixed by layering 50 µl of 50 % TCA directly on top of the incubation medium. The plates are incubated for a further 1 h at 2 °C.

2. The plates are then rinsed five times with tap water to remove solutes and allowed to completely dry.

3. Cells are subsequently stained with 200 µl SRB for 30 min and rinsed four times in 1 % glacial acetic acid.

4. When completely dry, add 200 µl/well of 10 mM Tris buffer, pH 10.5, to release unbound dye. Mix the resulting solution and read at 570 nm.

22.5
Crystal Violet Dye Elution

This assay is a protein binding dye assay in which cells are first fixed with formalin before staining with the crystal violet dye (Kueng et al. 1989; Scragg and Ferreira 1991)

▦ Materials

– PBS
– 10 % Formalin
– 0.25 % Aqueous crystal violet (Sigma, cat. no. C-3886): Prefilter through Whatman No. 1 paper.
– 33 % Glacial acetic acid

▦ Procedure

1. At the end of the incubation period remove the medium from the 96-well plates and rinse each well with 100 µl PBS.

2. Fix the cells by adding 100 µl of 10 % formalin for 10 min. Decant and allow the plates to air dry.

3. Add 100 µl of 0.25 % aqueous crystal violet for 10 min. Then rinse four times with tap water and allow to dry.

4. Solubilize the color by addition of 100 µl/well 33 % glacial acetic acid and read at 570 nm.

22.6
MTT Assay

The basis of this assay is that MTT, a soluble yellow dye, is metabolized by the enzyme succinate dehydrogenase in the mitochondria of metabolically active cells to a dark blue insoluble formazan product, which can be solubilized and the color produced measured (Mossman 1983; Alley et al. 1988; Vistica et al. 1991).

Materials

- MTT: tetrazolium dye: 3-(4,5-dimethylthiazol-2-yl)-2,5-diphenyl tetrazolium bromide (Sigma, cat. no. M-2128), is dissolved in PBS at a concentration of 5 mg/ml. This solution can be stored at 6 weeks at 4 °C in the dark (different authors recommend varying storage times). **Note:** MTT is believed to be mutagenic.
- Dimethyl sulfoxide (DMSO)

Procedure

1. At the end of the incubation period add 20 µl 5 mg/ml MTT to each well and incubate as before for a further 4 h.

2. Then carefully remove the medium from each well by pipetting, taking care not to disturb the formazan crystals.

3. Add 100 µl DMSO to each well and agitate to give a homogeneous color. Read at 570 nm.

Comments

Miniaturized in vitro toxicity tests provide a quick and clear means of assessing some of the toxicological potential of an agent. The range of measurable end points combined with the variety of cell types available gives these methods broad applicability. However, the type and parameters of the assay used to quantitate toxicity must be carefully chosen to give a meaningful result. The distinction between cytostatic and cytotoxic effects can be very difficult to ascertain with colorimetric endpoint assays. The timing of toxicant exposure to the cells can be important for agents with cell cycle specificity. The endpoint parameter chosen must be relevant to, but not interfere with, the toxic action of the test agent. The characteristics of the different endpoints assays have advantages and disadvantages, most notably in the areas of sensitivity and linearity, and a careful choice is therefore necessary for a relevant result. Once a suitable method has been investigated, a rigorous assessment is necessary to ensure relevance under the individual experimental conditions. Once such requirements have been satisfied, greater volumes of assays can be undertaken in a more accurate and reproducible manner using modern robotic technology. These systems are not sufficiently advanced to eliminate the need for other methods of toxicological assessment. However, appropriate application of these methods can give a lot of valuable information to the toxicologist.

Applications

- Scaling up of miniaturized assays

Two of the strengths of many in vitro methods of toxicity testing, especially those outlined earlier in the chapter, are the small size of the test and the rapidity of the result. These factors mean that these systems are amenable to scaling up, thereby providing a large volume of data, for example, about a series of related chemicals. On such a scale, the required effort becomes extremely labor intensive. However, recent advances in laboratory robotic systems have largely reduced this obstacle.

Robotic systems come to the fore when combined with other technologies such as combinatorial chemistry or screens of toxicity in a group of related agents (Gulakowski et al. 1991). They are also of particular use for large volumes of routine screening (QA/QC applications) to ensure the safety of a standard product. An example of such a use is routine endotoxin testing of water for injection (Lavelle and Barwick 1993). In the case of therapeutic drug development, with the anticipated low probability of finding a new suitable therapeutic agent from a complex lead chemical with many thousands of synthetic chemical permutations, a robotic system can give a very significant advantage in the initial toxicological screening leading to a reduction in the number of derivatives which go on to further, more extensive, tests (Floyd et al. 1996).

In practice the whole robotic unit is operated in a sterile airflow such as a laminar flow cabinet. The cell suspension, media and toxin dilutions are added to specified containers in the machine and the program controls all additions, incubations and can be coupled to a plate reader for automated results and data handling. The machinery accurately adds the specified quantities to the tissue culture plates in a very precise manner removing much of the laborious manual work involved in such tests.

References

Alley MC, Scudiero DA, Monks A, Hursey MC, Czerwinski MJ, Fine DL, Abott BJ, Mayo JG, Shoemaker RH, Boyle MR (1988) Feasibility of drug screening with panels of human tumour cell lines using a microculture tetrazolium assay. Cancer Res 48:589–601

Allison AC, Young MR (1973) Vital staining and fluorescence microscopy of lysosomes. In: Dingle JT and Fell HB (eds) Lysosomes in biology and pathology, North Holland, Amsterdam, p 600

Borenfreund E, Puerner JA (1983) A simple quantitative procedure using monolayer cultures for cytotoxicity assays. J Tissue Cult Methods 9: 7–9

Fiennes A, Walton J, Winterbourne D, McGlashan D, Hermon-Taylor J (1987) Quantitative correlation of neutral red dye uptake with cell numbers in human cancer cell cultures. Cell Biol Int 11:373–378

Floyd CD, Lewis CN, Whitttaker M (1996) More leads in the haystack. Chem Britain 32:31–35

Ford CHJ, Richardson VJ, Tsaltas G (1989) Comparison of tetrazolium colorimetric and [^{3}H]-uridine assays chemosensitivity testing. Cancer Chemother Pharmacol 24:295–301

Givens KT, Kitado S, Chen AK (1990) Proliferation of human ocular fibroblasts – an assessment of in vitro colorimetric assays. Investig Opthalmol Vis Sci 31:9

Green LM, Reade JN, Ware CF (1984) Rapid colorimetric assay for cell viability: Application to the quantitation of cytotoxic and growth inhibitory lympholines. J Immunol Meth 70:257

Gulakowski RJ, McMahon J, Stanley P, Moran R, Boyd M (1991) A semi-automated multiparameter approach for anti-HIV drug screening. J Virol Meth 33:87–100

Horakova K, Czikova S, Kernacova B (1978) The suitability of different quantitative methods for determination of the cytotoxic activity of agents in cell cultures. Neoplasia 25:309–315

Kueng W, Silber E, Eppenberger U (1989) Quantification of cells cultured on 96-well plates. Anal Biochem 182:16–19

Lavelle L, Barwick R (1993) Validation of the Biomek 1000 BioRobotics system for LAL endotoxin testing using the BioWhittakerkinetic-QCL system. Information bulletin A-1743 from Beckman Instruments Inc., Fullerton, California

Leprince P, Lefebure PP, Rigo P (1989) Cultured astroglia release a neuronotoxic activity that is not related to the excitotoxins. Brain Res 502:21–27

Martin A, Clynes M (1991) Acid phosphatase: endpoint for in vitro toxicity tets. In Vitro Cell Dev Biol 27A:183–184

Martin A, Clynes M (1993) Comparison of 5 microplate colorimetric assays for in vitro cytotoxicity testing and cell proliferation assays. Cytotechnology 11:49–58

Mossman T (1983) Rapid colorimetric assay for cellular growth and survival: Application to proliferation and cytotoxicity assays. J Immunol Meth 65:55–63

Scragg M, Ferreira L (1991) Evaluation of differential staining procedures for quantification of fibroblasts cultured in 96-well plates. Anal Biochem 198:80–85

Skehan P, Storeng R, Scudiero D, Monks A, McMahon J, Vistica D, Warren J, Bokesch H, Kenney S, Boyd, MR (1990) New colorimetric assay of anticancer drug screening. J Natl Cancer Inst 82:1107–1113

Triglia D, Braa S, Yonan C, Naughton G (1991) In vitro toxicity of various classes of test agents using the neutral red assay on a human three-dimensional physiologic skin model. In Vitro Cell Dev Biol 27A:239–244

Twentyman PR, Luscombe M (1987) A study of some variables in a tetrazolium dye (MTT) based assay for cell growth and chemosensitivity. Br J Cancer 56:279–285

Vistica DT, Skehan P, Scudiero D, Monks A, Pittman A, Boyd MR (1991) Tetrazolium-based assays for cellular viability: a critical examination of selected parameters affecting formazan production. Cancer Res 51:2515–2520

Chapter 23

Investigation of Anti-tumor Drug Accumulation and Efflux in Cell Culture

IRENE CLEARY* and MARTIN CLYNES

▦ Introduction

Resistance to chemotherapy represents a major factor in the failure of many cancer treatments. While resistance to chemotherapeutic agents is inherent in many human tumors, other tumors which are initially responsive to chemotherapy also develop resistant variants. Tumors inherently resistant to a variety of anti-tumor agents display minimal response to chemotherapy. In most cases the basis for this inherent resistance is undefined, although it probably involves a combination of kinetic and biochemical factors, including inability to effectively transport the drug into the tumor mass or to convert it into its active form. With acquired drug resistance, a population of cells initially sensitive to the drug develops resistant characteristics. Tumor cells, either partially or completely resistant to an anti-tumor agent, are, therefore, more likely to survive in a drug-containing environment. In an attempt to overcome this problem combination chemotherapy was introduced. Many tumors, however, can develop multidrug resistant (MDR) variants to a number of chemically unrelated anti-tumor agents. Given the lack of sustained response to chemotherapy in many human tumors there is considerable interest in understanding the mechanisms of MDR. Although a number of mechanisms have been described the most frequent determinant of MDR in many human tumors and cell lines appears to center on the ability of the cells to greatly decrease the cellular accumulation of the anti-tumor agent and thus prevent the drug from reaching its target site. It was first reported by Kessel et al. (1968) that anthracycline-resistant tumor cells accumulated less drug than the sensitive parental cell line. Initial observations led to the suggestion that the reduction in drug

* *Correspondence to* Irene Cleary, National Cell and Tissue Culture Centre, Dublin City University, Glasnevin, Dublin 9, Ireland; phone +353–1–7045701; fax +353–1–7045484; e-mail 75031256@raven.dcu.ie

accumulation in resistant cells was primarily due to differences in the cell membrane. However further studies showed that cellular resistance to antitumor agents, including the anthracyclines and vinca alkaloids, occurred in association with the activation of an energy dependent pump (Dano 1973), which was subsequently identified by Juliano and Ling (1976) and was designated P-glycoprotein or P-170. A number of other transport efflux pumps have since been identified in MDR cell lines, including the 190 kDa MDR-related protein MRP (Cole et al. 1992) and the 110 kDa lung resistance protein LRP (Scheper et al. 1993). These proteins act as energy-dependent efflux pumps for a variety of anti-tumor agents and thus reduce the cellular accumulation of the drug. Various techniques have been employed to determine the cellular accumulation, retention and efflux of anti-tumor agents in MDR cell lines. These procedures, while facilitating the evaluation of cellular drug levels, also represent useful tools for studying the effect of various agents on subcellular drug levels. A number of specific advantages and disadvantages are, however, attached to each of these methods. The techniques include radioautography, spectrofluorimetry, radiolabeled studies, fluorescent microscopy, high performance liquid chromatography (HPLC), and flow cytometry. Spectrofluorimetry, radiolabeled studies and fluorescence microscopy represent the more traditional approaches, while the latter two procedures are now widely applied for determining intracellular drug levels in cell culture.

23.1
Radioautography

One of the earlier techniques used to determine anti-tumor drug uptake in resistant tumor cells was radioautography. However, this techniques (for many applications) has long since been superseded by more accurate and less time consuming procedures. The earlier technique, which took approximately 3 weeks, involved labeling the cells with tritiated drug and determining the number of grains present in the nucleus of 200–1000 cells as counted by two or three independent observers. A number of earlier radiography studies were carried out to determine the cellular concentration of both radiolabeled daunorubicin and actinomycin D in Chinese hamster ovary cell lines (Biedler and Riehm 1970; Riehm and Biedler 1971).

▓ Materials

– Microscopic glass coverslips (Chance propper LTD)
– 35 mm petri dishes
– Radiolabeled drug (e.g., ^{3}H adriamycin, Amersham, UK)
– Methanol: 70 and 100 %
– Balanced salt solution
– Mayer's hematoxylin
– Kodak NTB liquid emulsion
– Kodak D-19 developer
– Kodak rapid fixer
– Dark room facilities

▓ Procedure

radiography

1. Cells are grown on glass coverslips in 35 mm petri dishes at 37 °C in 5 % CO_2 for 3 days and are in the exponential phase of growth at time of assay.

2. Cells are treated with the tritiated drug for specified time periods (actinomycin D, 2.66 and 5.32 µCi; daunorubicin, 0.305 µCi).

3. Coverslips are rinsed in complete medium for 1 min followed by 30 s in a balanced salt solution.

4. After fixation for 5 min each in 70 and 100 % methanol, the coverslips are allowed to dry and then rinsed for 30 min in a 37 °C water bath.

5. Coverslips are mounted on glass slides, allowed to dry for 2 days, dipped in Kodak NTB liquid emulsion (diluted 1:1 with distilled water) at 45 °C and then stored at 4 °C for 14 days.

6. After exposure, the preparations are developed in Kodak D-19 developer for 4 min at 18 °C, fixed for 8 min in Kodak rapid fixer and rinsed three times within 15 min with distilled water.

7. Slides are dried, stained in Mayer's hematoxylin for 8 min and rinsed in distilled water.

8. Depending on the density, the average number of grains per nucleus of 200–1000 cells was determined.

23.2
Spectrofluorimetry

A more accurate quantitative method for determining anti-tumor drug accumulation and efflux is spectrofluorimetry. Since a number of the anthracyclines including adriamycin and daunorubicin were found to be naturally fluorescent under ultraviolet illumination (Egorin et al. 1974), the cellular concentration of the drug can easily be determined by spectrofluorometric analysis. Early studies by Dano (1973) demonstrated decreased cellular levels of drug in resistant Ehrlich ascites tumor cells by determining the drug content in the medium and subtracting it from the total amount added to the cells. However, more recent procedures involve determining the quantity of drug directly extracted from the cells. Various parameters can be investigated by this method including drug accumulation, drug efflux, metabolic inhibition studies and circumvention studies.

Materials

- 6-well cluster plates (Costar)
- Adriamycin (Pharmacia, UK)
- Spectrofluorometer (e.g., Perkin Elmer LC50 luminescence spectrometer)
- Trypsin (0.25 %)/EDTA (0.01 %)
- 0.6 N HCl-methanol
- Ice-cold PBS
- Ice-cold ultrapure water
- Glucose-free medium (Gibco)
- Metabolic inhibitor: e.g., 10 µM antimycin or 10 mM sodium azide

Procedure

1. Sensitive and MDR cells are grown in 75 cm^2 flasks until approximately 80 % confluent and then trypsinized using a 0.25 % solution of trypsin containing 0.01 % EDTA.

    drug accumulation in adherent cells

2. To each well of the 6-well cluster plates, 1 ml of the cell suspension, at a concentration of $1 \times 10^5 - 5 \times 10^5$ cells/ml (concentration dependent on cell line), is added.

3. A further 3 ml of complete medium are added to the wells and the cells incubated at 37 °C in 5 % CO_2 for approximately 48 h.

4. The medium is then decanted and 4 ml of complete medium containing the anti-tumor agent (e.g., 10 μM adriamycin) are added to each of the wells, except the control wells to which 4 ml of complete medium are added.

5. The plates are incubated at 37 °C in 5 % CO_2 and after specified time periods the medium is removed and the cells are washed twice in ice-cold PBS.

6. Then, 2 ml of ice-cold ultrapure water are added to the wells which are then allowed to sit for 5 min to facilitate cell lysis (which can be viewed microscopically).

7. The adriamycin is then directly extracted from the cells by the addition of 2 ml 0.6 N HCl-methanol solution for 5 min.

8. The resulting solutions are transferred to universal containers and centrifuged at 4000 rpm for 10 min at 4 °C.

9. The supernatants are collected and the fluorescence determined using a luminescence spectrometer (Perkin Elmer LC50) with an excitation wavelength of 470 nm and an emission wavelength of 585 nm. The slit width for excitation and emission are 10 and 15 nm, respectively.

10. The concentration of drug present is quantitated from a linear standard curve prepared from the fluorescence of adriamycin concentrations.

Figure 23.1 illustrates the time course of adriamycin accumulation in the human lung parental SKMES-1 cells and its MDR variant SKMES-1/ADR. It also illustrates the effect of circumvention agents (verapamil and cyclosporin A) on drug uptake in the resistant cells.

Sample Calculation of Subcellular Drug Concentration

A standard curve is prepared from the fluorescence of known adriamycin concentration over a range of 0–1000 nM. The cellular concentration of adriamycin in each sample is quantitated from the linear standard curve (Fig. 23.2). A sample calculation is as follows: fluorescence value obtained = 28; adriamycin concentration from above graph =

Fig. 23.1. The time course of adriamycin accumulation in the parental SKMES-1 and resistant SKMES-1/ ADR cell lines. The effect of the circumvention agents verapamil (30 μg/ml) and cyclosporin A (10 μg/ml) on adriamycin accumulation in SKMES-1/ ADR cells

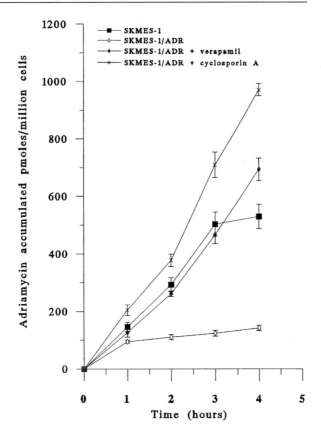

400 nM; drug extracted in total of 4 ml = 1.6 nmol/sample; total number of cells in 4 ml = 8×10^5 cells/well; subcellular adriamycin concentration = 2 nmol/10^6 cells.

Drug efflux from adherent cells can be monitored by preloading the cells with the drug (by inhibiting drug efflux) and determining the cellular concentration of drug when the cells are reintroduced into complete medium.

drug efflux in adherent cells

1. Sensitive and MDR cells are plated in 6-well cluster plates (as for the accumulation assay) and incubated for approximately 48 h at 37 °C in 5 % CO_2.

2. Following this incubation period the medium is decanted and 4 ml of glucose-free medium containing the drug (e.g., 10 μM adriamycin) and metabolic inhibitor (e.g., 10 μM antimycin A, 10 mM sodium azide) are added to each well with the exception of the control wells.

Fig. 23.2. Standard curve for the fluorescence of known adriamycin concentrations over a range of 0–1000 nM. The cellular concentration of adriamycin in each sample is quantitated from the standard curve

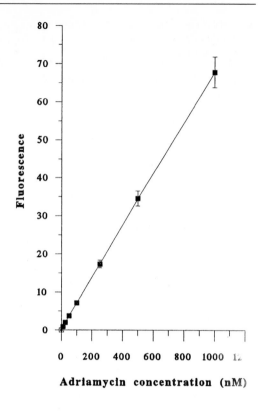

3. After 2 h the medium is removed and the cells are rinsed twice with complete medium.

4. To each of the wells 4 ml of complete medium are added and the cells are further incubated at 37 °C.

5. After specified time intervals the medium is decanted and the cellular drug extracted using 0.6 N HCl-methanol.

6. The resulting solutions are centrifuged and the fluorescence of the supernatants determined.

drug accumulation in nonadherent cells

1. Sensitive and MDR cells are seeded in volumes of medium at a concentration of 1×10^6 cells/ml.

2. Concentrated stock drug solution is added to give a final working concentration (e.g., 10 µM adriamycin) and the cells are incubated at 37 °C in 5 % CO_2.

3. After specific time periods after the addition of drug, a 1.5 ml aliquot of the cellular suspension is removed centrifuged and washed three times with ice-cold PBS using an Eppendorf microcentrifuge.

4. Control samples are removed at time 0, immediately before the drug is added.

5. Then, 2 ml of ice-cold ultrapure water is added to the cells to facilitate cell lysis and the drug is extracted directly from the cells by the addition of 0.6 N HCl-methanol.

6. Cell supernatants are removed and assayed spectrofluorometrically for drug content as described above.

Similar to adherent cells, drug efflux from nonadherent cells can be determined by preloading the cells with the drug and monitoring the decrease in the cellular concentration of the drug when the cells are reintroduced into complete medium.

drug efflux in nonadherent cells

1. Sensitive and MDR cells are preloaded with the drug in glucose-free medium containing a metabolic inhibitor for 2 h.

2. The drug containing medium is then removed, the cells are washed twice with complete medium by centrifugation and the cells are reseeded in complete medium.

3. After specific time periods 1.5-ml samples are removed and the cellular concentration of drug determined as described above.

4. Control samples are removed immediately after the preloading incubation period.

23.3
Radiolabeled Studies

Radiolabeled studies represents another commonly used technique employed in the study of anti-tumor drug accumulation and efflux. A number of the anti-tumor agents, including the anthracyclines, vinca alkaloids and epipophyllotoxins, can be radiolabeled and consequently the accumulation and efflux of these agents can be determined by monitoring the cellular concentration of the radioisotope through scintillation counting. Although relatively simple, radioactive detection presents some disadvantages including high investment costs and the inherent danger of working with radioisotopes. Strict guidelines must be followed at all times whilst handling and disposing of all reactive materials.

▤ Materials

- 6-well cluster plates (Costar)
- Radiolabeled drug (e.g., ^3H adriamycin or ^3H vincristine, Amersham, UK)
- Scintillation fluid (Ecolite, ICN, UK)
- Scintillation counter (e.g., Beckman LS 6500)
- 0.2 N NaOH
- 0.2 N HCl
- Ice-cold PBS

▤ Procedure

adherent cells

1. Sensitive and MDR cells are plated in 6-well cluster plates at a concentration of 1×10^5–5×10^5 cells/ml (concentration dependent on cell line) and incubated at 37 °C in 5 % CO_2 for approximately 48 h.

2. The medium is decanted and fresh medium containing the drug (e.g., ^3H-vincristine or ^3H-adriamycin) is added to the wells; the cells are incubated at 37 °C in 5 % CO_2.

3. For drug efflux experiments, a 60 min loading period is used followed by removal of medium, rinsing twice with complete medium and the addition of fresh medium.

4. After specified incubation periods the medium is decanted and the cells washed twice with ice-cold PBS.

5. The plates are then blotted dry using filter paper or allowed to air dry before the cells are solubilized by a 2-h incubation in 0.5 ml of 0.2 N NaOH at room temperature.

6. The cell lysates are neutralized by the addition of 0.5 ml of 0.2 N HCl, transferred to scintillation vials containing 5 ml of scintillation fluid and the radioactivity determined.

A number of alternative methods can be used to extract the radiolabeled drug from adherent cells including the following procedures:

- Lysing the cells with PBS containing 1 % Triton-X 100
- Digesting the cells with 0.5 N KOH at 70 °C for 2 h
- The cells can be harvested with a rubber scrapper and suspended in 0.9 ml ice-cold H_2O to facilitate cell lysis.

1. Nonadherent cells are seeded in volumes of medium at a concentration of 1×10^6 cells/ml.

2. Concentrated stock drug solution is added to give a final working concentration and the cells are incubated at 37 °C in 5 % CO_2.

3. For drug efflux experiments, a 60-min loading period is used followed by centrifugation to remove all extracellular drug and the addition of complete medium

4. After specified incubation periods 1.5-ml samples are removed and washed twice with ice-cold PBS by centrifugation.

5. Samples are then transferred to scintillation fluid and the radioactivity determined.

nonadherent
cells

23.4
Fluorescent Microscopy

Many quantitative studies on the accumulation and efflux of anti-tumor drug in intact cells have used spectrofluorometric and radiolabeled studies. These methods, however, do not allow examination of the intercellular compartmentalization of the drug in living cells nor do they allow observation of cellular heterogeneity. Fluorescent microscopy has been employed to determine the cellular localization of drugs in parental and MDR cells, as a number of the anthracycline anti-tumor agents are inherently fluorescent and can easily be visualised under ultraviolet illumination. This induces orange fluorescence at the site of drug localization. In general the drug is localized predominately in the nuclei of parental cells and in the cytoplasm of MDR cells. Confocal scanning microscopy is an optical microscopic technique offering significant advantages over conventional fluorescence microscopy, principally in terms of its greater resolution and elimination of autofluorescence. In laser scanning microscopy the sample is scanned by a diffraction beam of light, transmitted or reflected by the in-focus illuminated volume element. Alternatively, the fluorescence emission is excited within by the incident light and is focused, as in the first case, onto a photodetector. As the illumination spot is scanned over the sample, the electrical output from the detector is displayed at the appropriate spatial position on a monitor, thus building up a two-dimensional image (Shotton 1989).

▦ Materials

- Microscopic glass coverslips (Chance propper LTD)
- 35 mm petri dishes (Greiner)
- Microscope equipped with mercury lamp or confocal laser scanning system
- Ice-cold PBS

▦ Procedure

fluorescent microscopy

1. Microscope glass coverslips are washed with 70 % alcohol, flamed with a bunsen burner and placed in 35 mm petri dishes.

2. The cells are grown in 25 cm^2 flasks until approximately 80 % confluent, trypsinized and a single cell suspension of 1×10^6 cells/ml prepared.

3. To each of the petri dishes, 1 ml of the cell suspension is added; the dishes are incubated overnight at 37 °C in 5 % CO_2 to allow attachment of the cells to the coverslip.

4. The medium is decanted and the cells incubated with 1 ml of medium containing the drug for 2 h.

5. The drug-containing medium is decanted and the cells washed twice with ice-cold PBS.

6. The coverslips are then inverted onto clean glass slides and the edges sealed with silicon grease to prevent dehydration.

7. The cells can then be viewed for fluorescence under ultraviolet illumination using a microscope (Nikon) equipped with a mercury lamp or alternatively using a digital video confocal laser imaging system (Odyssey XL system).

23.5
High Performance Liquid Chromatography

A method that is now widely used in determining drug accumulation and efflux in cell lines is HPLC. This procedure has the advantage of allowing the detection of a number of anti-tumor agents once the conditions have been optimized. However, it is rather time consuming and less sensitive than some of the other methods already discussed. In addition a large number of cells are required to carry out the assay. The following protocol is used in the extraction and quantitation of the anti-tumor agent adriamycin by HPLC. The protocol described has been optimized for this drug, however a number of different extraction procedures for adriamycin and other anti-tumor agents have also been described in the literature. A number of the reagents used are toxic so extreme care must be taken throughout the procedure.

Materials

– Acetonitrile (Lab Scan, Analytical Sciences, Ireland)
– Silver nitrate (Sigma, UK)
– Phosphoric acid (Sigma, UK)
– HPLC system (e.g., Beckman System Gold)

Procedure

1. Cells are grown in 75 cm^2 flasks until approximately 80 % confluent and then trypsinized (adherent cells) or centrifuged (nonadherent cells) to obtain cell pellets.

   HPLC of adherent and nonadherent cells

2. The cell pellets are drained and stored at $-20\,°C$ until required.

3. Then, 100 µl ultrapure H$_2$O is added to the pellet to facilitate cell lysis and the cell suspension is transferred to 10 ml glass extraction tubes.

4. To each of the extraction tubes 300 µl of a daunorubicin internal standard (1–20 µg/ml) in methanol is added.

5. Each tube is then extracted by the addition of 1.3 ml acetonitrile and 100 µl of 33 % silver nitrate, vortexed for 1 min and centrifuged at 4000 rpm for 5 min at room temperature

6. From the top acetonitrate layer, 1.1 ml sample is removed, transferred to a glass autosample vial and evaporated to dryness under nitrogen.

7. Dried samples are reconstituted in 50 µl methanol and then loaded into the HPLC autosampler.

8. A number of different mobile phases for adriamycin extraction have been described, e.g., samples can be eluted with a mobile phase of 50 % acetonitrile, 35 % H_2O and 15 % 0.2 M phosphoric acid (Weibe et al. 1992) using a reverse phase C18 column using a flow rate of 0.33 ml/min.

9. The retention time for adriamycin is approximately 6–8 min and for daunorubicin approximately 13–17 min.

10. The cellular concentration of the drug is quantified by measuring the peak height of the drug and the internal standard. The peak heights of known standards are plotted against concentration to generate a standard curve and the peak height of the cellular extract is compared to the standard curve to determine the amount of adriamycin present.

An alternative method for extracting the drug from the cellular preparation involves extracting the adriamycin (and daunomycin internal standard) with 5 ml of chloroform:methanol (4:1), vortexing the resulting solution for 1 min and centrifuging it at $4000\,g$ for 20 min. The bottom chloroform layer is then removed, evaporated to dryness under nitrogen and reconstituted in methanol.

23.6
Flow Cytometry

Flow cytometry represents another procedure now used to determine anti-tumor drug accumulation and efflux in cell culture. Since agents such as adriamycin and daunorubicin emit fluorescence following excitation, intracellular fluorescence per cell can be measured by flow cytometry. The cellular accumulation of a number of compounds that are often used in the study of P-glycoprotein function can also be determined by flow cytometry. These compounds include rhodamine 123 (Twentyman et al. 1994) and calcein (Hollo et al. 1994), both of which act as substrates for the P-glycoprotein efflux pump. The following protocol has been used to determine the cellular levels of rhodamine 123 in sensitive and MDR adherent cells (Twentyman et al. 1994).

Materials

– Rhodamine 123 (Sigma, UK)
– Flow cytometer (e.g., Becton Dickenson FACscan)

Procedure

1. Cells in the exponential phase of growth are trypsinized and a single cell suspension of 2×10^5 cells is prepared and stored on ice until used.

2. Rhodamine 123 is added to the cells to produce a final concentration of 0.1–1 µg/ml and the cells are incubated at 37 °C in 5 % CO_2.

3. After specified time periods aliquots are removed and centrifuged to obtain cell pellets.

4. The supernatants are removed and the cells are resuspended in 0.5 ml of ice-cold complete medium.

5. The suspension is stored on ice until cells are analyzed by flow cytometry.

6. The intensity of fluorescence at 515–560 nm is measured using an excitation wavelenght of 488 nm.

7. The mean intensity of fluorescence is calculated from obtained histograms. From this the mean intensity of fluorescence in rhodamine untreated cells is subtracted and the mean intensity of rhodamine fluorescence is estimated.

For determining the intracellular levels of adriamycin, excitation is at 488 nm and the intensity of fluorescence is measured at 576–625 nm.

References

Biedler J, Riehm H (1970) Cellular resistance to actinomycin in chinese hamster cells in vitro: cross resistance, radioautography and cytogenetic studies. Cancer Res 30:1174–1184

Cole S, Bhardwaj G, Gerlach J (1992) Overexpression of a transporter gene in a multidrug resistant human lung cancer cell line. Science 258:1650–1654

Dano K (1973) Active outward transport of daunorubicin in resistant Ehrlich ascites tumor cells. Biochem Biophys Acta 323:466–483

Egorin M, Hildebrand R, Cumino E, Bachur N (1974) Cytofluorescence localisation of adriamycin and daunorubicin. Cancer Res 34:2234–2245

Hollo Z, Homolya L, Davis CW, Sarkadi B (1994) Calcein accumulation as a fluorometric functional assay of the multidrug transporter. Biochem Biophys Acta 1191:384–388

Juliano R, Ling V (1976) A surface glycoprotein modulating drug permeability in chinese hamster ovary cell mutants. Biochem Biophys Acta 455:152–162

Kessel D, Bottenrill V, Wodinsky I (1968) Uptake and retention of daunorubicin by mouse leukemic cells as factors in drug response. Cancer Res 28:938–941

Riehm H, Biedler J (1971) Cellular resistance to daunorubicin in chinese hamster cells in vitro. Cancer Res 31:409–412

Scheper R, Broxterman H, Scheffer G, Kaaijk P, Dalton W, Van Heijningen T, Van Kallen C, Slovak M, de Vries E, Van der Valk P, Meijer C, Pinedo H (1993) Overexpression of Mr*110,000 vesicular protein in non P-glycoprotein-mediated multidrug resistance. Cancer Res 53:1475–1479

Shotton D (1989) Confocal scanning optical microscopy and its application for biological specimens. J Cell Sci 94:175–206

Twentyman P, Rhodes T, Rayner S (1994) A comparison of rhodamine 123 accumulation and efflux with P-glycoprotein-mediated and MRP associated multidrug resistance phenotypes. Eur J Cancer 30A:1360–1369

Weibe V, Koester S, Lindberg M, Emshoff V, Baker J, Wurz G, de Gregario M (1992) Toremiferene and its metabolites enhances doxorubicin accumulation in estrogen receptor negative multidrug resistant human breast cancer cells. Investigational New Drugs 10:63–71

Chapter 24

Current Status of Genetically Engineered In Vitro Systems for Biotransformation

Johannes Doehmer*

Introduction

Biotransformation

Biotransformation is a complex enzyme system required in the metabolic activation of xenbobiotics, in which hydrophobic compounds are converted into hydrophilic ones which then may be excreted. The liver is the major organ for the metabolism of xenobiotics; however, other organs, e.g., lung, intestine, skin and brain, may participate in biotransformation as well. Metabolic activation is an essential prerequisite for many chemicals before they can be excreted. The formation of a reactive intermediate is the critical step in biotransformation. This intermediate is of toxicological relevance, as it can distort the physiological functions of a cell by reacting with DNA, RNA or protein, leading to cell death or malignant transformation of a cell into a cancer cell. The key enzymes in biotransformation are the monooxygenases cytochrome P450s, which transfer an atomic oxygen thereby activating a chemical in several ways. Although these cytochrome P450s are not substrate specific in a strict sense, there are preferences, e.g., cytochrome P450 1A1 for polycyclic aromatic hydrocarbons such as benzo[a]pyrene, or 1A2 for aromatic amines. For risk assessment, it is of fundamental interest to know which enzymes are involved and which metabolites are formed, as well as to understand the mechanism by which these metabolites became toxic to a cell.

* Johannes Doehmer, Institut für Toxikologie und Umwelthygiene, Technische Universität München, Lazarettstrasse 62, 80636 München, Germany; phone +49–89–3187–2973; fax +49–89–3187–3449; e-mail doehmer@gsf.de

Cell Cultures Genetically Engineered for Biotransformation Studies

In order to facilitate studies on biotransformation at the cellular and molecular level, several cellular systems have been genetically engineered for metabolic competence in reaction with xenobiotics. These systems differ in the recipient cell chosen for gene transfer of a recombinant expression vector and the kind of expression system. Each of these systems has its advantages and disadvantages as they differ in their genotypes and phenotypes. The challenge is to determine which is the most appropriate system for the experimental conditions at hand. Recently, several mammalian cell expression systems have been developed which combine expression of a specific biotransformation enzyme and suitable biological endpoints. The most advanced systems are the human lymphoblastoid cell line AHH1, engineered for expression of cytochrome P450s by an extrachromosomally replicating expression plasmid; or the V79 Chinese hamster cell line, engineered by an expression plasmid which stably integrated in the V79 genome. AHH1 cells have about tenfold higher levels of cytochrome P450 activity than V79 cells and are therefore more sensitive indicators in several tests. Further differences are in the way these cells are cultured and applied. For instance, V79 cells are fibroblasts which adhere to the surface provided, whereas AHH1 cells are lymphoblastoid cells that grow in suspension. This is a crucial difference in terms of application of these systems; for example, attached growing cells are readily used in a micronucleus assay.

This laboratory has developed the V79 based cytochrome P450 system which will be presented in more detail in the following sections. For information on the AHH1 cell system, the reader may request information from Gentest Corp., 6 Henshaw St, Woburn, MA 01801, USA.

V79 Cells Genetically Engineered for Cytochrome P450s

The V79 Chinese hamster cell line was established in the 1950s and has been the preferred mammalian cell line in toxicity testing ever since based on several phenotypic and genotypic characteristics. The cells grow with a doubling time of 12 h or less, have a stable diploid karyotype, are very robust, easy to passage, and do not have special medium requirements. V79 cells can be applied to cytotoxicity testing, mutagenicity testing, micronucleus assays, and sister chromatid exchange (SCE) assays.

V79 cells were known to be deficient in metabolic competence, especially for cytochrome P450s. Therefore, liver homogenate was used as a metabolic component in conjunction with the cells as the indicator component.

V79 cells genetically engineered for cytochrome P450 s combine the usefulness of toxicological testing with the absence of a cytochrome P450 background; thus they readily present a system for integrating cDNA encoding cytochrome P450 activity. The genetic engineering of V79 cells started in 1986 by expression of rat cytochrome P450 s and is currently being pursued in order to a yield a cell battery covering all major human hepatic cytochrome P450s.

The following cytochrome P450 V79 cell lines are available from the author's laboratory under a material transfer agreement:

- V79MZ parental

- V79MZr1A1, r1A2, r2B1, and r2E1

- V79MZh1A1, h1A2, h2A6, h2E1, h3A4, hOR

- V79MZf1A1

MZ denotes the V79 cell lines grown and characterized at the Institute of Toxicology, University of Mainz. This cell line differs from the V79NH cell line, grown and characterized at the GSF-Forschungszentrum, Institute for Toxicology, Neuherberg. V79NH cells have endogenous acetyltransferase activity, while V79MZ cells have no such activity. Furthermore, the prefix r denotes rat, h is for human, f for fish (scup fish). OR denotes human cytochrome P450 reductase, which has been coexpressed with cytochrome P450 s for improved activity, e.g., h3A4.

24.1
Cell Culture and Maintenance

Materials

- Culture flasks
- Humidified CO_2 incubator
- Inverted phase contrast microscope
- Dulbecco's modified Eagle's medium (DMEM): high glucose (4.5 g/l)
- Penicillin
- Streptomycin
- Geneticin (G418)
- Serum

▪ Procedure

V79 cells are grown in DMEM fortified with 4.5 g/l glucose and supplemented with 10 % fetal calf serum, penicillin (100 U/ml) and streptomycin (100 µg/ml). The cells are incubated at 37 °C, 90 % humidity, 5 %–10 % CO_2. Genetically engineered V79 cells are cultivated under the same conditions, except that the medium contains 400 µg/ml of G418 when grown for more than two passages in order to maintain selective pressure. Store cell stocks in liquid nitrogen at 1×10^6 cells/ml in DMEM with 10 % fetal calf serum and 10 % DMSO. Thaw cells by placing an ampoule in a 37 °C waterbath for 10 min. Transfer cells to a 75 cm² flask. Incubate in 10 ml of culture medium for at least 2 h to allow attachment and spreading of the cells, then exchange medium against fresh medium. **Note:** These cells grow extremely fast and may exhaust the medium within a short period of time if the cultures reaches over 80 % confluency. If this happens, cells begin to die, the pH indicator phenol red changes to yellow, and the cytochrome P 450 activity in this culture is lost. For best results, cells are seeded at 5 %–10 % confluency, fed every second day, and grown to a density of 80 %–90 %.

24.2
Application of V79 Cells in Toxicity Testing

▪ Materials

- 96-well culture plates
- Plate shaker
- Plate reader
- Petri dishes: 15 cm and 6 cm
- Dye for staining colonies
- Petri dishes: 3 cm
- Slides and cover slips
- Fluorescence microscope with special filter settings for Hoechst 33258

▪ Procedure

cytotoxicity testing

1. Suspend cells in culture medium (without G418) and seed into 96-well plates at 500 cells/well. Incubate for 16 h at 37 °C, 10 % CO_2/90 % air, 90 % humidity to allow the cells to attach to the plastic.

2. Add appropriate dilutions of the test compound, 12 replicate cultures for each dilution and controls. Divide plates into two groups in order to have the following exposure periods:

    – A 72 h exposure, followed immediately by processing and staining
    – A 24 h exposure, followed by 5 days culture without the test chemical

3. After exposure, process the cells as follows: Replace culture medium with neutral red medium, incubate for 90 min at 37 °C. Wash cells in PBS to remove excess dye. Fix in formaldehyde/CaCl2 solution for 30 s. Wash twice with PBS. Extract dye with ethanol/acetic acid solution, with vigorous shaking on a plate shaker.

4. Measure neutral red absorption at 546 nm in a plate reader, or at 405 nm if absorption has reached a maximum at 546 nm.

1. Day 1: Seed 1.5×10^6 cells in 30 ml of culture medium (without G418) into a 15 cm petri dish. Incubate for 18 h at 37 °C, 10 % CO_2/90 % humidity. **mutagenicity test**

2. Add 60 µl of test compound. If a solvent, e.g., DMSO, has been used, add an appropiate quantity to control cultures. Incubate for 24 h.

3. Terminate exposure by removing medium and replacing with fresh medium.

4. Day 4: Trypsinize the cells off the petri dishes. Resuspend in culture medium and subculture by seeding 3×10^6 cells into 15 cm petri dishes.

5. Day 8: Trypsinize the cells off the petri dishes.

6. Determine number of mutants and cloning efficiency as follows:

6a. To determine the number of mutants, resuspend cells in medium containing 6-thioguanine. Seed into 15 cm petri dishes at a density of 10^6 cells/dish (six replicate plates). Incubate for 10 days. Count the number of colonies on each plate.

6b. To determine cloning efficiency, resuspend cells in medium without 6-thioguanine at a density of 20 cells/ml. Seed 5 ml of cell suspension into 6 cm petri dishes (three replicates). Incubate for 9 days. Count the number of colonies on each plate.

7. Using the results from steps 6a and 6b, calculate the mutant frequency in test and control cultures. Subtract the control values from the test values to obtain the mutant frequency arising from exposure to the test compound.

micronucleus assay

1. Seed 10^5 cells in culture medium (without G418) containing appropriate dilutions of test chemical into 3 cm petri dishes. Incubate for 24 h.

2. Replace medium with fresh culture medium without test chemical. Incubate for 24 h.

3. Remove medium and treat cells briefly with hypotonic buffer. Fix cells with methanol/glacial acetic acid for a few minutes. Allow cells to dry.

4. Incubate cells with Hoechst 33258 solution for 1 h. Remove excess dye by washing three times with glycerin/H_2O for 30 min. Coat cells with the glycerin mounting solution and cover with cover slip.

5. Count micronuclei with a fluorescence microscope. Determine number of micronuclei per 500 cells.

Applications

V79 cells in metabolism studies

Ethoxyresorufin-O-Deethylase Activity

Resorufin derivatives are suitable substrates for several cytochrome P450s, e.g., ethoxyresorufin for CYP1A1 and 1A2; methoxyresorufin for CYP1A2; pentoxyresorufin for CYP2B1. The dealkylation by a cytochrome P450 yields resorufin, which can be measured spectrofluorometrically at an excitation of 550 nm and an emission at 585 nm. Resorufin is used as the standard. Activity can be checked on live cells or in a cell homogenate. Cells are grown in a 15 cm petri dish up to a confluency of 80–90%, starting from an inoculum of 2×10^6 cells with 2 or 3 days of incubation. Under these conditions cells have maximum activity. Cells lose activity if the culture is overgrown and the medium is exhausted. One 15 cm petri dish at 80% confluency yields more than 1 mg of total protein in cell homogenate. A cell homogenate is produced by scraping the cells from the dish into 5 ml ice cold PBS, collected in a 50 ml conical tube and centrifuged at top speed in a bench top centrifuge. The supernatant is removed and the pellet is shock-frozen in liquid nitrogen and kept at −80 °C for further processing. This is the best procedure for maintaining maximum activity in the homogenate. Any further handling by, e.g., sonication or freeze/thawing cycles lowers the activity.

For the enzyme assay 300–500 µg of total protein is needed, along with 1 ml reaction buffer containing 50 µmol Tris-HCl, pH 7.6; 5 µmol MgSO$_4$; 0.4 µmol NADPH. The reaction is started with 2 nmol of ethoxyresorufin at 37 °C for 5–30 min. The reaction is stopped by addition of 2.5 ml methanol.

Aryl Hydrocarbon Hydroxylase Activity

Cytochrome P450 1A1 and 1A2 activity can also be determined by measuring the fluorescence of hydroxylated metabolites of benzo[a]pyrene with excitation at 396 nm and emission at 522 nm; 3-OH-benzo[a]pyrene is used as reference. For this assay 300–500 µg of total protein as cell homogenate was added to 1 ml of reaction buffer containing 50 µmol Tris-HCl, pH 7.6; 3 µmol MgCl$_2$; 0.6 µmol NADPH. The reaction is started by the addition of 100 nmol benzo[a]pyrene and maintained for 30 min at 37 °C. The reaction is stopped by addition of 4 ml propanol/hexane, 1:3 (vol/vol).

Testosterone Hydroxylation

Testosterone is the substrate for many cytochrome P450s. Hydroxylation is cytochrome P450 specific at certain sites, e.g., 6β-hydroxylation for 3A4. Hydroxylated metabolites are detected and measured by reverse-phase HPLC with UV detection. The assay is performed with 1–3 mg total protein in 50 mmol Hepes buffer, pH 7.4; 5 mmol MgCl$_2$; 38 mmol KCl; 1 µmol NADPH. After a pre-incubation for 5 min, the reaction was started by addition of 500 µmol testosterone and incubation for 10 min at 37 °C. The reaction is stopped by adding 100 µl 120 mmol NaOH. Thereafter, 0.65 µg corticosterone is added as an internal standard. Metabolites and internal standard are extracted with extraction columns (Varian Chem Elut, Varian, Harbor City, USA) and eluted with dichloromethane, which is evaporated under nitrogen to dryness. The residue is dissolved in 20 µl methanol. Chromatic separation of the metabolites may be carried out with a Kromasil C18 reverse-phase column (250×3 mm, particle size 5 µm; Knauer, Berlin, Germany). The metabolites are detected at

254 nm. Product elution is performed at ambient temperature and at a flow rate of 0.6 ml/min with a gradient as follows: starting with 30 % solvent B followed by a linear increase to 40 % solvent B for 25 min and 65 % solvent B for 15 min (solvent A, 93:7 H_2O:tetrahydrofuran; solvent B, 93:7 methanol:tetrahydrofuran).

References

Doehmer J, Schneider A, Faßbender M, Soballa V, Schmalix WA, Greim H (1995) Genetically engineered mammalian cells and applications. Toxicol Lett 82/83:823–827

Gonzalez FJ, Crespi CL, Gelboin HV (1991) cDNA-expressed human cytochromes P450s: a new age of molecular toxicology and human risk assessment. Mutat Res 247:113–127

Jacob J, Raab G, Soballa V, Schmalix WA, Grimmer G, Greim H, Doehmer J, Seidel A (1996) Cytochrome P450-mediated activation of phenanthrene in genetically engineered V79 Chinese hamster cells. Environ Toxicol Pharmacol 1:1–11

Schneider A Schmalix WA Siruguri V deGroene EM Horbach GJ Kleingeist B Lang D Böcker R Belloc C Beaune P Greim H Doehmer J (1996) Stable expression of human cytochrome P450 3A4 in conjunction with human NADPH-cytochrome P450 oxidoreductase in V79 Chinese hamster cells. Arch Biochem Biophys 332:295–304

Part V

Industrial Application
of Animal Cell Culture

Chapter 25

Scale-Up of Animal Cell Cultures

Eunan McGlinchey*, Donnacha O'Driscoll, and Martin Clynes

Introduction

The techniques of large-scale animal cell culture technology have been in use for almost 30 years now and have become increasingly important in the development and production of biological products of commercial interest (see Arathoon and Birch 1986). Some of the first products produced by large-scale cell culture methods included human vaccines (e.g., polio, rubella), veterinary vaccines (e.g., foot and mouth disease, FMD) and interferon-α. More recently, an increasing number of recombinant DNA-derived products have been coming onto the market including cytokines, hormones and vaccines along with monoclonal antibodies for diagnostic and therapeutic use.

Once developed on a small scale these processes must be scaled up to a suitable level for production of the compound of interest. This usually involves a number of different technologies as each time the producer cells must be taken from the working cell bank, normally a small vial stored in liquid nitrogen, and scaled up to production level, which can be several hundred liters or up to 10 000 l in volume (see Griffiths 1990, 1992; Lubiniecki 1990; Freshney 1992).

Animal cells are normally classified by their requirement for anchorage-dependent growth or by their ability to grow in suspension culture (anchorage independent cells). Cells which grow in suspension are readily scaled up to production level using similar fermentation vessels as designed for microbial cells along with relevant modifications for the more shear sensitive animal cells. Although animal cells are relatively fragile, scale-up has not been hindered by this and many different types of bioreactor vessels have been employed and adapted to suit particular needs.

* *Correspondence to* Eunan McGlinchey, National Cell and Tissue Culture Centre, Dublin City University, Glasnevin, Dublin 9, Ireland; phone +353–1–7045719; fax +353–1–7045484/5988; e-mail Mcglince@ccmail.dcu.ie

Anchorage-dependent cells need a surface for attachment in order to proliferate and strategies for scale-up of anchorage-dependent cells must take this into account. These cells can be grown in either static cultures (e.g., hollow fibers) or in suspension culture (e.g., stirred tank bioreactor, on microcarriers), but generally suspension cultures are preferred due to their relative ease of scale-up.

Animal cells can, however, be sensitive to mechanical agitation as well as gas sparging and certain additives may be used to help protect cells in suspension culture. A wide variety of both natural and synthetic additives are in use including serum, methylcellulose and Pluronic F68 (Wu 1995). Serum has additional biological effects on cells since it is a source of growth factors, hormones and other proteins necessary for cell growth and proliferation.

The major concern with large-scale cell culture is the need to maintain sterility for a long period of time as microbial contamination can prove very costly in terms of time and materials, particularly at larger volumes. Thus cultivation of animal cells often involves more stringent criteria for technical equipment, culture conditions, and process operation than does traditional microbial fermentation. Animal cell fermentations may last up to several months and asepsis must be maintained throughout. Mycoplasma screening of master cell banks is particularly important since these are difficult to detect in culture yet can have adverse effects on cell growth. Antibiotics can be very effective although they may mask low level infections for some time or can give rise to antibiotic resistance. Antibiotics are often used in large-scale processes although there is a trend to move away from these, particularly from a regulatory view point.

When scaling-up cells two approaches can be used in order to achieve the desired production scale:

- Increase the number of units already in operation (multiple-process system), e.g., scale-up by increasing the number of spinner flasks or roller bottles.

- Increase the size of the culture system without increasing the number of vessels used (unit-process system), e.g., scale-up in a large fermentation vessel.

Where the decision is made to scale-up with a unit-process system then it has to be decided whether to run the system in a batch or continuous perfusion mode. Batch cultures allow the seeded cells to grow to their final density without changing the growth media while perfused cultures have fresh media added continuously and waste media is removed. Batch cultures are simple and easy to run whereas perfusion cultures are more complex to set up and maintain. Perfusion cultures are normally run on a small scale (e.g., 10 l) but have many claimed advantages for example low capital cost, higher productivity, less work between runs and more rapid product processing.

Anchorage-Independent Cells

To start a cell culture from frozen stocks, the usual method is to thaw the cells into a small tissue culture flask – most often plastic disposable flasks are used which have surface areas ranging from 25–200 cm^2. Generally, flasks are made of polystyrene and have a negative surface charge on the culture surface which promotes attachment of cells. The flasks are typically closed systems, but flasks are also available with 0.2 µm filters incorporated into the cap, which allows exchange of gases between the flask and the incubator environment.

Anchorage-independent cells may settle and adhere slightly to the base of the flask and a quick shake is all that is required to dislodge cells from the flask surface. Other cells may adhere more strongly and will have to be removed by enzymatic treatment.

Increases in the number of cells on a small scale can be achieved by using larger flasks or increasing the number of flasks. Anchorage-independent cells may also be grown in flexible cell culture bags (e.g., SiCulture Bag, TC Tech Corp.) which allow diffusion of gases and can improve pH control within the culture. Culture bags can vary in size from 100 ml up to a few liters, thus allowing for easy, moderate scale-up of anchorage-independent cells.

Suspension Cultures

Anchorage-independent cells may also be easily grown in suspension using the spinner flask. Cells are suspended in the culture media and kept in suspension by gentle agitation with a magnetic bar, usually suspended from above. The flask normally has side arms for use as additional ports or for gassing the culture. Vessels are normally made of glass and modifications of the basic design are available from different suppliers. The size of spinner flasks varies from 100 ml to 20 l and the flask rests on special magnetic stirring modules with variable speed control (between 10 and 300 rpm).

Cells can be grown up to densities of $1-2\times10^6$/ml. Typical cells types to be grown in suspension are hybridoma cells, Vero and BHK cells. In order to reduce clumping of cells grown in suspension, special commercially available media are available which have a reduced calcium and magnesium ion concentration.

Anchorage-Dependent Cells

Anchorage dependent cells are started from frozen cultures and thawed into flasks, as in the case of anchorage-independent cells above. Anchorage dependent cells attach strongly to the surface of the flask and must be removed by enzymatic disaggregation, e.g., trypsin treatment.

Often the next stage in the scale-up of anchorage-dependent cells is the use of roller bottles. These are available in glass or in various plastics and allow an increased surface area for cell adhesion. Cells attach to the inner surface of the bottle while the bottle is rotated gently on a specialized roller apparatus generally at speeds of 0.5–5 rpm. The available media in the roller bottle is distributed to all of the attached cells as the bottle rotates, thus ensuring no cells are left without media for long.

There are variations in the design of roller bottles available which increase the surface area and thus the yield of cells per bottle. The usual roller bottle has a smooth surface and an area of 850 cm^2 – but longer roller bottles are available with double the length and thus double the surface area. Other variations include corrugated surfaces (expanded surface roller bottle, Bibby Sterilin Ltd.) or pleated surfaces (Falcon, Becton Dickinson) which can double the surface area available for growth.

Large numbers of roller bottles are used in vaccine manufacture and may be handled either manually or, more recently, robotically within a class 100 laminar airflow environment, thus reducing the risk of infection from human operators.

The Nunc multitray unit is another option for scale-up of anchorage dependent cells. Cells attach to a flat square surface which is layered with media as with a flask culture. There are normally ten chambers per multitray unit giving an overall area of 6000 cm^2, but larger units are also available. The culture is static and must be incubated in a large incubator or hot room for temperature control.

The Costar CellCube is available for scale-up of anchorage dependent cells under high density conditions in a perfusion mode. The simplest unit consists of a single cube composed of numerous parallel plates (made of styrene) through which media is perfused thus allowing cells to grow and proliferate on both sides of the plates. The surface area available for cell growth is variable ranging from 21 250 cm^2 up to 340 000 cm^2 depending on the cube size used and the number of cubes interconnected. The pH and dissolved oxygen levels within the system are monitored and controlled.

Cells which are anchorage dependent may be grown in suspension cultures by using microcarrier technology. Microcarriers are small spherical beads which act as a support for anchorage dependent cells while still being kept in suspension by gentle agitation. At laboratory scale microcarriers are used with spinner flasks and at larger scale can be used in bioreactor systems. The growth of cells on small spheres in suspension was first introduced by van Wezel using an ion-exchange resin but since then there have been many types of microcarriers developed for use such as dextran (e.g., Cytodex, Pharmacia), cellulose (e.g., DE-53, Whatman) or polystyrene (e.g., Biosilon, Nunc). The advantage of microcarrier technology is that it enables anchorage dependent cells to be grown in suspension cultures thus allowing easier scale-up. Other advantages include improved control of culture parameters, increased production yields, reduced requirement for serum and lower risk of contamination. Microcarrier technology has found many applications in cell culture including its use in the production of cell derived products and for virus production.

Materials

- Biosafety cabinet
- Incubator
- Microscope
- Autoclave
- Roller bottle apparatus, spinner apparatus
- Roller bottles (glass or plastic), spinner flasks
- Cell culture media: The cell culture media used depends on the cell type to be cultured and is usually advised by the supplier of the cell line. Cell culture media can be bought from suppliers as a ready to use solution, as a 10× concentrate or as a powder.
- Animal cells and cell culture materials: i.e., microcarriers, siliconizing agent, EDTA(e.g., Sigma cat. no. EDS), PBS-A (Ca/Mg-free, e.g., Oxoid cat. no. BR12a), trypsin (Gibco cat. no. 043–05090), DMSO, 10 % CO_2 gas, sterile disposable plastics (flasks, pipettes, etc.) and filters

25.1
Culture of Anchorage-Dependent Cells

Anchorage-dependent cells proliferate in culture to eventually form a continuous layer on the surface into which they are inoculated. When the surface is covered completely by a monolayer of cells, the culture is termed "confluent". When cells reach approximately 80–100 % confluency in a culture vessel, they should be subcultured and reseeded at a lower density. Subculturing is achieved by adding trypsin (which causes adherent cells to become rounded and detach from the surface). The process of subculturing by trypsinization is often referred to as "passaging". Each trypsinization corresponds to a single passage, and consequently, passage number gives an indication of the age of a cell stock. In general master cell stocks (often referred to as a master cell bank) should be frozen and used to generate secondary stocks (or working cell bank). Cultures often can be discarded once 10–15 passages have elapsed since thawing, since cell characteristics tend to vary with age. Maintenance of working stocks within a given range of passages should reduce experimental error caused by ageing cell populations.

▓ Procedure

Flasks

thawing and inoculation

1. Remove the vial of cells with care – allow excess liquid nitrogen to drain from the surface of the vial, as liquid nitrogen vaporizes rapidly with temperature and this can cause vials to shatter.

2. Thaw the vial in warm water (ideally at 37 °C).

3. When fully thawed, sanitize the surface (using alcohol) and transfer cells immediately into a universal containing 5 ml cell culture media.

4. Spin the cell suspension at 1000 rpm. for 5 min.

5. Resuspend in 5–7 ml of growth media. Place in 25 cm^2 flask, label and incubate at 37 °C.

6. Check the viability of cells under a microscope. After 24 h replace the medium to remove dead cells.

A solution of phosphate buffered saline (PBS-A) containing 0.25 % tryp- **passaging cells**
sin w/v and 0.02 % w/v EDTA is used for cell passaging. The trypsin may
be prepared as follows: To 88 ml sterile PBS-A add 10 ml of sterile trypsin
solution (2.5 % w/v) and 2 ml of sterile EDTA solution (1 % w/v). The
resulting trypsin-EDTA solution should be divided into 20 ml aliquots
and stored at $-20\,°C$ until required.

1. Decant the growth medium from the flask of cells.

2. Add approx. 2 ml of trypsin-EDTA solution. Close the flask and rotate
   gently for a few seconds (the purpose of this step is to remove any
   remaining traces of serum from the culture flask as trypsin is inactivated by serum).

3. Pour off waste trypsin and add a further 4 ml trypsin-EDTA to the
   culture flask.

4. Seal the flask and incubate at $37\,°C$ until the cells have detached
   (3–10 min, depending on the cell line).

5. Add an equal volume of growth medium containing serum. Rotate the
   flask gently (as before) and transfer the contents to a universal container.

6. Centrifuge the suspension at 1000 rpm for 5 min.

7. Carefully decant the supernatant and resuspend the pellet in fresh
   medium.

8. Count the cells using a hemocytometer and then seed into a new culture flask(s) at the required density.

9. Depending on the cell type, a single flask can yield enough cells to
   freshly inoculate approximately two to five flasks (referred to as the
   cell split ratio).

Subconfluent (75–90 %) exponential cultures should be used to give a **freezing cells for**
healthy cell suspension of 5×10^6 cells/ml or more. **storage**

1. Prepare a 10 % DMSO stock solution by adding 1 ml DMSO to 9 ml
   serum; filter sterilize.

2. Trypsinize and count the cells as described above (either flask or roller bottle cultures may be used). Centrifuge the cell suspension again
   and resuspend the pellet in serum to the desired cell density.

3. Mix together equal volumes of cell suspension and DMSO (10 % v/v). Add the DMSO solution slowly to the cell suspension (drop by drop for the first few ml).

4. When all the DMSO has been added, mix well and dispense 1–1.5 ml of the cell suspension into the vials for freezing. Label each vial and place in the neck of the liquid nitrogen freezer for 3–5 h initially. The vials are then placed into the freezer (e. g., Taylor-Wharton), and later one vial should be thawed to check for viability and sterility.

Roller Bottles

Cell culture in roller bottles follows the same principle as that of flasks. A much larger surface area is available for growth and therefore a greater number of cells can be generated.

cleaning   It is important that the roller bottles are thoroughly cleaned to remove any residues from previous cultures.

1. Soak the bottles in a warm solution of detergent for approximately 1 h and then scrub vigorously using a brush.

2. Rinse well (three times) in tap water followed by rinsing in purified water (three times). Invert the bottles and air dry at 37 °C in a hot room or incubator.

Alternatively glassware can be washed in a glass-washer machine with appropriate feeds of tap water and purified water.

sterilization   Roller bottles can be sterilized by autoclaving or using a dry heat oven.
To autoclave, loosen the cap on a clean dry roller bottle and cover with tin foil. Autoclave the complete roller bottle at 121 °C for 20 min. When cool, retighten cap on the roller bottle. The bottle and cap may also be autoclaved separately and then aseptically reassembled before use.
Sterilization in a dry heat oven should be carried out at the appropriate cycle, e.g., 180 °C for 3 h. Some caps may not be able to withstand the high temperatures in an oven and may be sterilized separately by autoclaving.

inoculation   1. Allow the growth medium and roller bottle to warm at 37 °C for at least 30 min beforehand. This will help prevent cells clumping and ensure an even spread of the monolayer throughout the roller bottle.

2. Trypsinize the cells and perform a cell count. A total of approximately $7-10\times10^6$ cells are needed to set up one roller bottle, though this will vary depending on the cell line used.

3. Add the cell suspension to 100 ml growth medium + serum; mix and transfer carefully into the roller bottle.

4. Place on a roller apparatus at 0.25 rpm for the first 24 h.

5. After 24 h check that the cells are attached and increase speed to 0.75 rpm. An increase in cell numbers can be monitored regularly by means of an inverted microscope.

25.2
Culture of Anchorage Independent Cells (Suspension Cultures) Using Spinner Flasks

Spinner flasks should be cleaned thoroughly before use (as for roller bottles – see above).

The following protocol is based on the use of a 250 ml vessel with a 100 ml working volume. For larger or smaller vessels the quantities given can be adjusted accordingly. Also the parameters given, such as trypsinizing times, vessel rpm, and inoculation concentrations, are guidelines based on experience with various cell lines. Optimum conditions for particular cell types can be established by performing a few trial runs.

▓ Procedure

Preparation of Spinner Flasks

Note: Use a fume cupboard.

To a clean, dry culture vessel, add 10 ml siliconizing agent, e.g., Sigmacote (Sigma, cat. no. SL-2). Rotate the vessel to ensure the entire inner surface has been coated. Drain off any surplus fluid and leave the vessel to air dry in the fume hood for at least 6 h. When dry, rinse three times with purified water. Autoclave the vessel at 121 °C for 20 min.

Free Cell Suspension

inoculation

1. Prepare a cell suspension for inoculation (e.g., from flask cultures) and dilute the cells to give a final concentration of 10^7 cells per 100 ml of growth medium.

2. Rinse the sterile spinner flask with about 20 ml of culture medium.

3. Aseptically add the cells to the vessel along with 90 ml of medium +5% fetal calf serum.

4. Place the spinner flask on a spinner apparatus at 15 rpm for 24 h and then increase agitation to 40 rpm.

5. The increase in cell numbers can be followed by removing a 1 ml sample from a well mixed spinner flask and performing a cell count. Cells must first be centrifuged and then resuspended in trypsin to disaggregate the clumps. Spin again, resuspend in medium +5% fetal calf serum and perform a cell count.

refeeding and harvesting

Feeding a free cell suspension culture is troublesome because the large volume of cells must be centrifuged to remove them from the spent medium.

1. Under sterile conditions distribute the cell suspension from the vessel into 50 ml sterile tubes.

2. Centrifuge the tubes at 1000 rpm for 5 min.

3. Pour off the supernatant and resuspend the pellet in 20 ml preheated cell culture medium. A 10 ml pipette can be used to pipette the suspension up and down order to break up any cell clumps.

4. Transfer the resuspended cells from each tube back into the culture vessel and bring the volume back to its original level with culture medium. If necessary, gas the cells with 10% CO_2 air (sterile filtered) for approximately 1 min and return them to the incubator.

Microcarrier Cell Suspension

preparation of microcarrier beads

1. Add 0.3 g of the microcarrier beads (Cytodex 3) to a siliconized glass bottle containing 30 ml of PBS-A (Mg, Ca-free).

2. Allow the suspension to stand at 37 °C for 2–3 h with occasional agitation.

3. Remove the supernatant and replace with fresh PBS-A (approximately 10 ml).

4. Allow the beads to settle again and remove the PBS-A. Replace with another 10 ml of PBS-A and then autoclave the washed microcarrier suspension.

5. After sterilization, aseptically remove the PBS-A and replace with pre-warmed medium +5 % fetal calf serum. Allow the microcarriers to settle, remove the supernatant and resuspend the microcarriers in 10 ml fresh medium. Then transfer the microcarriers to the sterile siliconized spinner flask.

inoculation of spinner flask

1. Prepare a cell suspension for inoculation (e.g., from flask cultures) and dilute the cells to give a final concentration of 10^7 cells per 100 ml growth medium.

2. Add the cell suspension carefully to spinner flask containing the microcarriers followed by 40 ml growth medium.

3. Stir at 30 rpm for 1 min with 45 min intervals to allow cell attachment to the microcarriers.

4. After approximately 12 h aseptically add another 50 ml growth medium and agitate the suspension at 60 rpm continuously.

refeeding the microcarrier suspension

Since microcarriers with attached cells are significantly heavier than free cells, the beads will settle on standing in a short space of time without harming the cells.

Allow the culture vessel to stand in a sterile cabinet for 4–8 min until the microcarriers have settled. Carefully decant the supernatant and then bring the volume back to its original level with culture medium. If required the cells are gassed with 10 % CO_2 air (sterile filtered) for approximately 1 min and returned to the incubator.

counting

Counting of microcarrier cultures is based on the method of Sanford et al. (1951). The culture sample is placed in a hypertonic solution causing the cells to burst. Crystal violet is incorporated into the solution to stain freed nuclei. The nuclei can then be counted using a hemocytometer. The microcarrier beads are too large to enter the counting chamber and do not interfere with the count.

1. Remove a 1 ml sample from a well mixed spinner flask.

2. Spin at 1500 rpm for 5 min.

3. Remove supernatant and resuspend the pellet of microcarriers in 200 µl 0.1 % (w/v) crystal violet in 0.1 M citric acid.

4. Mix the suspension well and incubate at 37 °C for 1 h.

5. After incubation pipette the suspension gently up and down to dislodge the nuclei. Load the sample onto a prepared hemocytometer and count the stained nuclei as for whole cells.

An estimate of the viable cell count can also be achieved by trypsinizing and staining the cells.

1. Remove 0.5 ml of sample from a well mixed flask.

2. Add 0.5 ml trypsin to the sample and mix well.

3. Allow to stand at room temperature for 5 –10 min (mixing the solution every 2 min with a pipette).

4. Add 0.5 ml trypan blue stain and count the cells under an inverted microscope using a hemocytometer.

25.3
Culture of Anchorage Independent Cells Using Roller Bottles

Spinner flasks and a stirrer apparatus are expensive. If, however, a laboratory is equipped for the use of roller bottles, such equipment can be used as a cheaper but effective way to grow microcarrier suspension cultures.

▥ Procedure

preparation of roller bottles

1. Prepare a clean glass roller bottle as described. Siliconize the bottle to prevent adhesion of microcarriers.

2. Add approx. 500 ml PBS-A to the roller bottle and then add 6 g Cytodex 3 microcarrier beads.

3. Mix the microcarriers every 30 min for about 3 h to allow the microcarriers to swell.

4. After swelling, decant off the PBS-A and add fresh PBS-A to bring the final volume back to 500 ml.

5. Autoclave the roller bottle and microcarrier suspension (as a liquid cycle).

6. After autoclaving, aseptically rinse the microcarriers three times in the culture media. This is achieved by allowing the microcarriers to settle then decanting off the liquid and adding fresh medium. The final volume should be 500 ml.

Before inoculation of cells, the roller bottle and microcarrier suspension should be preheated to 37 °C.

inoculation of cells

1. Gas the roller bottle with a 10 % CO_2 in air mixture for approximately 1 min (using filter sterilized gas).

2. Inoculate the roller bottle with a prepared cell suspension to give a final concentration of $1-2 \times 10^5$ cells/ml.

3. Place the roller bottle on a roller apparatus at 4–6 rpm and incubate at 37 °C.

4. Count cells as described above for spinner flask cultures.

References

Arathoon WR, Birch JR (1986) Large-scale cell culture in biotechnology. Science 232: 1390–95

Freshney RI (1987) Culture of animal cells: a manual of basic technique, 2nd edn. Alan R Liss, New York

Griffiths B (1992) Scaling-up of animal cell cultures. In: Freshney RI (ed) Animal cell culture: a practical approach. IRL, Oxford, pp 47–93

Griffiths JB (1990) Scale-up of suspension and anchorage-dependent animal cells. In: Pollard JW, Walker JM (eds) Methods in molecular biology, vol 5. Humana, New Jersey, pp 49–63

Lubiniecki AS (1990) Large-scale mammalian cell culture technology, vol 10. Bioprocess Technology, Dekker, New York

Sanford KK, Earle WR, Evans VJ, et al. (1951) The measurement of proliferation in tissue cultures by enumeration of cell nuclei. J Natl Cancer Inst 11: 773–95

Wu J (1995) Mechanisms of animal cell damage associated with gas bubbles and cell protection by medium additives. J Biotechnol 43:81–94

▓ Suppliers

General Cell Culture Materials

Company	Address	Phone	Fax
Becton Dickinson UK Ltd.	Between Towns Road, Cowley Oxford OX4 3LY	+44 1 865 748844	+44 1 865 781523
Bibby Sterilin Ltd.	Tilling Drive, Stone Staffordshire ST15 OSA UK	+44 1 785 812121	+44 1 785 813748
Costar UK Ltd.	10, The Valley Centre Gordon Road High Wycombe, Bucks HP13 6EQ UK	+44 1 494 471207	+44 1 494 464891
HyClone Europe Ltd.	Nelson Industrial Estate Cramlington Northumberland NE23 9BL, UK	+44 1 670 734093	+44 1 670 732537
Life Sciences	Sedgewick Rd. Luton Beds. LU4 9DT UK	+44 1 582 597676	+44 1 1582 597676
Life Technologies Ltd. (Gibco BRL)	3 Fountain Drive Inchinnan Business Park Paisley PA4 9RF Scotland	+44 141 8146100	+44 141 8146317
Nalge (Europe) Ltd.	Foxwood Court, Rotherwas Hereford HR2 6JQ UK	+44 1432 263933	+44 1432 351923
Pharmacia	Uppsala Sweden	+46 181 65000	+46 181 43820

Company	Address	Phone	Fax
Sigma-Aldrich Co. Ltd.	Fancy Road, Poole Dorset BH12 4QH UK	+44 1202 733114 (free) 0800 373731	(free) 0800 378785
TC Tech Corporation	25885 Birch Bluff Road Minneapolis, MN 55331, USA	+1 612 470 5938	+1 612 470 4106
Techne (Cambridge) Ltd.	Duxford Cambridge CB2 4PZ UK	+44 1223 832401	+44 1223 836838
Taylor-Wharton	4075 Hamilton Blvd. Theodore, AL, USA 36590–0568	+1 334 4438680	+1 334 4432250

Fermentation Equipment

Company	Address	Phone	Fax
Applikon Ltd.	Station Drive Bredon Tewkesbury Gloucester GL20 7HH UK	+44 1 684 772425	+44 1 684 772168
B.M. Brownes UK Ltd. (UK agents for B. Braun)	Pincents Kiln Industrial Park Pincenys Kiln Calcot Berkshire RGJ1 75B UK	+44 1 734305333	+44 1 734305111
Bioengineering AG	Sagenrainstrasse 7 CH 8636 Wald Switzerland	+41 55 256 8 111	+41 55 256 8 256
Biolaffite	10 Rue De Temara F-78100 St German-En-Laye France	+33 1 30615260	+33 1 306 15234

Company	Address	Phone	Fax
Life Sciences (UK agents for Bio-laffite)	Sedgewick Rd. Luton Beds. LU4 9DT UK	+44 582 597676	+44 1582 597676
Infors U.K. Ltd.	Fortune House 10 Bridgeman Tce. Wigan WN1 1SX UK	+44 1 942 825025	+44 1 942 820412
New Brunswick Scientific	Edison House 163 Dixons Hill Rd. North Mymms Hatfield Herts AL9 7JE UK	+44 1707 275733	+44 1707 267859

Chapter 26

Purification of Monoclonal Antibodies Produced by Animal Cells in Culture

Conor O'Dea[*,1,2], Suzanne O'Connor[1], and Martin Clynes[1]

Introduction

Monoclonal antibodies are the most common proteins produced by animal cells in culture. They have numerous applications as diagnostics, therapeutics and as basic research tools. The purification of an antibody from the cell culture supernatant or ascites is necessary for most applications. Two things will determine what is the best purification strategy. Firstly, the nature of the starting material, e.g., type and level of impurities and class of antibody; secondly, the level of purity required at the end of the separation process.

The Starting Material

Monoclonal antibodies are produced in a wide range of different culture systems. Harvests derived from suspension cultures and stirred bioreactors show antibody concentrations of 0.01–0.1 mg/ml. Harvests from high cell density bioreactors can have antibody concentrations up to 10 mg/ml. The starting concentration will have an impact on the sizing and operation of the purification steps. If the concentration is too low a volume reducing step may be required prior to purification. Binding characteristics tend to be better at higher concentrations. Depending on whether the cell culture system generates much cell debris, a clarification step, e.g., centrifugation or filtration, may be required upstream of any

* *Correspondence to* Conor O'Dea: phone +141–946–9999; fax +141–946–0000; e-mail co'dea@q-one.co.uk

[1] Conor O'Dea, Suzanne O'Connor, Martin Clynes, National Cell and Tissue Culture Centre, Dublin City University, Glasnevin, Dublin 9, Ireland

[2] Conor O'Dea, current address: Q-1 Biotech Ltd., Todd Campus, West of Scotland Science park, Glasgow G21 0XA

purification. For the purpose of this chapter, it will be presumed that the starting material has already been clarified.

The amount of contaminating protein will vary depending on whether there has been much disruption of the cells, releasing intracellular proteins into the media. Even with protein-free media there will always be some intracellular proteins leaking from cells. Fetal calf serum and growth factors such as insulin and transferrin are also a source of contaminating protein. Other contaminants, e.g., antibiotics, dyes, and DNA, may need to be separated from the antibody of interest.

One advantage of using animal cells instead of mouse ascites is that one does not have to worry about copurifying host antibodies, although another problem may arise from the presence of bovine IgGs in fetal calf serum.

It is also necessary to know the class of antibody to be purified. For example, although protein A affinity chromatography is a very useful tool in purifying monoclonal antibodies, it will not bind antibody classes IgA and IgM. Within the IgG class, various epitopes will bind at different relative affinities. There are also variations from species to species, for example mouse IgG_1 has a low binding affinity whereas human IgG_1 has a high affinity.

How Pure Is Pure?

Once we have some idea as to what our starting material may consist of, the second question to address is what degree of purity is required? This will vary depending on the specific application of the antibody. In some cases purification of an antibody is not necessary. If only specificity is required then each batch of antibody is tested for titer and used directly without further processing. A single step ammonium sulfate precipitation is sometimes all that is required to obtain antibody with an adequate degree of purity. In the case of therapeutic grade antibodies a very high level of purity would be required. This may include a protein A affinity column followed by an anion and/or a cation exchange column. In addition, a gel filtration column to remove aggregates may be placed at the end of the process. The process would need to be validated for the removal of all contaminating proteins, DNA, virus and leachables from any of the columns (Sofer 1991).

Development of a Purification Method

It is hoped that within the selection of protocols presented in this chapter, the reader will find at least one method which gives the desired result. However, as no two monoclonals are exactly alike, some method development may be necessary. Many of the protocols outlined in this chapter include a variety of buffer loading and elution conditions to help assist in the development process. Also, the flow rates, column sizing, etc. may need to be optimized. Use the manufacturers guidelines as a starting point. Since the stability of antibodies will vary, the conditions will need to be such that the activity of the antibody remains intact after the purification. Development of quantitative assays for the activity of the antibody is necessary prior to developing a good purification process.

It is an obvious advantage to have a knowledge of proteins and the techniques used in their purification. A good starting point is the book by Scopes (1994) on protein purification. For an extensive book on antibody purification techniques and method development refer to Gagnon (1996).

Getting Started

This chapter introduces the types of protocols one can apply to the purification of monoclonal antibodies from cell culture media. As there is a very broad range of techniques which can be used, we will concentrate on a selection of some of the most commonly applied methods. These protocols should be applicable to the purification of most antibodies. Precipitation is a simple method if the level of purity required is not that high (Curling 1980). Protein A is usually the most popular and useful technique for achieving high purity in a single step (Beyzavi 1990). More recently, other non-affinity steps have become available, including ABx Bakerbond from Baker, Inc. (Nau 1989), or hydroxyapatite (Bukovsky and Kennett 1987). If a high degree of purity is required a wide range of method combinations can be applied. All supernatants should be clarified of cells and cell debris before starting any purification method.

▧ Materials

general equipment

A typical column chromatography system consists of a peristaltic pump, a range of column sizes, monitors for UV and conductivity, a gradient mixer, a chart recorder and a fraction collector. Many manufacturers, e.g., Pharmacia, provide a complete system, although all the components can also be bought separately. Examples of small chromatography columns are the Pharmacia C and XK range, or the Econo-chromatography column range from Biorad Ltd.

Currently there are a number of automated purification systems on the market; examples include the Bio-cad system from PerSeptive Biosystems Inc, and the Biopilot system from Pharmacia. These systems allow for rapid process development by allowing multiple methods to be performed automatically in sequence. However, chromatography can also be performed on the most basic equipment. The dimensions and scale of the columns will depend on the particular application.

A range of analytical equipment will also be required to monitor the activity and purity of the protein during purification. Electrophoresis is the easiest way to monitor protein purity using SDS-PAGE or isoelectric focusing. A complete system comes with tank assembly, power pack and sandwich plates, available from Bio-Rad. The Pharmacia Phast system allows for rapid running and development of pre-cast gels. A simple method for quantifying the concentration of pure antibody is by measuring the OD at A_{280}; 1 mg/ml of antibody (IgG) gives an OD of approximately 1.4.

▧ Procedure

Ammonium Sulfate Precipitation

A single step precipitation may allow up to 65 % purity, with yields between 50 % and 80 %. The technique is simple and requires no specialized equipment. Drawbacks may include loss of antibody activity due to denaturation, low antibody purity and poor reproducibility. The method below is a typical protocol. It can be modified by varying the salt concentration, temperature and incubation times to improve results.

precipitation

1. Saturated ammonium sulfate is 4.1 M at 25 °C. Add 761 g to 1 l of distilled water and adjust the pH to 7.0 with $2NH_2SO_4$. **Note:** Handle the acid with care, use eye protection.

2. Slowly add ammonium sulfate to the antibody containing culture supernatant. Add sufficient ammonium sulfate to give a final concentration of 45 % saturated ammonium sulfate (i.e., 45 volumes saturated ammonium sulfate + 55 volumes antibody).

3. Stir at room temperature for 30 min; collect precipitate by centrifugation (2500 g for 15 min) and wash the precipitate once in 45 % saturated ammonium sulfate.

4. Redissolve the precipitate in PBS to the original volume and reprecipitate at 40 % saturated ammonium sulfate.

5. Wash the precipitate once in 40 % ammonium sulfate and redissolve in a minimal volume of PBS or other buffer as required. Dialyze at 4 °C against two changes of buffer. Clarify by centrifugation.

Note: Allow room in the dialysis tubing for expansion of the antibody solution.

Note: If the antibody is to undergo further chromatography purification, then dialyze the antibody into the loading buffer. Otherwise store the antibody in 0.02 % sodium azide, if appropriate.

Affinity Chromatography

Protein A binds to the Fc portion of the antibody. As this is the most conserved part, protein A can be used for the purification of a wide range of monoclonals. One can achieve high antibody purity, approximately 95 %, in one simple step and it is a relatively easy method to develop (Pepper 1990 and Beyzavi 1990). However, protein A is expensive, it may decrease the activity of the antibody and will bind nonspecific IgGs. Leaching of protein A must be taken into account when purifying therapeutic grade antibodies.

Monoclonal antibodies have different binding affinities for protein A. Two protocols are described for antibodies with high and low affinities for protein A. An example of antibodies with relatively high affinity for protein A are mouse IgG_{2a} and IgG_{2b} and most human IgGs. In these cases the following protocol is worth trying.

1. Adjust the pH of the supernatant to 8.0 with 1.0 M Tris HCl, pH 8.0. Size the column according to its capacity and the amount of antibody to be purified.

antibodies with high relative affinity for protein A

2. Equilibrate the column with ten column volumes of 10 mM Tris HCl, pH 8.0.

3. Load the column at a flow rate that is recommended by the manufacturer. Do not exceed the capacity of the column, which may decrease with time, especially if correct storage conditions are not followed.

4. Wash the column with ten column volumes of 10 mM Tris HCl, pH 8.0.

5. Elute the column with 100 mM glycine HCl, pH 3.0, and collect the eluant in small fractions.

6. Immediately readjust the pH of the fractions back to neutral pH by adding approximately 10 % 1 M Tris HCl, pH 8.0.

7. The pH of the column should also be readjusted back to a neutral pH as quickly as possible.

8. Strip the column by washing with five column volumes of an appropriate regeneration buffer, e.g., HCl pH 1.5, followed by washing the column with five bed volumes of PBS for ProSep A or 50 mM acetic acid followed by PBS for Sepharose based protein A.

9. Store the column in one of the following agents: 0.02 % sodium azide; 0.01 % thiomersal; benzoic acid; 20 % ethanol.

antibodies with low relative affinity for protein A

The following loading buffers can be used if the expected affinity of the antibody for protein A is low, e.g., mouse IgG$_1$ and rabbit IgGs.

1. The clarified supernatant should be dialyzed diafiltered or diluted fivefold with one of the following load buffers:

   – 0.1 M Tris HCl, 3 M NaCl, pH 8.9
   – 1 M Glycine/NaOH +0.15 M NaCl, pH 8.6
   – 0.1 M Borate/0.15 M NaCl, pH 8.5

2. Equilibrate the column with five bed volumes of loading buffer.

3. Load the column and wash with five bed volumes of loading buffer or until the UV absorbance is back to baseline.

4. Use a stepwise pH gradient from 0.1 M citrate, pH 6.0 to 3.0.

5. Strip the column by washing with five column volumes of an appropriate regeneration buffer, e.g., HCl pH 1.5, followed by washing the column with five bed volumes PBS for ProSep A or 50 mM acetic acid followed by PBS for Sepharose based protein A.

6. Store the column in one of the following agents: 0.02 % sodium azide; 0.01 % thiomersal; benzoic acid; 20 % ethanol.

Note: The protein A performance will vary depending on the manufacturer with regard to the binding capacity and the flow rates one can use.

Note: It is usual for the affinity columns to be performed at 4 °C, biological columns are less stable due to protease digestion.

Note: Storage reagents such as sodium azide may interfere with some assays or coupling reactions. It should be removed by dialysis or gel filtration.

Protein G is an affinity resin similar to protein A. Although it is not as popular as protein A it is useful in that it binds mouse IgG_1, rat IgGs and human IgG_1 with stronger affinity (Akerstrom and Bjorck 1986; Bill 1995). Use as described above for protein A but with the following buffers:

Protein G

- Load/wash buffer: 50 mM sodium phosphate
- Elution buffer: 0.1 M glycine HCl, pH 2.7
- Immediately readjust the pH back to neutral by adding approximately 10 % of 1 M Tris HCl, pH 8.0.
- Regenerate and store as described for protein A.

Ion Exchange Chromatography

Ion exchange chromatography can be applied to the purification of antibodies as described by Kent (1994). It can also be used in conjunction with other methods to remove residual contaminants. While there are a variety of different cation, anion or mixed ion exchangers, only two popular protocols are described here.

DEAE anion exchanger can be used in addition to a capture step (protein A or Bakerbond ABx) to remove negatively charged species, e.g., DNA and endotoxins from the antibody pool. The antibodies are allowed to bind to the DEAE by raising the starting pH above the pI of the antibody. Elution takes place by increasing the ionic strength of the column elution buffer which can be done in a stepwise or gradient manner. Alternatively, a pH gradient can be applied to the column, going from a higher to a lower pH. Optimal running conditions will have to be determined for

DEAE anion exchange

each antibody. DEAE exhibits poor capacity for strongly basic antibodies. The following is an example of a separation protocol in which the antibody is bound to the column and eluted under conditions which separate it from the contaminants. The buffers are as follows:

- Load and wash buffer: 50 mM Tris Acetate, pH 8.6
- Elution Buffer: (a) 50 mM Tris acetate; 300 mM NaCl; pH 8.6; (b) 25 mM sodium acetate; 15 mM sodium chloride; pH 4.5
- Strip buffer: 50 mM sodium acetate; 1 M NaCl; pH 4.0

1. Pour a column using approximately 1 ml of resin for every 5 mg of antibody.

2. Strip the column with three bed volumes of strip buffer.

3. Equilibrate the column with ten bed volumes of load and wash buffer.

4. The antibody solution should be dialyzed/diafiltered against the loading buffer.

5. Load the antibody onto the column and wash with ten bed volumes of load and wash buffer.

6. Elute the column by increasing the salt concentration in the wash buffer. A gradient can be set up going from the load buffer to 50 mM Tris acetate, 300 mM NaCl, pH 8.6. The linear gradient can be converted to a stepwise gradient once the salt concentration for elution is determined. Alternatively, the column can be eluted by washing with an elution buffer which lowers the pH of the column, e.g., 25 mM sodium acetate, 15 mM NaCl, pH 4.5, which acts to generate a pH gradient.

7. Monitor the UV trace for breakthrough and eluted fractions. If the antibody solution is not very pure then an SDS PAGE gel can be used to determine the presence of the antibody peak.

8. Strip the column by washing with four column volumes of strip buffer. Monitor the strip fraction for the presence of antibody.

ABx mixed ion exchange chromatography

ABx Silica Bakerbond from Baker, Inc. is a mixed ion exchanger developed by Baker for the purification of antibodies (Nau 1989). Although the resin is significantly cheaper than protein A, good results can be obtained by a single step purification. More development work to find the correct purification conditions may be required. As a starting point it is worth trying these running conditions:

- Load buffer; 25 mM MES, pH 6.0
- Elution buffer: (a) 0.5 M $2NH_4SO_4$; 20 mM KH_2PO_4; pH 7.0; or (b) 1 M NaOAc, pH 7.0
- Strip buffer: 1 M NaOAc, pH 3.0

Purification of Monoclonal Antibodies by Hydroxyapatite

This procedure requires no modification of the sample before loading (Bukovsky and Kennett 1987). The material is cheap and the yields are generally good. However, samples containing fetal bovine serum are usually contaminated with albumin. This can be removed by ammonium sulfate precipitation (see above) prior to the hydroxyapatite column. The resin is also susceptible to biodegradation. This method works well for IgM and for subclasses of IgG that have a low affinity for the affinity columns.

The buffers are as follows: hydroxyapatite
- Equilibration buffer: 0.05 M MES; 1 mM potassium phosphate; pH 6.0
- Elution buffer: 0.05 M MES; 1 mM potassium phosphate; pH 6.0. Add NaCl to a final concentration of 0.5 M.
- Strip buffer: 0.05 M MES; 1 mM potassium phosphate; pH 6.0.

1. Equilibrate column with ten column volumes of equilibration buffer.

2. Load 2 % of the column volume of the unequilibrated sample.

3. Wash with two column volumes of equilibration buffer.

4. Elute in a ten column volume linear gradient using elution buffer.

If the antibody does not elute continue as follows.

5. Wash with a two column volume linear gradient of equilibration buffer.

6. Elute in a ten column volume linear gradient using strip buffer.

7. Strip with five column volumes of strip buffer.

Note: If the antibody elutes at relatively high sodium chloride concentration substitute the equilibration buffer with 0.05 M Hepes, pH 7.0, or even BICINE, pH 8.0.

Note: If the antibody elutes only in phosphate, then loading in the presence of sodium chloride will reduce the amount of contaminants binding to the column.

484 CONOR O'DEA, SUZANNE O'CONNOR, MARTIN CLYNES

Hydrophobic Interaction Chromatography

Antibodies are generally hydrophobic in nature compared to other likely protein contaminants. This can be exploited to purify antibodies. In the presence of high salt concentrations the hydrophobic interaction between antibody and the hydrophobic ligand on the column is enhanced and the antibody binds. The antibody can be eluted by simply washing with low salt buffers.

Hydrophobic interaction chromatography (HIC) is useful for all types of monoclonal antibodies including IgMs. Purity of up to 90 % is possible in a single step. The method is also effective in removing nonspecific antibodies. The biggest drawback of HIC is its tendency to affect the activity of the antibody. Also, development of a separation procedure is not as simple as with the affinity resins. There is a wide range of hydrophobic media resins which can be used. A general rule is to select the most hydrophobic resin which does not denature the antibody. This should give better resolution with lower salt concentrations. There is a good overview of the application of HIC to the purification of monoclonal antibodies in the book by Gagnon (1996). The following is a typical HIC procedure for media less hydrophobic than phenyl:

hydrophobic interaction chromatography

The buffers are made up as follows:

– Buffer A: 0.05 M sodium phosphate; 1.5 M ammonium sulfate; pH 7.0
– Buffer B: 0.05 M sodium phosphate, pH 7.0

1. Equilibrate the column with ten column volumes of buffer A.

2. Load approximately < 5 % of a column volume of unequilibrated sample. Load less than 4 mg of antibody/ml of resin.

3. Wash with five column volumes buffer A:

4. Elute with ten column volumes in a linear gradient to 100 % buffer B. Collect fractions no greater than 0.5 bed volumes and analyze for the presence of antibody.

4. Strip with five column volumes buffer B.

Note: Variation in the media ligand, salt concentrations and pH during loading and elution may lead to better results.

Note: Usually media with hydrophobicity higher than the phenyl ligand will not be suitable for purification of monoclonals.

References

Akerstrom B, Bjorck L (1986) A physiochemical study of protein G, a molecule with unique immunoglobulin G-binding properties. J Biol Chem 261: 10240–10247

Beyzavi K, Wood H C (1990) Purification of mouse IgG$_1$ on protein A and the measurement of contaminating protein A. Separ Biotechnol 2: 452–61

Bill E, Lutz U, Karlsson BM, Sparrman M, Allgaier H (1995) Optimisation of protein G chromatography for biopharmaceutical monoclonal antibodies. J Mol Recog 8:90–4.

Bukovsky J, Kennett RH (1987) Simple and rapid purification of monoclonal antibodies from cell culture supernatant and ascites fluid by hydroxyapatite chromatography on analytical and preparative scales. Hybridoma 6: 219–228

Curling J (1980) Methods of plasma protein fractionation. Academic Press, London

Gagnon P (1996) Purification tools for monoclonal antibodies. Validated Biosystems, Tucson, Arizona, pp 103–126

Harlow E, Lane D (1988) Antibodies: a laboratory manual. Cold Springer Harbor, Cold Springer Harbor, New York, pp 288–318

Kent UM (1994) Purification of antibodies using ion exchange chromatography. Methods Mol Biol 34:23–7

Nau D (1989)Chromatographic methods for antibody purification and analysis. Biochromatography 4:4

Pepper D (1990) In: Laboratory methods in immunology, vol. II, CRC, Boca Raton, p169

Scopes R (1994) Protein purification principals and practice. Springer, Berlin Heidelberg New York

Sofer G, Nystrom LE (1991)Process chromatography: a guide to validation. Academic Press, San Diego

Suppliers

Pharmacia Biotech
Chromatography media e.g., protein A, ion exchange, including DEAE Sepharose CL-6B, hydrophobic interaction chromatography and others. They also supply all chromatography instrumentation and electrophoresis equipment.
Head office, Uppsala, Sweden; fax (Sweden) 46(0)18143820
Fax (UK) 01727–814001

Bio-rad
Chromatography media including hydroxyapatite, chromatography instrumentation and electrophoretic equipment
Head office: 2000 Alfred Nobel Drive, Hercules, California 94547, USA
Fax (USA)1–800–879–2289 or fax (UK) 01442–259118

Bioprocessing Ltd.
Protein A and other chromatography media.
Head office: Medomsley Rd, Consett, Co. Durham, DH8 6TJ, UK
Fax:(UK) 01207–500944

J.T. Baker, Inc.
Chromatography media including Bakerbond ABx
Head office 222 Red School Lane, Phillipsburg, New Jersey 08865, USA
Fax (USA) 908 859–9318

Biosafety Assessment of Cultured Animal Cells Used for Biopharmaceutical Production

Carl N. Martin* and David W. Birch

▒ Introduction

In 1973, genetic engineering was born following the pioneering work of Cohen and co-workers in California, who cloned a synthetic DNA molecule into a bacterium such that the constructed molecule was passed to daughter cells. Following these initial experiments, there has been an explosion of interest in the use of recombinant DNA technology using bacteria, yeast and continuous mammalian cell lines. It is now possible to produce vaccines, hormones, cytokines, blood factors and monoclonal antibodies using developments of these pioneering techniques.

Hybridoma technology, first developed by Köhler and Milstein in 1975, has also been widely used to produce monoclonal antibodies which can target specific tissues for diagnostic or therapeutic purposes.

There has been recent excitement in the potential to treat genetically inherited diseases by combining rDNA technology with a delivery system, such as a replication incompetent adenovirus or murine retrovirus, to allow incorporation of the gene into target cells of the recipient.

The production of these biological therapeutics and prophylactics (biopharmaceuticals) has a number of risks associated with it. Of concern here is the risk that mammalian cells used in the production of many biopharmaceuticals (see Fig. 27.1) may be contaminated by endogenous retroviruses or may become contaminated by exogenous, adventitious viruses or mycoplasma.

To address these concerns, a strategy has been developed by regulatory authorities which involves three independent approaches. Firstly, all starting materials (cell banks) and additives are thoroughly tested. Secondly, the purification process which is designed to produce chemi-

* *Correspondence to* Carl N. Martin, Corning Hazleton, Otley Road, Harrogate, HG3 1PY, England; phone +44–1423–500011; fax +44–1423–569595; e-mail cmartin@hazle.co.uk

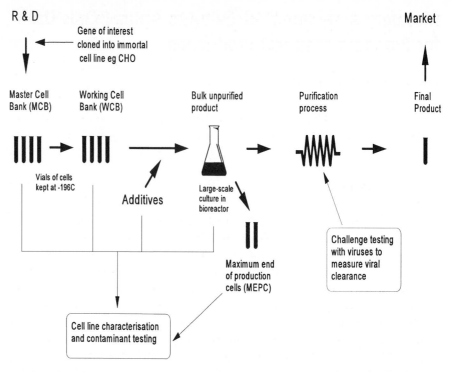

Fig. 27.1. Testing of the various manufacturing stages of a rDNA biopharmaceutical manufactured in CHO cells

cally pure product must be validated for its ability to inactivate or remove viruses. Finally, batches of unpurified product are routinely tested. This chapter concentrates on the methodology employed in the tissue culture assays which are used to assay the various cell banks and bulk unpurified culture fluids (Fig. 27.1) for the presence of viruses. Plaque assays and tissue culture infectivity dose$_{50}$ (TCID$_{50}$) assays used in the purification process validation are also described.

Cell Bank and Additive Testing

Cell banks must be thoroughly tested for the presence of endogenous retroviruses and exogenous adventitious viruses. The types of virus potentially present will depend on the host range of the producer cells being tested and the species of origin of these cells and any biological additives. This chapter cannot hope to cover the wide variety of assays needed to cover all eventualities, thus we have concentrated on providing a number of the more widely used tests (Schmidt 1979).

The Detection of Murine Retroviruses

Rodent cell lines are extensively used in the production of diagnostic and therapeutic biopharmaceuticals, hence manufacturers must address the concerns arising from the potential presence of retroviruses in the cell banks.

The members of the Retroviridae family are of great importance when considering the safety of biopharmaceuticals. The group includes both HIV and HTLV – viruses responsible for human diseases (although these viruses are of no concern to rodent cell lines). Retroviruses are able to integrate into the host genome and hence alter the pattern of gene expression. Once integrated they are able to pass vertically to daughter cells and so spread the infection without the need for virus replication. As a group they demonstrate a wide range of interactions with their hosts, ranging from benign infection to fatal disease.

Despite the breadth of biological effects, all retroviruses exhibit a similar morphology and life cycle. All are enveloped with a genome of RNA. The envelope displays the env protein which serves as the acceptor molecule, binding to receptor molecules on the cell surface and allowing the virus entry into the host cell. The virus then uncoats and transcribes a DNA copy of the RNA genome using the enzyme reverse transcriptase which is present in the mature virus particle. The RNA template is digested and a complementary DNA strand is synthesized. The double-stranded DNA molecule then circularizes and integrates into the host cell genome, forming a provirus. Proviral integration takes place in mitotically active cells and between one and 30 proviral sequences may be integrated. The provirus can then be transmitted vertically as a stable component of the host cell genome, often maintained in an unexpressed state by methylation.

Derepression of the integrated provirus is a key step in the retrovirus life cycle. It can be brought about by various chemical stimuli, including mitogens and halogenated pyrimidines. In vitro culture of the host cells can also lead to derepression, and late passage cells are more likely to express virus particles than those at early passage. Transcription of the provirus is followed by translation, and the retrovirus proteins assemble with genomic RNA to form a mature particle. This can then be released from the cell to start a new round of infection.

The family Retroviridae is subdivided into three groups: the Oncovirinae, Lentivirinae and Spumavirinae.

- Oncovirinae (oncos meaning tumor): Members of this family are also known as RNA tumor viruses because of their ability to induce cellular transformation.
- Lentivirinae (lente meaning slow): This family, which includes HIV, is associated with slowly progressive diseases.
- Spumavirinae (spuma meaning foam): Spumaviruses have been isolated as contaminants of primary tissue culture cells. These viruses derive their name from the characteristic foaming degeneration they induce in cultured cells.

All rodent retroviruses belong to the oncovirinae and are subdivided into several groups according the morphology of the virus particle as determined by electron microscopy.

There are two types of intracellular A-type particles in murine cells, intracisternal A-type particles (IAPs) and cytoplasmic A-type particles (CAPs). IAPs develop by budding into the cisternae of the endoplasmic reticulum where they remain sequestered. IAPs are noninfectious and often constitute the most numerous retrovirus particles in a rodent cell. Cytoplasmic A-type particles are precursors to the mature B-type particles. The immature particles develop their inner core in the cytoplasm and acquire their envelope as they bud from the plasma membrane. The only infectious murine example is the mature form of mouse mammary tumor virus.

C-type particles form at the surface membrane. All murine leukemia viruses belong to this group

Cytoplasmic A-type particles have also been found in Chinese hamster cells. Typically, these particles are found associated with centrioles and are non-infectious. Syrian hamster cells have been found to contain retroviral particles which are referred to as R-type particles. These particles are intracisternal and are noninfectious.

The murine genome contains especially large numbers of proviral related sequences. Sequence analysis reveals at least 1000 copies per haploid genome, many of which contain multiple stop codons and are not expressed (Lueders 1991). Electron microscopic analysis of hybridoma cell lines often reveals both A-type (predominantly IAPs) and C-type particles, with IAPs forming the predominant type. All murine hybridoma cell banks probably contain IAPs, often many thousands per cell, but there is no evidence that any are infectious.

C-type particles are usually present in smaller numbers than A-type particles. Unlike A-type particles, some C-types are infectious, and are classified according to the host range they exhibit.

All Chinese hamster ovary (CHO) cell lines probably produce retrovirus particles. As with retroviruses of murine cells, they can be classified as either A-type or C-type. However, unlike their counterparts in murine cells, the CHO A-type particles assemble in the cytoplasm. No infectious A- or C-type particles have been found in CHO cells.

C-type particles are classified according to the host range they exhibit:

- Xenotropic viruses are able to grow in nonmurine cells but are unable to grow in murine cell lines. Some xenotropes exhibit very wide host ranges.
- Polytropic viruses are formed by recombination between exogenous ecotropic retrovirus and nonecotropic endogenous env sequences (Stoye and Coffin 1987). These viruses have been isolated in spontaneous lymphomas of AKR mice and malignant tissue of other mouse strains. A characteristic of many polytropic viruses is the ability to induce focus formation on cultured mink lung cells, hence their designation as mink cell focus-forming (MCF) viruses. They can infect murine cells, but have limited infectivity to non-murine cells, and do not produce syncytia in the XC plaque assay. MCF viruses are detectable using the mink S^+L^- and MCF assays.
- Amphotropic viruses can infect wide host range including murine and non-murine cells. They differ from the MCF viruses in their wider nonrodent infectivity and differences in the structure of the env gene. Amphotropic murine retroviruses have only been recovered from wild mice from California and not in laboratory strains. Their mode of transmission does not appear to include inheritance through the germline. Amphotropic retroviruses which can infect human cells can be generated in culture and have been used as vectors for human gene therapy.
- Ecotropic viruses have a limited host range, only replicating in cells of mouse or rat origin. A murine cell may contain several copies of ecotropic proviral DNA, but all are essentially the same and only one strain of ecotrope is expressed.

A number of virus detection assays are available to evaluate the retrovirus status of a rodent cell line and the key ones are described (Toyoshima and Vogt 1969; Morse and Hartley 1986).

The S^+L^- assay is a sensitive method for the detection of a wide range of xenotropic, polytropic and amphotropic retroviruses. It utilizes intracellular complementation between an exogenous retrovirus and an endogenous provirus carried by the indicator cell line.

There are two functional categories of rodent retroviruses. Tumor viruses are designated r^-t^+, since they can enter and transform (t^+) certain cell types but cannot replicate (r^-). Helper viruses are designated r^+t^- as they are replication competent (r^+) in cells but do not cause transformation (t^-).

The cell lines used in the S^+L^- assay carry the genome of a sarcoma virus which is r^-t^+. When infected with a r^+t^- virus, complementation occurs, the sarcoma genome is "rescued" and a focus is formed.

Various S^+L^- cell lines are available. The most suitable cell line for the detection of murine xenotropic retroviruses is the mink lung cell (S^+L^- MiCl1) which carries the Moloney murine sarcoma virus genome. This cell line is more sensitive to xenotrope infection than the feline S^+L^-(PG4) cell line, while the PG4 cell line is more sensitive for detecting amphotropic retroviruses. The feline S^+L^- line is particularly useful in gene therapy protocols when assaying for amphotropic replication competent retroviruses (RCR). These may arise following recombination between the therapeutic viral vector and endogenous viral sequences present in cell lines used to replicate the vector.

The XC plaque assay is used for the detection of ecotropic murine retroviruses. Culture supernatant taken from producer cells is inoculated onto SC-1 mouse cells. These cells support the growth of a wide range of ecotropic murine retroviruses without producing any cytopathic effect. After incubation, the SC-1 cells are given a controlled dose of UV irradiation to kill the cells without inactivating the virus. The cells are then overlaid with XC cells, a continuous Wistar rat tumor cell line infected with Rous sarcoma virus. If ecotropic retroviruses are present they can provide a helper function to the sarcoma virus and syncytium formation occurs.

The irradiation of the SC-1 cells makes syncytium formation easier to visualize by preventing overgrowth of these cells when the XC cells are added. By using an appropriately controlled dose of UV irradiation, the loss of retroviruses is minimized.

The Detection of Adventitious Viruses

The term "adventitious virus" covers any virus which could infect the producer cell line but excludes retroviruses. As the range of potential infecting viruses may be very large, there is a need for the application of general screening assays which are sensitive to a broad spectrum of agents (Poiley 1990). In some circumstances, a specific virus may be identified as posing a potential hazard and for these agents application of sensitive and specific assays are required.

The assay described involves the inoculation of multiple dishes of indicator cells with test cell supernatant or an homogenate prepared from these cells. Samples are observed each working day for cytopathic effect over a 2-week period and are tested on two occasions for hemadsorption and hemagglutination. All negative cultures are passaged onto fresh cultures of indicator cells and observed for a further 2 weeks. The choice of indicator cell lines will depend on the species of origin of the producer cells, their host range and any biological additives used in the production process. Indicator cells routinely used include continuous cell lines and primary embryonic cultures of human, lapine, canine, porcine, bovine, murine, simian and avian origin. Hemadsorption and hemagglutination are carried out using guinea pig, chicken and human group O red blood cells. Red blood cells from other species may be used as circumstances dictate. The assay is applied to cells from different species by careful choice of the indicator cells used. Where the producer cells have been in contact with bovine serum, it is usual to include an immunoassay for noncytopathic effect forming bovine viral diarrhea virus (BVDV) though any relevant virus may also be assayed in this way.

27.1
Mink S⁺L⁻ Focus-Induction Assay for the Detection of Murine Xenotropic Retroviruses

This assay is the most sensitive assay for the detection of murine xenotropic retroviruses (MXRV) (Peebles 1975).

Materials

All incubations are carried out at 37 °C in a humidified 5% CO_2 incubator.

- Mink S⁺L⁻ cells (ATCC CCL 64.1)
- Mink lung cells (ATCC CCL 64)

Murine xenotropic retroviruses are not readily available from any repository or commercial supplier but can easily be obtained from research laboratories working in the field.

- Dulbecco's modified Eagle medium (DMEM) complete growth medium for mink S$^+$L$^-$ cells: supplemented with 10 % (v/v) fetal calf serum; 1x nonessential amino acids; 2.5 µg/ml fungizone; 50 µg/ml gentamycin (Life Technologies Ltd.)
- Minimal essential medium (MEM) complete growth medium for mink lung cells: supplemented with 10 % (v/v) fetal calf serum; 1× nonessential amino acids; 2.5 µg/ml fungizone; 50 µg/ml gentamycin (Life Technologies Ltd.)
- Polybrene (hexadimethrine bromide): prepared as a 2 mg/ml solution in purified water) (Sigma). **Note:** Polybrene is a polycation and aids viral infectivity. Its mode of action is not fully understood but it is thought to act at two levels; firstly by reducing electrostatic repulsion due to its neutralizing effect on both viral and cellular negative charges and, secondly, by the inactivation of polyanionic viral inhibitors.
- Test material: supernatant from a culture of cells from the bank to be tested
- Positive control: three separate controls consisting of murine xenotropic retrovirus diluted with complete medium containing 2 µg/ml Polybrene to yield between approximately 2, 20 and 200 focus-forming units (ffu) per 0.2 ml inoculum, respectively
- Negative control: complete medium containing 2 µg/ml Polybrene

▓ Procedure

direct assay
1. Mink S$^+$L$^-$ cells are seeded into appropriately labeled 60 mm tissue culture dishes at 2.4×10^5 cells/dish in DMEM complete growth medium (4 ml/dish). The dishes are then incubated (three dishes for each test sample and three dishes for each control).

2. The following day, the medium is removed by aspiration and the cultures refed with complete growth medium containing 2 µg/ml Polybrene.

3. After 2-h incubation, 0.2 ml of control and test material are inoculated onto the appropriate dishes in triplicate and further incubated.

4. The following day, the culture medium is removed and all dishes refed with 4 ml/dish complete growth medium and incubated.

5. Cultures are refed 4 days after inoculation with 4 ml/dish complete growth medium if cells are not confluent.

6. The assay is scored, using an inverted microscope, when foci are observed in the positive control.

Note: The assay can be made more sensitive by amplifying any virus present. This is done by prior cultivation of the test material with host cells as described below. The test material and positive and negative controls are as above.

1. Mink lung cells (which are known to support the growth of xeno-tropic rodent retroviruses) are seeded into appropriately labeled culture vessels at 2.4×10^5 cells/25 cm^2 tissue culture flask in MEM complete growth medium (5 ml/flask) and incubated (two flasks for each test sample and two flasks for each control).   **amplified assay**

2. The following day, the medium is removed by aspiration and the cultures refed with complete growth medium containing 2 µg/ml Polybrene.

3. After 2-h incubation, 0.2 ml of test material, positive controls and negative control are inoculated into appropriate culture vessels in duplicate. Culture vessels are then further incubated.

4. The following day, culture medium is removed and all culture vessels refed with complete growth medium.

5. After a further 2 days incubation, the cells are subcultured at a split ratio of 1:4.

6. Cells are further incubated until confluent (usually 2–3 days) and then second subculture is performed; growing cultures are refed as necessary during this period to maintain healthy cultures.

7. Once the cells are 70–100 % confluent following the second subculture, culture medium from each vessel is removed and clarified by centrifugation at 1000 g for 10 min at 20 °C. Aliquots (1 ml) of clarified supernatant from each sample are stored at −70 °C or below until it is tested in the mink S$^+$L$^-$ focus induction assay as described earlier.

Acceptance criteria: The assay is acceptable if: (a) the mean focus count of the positive control batch used in the assay is less than 1 log different from the original titration of the control batch; (b) there is an increase in the number of foci obtained in the positive control following passage in mink lung cells compared with the number initially inoculated; (c) no foci are observed in the negative controls.

Evaluation criteria: The assay is considered positive if at least one focus is observed on any of the triplicate dishes plated after treatment with the test material in either the direct or amplified assays.

Results

A set of example results are presented in Table 27.1. In this case the test material was found to be contaminated at a very low level.

If foci are detected in the test material dishes in the direct assay, the number of retrovirus present per ml in the original sample can be determined by multiplying the mean foci per dish by the dilution factor and the reciprocal of the inoculation volume. Where more than one dilution yields foci, then the dilution yielding between 10 and 100 ffu per 60 mm

Table 27.1. Direct assay results

Sample (0.2 ml)	Dilution	Foci/dish	Mean foci/dish
Test sample	Neat	0, 0, 0	0.0000
Positive control (~200 ffu MXRV)	Neat	184, 206, 210	200
Positive control (~20 ffu MXRV)	10^{-1}	20, 26, 23	23
Positive control (~2 ffu MXRV)	10^{-2}	2, 3, 2	2
Negative control	–	0, 0, 0	0.0000

MXRV, murine xenotropic retrovirus; ffu, focus forming units.

Table 27.2. Amplified assay results

Sample (0.2 ml)	Dilution	Foci/dish	Mean foci/dish
Amplified test sample	Neat	150, 165, 172	162
Positive amplified control containing ~200 ffu MXRV prior to amplification	Neat	TNTC	TNTC
Positive amplified control containing ~20 ffu MXRV prior to amplification	10^{-1}	TNTC	TNTC
Positive amplified control containing ~2 ffu MXRV prior to amplification	10^{-2}	TNTC	TNTC
Negative control	–	0, 0, 0	0.0000

MXRV, murine xenotropic retrovirus; TNTC, too numerous to count; ffu, focus forming units.

plate should be used for the most accurate calculation. From Table 27.1 it can be seen that the test sample was negative. Table 27.2 demonstrates the increased sensitivity of the amplified assay, i.e., the sample was positive only after amplification. The amplified assay is not quantitative and the numbers of foci bear no calculable relationship to the number of retroviruses in the original test sample.

Troubleshooting

The mink S^+L^- assay is temperature sensitive and the rate of cell growth and focus formation is severely affected by periods of interrupted incubation. All manipulations should be kept to a minimum.

The mink S^+L^- cells have a finite passage of approximately 20 (at a split ratio of 1:6) after which the cells tend to form spontaneous foci. The cells also tend to form ridges when the cells begin to become fully confluent or overgrow at the end of an assay. Where this is a problem the cells may be refed with complete growth medium containing 2% (rather than 10%) (v/v) fetal calf serum.

27.2
Feline S^+L^- Focus-induction Assay for the Detection of Murine Amphotropic Retroviruses

This assay is the most sensitive for the detection of murine amphotropic retroviruses (Bassin et al. 1982; Haapla et al. 1985).

Materials

- Feline S^+L^- cells (ATCC CRL 2032)
- McCoys 5A complete growth medium containing 5% (v/v) fetal calf serum; 1x nonessential amino acids; 2.5 µg/ml fungizone; 50 µg/ml gentamycin (Life Technologies Ltd.)

▪ Procedure

direct assay

1. Feline S⁺L⁻ cells (PG4) are seeded into appropriately labeled 60 mm tissue culture dishes at 2.0×10^5 cells/dish in McCoys complete growth medium with 4 µg/ml Polybrene (4 ml/dish). The dishes are then incubated (three dishes for each test sample and three dishes for each control).

2. The following day, the medium is removed by aspiration and the cultures inoculated with 1.0 ml of test material or controls.

3. After 1 h, 4 ml/dish complete growth medium are added and dishes are then incubated until cells are confluent.

4. The assay is scored, using an inverted microscope, when foci are observed in the positive control.

Note: The assay can be made more sensitive by amplifying any virus present. This is done by prior cultivation of the test material with PG4 cells as described below.

amplified assay

Steps 1–3 are performed as for the direct assay except that three flasks per sample and per control are seeded.

5. When the cells are confluent they are subcultured at a split ratio of 1:4 and these cultures are then incubated. If at any time the positive control cells show signs of excessive focus formation, they are scored without further passage.

6. Foci form in these flasks and the assay is scored when cells are confluent.

27.3
XC Plaque Assay for the Detection of Ecotropic Murine Retroviruses

This assay is used for the detection of murine ecotropic retroviruses (Rowe et al. 1970). An S⁺L⁻ assay using murine rather than mink S⁺L⁻ cells can also be used.

Materials

- Crystal violet (0.13 % w/v) / formaldehyde (11.2 % v/v). All components are available from Sigma.
- UV lamp (Panasonic germicidal ultraviolet G15T8 tubes housed in Luxo angle poise lamp). **Note:** Suitable personal protective equipment must be used, i.e., face shield and thick gloves.
- SC-1 cells (ATCC CRL 1404)
- XC cells (ATCC CCL 165)
- Rauscher murine leukaemia virus (ATCC VR 1413)
- MEM complete growth medium for SC-1 and for XC cells: MEM containing 2 %–20 % (v/v) fetal calf serum; 1x nonessential amino acids; 2.5 µg/ml fungizone; 50 µg/ml gentamycin (Life Technologies Ltd.)
- Test material: Supernatant from a culture of cells from the bank to be tested
- Positive control: Three separate controls consisting of murine ecotropic retrovirus diluted with complete growth medium containing 8 µg/ml Polybrene to yield between approximately 2, 20 and 200 syncytium (plaque) forming units (pfu) per 0.2 ml inoculum, respectively
- UV control: SC1 cells treated with UV and observed at the end of the assay to ensure the dose of UV used was lethal
- XC control: XC cells untreated and observed at the end of the assay to ensure adequate growth of this cell type (difficult to observe in treated cultures due to the presence of irradiated SC1 cells)
- Negative control: Complete growth medium containing 8 µg/ml Polybrene

Procedure

1. SC-1 cells are seeded into appropriately labeled dishes at 1.5×10^5 cells/60 mm dish in MEM complete growth medium and incubated (three dishes for each test sample and three dishes for each control). Three dishes are labeled "UV control".

    direct XC assay

2. The following day (day 2), the medium is removed by aspiration and 4 ml MEM complete growth medium containing 8 µg/ml Polybrene is added to all cultures. After 2 h, 0.2 ml of test material and controls are inoculated onto the appropriate dishes in triplicate and further incubated.

3. On day 3 the medium is removed by aspiration and all dishes refed with complete growth medium.

4. On day 6, the medium is removed and all cultures are exposed to a lethal dose of UV (320–360 µW/cm^2 for 25 s). All dishes, except the UV controls, are overlaid with 4 ml XC cell suspension at 3.8×10^5 cells/ml. In addition, three fresh dishes, labeled as XC control, are seeded with 4 ml of the XC cell suspension. The UV control dishes are refed with complete growth medium.

5. All dishes are incubated and when the XC control dishes are confluent and plaques are seen in the positive controls, the medium is removed from all cultures and the cell sheets washed, stained with crystal violet/formaldehyde for 1 min and then washed with tap water and drained.

6. When dry, the cell sheets are scored for syncytium (plaque) formation using an inverted microscope.

amplified XC assay

1. SC-1 cells are seeded into appropriately labeled 25 cm^2 culture flasks at 1.5×10^5 flask in 4 ml MEM complete growth medium and incubated.

2. The following day, the medium is removed by aspiration and the cultures refed with complete growth medium containing 8 µg/ml Polybrene. After 2 h incubation, 0.2 ml of test material, positive controls (containing approximately 2, 20 and 200 pfu) and negative control are inoculated into appropriate culture vessels in duplicate. Culture vessels are then further incubated.

3. The following day the inoculum is removed and all culture vessels refed with complete growth medium and returned to the incubator. When cells are confluent, (usually after 2 days) they are subcultured at a split ratio of 1:4 and, once cells have again reached confluence, a second subculture is performed. Growing cultures are refed as necessary to maintain healthy cultures.

4. Once the cells are 70–100 % confluent, culture medium from each vessel is removed and clarified by centrifugation. Aliquots (1 ml) of clarified supernatant from each sample are stored at −70 °C or below until tested in the direct XC assay as described earlier.

– Test material: Supernatant from a culture of cells from the bank to be tested.

– Positive controls: Three separate controls consisting of murine ectopic retrovirus diluted with complete growth medium containing 8 µg/ml Polybrene to yield between approximately 2, 20, and 200 syncytium (plaque) forming units (pfu) per 0.2 ml inoculum respectively.
– Negative control: Complete growth medium containing 8 µg/ml Polybrene.

Acceptance criteria: The assay is acceptable if: (a) the mean plaque count of the positive control batch used in the assay is less than 1 log different from the original titration of the control batch; (b) there is an increase in the number of plaques obtained in the positive control following passage in SC-1 cells compared with the number initially inoculated; (c) no plaques are observed in the negative controls. In addition the UV control must show that the SC-1 cells have been lethally irradiated and that cells are confluent in the XC controls.

Evaluation criteria: The assay is considered positive if at least one plaque is observed on any of the triplicate dishes plated after treatment with the test material in either the direct or amplified assays.

Troubleshooting

● It is imperative that the UV lamp is correctly calibrated so that the correct dose of UV irradiation is delivered to the SC-1 cells. This should be established by initial studies using a range of doses under experimental conditions.

● When the XC cells are added to the SC-1 cells, they must be added slowly, down the side of the dish, minimizing any disturbance to the SC-1 cell sheet.

● The irradiation of the SC-1 cells makes syncytium formation easier to visualize preventing overgrowth of these cells when the XC cells are added. By using an appropriately controlled dose of UV the loss of retroviruses is minimized.

● Both types of cells used in the assay have a finite number of passages. The SC-1 cells should not be taken above passage 104 (at a split ratio of 1:4). The XC cells should not be taken above passage 89 (at a split ratio of 1:4).

27.4
In Vitro Evaluation of Adventitious Viruses in Cell Cultures with Immunological Detection of Bovine Viral Diarrhea Virus

▦ Materials

- MDBK (BVDV free) cells (ATCC CCL 22)
- Vero cells (ATCC CCL 81)
- MRC-5 cells (ATCC CCL 171)
- Parainfluenza 3 (PI3) virus (ATCC VR 281): positive control virus for cytopathic effect (CPE), hemadsorption and hemagglutination
- Bovine viral diarrhea virus (BVDV) (ATCC VR 524): positive control virus for BVDV immunoassay, New York strain (noncytopathic)
- MEM complete growth medium for MDBK (BVDV free) cells: MEM containing 2%–20% (v/v) γ-irradiated fetal calf serum; 1× nonessential amino acids; 2.5 µg/ml Fungizone and 50 µg/ml gentamycin (Life Technologies Ltd.)
- MEM complete growth medium for Vero and MRC-5 cells: MEM containing 2%–20% (v/v) fetal calf serum; 1x nonessential amino acids; 2.5 µg/ml fungizone; 50 µg/ml gentamycin (Life Technologies Ltd.)
- MEM containing 5% (v/v) tryptose phosphate broth (diluent for positive control virus dilutions) (Life Technologies Ltd.)
- Hanks balanced salt solution (HBSS) without phenol red (Life Technologies Ltd.)
- Phosphate buffered saline (PBS) (Oxoid/Unipath)
- Acetone (Rathburn Chemicals Ltd.)
- IPEX-BVDV immunostaining kit (Central Veterinary Laboratory)
- Mammalian erythrocytes, e.g., human type O (Regional Blood Transfusion Services); 1 day old chicken and guinea pig type O (TCS Biologicals Ltd.)
- Hemagglutination buffer (Oxoid/Unipath)
- Test material: See protocol above
- Positive control for the adventitious virus assay: parainfluenza 3 virus diluted in MEM containing 5%(v/v) tryptose phosphate broth to achieve a titer of approximately 1×10^4 pfu/ml
- Positive control for the immunoassay for noncytopathic BVDV: BVDV, New York strain, added to the cell sheets in tenfold serial dilutions
- Negative control (both assays): MEM containing 5% (v/v) tryptose phosphate broth

Procedure

A short protocol is shown in Fig. 27.2.

The following is carried out for each indicator cell line (in which the test material is the supernatant from a cell culture; it is a regulatory requirement that these cells also be included as indicator cells, a somewhat circular exercise of dubious scientific value. In this case these test material indicator cells, TMIC, are also treated as described for the other indicator cells).

adventitious virus assay

1. On day 3 indicator cells are seeded into a total of 36 wells (6×6-well plates) per indicator cell line. Two 6-well plates are labeled test material, two plates positive control and two plates negative control. All plates are labeled with the cell type.

2. On the day of inoculation, the inoculum is prepared as follows: For test cells which grow as a monolayer culture, culture supernatant is removed from the cells and centrifuged at approximately 1000 g for 10 min at 20 °C. The supernatant is removed and used as the test article inoculum. For test cells which grow as a suspension culture, the cells are centrifuged at approximately 1000 g for 10 min at 20 °C. The supernatant is removed and used as the test article inoculum. It may in some cases be advantageous to prepare an inoculum consisting of supernatant plus cells which have been disrupted by freeze-thawing. This sample may be used simultaneously in the BVDV assay or an aliquot may be frozen at −70 °C and stored for use in that assay at a later date.

3. For each indicator cell type, the medium is aspirated from subconfluent cultures and the cell sheets are washed twice with 2 ml HBSS without phenol red. TMIC are not washed. The final washing is aspirated from the wells and 1 ml of supernatant from the test article cells is inoculated onto two plates (a total of 12 wells). TMIC are simply refed with 2 ml/well complete growth medium. Appropriate positive and negative controls are similarly inoculated onto duplicate plates of indicator cells. Plates are incubated at 37 °C for 60 min.

4. The inoculum is aspirated from the wells and cells refed with 2 ml/well of the appropriate complete growth medium. Plates are incubated and observed each working day for CPE for 14 days.

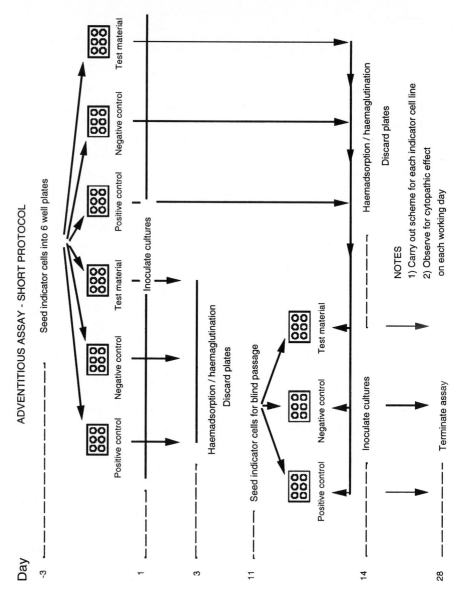

Fig. 27.2. Adventitious assay: a short protocol

5. On day 3, hemagglutination and hemadsorption assays are carried out on one plate inoculated with test material, one plate inoculated with positive control and one plate inoculated with negative control. In addition, one of the duplicate plates of the TMIC is also tested for hemagglutination and hemadsorption.

6. For each indicator cell type: medium from each well of one plate is pooled into a sterile centrifuge tube and spun at approximately 1000 g for 10 min at 20 °C. The clarified supernatant is used for the hemagglutination assay using 1 day old chick, guinea pig and human type O blood. Duplicate 96-well plates, containing 50 μl hemagglutination buffer, 50 μl culture supernatant and 50 μl of the appropriate blood are prepared. One plate is placed in the refrigerator at 4 °C and one incubated at 37 °C for 60 min. At the end of each incubation, the plates are observed for hemagglutination and the results recorded.

7. For each indicator cell type: the cell sheets are washed with HBSS without phenol red and, after aspirating the final wash, 1 day old chick, guinea pig and human type O blood are separately added to duplicate wells. Firstly, the plate is placed in the refrigerator for 30 min and then incubated at 37 °C for 30 min. At the end of each incubation period the cell sheets are observed for hemadsorption and the results recorded.

8. If the TMIC grow as monolayer cultures, the hemadsorption and hemagglutination assays are performed as above. If, however, the TMIC cells grow as a suspension culture, the cells from duplicate wells are transferred to three separate sterile centrifuge tubes and spun at approximately 1000 g for 10 min at 20 °C to pellet the cells. The supernatants are pooled into a fresh sterile centrifuge tube and spun at approximately 1000 g for 10 min. The supernatant is then used in the hemagglutination assay as described above. The cell pellets are resuspended in 1–3 ml of HBSS without phenol red and centrifuged at approximately 1000 g for 10 min at 20 °C. The wash in HBSS without phenol red is repeated twice. Cell pellets are separately resuspended in 1 day old chick, guinea pig and human type O blood solutions and the suspension returned to the original wells. The plates are then treated as described above for hemadsorption.

9. If the test article inoculated indicator cells and TMIC give a negative result for hemadsorption or hemagglutination, the corresponding wells in the duplicate plate are refed with the appropriate growth

medium and the hemagglutination and hemadsorption assays repeated at day 14.

10. If the test article inoculated indicator cells and TMIC give a negative result for all endpoints, including CPE, on day 14; supernatant samples from the test article inoculated indicator cells and negative control indicator cells are used to inoculate fresh indicator cells of the corresponding type. Fresh positive controls are also set up. These cells are observed each working day for CPE for a further 14 days.

Acceptance criteria: The assay is considered acceptable if the positive control virus causes CPE in all the indicator cells used; the positive control virus causes hemagglutination and/or hemadsorption with at least one type of erythrocyte at one or more temperatures at one or more time points for at least one indicator cell line; the negative control inoculated indicator cells show normal morphology and the negative control inoculated indicator cells show no evidence of hemagglutination and hemadsorption.

Evaluation criteria: The assay is considered positive if any indicator cells inoculated with test material show a CPE or hemagglutination or hemadsorption with any of the blood types assayed.

immunoassay for noncytopathic BVDV

This assay may be performed at the same time as the assay in 6-well plates using the same inoculum. Alternatively, a portion of the inoculum for the main assay may be frozen at $-70\,°C$ or below after preparation and used in the BVDV assay at a later date.

1. MDBK (BVDV free) cells are seeded by adding 0.1 ml of a 1.0×10^5 cells/ml cell suspension to the appropriate number of 96-well plate wells. Cells are then incubated.

2. When cells are 70–90 % confluent (usually 3 days after seeding), the medium is aspirated from each well and the cells washed once with HBSS without phenol red.

3. After aspirating the final wash, replicate wells are inoculated separately with either positive control, negative control or test sample supernatant.

4. Plates are incubated for 60 min and the inoculum is then aspirated and the cells refed with complete growth medium.

5. Three to 5 days later, the medium is aspirated from each well and the cells washed with PBS. The final wash is aspirated and cells fixed by

the addition of 20 % (v/v) acetone in PBS. Plates are then drained and air dried.

6. The plates may then be stored at -20 °C in a sealed bag until required for staining.

7. Full details of the immunostaining procedure is provided in the IPEX BVDV staining kit.

Acceptance criteria: The assay is considered acceptable if MDBK cells inoculated with positive control show cytoplasmic staining after the addition of anti-BVDV antibodies; MDBK cells inoculated with negative control exhibit no cytoplasmic staining after the addition of anti-BVDV antibodies.

Evaluation criteria: The assay is considered as positive if MDBK cells inoculated with test article show cytoplasmic staining after the addition of anti-BVDV antibodies.

Troubleshooting

- The cell monolayers should be approximately 50 % confluent when inoculated with test material to increase detection as many types of potential contaminants as possible.

- It is important not to overinfect the positive control cultures with cytopathic positive control virus, otherwise all the cells will have detached by the second hemagglutination and hemadsorption.

27.5
Process Validation: Viral Quantitation with Plaque Assay

The assays used in cell bank testing only evaluate relatively small aliquots of the test material for the presence of possible contaminants. Even if the results of these tests are negative, there remains the possibility that the cell bank contains virus not detected due to the statistical limitations of sampling. It is also possible that viruses may not be detected by the particular indicator cell lines utilized. For these and other reasons testing of the cell banks can never demonstrate absolute freedom from contamination. To give a high level of confidence in the safety of the final product, it is therefore necessary to demonstrate that any undetected viral contamination will be cleared or inactivated during purification of the product.

Selected areas of the manufacturers purification process (usually a scaled down version) are deliberately spiked with virus of high titer. Samples are taken during the purification procedure and assayed for the presence of infectious virus using quantitative virus assays.

Choice of the actual viruses used is based on experience and a knowledge of the relevant regulatory guidelines. It is not practical to give details of all possible assays here and so a selection of representative studies have been presented.

Prior to starting quantitative assay, it is most important to determine the cytotoxicity and assay interference which may be caused by the test solution. There will be column fractions, virucidal mixes, etc., resulting from the purification process.

cytotoxicity

A cytotoxicity assay is performed by adding serial dilutions of the test solution to the appropriate indicator cell monolayers (i.e., those cells intended for use in the adventitious assay). The cells are then cultured using medium and conditions appropriate to the particular cell line. The cells are observed daily for CPE and the maximum concentration of the test solution which does not cause toxic effect noted.

viral interference

A concentration of test solution which is noncytotoxic may still interfere with viral infectivity. Interference may take place because test solutions alter cell surface receptors, or because they change the electrostatic charges on the surface of cell and virus. To investigate this, serial dilutions of the test solution are prepared and known amounts of virus added to replicate portions of each dilution of test solution. The numbers of plaques which are formed at different concentrations of the test solution are then compared, and the highest concentration of test solution which does not cause cytotoxicity or assay interference is then selected for use in the assay.

▦ Materials

- MDBK (BVDV free) cells (ATCC CCL 22)
- BT cells (ATCC CRL 1390)
- CV-1 cells (ATCC CCL 70)
- Vero cells (ATCC CCL 81)
- MRC-5 cells (ATCC CCL 171)
- Infectious bovine rhinotracheitis virus (IBRV), LA strain (ATCC VR 188), plaqued on MDBK (BVDV free) cells

- Bovine viral diarrhoea virus (BVDV), NADL strain (ATCC VR 534) plaqued on MDBK (BVDV free) cells and plaqued on BT cells
- Parainfluenza 3 (PI3) virus, SF-4 strain (ATCC VR 281) plaqued on CV-1 cells
- Reovirus type 3 (REO3), Abney strain (ATCC VR 232), plaqued on CV-1 cells
- Simian virus 40 (SV40), Pa-57 strain (ATCC VR 239), plaqued on CV-1 cells
- Poliovirus, Chat strain (ATCC VR 192), plaqued on Vero cells
- Herpes simplex type 1 virus (HSV-1), strain HF (ATCC VR 260), plaqued on Vero cells
- Cytomegalovirus (CMV), AD-169 strain (ATCC VR 538), plaqued on MRC-5 cells
- MEM complete growth medium for MDBK (BVDV-free) cells: MEM containing 10 % (v/v) γ-irradiated fetal calf serum; 1x nonessential amino acids; 2.5 µg/ml fungizone; 50 µg/ml gentamycin (Life Technologies Ltd.)
- DMEM complete growth medium for BT cells: DMEM containing 10 % (v/v) horse serum; 1x nonessential amino acids; 2.5 µg/ml fungizone; 50 µg/ml gentamycin (Life Technologies Ltd.)
- MEM complete growth medium for CV-1, Vero and MRC-5 cells: MEM containing 10 % (v/v) fetal calf serum; 1x nonessential amino acids; 2.5 µg/ml fungizone; 50 µg/ml gentamycin (Life Technologies Ltd.)
- MEM containing 5 % (v/v) tryptose phosphate broth (virus diluent) (Life Technologies Ltd.)
- DEAE dextran: 5 mg/ml prepared in HBSS without phenol red (Sigma)
- HBSS without phenol red (Life Technologies Ltd.)
- Basal medium Eagles (2×) + (4×) L-glutamine; 4 % (v/v) serum + Earles salt (Life Technologies Ltd.)
- Agarose solution: 1.8 % (w/v) prepared in purified water (ICN Biomedicals Ltd.)
- Neutral red solution: 3.3 g/l (Sigma)

The required volume of overlay medium is prepared by adding 1.8 % (w/v) agarose in purified water to an equal volume of basal medium Eagles (2×) containing L-glutamine (4×), Earles salts +4 % (v/v) serum (2× supplemented basal medium). The final concentrations of the overlay constituents are 0.9 % (w/v) agarose and 1× supplemented basal medium.

Note: Antibiotics may be added if required to all buffers to a final concentration of 50 µg/ml gentamycin and 2.5 µg/ml Fungizone (Life Technologies Ltd.)

- Test material: Manufacturers process samples
- Positive control: The viruses are listed at the beginning of this section. Serial tenfold dilutions are added to the cell sheet in order to obtain an accurate viral titer
- Negative control: Medium used to dilute the positive control, i.e., minimum essential medium +5 % tryptose phosphate broth

▦ Procedure

Quantitative viral plaque assays are performed by inoculating confluent monolayers of indicator cells with test material and overlaying with a basal medium/agarose mixture. Because the cells are covered by a solid overlay not liquid media, areas of viral infection develop in a radial direction and form discrete areas of infection. If the test material is assayed after serial dilution, one of the dilutions will contain a countable number of plaques, e.g., 10–100 plaques, from which the viral titer of the sample can be determined.

Prior to performing an assay it is advisable to have performed cytotoxicity and assay interference assays.

1. Appropriately labeled 60 mm Petri dishes are seeded with indicator cells at a concentration of $4.0–6.0 \times 10^5$ cells per dish in 4 ml of the appropriate complete growth medium. In general, fast growing cells (e.g., MDBK [BVDV-free] cells) are seeded at the lower end of the range and slower growing cells (e.g., CV-1 cells) at the upper end of the range. Dishes are placed on appropriately labeled trays which are placed at 37 °C in a humidified, 5 % CO_2 incubator.

2. When cells are approximately 80 %–100 % confluent, the medium is aspirated from each dish. For poliovirus and SV-40 assays, the cells are first washed with HBSS without phenol red and then treated with 4 ml of a 50 mg/ml solution of DEAE dextran in HBSS without phenol red. The cells are incubated for 90 min, after which the DEAE dextran is aspirated.

3. The cells used in all assays are washed twice with HBSS without phenol red and the final wash aspirated prior to inoculation.

4. Triplicate dishes are separately inoculated with 1.0 ml of the appropriate dilution of test sample or control. Dilutions are made in virus diluent. All dishes are then incubated for a minimum of 1 h and a maximum of 2 h.

5. After incubation, the inoculum is aspirated and the cells in each dish overlaid with 4 ml of overlay medium. After the overlay has been applied, the dishes are refrigerated until the overlay medium has hardened (typically 5–10 min).

6. The dishes are then incubated. Additional overlay medium may be applied to the existing overlay as described above if the cells need to be incubated for in excess of 2 weeks for the plaques to develop.

7. When plaques are fully developed in the positive controls, cultures may be stained with neutral red to aid plaque visualization. Neutral red is added to overlay medium to achieve a final neutral red concentration not exceeding 0.0005 % (w/v) per 4.0 ml of overlay medium. This overlay medium is added to the dishes which are then refrigerated until the agarose has hardened. Dishes are then incubated. Typical times for each type of plaque to appear are shown in Table 27.3.

8. Plaques are counted using an inverted microscope.

9. The titer of the sample or control is calculated by multiplying the mean plaque count per dish by the dilution performed on the sample or control. The dilution giving a count of between 10 and 100 plaques is used in the calculation.

Table 27.3. Viral growth characteristics

Indicator cells	Virus	Time for plaques to appear postinoculation (days)
MDBK (BVDV free)	IBRV	3–5
MDBK (BVDV free) or BT	BVDV	3–5
CV-1	PI3	10–12
	REO3	6–8
	SV-40	12–16
Vero	Polio	3–5
	CMV	12–16
MRC-5	HSV-1	3–5

Assay acceptance criteria: The assay is considered valid if the titer of the positive control used in the assay is less than 1 log different from the expected titer of that virus batch and if no plaques are observed in the assay negative control.

Assay evaluation criteria: The test sample is considered positive and the titer of the sample calculated if 1 or more plaques are observed in the test sample.

Results

The results are shown in Table 27.4.

Table 27.4. Results of the plaque assay

Sample	Incubation time (min)	Assay dilution	Spiking dilution	Plaques/ dish[a]	Mean plaques/ dish	Extrapolated titer (PFU/ml)	RI (log$_{10}$)[b]
Assay negative control	0	Neat	NA	0,0,0	0	0	NA
Assay positive control	0	10^{-3}	1:50	TNTC	TNTC	–	NA
		10^{-4}		TNTC	TNTC	–	
		10^{-5}		26,29,29	28.0	1.4×10^8	
		10^{-6}		3,3,2	NC	–	
1	0	Neat	1:50	TNTC	TNTC	–	2.25
		10^{-1}		TNTC	TNTC	–	
		10^{-2}		TNTC	TNTC	–	
		10^{-3}		15,13,19	15.7	7.9×10^5	
		10^{-4}		1,1,1	NC	–	
		10^{-5}		0,0,0	NC	–	
		10^{-6}		0,0,0	NC	–	
2	15	Neat	1:50	TNTC	TNTC	–	2.30
		10^{-1}		TNTC	TNTC	–	
		10^{-2}		TNTC	TNTC	–	
		10^{-3}		13,17,1 2	14.0	7.0×10^5	
		10^{-4}		2,1,0	NC	–	
		10^{-5}		0,0,0	NC	–	
		10^{-6}		0,0,0	NC	–	

Table 27.4. (Continue)

Sample	Incubation time (min)	Assay dilution	Spiking dilution	Plaques/ dish[a]	Mean plaques/ dish	Extrapol- ated titer (PFU/ml)	RI ($\log_{10}$)[b]
3	30	Neat	1:50	TNTC	TNTC	–	3.87
		10^{-1}		33,41,37	37.0	1.9×10^4	
		10^{-2}		4,3,2	NC	–	
		10^{-3}		0,0,0	NC	–	
		10^{-4}		0,0,0	NC	–	
		10^{-5}		0,0,0	NC	–	
		10^{-6}		0,0,0	NC	–	
4	60	Neat	1:50	TNTC	TNTC	–	3.87
		10^{-1}		35,38,40	37.7	1.9×10^4	
		10^{-2}		5,3,3	NC	–	
		10^{-3}		0,0,0	NC	–	
		10^{-4}		0,0,0	NC	–	
		10^{-5}		0,0,0	NC	–	
		10^{-6}		0,0,0	NC	–	
5	180	Neat	1:50	TNTC	TNTC	–	4.27
		10^{-1}		14,12,1 9	15.0	7.5×10^3	
		10^{-2}		2,2,1	NC	–	
		10^{-3}		0,0,0	NC	–	
		10^{-4}		0,0,0	NC	–	
		10^{-5}		0,0,0	NC	–	
		10^{-6}		0,0,0	NC	–	
6	300	Neat	1:50	TNTC	TNTC	–	4.26
		10^{-1}		16,17,13	15.3	7.7×10^3	
		10^{-2}		3,1,1	NC	–	
		10^{-3}		0,0,0	NC	–	
		10^{-4}		0,0,0	NC	–	
		10^{-5}		0,0,0	NC	–	
		10^{-6}		0,0,0	NC	–	

NC, not calculated; TNTC, too numerous to count (on triplicate dishes); NA, not applicable; RI, reduction index; PFU, plaque forming units.

[a] Using a 1 ml inoculum volume.

[b] Calculated using the titer of assay positive control.

▨ Troubleshooting

- The dishes must be well drained before the agarose overlay is added to minimize the possibility of ill defined plaques. The cell sheet, however, must not be allowed to dry out.

- The agarose overlay solution must be cooled to approximately 37 °C before adding it to the cell sheet. The overlay medium should not be prepared too far in advance of its use otherwise it will solidify.

- Avoid introducing air bubbles into the overlay medium as they may obscure plaques below them.

- The dishes must not be left in the refrigerator for longer than is necessary to set the agarose overlay. However, the overlay must be fully set before transfer into the incubator. The agarose medium usually takes 5–10 min to solidify at 4 °C.

27.6
Process Validation with TCID$_{50}$ Assay

The tissue culture infectivity dose 50 (TCID$_{50}$) assay is a relatively simple assay used to calculate the dilution of virus required to infect 50 % of inoculated cultures. These types of assays require that the test material is assayed over a wide range of dilutions using several replicates per dilution to establish a titer for that sample.

Numerous indicator cell/virus combinations are possible. Two examples are provided in this section.

▨ Materials

- BT cells (ATCC CRL 1390)
- LLC-PK1 cells (ATCC CL 101)
- Bovine viral diarrhea virus (BVDV), NADL strain (ATCC VR 534) assayed using BT cells
- Porcine parvovirus (PPV), CVL 1234/6–1 strain (Central Veterinary Laboratory, UK), assayed using LLC-PK1 cells
- DMEM complete growth medium for BT cells: DMEM containing 10 % (v/v) horse serum, 1× nonessential amino acids; 2.5 µg/ml fungizone; 50 µg/ml gentamycin (Life Technologies Ltd.)

- M199 complete growth medium for LLC-PK1 cells: Medium 199 with Earles salts and sodium bicarbonate but without L-glutamine (Sigma) containing 5 % (v/v) fetal calf serum; 1× nonessential amino acids; 2.5 µg/ml Fungizone; 50 mg/ml gentamycin (Life Technologies Ltd.)
- MEM containing 5 % (v/v) tryptose phosphate broth (virus diluent) (Life Technologies Ltd.)
- Test material: Manufacturers process samples
- Positive control: The viruses are listed at the beginning of this section. Appropriate serial dilutions are added to the cells in order to titrate out the virus present and obtain an accurate viral titer
- Negative control: Medium used to dilute the positive control, i.e., minimum essential medium +5 % tryptose phosphate broth

Procedure

TCID$_{50}$ Assay

1. The appropriate indicator cells are seeded into the required number of wells of suitably labeled assay plates (either 500 µl/well for 24-well plate wells or 50 µl/well for 96-well plate wells) at a concentration of 1×10^5 cells/ml in the appropriate growth medium. It is usual to use an even number of wells, i.e., 4 or 8 for each dilution. The greater the number of wells employed the greater the statistical power.

2. An aliquot (500 µl/well for 24-well plate wells or 50 µl/well for 96 well plate well) of undiluted or appropriately diluted test material or control is added to the appropriate number of wells.

3. All cultures are incubated at approximately 37 °C in a humidified, 5 % CO$_2$ incubator.

4. All cultures are observed for CPE each working day after inoculation and when there is no further increase in the number of wells exhibiting CPE the assay is scored using an inverted microscope.

Acceptance criteria: The assay is considered valid if the titer of the positive control used in the assay is less than 1 log different from the expected titer of that virus batch and if no CPE is observed in any well of the negative control.

Evaluation criteria: The TCID$_{50}$ titer of the each sample is calculated using a simplified version of the Spearman-Kärber formula:

$$m = x_k + d/2 - d\Sigma\tilde{o}p_i.$$

The standard deviation of m (s_m) can be calculated by the following formula:

$$s_m{}^2 = d^2 \Sigma \tilde{o}(p_i[1-p_i]/[n_i-1]),$$

where m is the logarithm of the titer in the test volume (usually then converted to log titer/ml); x_k is the logarithm of the smallest dose which induces the infection of all cultures; d is the logarithm of the dilution factor; p_i is the number of positive wells per total wells for each dilution; s_m is the standard deviation; n_i is the number of cultures tested per dilution.

example of positive control

Worked Example of the Extrapolated $\log_{10}$ TCID$_{50}$ Unit/ml Titer of the Positive Control

The smallest dose which infected all cell cultures was the 10^{-4} dilution, x_k $=-4.0$. As the dilution factor was 10, $d=1$ and $d/2=0.5$. The number of positive wells/total wells for the unaffected dilution adjacent to the first effect dilution $p_1=8/8=1.0$, and $p_2=7/8=0.875$; so $\Sigma \tilde{o} p_i=1.875$ (if other dilutions had shown effects these would have been included in the summation also).

Applying the simplified version of the Spearman-Kärber formula:

$$m = -4 + 0.5 - (1 \times 1.875)$$

$$m = -5.375$$

The TCID$_{50}$ is $10^{5.375}$ or 5.375 $\log_{10}$. This value is per 50 ml; the extrapolated $\log_{10}$ TCID$_{50}$ titer/ml is obtained by adding 1.301 ($\log_{10}$ 20, as the inoculum volume was 1/20th of an ml) and by adding 1.0 ($\log_{10}$ 10, as there was a tenfold virus spiking dilution). When all these values have been added in, the extrapolated $\log_{10}$ TCID$_{50}$ unit/ml titer is 7.68 (to two decimal places).

The standard deviation of the titer is calculated as follows:

As $p_1=1.0$, $p_2=0.875$ and n in all dilutions is 8, then

$$p_1(1-p_1)/(n_1-1) = 1.0(1-1)/(8-1) = 0$$

$$p_2(1-p_2)/(n_2-1) = 0.875(1-0.875)/(8-1) = 0.015625$$

$$s_m{}^2 = 1^2 \times 0 + 0.015625$$

$$s_m{}^2 = 0.015625$$

$$s_m = 0.13 \text{ (to two decimal places)}$$

The extrapolated assay positive control titer is 7.68 ± 0.13 $\log_{10}$ $TCID_{50}$ unit/ml.

Results

Typical results are shown in Table 27.5.

Table 27.5. Results of the $TCID_{50}$ assay

Sample	Incubation time (min)	Initial dilution to prevent cytotoxicity	Spiking dilution	Assay dilution[a]							Extrapolated titer[b]	RI ($\log_{10}$)[c]
				N	−1	−2	−3	−4	−5	−6		
Assay negative control	NA	NA	NA	0	NA						0	NA
Assay positive control	0	NA	1:10	NA	8	8	8	8	7	0	7.68	NA
1	0	1:4	1:10	8	8	8	8	8	6	0	8.15	−0.47
2	15	1:4	1:10	8	8	8	8	8	4	0	7.90	−0.22
3	30	1:4	1:10	8	8	8	8	8	5	0	8.03	−0.35
4	60	1:4	1:10	8	8	8	8	8	3	0	7.78	−0.10
5	180	1:4	1:10	8	8	8	8	8	2	0	7.65	0.03
6	300	1:4	1:10	8	8	8	8	8	2	0	7.65	0.03

CPE, cytopathic effect; RI, reduction index.
[a] Number of wells showing CPE in 8 wells of a 96 well plate, using 50 ml inoculum per well.
[b] $\log_{10}$ $TCID_{50}$ units/ml.
[c] Calculated using the titer of assay positive control.

Troubleshooting

- Although only two examples have been given, the method described lends itself to a wide range of viruses and indicator cells. For parvoviruses, however, a second or passage inoculation is required once the cells from the initial inoculation have become confluent (typically 3–4 days after the initial inoculation). Appropriate volumes of supernatant from the primary inoculation are transferred (passaged) onto freshly seeded cells. These cells are then incubated and scored for CPE. This procedure is required because parvovirus induced CPE is most pronounced in cells which are actively dividing.

- As the medium in the wells on the outside edges of 96-well plates tends to dry out more quickly it may be necessary to add more medium to the wells to prevent them drying out.

- Most CPE is easy to visualize; the cells, however, may be fixed and stained, e.g., 0.13 % (w/v) crystal violet/11.2 % (v/v) formaldehyde solution, to aid visualization.

References

Bassin RH, Ruscetti S, Ali I, Haapala DK, Reins A (1982) Normal DBA/2 mouse cells synthesize a glycoprotein which interferes with MCF virus infection. Virology 123:139–151

Haapla DK, Robey WG, Oroszlan SD, Tsai WP (1985) Isolation from cats of an endogenous type C virus with a novel envelope glycoprotein. J Virol 53:827–833

Kohler G, Milstein C (1975) Continuous cultures of fused cells secreting antibody of predefined specificity. Nature 256:495–497

Lueders KK (1991) Genomic organization and expression of endogenous retrovirus-like elements in cultured rodent cells. Biologicals 19:1–7

Morse HC III, Hartley JW (1986) Murine leukemia viruses. In: Bhatt P, Jacoby R, Morse H III, New A (eds) Viral and mycoplasma infections of laboratory rodents. Academic, Orlando FL, pp 349–388

Peebles PT (1975) An in vitro focus-induction assay for xenotropic murine leukemia virus, feline leukemia virus C and the feline-primate viruses. Virology 67:288–291

Poiley JA (1990) Methods for the detection of adventitious viruses in cell culture used in the production of biotechnology products. In: AS Lubiniecki (ed) Large-scale mammalian cell culture technology. Marcel Dekker, New York

Rowe WP, Pugh WE, Hartley JW (1970) Plaque assay technique for murine leukaemia viruses. Virology 42:1136–1139

Schmidt NJ (1979) Cell culture techniques for diagnostic virology. In: Lennette EH, Schmidt NJ (eds) Diagnostic procedures for viral, rickettsial and chlamydial infections 5th (edn). American Public Health Association, Washington, DC, pp 51–100

Stoye JP, Coffin JM (1987) The four classes of endogenous murine leukaemia virus: structural relationships and potential for recombination J Virol 61:2659–2669

Toyoshima K, Vogt PK (1969) Enhancement and inhibition of avian sarcoma viruses by polycations and polyanions. Virology 38:414–426

▓ Suppliers

Media and Supplements

Life Technologies Ltd.
3 Fountain Drive
Inchinnan Business Park
Paisley PA4 9RF
Scotland

Sigma-Aldrich Company Limited

Fancy Road
Poole
Dorset BH17 7NH
UK

Cells and Viruses

American Type Culture Collection
123012 Parklawn Drive
Rockville
Maryland 20852
USA

Central Veterinary Laboratory
New Haw
Addlestone
Surrey KT15 3NB
UK

Tissue Culture Plasticware

Falcon Cell Culture Products
Becton Dickinson UK Limited
Between Towns Road
Cowley
Oxford OX4 3LY
UK

UV Detector (Model: J 225)

UVP Inc.
5100 Walnut Grove Avenue
P.O. Box 1501
San Gabriel
California 91778
USA

Mammaliam Erythrocytes

TCS Biologicals Ltd
Botolph Claydon
Buckingham
MK18 2LR
UK

Regional blood transfusion centers

Oxoid Hemagglutination Buffer

Unipath Ltd
Wade Road
Basingstoke
Hants RG24 OPW
UK

IPEX-BVDV Kit

Central Veterinary Laboratory
New Haw
Addlestone
Surrey KT15 3NB
UK

Low Gelling Temperature Agarose

ICN Biomedicals Ltd.
Eagle House
Perigrine Business Park
Gomm Road
High Wycombe
Bucks HP13 5BR
UK

Biochemical Reagents

Sigma-Aldrich Company Limited
Fancy Road
Poole
Dorset BH17 7NH
UK

Rathburn Chemicals Ltd.
Walkerburn
Scotland

Chapter 28

Cytokine Cell Culture Assays

ALICE REDMOND*, JACQUELINE QUINN, AND SUSAN MOLLOY

Introduction

Cytokines are defined as secreted regulatory proteins that control the survival, growth, differentiation and effector function of tissues. Cytokines encompass those families of regulators known as growth factors, colony-stimulating factors, interleukins (ILs), lymphokines, monokines and interferons (IFNs). Measuring cytokines is an important facet of understanding the complex interactions of the immune response mediated by these multifunctional, peptide, T cell products. For example cytokines regulate the activation, growth and differentiation of many cell types and have been implicated in diseases such as arthritis, allergy and leukemia (Aria 1990). In some cases, cytokines are involved in the pathology of the disease, while in other cases they may be used clinically as drugs, such as IFN, granulocyte/macrophage colony-stimulating factor (GM-CSF) and erythropoietin. There is a broad range of both cell culture assays available for cytokines (Whiteside 1994; Aggarwai 1991) and commercial ELISA kits for detecting cytokines (Ledur 1995; Mire-Sluis 1995).

In this chapter we will summarize the assay systems (assay protocols, sources of reagents and cell lines) that have been used effectively for the analysis of cytokines (Tables 28.1, 28.2). Many of these systems have been reported in the literature (Table 28.3).

* *Correspondence to* Alice Redmond, Novartis Ringaskiddy Ltd., Ringaskiddy, County Cork, Ireland; phone +353–021–378601; fax+353–021–378790; e-mail redmond@pharma.Novartis.com

Table 28.1. Cell lines used in the analysis of cytokines

Cytokine	Cell line assay	Cytokine	Cell line assay
IL-1	T lymphoma cell line EL-4 T Fibroblasts WISH	GM-CSF	Bone marrow leukemia cells
IL-2	T lymphocyte line CTLL Jurkat EL-4 MLA144	LIF	M1 myeloid leukemia cell line
IL-3	FDCP-1/FDCP-2 32DC1 AC2 Ea.3.123 AML193 Mo7e	TNF	L-929 cells PK(15) WEHI 164 cells L-M fibroblasts
IL-4	PBMNC CCL-185 CDC-35 T- cells MO7e	Lymphotoxin	L-929 cells ML-19 cells
IL-5	TF-1	IFNγ	L cells
IL-6	SKW6-CL4 B13.29 MH60.BSF2	FGF	Vascular endothelial cells
IL-7	IxN/2b DW24	HGF	CCL-64
IL-8	Neutrophils	GCSF	T cell depleted normal bone marrow
IL-10	D36	Oncostatin M	Melanoma cell line A-375
IL-11	B9–11	TGF-β	NRK cells CCL-64 TF-1
IL-15	CTLL-2	IFNα	WISH cells L cells
IL-16	T cells	IFNβ	A549
CSF	Mouse bone marrow		

IL, interleukin; GM-CSF, granulocyte/macrophage colony-stimulating factor; LIF, leukemia inhibitory factor; TNF, tumor necrosis factor; IFN, interferon; HGF, hepatocyte growth factor; GCSF, granulocyte colony-stimulating factor; FGF, fibroblast growth factor; TGF, transforming growth factor; CSF, colony-stimulating factor.

Table 28.2. Summary table of recognized producers of a selection of cytokines

Cytokine	Cell line producer	Cytokine	Cell line producer
IL-1	Monocyte and macrophage cell lines NK cells B cell lines Neutrophils	GCSF	Bladder cell line 5637 SCC CHU2
IL-2	T cells NK cells LAK cells	GM-CSF	Activated T cells Macrophages Fibroblasts Endothelial cells Mo cell line
IL-3	Activated T lymphocytes WEHI3b	LIF	Alloreactive T lymphocytes Bladder cell line 5637 Melanoma SEKI Colo-16 TRHP-1
IL-4	T cells	TNF	A431 cells KB cells Macrophages T lymphocytes
IL-6	TCL-NA1 CeSS U937 P388D1 MG63 T24 A549 SK-MG-4 U373	Lymphotoxin	T lymphocytes Natural killer cells Myeloma cells
IL-7	SK-HEP-1 IMR-90	PDGF	Fibroblasts A172 T24 GepG-2
IL-8	T cells	FGF	Melanoma Glioma Glioblastoma
IL-16	CD8[+] T cells	IL-12	Human Langerhans cells
HGF	Fibroblasts Smooth muscle Leukocytes	IL-10	T cells

Table 28.2. (Continue)

Cytokine	Cell line producer	Cytokine	Cell line producer
VEGF	Renal cell carcinoma	VPF	U937 GS-9L glioma cell line
TGF-β	Osteoblasts Fibroblasts Lymphocytes	EGF	Brain, kidney Adrenal medulla Salivary gland
IFNα	Buffycoat leukocytes Namalwa cells B lymphocytes	Oncostatin M	U-937 T lymphocytes
IFNβ	Fibroblasts Human stem cells	Erythropoietin	Adult kidney
IFNγ	Fibroblasts Natural killer cells	TGFα	Placenta Platelets

IL, interleukin; GM-CSF, granulocyte/macrophage colony-stimulating factor; LIF, TNF, tumor necrosis factor; IFN, interferon; HGF, hepatocyte growth factor; GCSF, granulocyte colony-stimulating factor; FGF, fibroblast growth factor; TGF, transforming growth factor; EGF, epidermal growth factor; PDGF, platelet-derived growth factor.

Table 28.3. Summary of the cell culture assay protocols used in the analysis of cytokines

Assay type	Reference	Cytokine
Virus challenge assay	Peskta (1986)	IFNα, β, γ
Radioisotopes/scintillation counting	Page (1996) Borset (1996) Jeannin (1996) Higuchi (1995) Antar (1995)	IL-1, 2, 3, 4, 7, 10, GM-CSF, TGF-α, HGF, lymphotoxin
MTT	Lu (1994) Corit (1994)	IL-3, 6, 11, TNF
Immunoglobulin production assay/ ELISA	Hang (1994)	IL-6
Boyden chamber/microscopy chemotactic index	Clemens (1987)	IL-8
P-nitrophenyl-β-D-glucuronide Spectrophotometer (405 nm)	Clemens (1987)	IL-8
Soft agar assay	Hill (1983) Courtney (1978)	CSF, G-CSF, GM- CSF, LIF, TGF-β

Table 28.3. (Continue)

Assay type	Reference	Cytokine
Staining with crystal violet, spectrophotometer	Freshney (1997)	TNF
Iodine labeled cytokines	de Jong (1996)	IL-15, 2
Hybridoma proliferation assay	Marshall et al. 1996	IL-6
Methotrexate recombinant indicator cell line assay	Lieonart (1996)	IFN-α
Fluorescent dye assay e.g., ethidium homodimer 1	Levesque 1996	TNF-α
Fluorescent dye e.g., alamar blue	Shanan (1994)	TNF
^{51}Chromium release cytotoxicity	Ruff (1981)	TNF
Neutralizing antibodies	Cagliero (1995) Randall (1996) Pauli (1994)	TGF-β, TGF-α
Chemotaxis assay	Taub (1996)	IL-8
IGF transfection assay	James (1996)	IGF
Fluoroscent immunostaining	Watari(1996)	IL-1β
2'-5' Oligoadenylate synthetase assay	Hamiliton (1996)	IFNα and IFNβ

IL, interleukin; GM-CSF, granulocyte/macrophage colony-stimulating factor; HGF, hepatocyte growth factor; TNF, tumor necrosis factor; CSF, colony-stimulating factor; G-CSF, granulocyte colony-stimulating factor; TGF, transforming growth factor; IFN, interferon.

28.2
Determination of Interferon Activity

There are many and varied assay systems for IFNs, most of which focus on their antiviral properties. While this protocol will focus on this type of assay, another very effective system involves measuring the inhibition of viral RNA synthesis. This is a qualitative assay which provides a direct measure of the inhibition of viral protein synthesis by radiolabeling proteins being synthesised in virus infected cells. IFN-treated cells are then analyzed by polyacrylamide gel electrophoresis and the use of radiolabeled monoclonal antibodies for use in radioimmunoassays.

This assay system uses L cells as indicator cells and vesicular stomatitis virus (VSV) as the challenge virus and is based on the inhibition by

IFN of the cytopathic effect (CPE) caused by VSV infection. The procedure is as indicated below. The critical points pertaining to this type of assay are as follows:

The cell density of indicator cells should be chosen so as to give good (CPE) but low IFN sensitivity. Low cell density will give CPE but low IFN sensitivity, whereas high cell density will give IFN sensitivity but tends to make the CPE difficult to observe. In the assay system described here, 3×10^4 cells/well are found to be appropriate.

The dose challenge virus should be sufficient so that complete CPE is produced with certainty in control cultures not treated with IFN; too low virus doses will cause low IFN sensitivities. The normal dose is 100 $TCID_{50}$ (50 % tissue culture infectious dose) of VSV as the challenge. The IFN sensitivity of the assay system is that one experimental unit/ml (protection of 50 % of the cell culture from viral CPE) equals 1–5 IU/ml.

The incubation time of cells with the IFN samples before virus challenge may be shortened to 8 h; this is reported not to have any effect on the sensitivity of the assay.

Procedure

IFN assay

1. In each well of a 96-well microtiter plate (e.g., Costar), inoculate 3×10^4 cells in 100 μl of growth medium (GM): MEM-G containing 60 μg/ml kanamycin (Gibco) supplemented with 1 g/l glucose (final 2 g/l); 5–10 % (v/v) fetal calf serum.

2. After 10–20 h incubation at 37 °C in a 5 % CO_2 incubator, drain the GM by aspiration or shaking. To the individual wells(100 μl/well), immediately add IFN samples serially diluted in MM: MEM-G containing 60 μg/ml kanamycin (Gibco) supplemented with 1 g/l glucose (final 2 g/l); 0.5 % (v/v) fetal calf serum.

3. Incubate the plates at 37 °C for 8–20 h in a 5 % CO_2 incubator.

4. Aspirate the medium by aspiration or shaking as described above.

5. Immediately, add VSV in MM, containing approximately a tissue culture infectivity dose (TCID) of 100 in 100 μl, to each well.

6. At 20–30 h postinfection, access the CPE of VSV in each well either by observation under a microscope, or by staining the cells with crystal violet (Sigma). The extent of the CPE is scored from 0 to 4: 0 means no CPE or less than a tenth of the cell monolayer killed; 1 means a

quarter killed; 2 means about a half destroyed; 3 means about three quarters; and 4 means over nine tenths of the monolayer destroyed.

7. The reciprocal of the dilution of the IFN sample gives 50 % protection of the cell monolayer from CPE (a CPE score of 2)and is taken to be the titer, expressed as experimental U/ml. This should be normalized to IU/ml to correct for variations in IFN sensitivity of the assay system from one assay to another. For this purpose, the international reference mouse IFNα preparation (NIH-G002–904–511) or equivalent must be titrated as the internal standard in each assay in parallel with the test samples. IFN titer corresponding to the endpoint can be obtained graphically from semi-log plots (Fig. 28.1) or by linear regression calculations, using only those points that fall in the linear point of the curve.

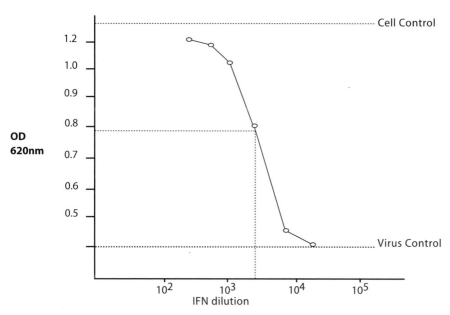

Fig. 28.1. Typical titration curve obtained from a cytopathic effect reduction (CPER) assay. The IFN preparation has a titer of 3400 U/ml

28.2
Radioisotope Assays: Incorporation of [³H]Thymidine for Assay of Interleukin-1

Measurement of the incorporation of radiolabeled metabolites is a frequently used endpoint for assays of intermediate and short-term duration. Measurement of [³H]thymidine incorporation into DNA and [³H]uridine incorporation into RNA are two of the most commonly used methods. In short-term assays, which do not include a recovery period, there are a number of disadvantages that should be taken into account. These include that responses noted may relate to changes in size of the intracellular pool of nucleotides rather than changes in DNA synthesis, as some agents may cause increase uptake of exogenous [³H]thymidine due to a transfer to the "salvage" pathway and continuation of DNA synthesis in the absence of [³H]thymidine incorporation can occur.

A modification of the EL-4 cell line was reported using the subclone EL-4NOB.1. It was derived by the maintenance of EL-4 cell line for 4 weeks at high cell density. El-4NOB.1 is maintained by RPMI1640 supplemented with 5 % FCS and is grown to a maximum cell density of 5×10^5 cells/ml before subculturing. EL-4 NOB.1 produces IL-2 in response to IL-1 alone and there is no requirement for mitogen or calcium ionophore; the assay may be performed in two stages as for the IL-4 cell line or as a 24 h one stage assay. The one stage assay is described here.

■ Procedure

radio-
immunoassay

1. Wash EL-4NOB.1 cells twice by centrifugation at 100 g for 10 min and resuspend at 2×10^8 cells/ml.

2. Add 100 µl EL-4NOB.1 cell suspension to the wells of a 96-well flat bottomed microtiter plate together with 50 µl of appropriate dilutions of the test sample.

3. Wash CTLL-2 cells free of growth medium and suspend them at 8×10^4 cells/ml. Add 50 µl of cells to each well and incubate at 37 °C for 20 h.

4. Pulse the culture with tritiated thymidine (0.5 µCi/10 µl/well; sp. act. 45 Ci/mmol) and incubate for 4 h at 37 °C. Harvest the cultures onto glass fiber filter discs using an automated cell harvester, dry the filter disks and access the thymidine incorporation by liquid scintillation counting. Refer to Fig. 28.2 for a diagrammatic representation of the thymidine assay compared to an MTT assay.

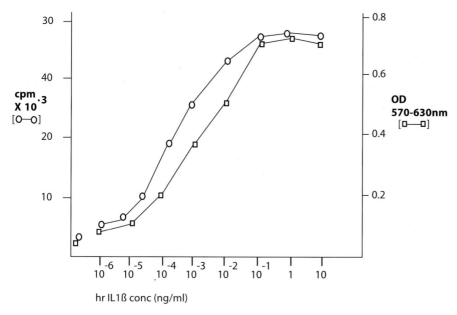

Fig. 28.2. EL4 NOB.1/CTLL-2 assay (two stage) for IL-1. Cell proliferation was measures by[³H]thymidine uptake during the last 4 h of culture or by colorimetric assay with MTT

Results

The quantity of tritiated thymidine used can vary from 0.5 to 1.5 µCi/10 µl/well; sp. act. 45 Ci/mmol.

One unit of IL-1 is defined as that amount of cytokine that produces 50 % of the maximal biological response.

28.3
MTT Assay

The MTT assay involves the use of a tetrazolium salt (3, (4,5-dimethylthiazol-2-yl)-2,5-diphenyl tetrazolium bromide) which is metabolized to a colored formazan salt by the mitochondrial enzyme (succinate dehydrogenase) in living cells. This assay has been reported as suitable for the assaying of a number of cytokines, e.g., IL-3, IL-6 and GMCSF.

■ Procedure

MTT preparation

Add 250 mg MTT to 50 ml PBS and place on a magnetic stirrer until dissolved. Filter the MTT solution through a 0.2 μm Nalgene 150 ml filter unit and store in the base container at 4–8 °C in the dark. Use the solution within 1 week; it may be stored at −20 °C for up to 3 months.

assay

1. Dispense 100 μl culture media (without IL-6) to the wells of a 96-well microtiter plate leaving the first vertical row, 1A – 1H, empty. Dilute a standard batch of IL-6 and samples in solution to a concentration of 1.5 ng/ml with culture media.

2. Dispense the samples, standards and blanks as follows:

3. Pipette 200 μl sample starting solution to wells 1B–1D.

4. Pipette 200 μl standard solutions to wells 1E–1G.

5. Add 200 μl culture media to 1A and 1H (blank wells).

6. Using an 8-channel pipetter, withdraw 100 μl from rows 1A–1H and perform a 1:2 serial dilution across the plate in the previously dispensed media, mixing the new volume with the 100 μl contained in each well.

   Note: This mixing step should involve at least five pipetting actions for each vertical row, excluding the first. This helps to improve reproducibility of the assay.

cell preparation

1. Transfer the contents of one 75 cm^2 flask to a 50 ml centrifuge tube.

2. Fill to the 50 ml mark with PBS. Centrifuge the cells for 10 min at 1000 rpm.

3. Pour off the supernatant into a sterile waste bottle. Add 50 ml PBS to the centrifuge tube and centrifuge for 10 min at 1000 rpm to wash the cells; repeat this step four times to complete washing. The increased washing time can reduce blank levels and improve the reproducibility of the assay.

4. Pour off the supernatant into a sterile waste bottle and resuspend the pellet in 5 ml culture media.

5. Perform a cell count and adjust the cell concentration to 5×10^4 cells/ml.

6. Dispense 100 µl of cell suspension (5×10^4 cells/ml) to all wells. Each well contains 5×10^3 cells/ml. Incubate the plate for 72 ± 1 h at 37 °C and 5 % CO_2.

7. At the end of the incubation period remove the plate to the laminar air flow and add 10 µl MTT stock solution to each well.

8. Return the plate to the incubator for 5 h (± 20 min). Remove the plate after 5 h to the laminar air flow and add 50 µl SDS stock solution to each well. Return the plate to the incubator overnight for a period of 16–24 h. Read the plate on the ELISA plate reader at 570 nm. Plates should be read within 30 min of their removal from the incubator. Prior to reading the plate, wipe the undersurface with a tissue moistened with 70 % isopropyl alcohol (IPK) to remove dust or fingerprints.

9. Calculate the potency of the sample against the sample using a 4-parameter fit curve of standard VSV sample comparing IC_{50} (inhibitory concentration 50 %) values or, alternatively, using a parallel line bioassay. Refer to Fig. 28.2 for a diagrammatic representation of results from the MTT assay.

28.4
Cytotoxicity Assays

There are a wide variety of cytotoxicity assay available that have all been developed as automated cytotoxicity assays in microtiter plates. Interpretation of the significance of results is dependent upon distinguishing between assays which measure cytotoxicity and assays that measure survival. Cytotoxicity assays measure agents induced alterations in metabolic pathways or structural integrity which may or may not be related directly to death, whereas survival assays measure the end results of such metabolic alterations which may be either cell recovery or cell death. Theoretically the only reliable index of survival in proliferating cells is the demonstration of reproductive integrity, as shown by clonogenicity. Metabolic parameters may also be used as a measure of survival when the cell population has been given time for metabolic recovery following drug exposure. Before choosing a cytotoxic assay for the estimation of cytokine activity look carefully at the mode of action and expected response of the cytokine under study. The example given here is specific for TNF and lymphotoxin (LT) but can be applied to a broad range of cytokines.

▦ Procedure

assay summary   TNF/LT samples are incubated for 1–3 days with monolayers of L929 cells, a murine cell line that grows in monolayer. Any cells that are killed detach from the plastic. At the end of the assay, the remaining adherent viable cells are stained with crystal violet. The amount of dye taken up is proportional to the number of residual cells and can be quantitated spectrophotometrically with a 96-well plate reader. A 1-day or 3-day assay can be used.

Setting Up the 3-Day Assay

cell preparation

1. Pour off the culture medium from the flask and wash the cells once with Dulbecco's PBS.

2. Add 2 ml of trypsin EDTA solution (Gibco) diluted 1:4 with PBS.

3. Incubate the cells at 37 °C for the duration required to remove the cells from the base of the flask. Tap the flask gently to obtain a single cell suspension.

4. Add an equal volume of cell culture medium to neutralize the trypsin.

5. Using a hemocytometer count the cells; centrifuge the cells at 500 rpm for 5 min. Resuspend the cells at a concentration of 10^5/ml in culture medium.

6. Using an eight channel micropipette plate 75 µl of the cell suspension into each well. Leave the first row of eight blank. Shake the plate in a figure-eight motion for 30 s to ensure even distribution of the mono-layer. While preparing the sample place the 96-well plate in the incubator at 37 °C 5 % CO_2.

sample preparation

1. Filter sterilize the sample using a 0.22-µm syringe filter. The exact range of dilutions depends upon the nature of the test sample and sensitivity of the target cells. It is suggested to make a series of dilutions in the range of 50 000–1 000 000 for high titer (e.g., TNF serum from rodents with endotoxic shock). Add 500–10 000 for unconcentrated supernatants for tissue culture cells. On each plate include a positive and negative control. The positive control is the cells incubated with an international reference standard or a laboratory reference standard. Add 75 µl to each well.

2. Following addition of the samples shake the plate in a figure-eight motion for 30 s to ensure even distribution of the drug substance. Incubate the plate at 37 °C 5 % CO_2.

Modifications for the 1-Day Assay

1. Plate out the cells at 5×10^5 cells/ml. Shake the plate in a figure-eight motion for 30 s to ensure even distribution of the cell monolayer. It is a requirement of the 1 day assay of the cells that they are attached before addition of the drug substance.

cell line preparation

2. Incubate the cells at 37 °C 5 % CO_2 for 4 h to allow for the establishment of the monolayer. Some researchers have found better assay reproducibility if the cell cultures are plated out a day before they are required at a cell density of 3×10^5 cells cells/ml.

1. Dissolve actinomycin D (Sigma) in PBS at 20 µg/ml (it may be necessary to warm the solution to 37 °C to obtain complete solubilization). The actinomycin D may be stored at −20 °C in 5 ml portions for 3–6 months.

sample preparation

2. Dilute the test samples in culture medium containing 2 µg/ml actinomycin D to give a final culture concentration of 1 µg actinomycin D when added to the cells.

Note: Actinomycin D can be substituted with emetine, cycloheximide, and mitomycin C. The increased susceptibility of L929 cells in the presence of these inhibitors of protein and RNA synthesis is possibly due to their interference with repair mechanisms.

Note: Perform the following operation in a fume cupboard; wear disposable gloves.

termination of the 1- and 3-day assays

1. Initially prepare a 1 % aqueous solution of crystal violet; filter the solution through No 1 Whatman filter paper to remove any undissolved matter. Take care not to scratch the base of the 96-well plate. A few days postpreparation a precipitate may re-form thus it is advisable to check the nature of the 1 % crystal violet solution and to filter if necessary.

2. After incubation invert the plate over the sink and flick out the culture medium, with the plate facing upwards.

3. Immerse it in a 5 % formaldehyde (prepared in PBS) in a plastic container; allow one row to fill at a time to avoid bubbles. Leave for 5 min to allow the cells to fix to the plastic.

4. Wash the plate three times with water or PBS. Post-staining, wash the plate in excessive volumes of water until no blue/violet dye is evident. Wipe the undersurface of the plate to remove residual water, leave the plate inverted on tissue paper to dry. The plate can be read dry, but unequal distribution of dye can give erroneous readings.

5. To obtain an even distribution of dye add 100 μl of 33 % aqueous glacial acetic acid to each of the wells including the blank row. Shake the plate for 60 s to ensure complete solubilization.

6. Read the plate at 540 nm on a standard ELISA reader. Blank the reader against the eight blank wells on the plate.

7. For each sample work out the percent cytotoxicity from the formula $100\,(a-b)/a$, where a and b are the mean absorbances of the triplicate wells with culture medium alone and test sample, respectively.

8. Calculate the LD_{50}.

28.5
Clonogenic Assays

One of the most generally accepted methods of assaying for survival and stimulating effect is the measurement of the ability of the cells to form colonies in isolation. This is usually achieved by simple dilution of a single cell suspension and determination of survival by counting colonies. A lower threshold must be established with the doubling time of the cells being studied and the total duration of the assay. A good point to aim for is five to six doublings. Colonies of this size can be microscopically determined with ease. Growth in soft agar is a very useful assay system for cell lines which have a comparatively high plating efficiency.

The assay described in the next protocol is a general clonogenic assay that can be used for a number of cytokines. It was developed by Hill (1983) from an original soft agar assay established by Courtney (1978).

Procedure

1. Prepare soft agar 2.5 % (w/v) low gelling temperature agarose in purified water. Boil to dissolve the agar and allow to cool to 60 °C. Prepare the appropriate medium +10 % FCS and place at 37 °C to warm. Mix 2 ml medium with 8 ml medium, to give a final concentration of 0.5 % agarose. Keep at 37 °C until required.

2. Dilute packed and washed August rat red blood cells 1:8 with growth medium +10 % FCS.

3. Trypsinize a stock culture of cells in the standard manner to give a single cell suspension. Prepare a dilution of cells to give a concentration of 6×10^4 cells in 1.8 ml.

4. Mix the diluted cell suspension with 0.6 ml diluted August rat red blood cells and 3.6 ml medium+10 % FCS +0.5 % agarose. The final volume is 6 ml.

5. Add the required volume of test cytokine. Dispense 1 ml volumes of the mix into 2 ml polypropylene tubes; when sealing the tubes ensure a gas permeable seal has been created.

6. Plunge the tubes into an ice bath to set the agar. Incubate at 37 °C in a CO_2 incubator. Feed every 6 days by the addition of 1 ml of medium +10 % FCS + the appropriate cytokine dilution to the top layer of agar.

7. After approximately 3 weeks examine the agar for colonies by placing the tubes on a gridded petri dish and covering with the inverted lid to compress the agar. Examine under low power microscopy. Score the plates for colonies when control colonies have reached a predetermined size. This is usually more than 50 cells for cell lines, but a size of more than 20–30 cells has been used when the population has a slow growth rate and a low plating efficiency.

Troubleshooting

- As with all bioassays the endpoint obtained in a given experiment is subject to variations due to difference in reagents used, particularly the cell lines and the cytokine stock solutions. In order to ensure uniformity and reproducibility, all cytokines in use in a laboratory should be titrated and compared against standards provided free of charge by the World Health Organization, the National Institute of Biological Standards and Control and the National Institute of Health Bethesda, Maryland. It is strongly recommended that the relationship between laboratory units and the reference standards are established and re-tested at selected intervals (at least once yearly). It is of the utmost importance to verify the stability of the laboratory standard over a specified time frame against the international reference standard (Meager 1988). Reagents (media type, source of critical reagents) used in cytokine assays should be standardized to ensure reproducibility of assay systems.

- The correct storage temperature for samples is often a difficult parameter to decide. A good rule of thumb is that most cytokines are not stable above refrigeration temperature, thus exposure at these temperatures should be minimized as far as is possible. A more diffi-cult decision is posed when deciding whether to store samples at 2–8 °C or less than −20 °C. As a general rule, cytokine samples con-taining serum or other proteins in high concentrations are stable at 2–8 °C while highly pure preparations containing low amounts of protein (<1 mg/ml) are often unstable at this temperature. A temper-ature of −80 °C is often recommended for freeze dried highly pure preparations. If the chosen storage temperature is 2–8 °C keep in mind that storage at this temperature does not prevent growth of mold contaminants; thus it is advised to filter sterilize prior to stor-age. Frozen samples should not be thawed too rapidly, repetitive freezing and thawing of the same sample should be avoided. Care should be taken to ensure that all cytokine standards (IRS and LRS), samples pending analysis and all temperature susceptible laboratory reagents are stored in fridges and freezers that are frequently cali-brated, maintained and monitored.

- The design of assays should be carefully considered in relation to the type and sources of the test material the number of samples and their volumes and the degree of statistical confidence required. As stated above all assays should be carefully calibrated against an IRS and using the LRS. When microtiter plates are used for assays it is recom-mended to have blanks, positive and negative controls on each plate. It is recommended that the test is at least performed in triplicate, where sample volume allows. In fact, repeating assays is good scien-tific practice. It confirms not only the reproducibility of the assay method, but also provides further quantitative data from which, when analyzed together or in successive assays, the geometric mean titers can be calculated, leading to increased statistical validity of the results.

- It should also be noted that there is great variation in the same cell line from different sources. It is thus recommended at the onset to test a validated preparation of the cytokine under study with the same cell line from different sources and select the most reactive cell line on the basis of the analytical results.

- It is of critical importance at the assay development stage to ascertain the stability of the cytokine assay with passage number of the appropriate cell line. It is advisable to choose a certain passage range and establish the given stability and reproducibility of the assay under those conditions. It is recommended to perform all subsequent assays in the given cell passage range. Overall, best results are obtained with healthy cells in the log phase of growth. This can be achieved by adhering to a strict subculturing routine, seeding the cells at consistent time intervals and cell densities.

- Many cells line that are reported in the literature as being dependent on one cytokine may be also stimulated by other cytokines. This is not a problem when one is dealing with a known purified cytokine, but if an investigator has isolated an autocrine substance and is attempting to ascertain the identity of the material by using a reported dependent cell line, the investigator may be sent on the wrong trail. If this information is not available from the literature it is advisable to perform assays with various cytokine standards on the designated cell line to establish a baseline of cytokine activity for the potentially useful cell line. This may save a substantial amount of time trying to understand obscure results!

- Background values in cytokines assays tend to be high; values should be at least two to three times the background before a positive is scored. For example, many experimenters find that with IL-3 dependent cell lines the negative control values rise either suddenly or over a period of time. This may be due to infection with mycoplasma or, if cultures are allowed to overgrow, the selecting out of cells with different characteristics such that the culture gradually becomes significantly different from the original one. The remedy for persistent high background is to reclone the cell line.

References

Aggarwal BB, Gutterman JU (1987) Human cytokines. Blackwell Scientific Publications, Oxford

Arai K I, Lee F, Miyajima A, Miyatake S, Arai N, Yokota T (1990) Cytokines: co-ordinators of immune and inflammatory response, Annu Rev Biochem 59:783–836

Borset M, Waage A, Sundan A (1996) Hepatocyte growth factor reverses the TGF-β induced growth inhibition of CCl-64 cells. J Immunol Methods 189:59–64

Cagliero E, Roth T, Taylor AW, Lorenzi M (1995) The effect of high glucose on human endothelial cell growth and gene expression are not mediated by TGFβ. J Lab Invest 73: 667–673

Clemens MJ, Morris AG, Gearing AJH (1987) Lymphokines and interferons. A practical appraoch. IRL, Oxford

Cortii A, Poiesi C, Merli S, Cassani G (1994) Tumor necrosis factor (TNF) α quantification by ELISA and Bioassay: effects of TNFα-soluble receptor (p55) complex dissociation during assay incubations. J Immunol Methods 177:191–198

Courtney VD, Mills J (1978) The use of a clongenic assay system for the analysis of IFN. Br J Cancer 37:261–269

De Jong LO, Farner NL, Widmer MB, Giri JG (1996) Interaction of IL-15 with shared IL-2 receptor β and γc subunits. J Immunol 156:1339–1348.

Freshney (1992) Animal cell culture: a practical approach. IRL, Oxford

Garrigue-Antar P, Barbieux I , Liebeau B , Boisteau O, Gregoire C (1996) Optimisation of the CCL64 based bioassay for TGF-β. J Immunol Methods 189:269–274

Hamilton JA, Whitty GA, Kola I, Hertzog PJ (1996) Endogenous INFαβ suppress colony stimulating factor CSF 1 stimulated macrophage DNA synthesis and mediates inhibitory effects of lipopolysaccharide and TNF-α. J Immunol 156:2553–2557

Hang L Svanborg C, Andersson G (1994) Determination of IL-6 in human urine and epithelial supernatants. Int Arch Allergy Immunol 105: 397–403

Higuchi M, Singh SM, Aggarwal BB (1995) Characterisiation of the atopoic effects of human tumor necrosis factor: development of a highly rapid and specific bioassay for human tumor necrosis factor and lymphotoxin using target cells. J Immunol Methods 178:173–181

Hill BT, Whelan RDH (1983) Modifcation of classic clonogenic assay systems. Cell Biol Int Rep 7:617–622

James P (1996) IGFBP-5 in muscle differentiation. J Cell Biol 133:681–688

Jeannin P, Delnestte Y, Seveso M, Life P, Bonnefoy J-V (1996) IL-12 synergises with IL-2 and other stimuli in inducing L-10 production by human T cells. J Immunol 156: 3159–3164

Ledur A, Fitting C, David B, Hamberger C, Cavaillon J-M (1995) Variable estimates of cytokine levels produced by commercial ELISA kits results using international reference standards. J Immunol Methods 186 171–179

Levesque A, Paquet A, Page M (1995) Improved fluorescent bioassay for the detection of tumor necrosis factor activity. J Immunol Methods 178 71–76

Lieonart R, Naf D, Browning H, Weissmann C (1996) A novel Quantitative bioassay for type 1 interferon using a recombinant indicator cell line. Biotechnology 8:1263–1266

Lu Z-Y, Zhang X-G, Gu Z-J, Yasukawa K, Amiot M, Etrillard M, Bataille R, Klein B (1994) A highly sensitive bioassay for human interleukin -11. J Immunol Methods 173:19–26

Marshall JS, Leal-Berumem I, Nielsen L, Glibetic M, Jordana M (1996) Interleukin (IL) 10 inhibits long term IL- 6 production but not performed mediator release from rat peritoneal mast cells. J Clin Invest 97: 1122–1128

Meager A (1987) Biological standardisation of cytokines. Dev Bio Standard 69:199–206

Mire-Sluis A, Gaines-Das R, Thorpe R (1995) Immunoassays for detecting cytokines: What are they really measuring? J Immunol Methods 186 157–160

Page LA, Toop MS, Thorpe R, Mire-Sluis AR(1996) An antiproliferative bioassay for interleukin 4. J Immunol Methods 189 129–135

Pauli U, Bertonin G, Duerr M, Peterhans E (1994) A bioassay for the detection of tumor necrosis factor from eight different species: evaluation of neutralisation rates of a monoclonal antibody against TNF-α. J Immunol Methods 171:263–265.

Pestka S (1986) Interferons from 1982 to 1986. Methods Enzymol 119:3–14

Randall L A, Wadhwa M, Thorpe R, Mire-Sluis AR (1994) A novel sensitive bioassay for transforming factor β. J Immunol Methods 171:263–265

Ruff MR, Gillford, GE (1981) In: Pick E (ed) Lymphokines. Academic Press, New York, p 235

Shanan TA, Siegel PD, Sorenson WG, Kuschenr WG, Lewis DM (1994) A new sensitive bioassay for TNF. J Immunol Methods 175:181–187

Taub DD, Anver M, Oppenheim JJ, Longo DL, Murphy WJ (1996) Lymphocyte recruitment by interleukin-8 (IL-8). J Clin Invest 97:1931–1941

Watari K, Mayani H, Lee F, Dragowska W, Lansdorp P M, Schrader JW (1996) Production of interleukin 1β by human hematopoietic progenitor cells. J Clin Invest 97:1666–1674

Whiteside L (1994) Cytokine measurements and interpretation of cytokine assays in human disease J Clin Immunol 14:327–39

▓ Suppliers

Amercian Type Culture Collection
12301 Parklawn Drive
Rockville, Maryland 20852
USA

Amersham International PLC
2636 South Clearbrook Drive
Atlington Heights, Illinois 60005
USA

European Culture of Animal Cell Cultures
Porton Down
Salisbury SP4 OJG
UK

Genzyme (Koch-light)
Rookwood Way
Haverhill, Suffolk CP98PB
UK

Gibco
P.O. Box 35
Trident House
Renfrew, Paisley PA34EF
Scotland

World Health Organization
Washington, DC
USA

National Institute of Health
Research Resources Branch
Bethesda, Maryland 20205
USA

National Institute for Biological
Standards and Control
Blanche Lane
South Mimms
Potters Bar, Hertfortshire EN6, 3QG
UK

Sigma Chemicals Co. Ltd.
Fancy Road
Poole, Dorset BH177NH
UK

Sterlin Ltd.
Sterlin House
Clockhouse Lane
Feltham, Middlesex TW148QS
UK

Corning Ltd., Stone
Staffs ST150BG
UK

Millipore Company
11–15 Peterborough House
Harrow, Middlesex HA1 2YH
UK

Flow Laboratories Ltd.
P.O. Box 17
Second Avenue Ind. Estate
Irvine, Ayrshire, KA128NB
UK

◼ Abbreviations

ATCC	American-type culture collection
CPE	Cytopathic effect
DMEM	Dulbecco's modified Eagle's medium
EDTA	ethylenediamine tetra-acetic acid
EGF	Epidermal growth factor
FCS	Fetal calf serum
FGF	Fibroblast growth factor
GMCSF	Granulocyte macrophage stimulating factor
GM	Growth medium
HGF	Hepatocyte growth factor
IL	Interleukin
INF	Interferon
IRS	International reference standard

LRS	Laboratory reference standard
LAF	Laminar air flow
MEM	Minimal essential medium
MTT	dimethylthiazole diphenyltetrazolium
OD	Optical density
PBS	Phosphate buffered serum
PDGF	Platelet- derived growth factor
TGF	Transforming growth factor
TNF	Tumor necrosis factor
TCID	50 % Tissue culture infectious dose
VSV	Vesicular stomatitis virus

Part VI

Genetic Manipulation and Analysis of Human and Animal Cells in Culture

Chapter 29

Antisense Methods in Cell Culture

Daragh Byrne*, Carmel Daly, and Martin Clynes

Introduction

The notion of using specific oligonucleotides for the modulation of gene expression surfaced nearly two decades ago when DNA oligonucleotides were used to inhibit Rous sarcoma virus replication and cell transformation (Zamenick 1978). The basic idea behind the use of antisense molecules is quite simple. An oligonucleotide chain containing bases complementary to a given target (sense) sequence can bind to this target sequence, and thus inhibit its processing, through the interactions defined by the Watson and Crick base pairing rules. These rules state that a guanosine must always base pair with a cytosine, while a tyrosine must base pair with an adenosine. Thus, once the sequence of a target gene is known, an antisense molecule can be designed which specifically binds to this gene alone. The initial antisense molecules were oligodeoxynucleotide sequences targeted to mRNA molecules. However, recent advances in molecular biology have widened the spectrum of antisense compounds to include RNA oligos targeted against RNA, DNA oligos targeted against DNA (termed antigenes) and either DNA or RNA oligos which bind to and inhibit the function of peptides and proteins (termed aptamers) (for reviews see Ma and Calvo 1996; Scanlon et al. 1995; Wagner 1994, 1995; Zhang 1996).

Although the basic concept is simple, the actual mechanisms which lie behind the antisense mediated inhibition of gene function are quite complicated and are still somewhat of a mystery. The main mechanisms of inhibition of the antisense molecules appears to be the steric inhibition of the transcriptional and translational machinery as they progress

* *Correspondence to* Daragh Byrne, National cell and Tissue Culture Centre, Dublin City University, Glasnevin, Dublin 9, Ireland; phone +353–1–7045728; fax +353–1–7045484; e-mail 94970254@tolka.dcu.ie

along the coding sequences and the induction of cellular enzymes, especially RNase H and adenosine deaminase, which selectively destroy the RNA in DNA/RNA and RNA/RNA hybrids respectively. In addition, the interference with crucial secondary structures may play a role. However, it is becoming apparent that not all the effects seen with antisense molecules can be accounted for by these mechanisms alone. The increasing appearance of non-specific side effects further complicates the issue and there is increasing doubt by some as to the actual existence of a true antisense effect (Stein 1995; Antisense R&D, editorial, 1994). Despite lingering doubts, there is continuing research on a large scale into the use of antisense molecules for both research and therapeutic purposes. Indeed, there are many ongoing clinical trials into their use against a wide variety of genetic and viral diseases, especially cancer and AIDS.

There are a number of important considerations when designing an antisense molecule. Firstly, there is the target site, which must be specific to the gene and conform to certain rules governing secondary structure and base composition (Wagner 1995). However, there are no set rules for deciding on the actual sequence site to which the antisense is designed, or for determining the optimum sequence length. When the antisense is part of an expression vector, it is common for all, or a large portion, of the target gene sequence to be incorporated into the vector in reverse, but short antisense sequences (<100 base pairs) are also frequently employed. As regards antisense oligonucleotides, it is very much a hit and miss affair when trying to determine the optimum target site and length. Generally, a good starting point is to design a range a 20 base pairs oligonucleotides, targeting various different sites in the gene, including the 5' UTR (untranslated region), the start codon region, the stop codon region and the 3' UTR, as well as the coding region itself. These oligos should be screened against a gene bank database to ensure that they are unique to the gene in the species in question. The appropriate sequences should then be assayed for their ability to down-regulate the appropriate mRNA at a cellular level in order to determine the most effective target sequences. The most potent antisense molecules can then be fine tuned by lengthening or shortening or shifting the target region laterally along the sequence.

Then one must choose the type of base analogue (whether DNA or RNA) if using synthetic oligos, since natural phosphodiester bonds may be too quickly degraded in a cellular environment. There are now many such analogues, the most popular and widely studied being phosphorothioates (Wagner 1994, 1995). The pharmacology and metabolism of the nucleotides and their breakdown products is a critical consideration if the eventual aim is for the therapeutic administration of the antisense to patients (Bennett et al. 1995; Sharma and Naraynan 1995; Crooke 1995).

When designing experiments to examine the effects of antisense molecules, one must employ very strict controls and perform rigorous testing to ensure that the results obtained are indeed due to the effect of the antisense reagent (Sharma and Naraynan 1995; Stein 1995; Wagner 1995). This usually involves the demonstration of a quantitative reduction in the levels of mRNA and of protein levels, with related genes or gene products unaffected.

Finally, one must consider the method of delivery of the oligonucleotides into the cells, as this will depend on the type of antisense employed, the cell type which is being targeted and whether the cell is in an in vitro or an in vivo environment (Bergan et al. 1993; J Drug Targeting, editorial, 1995; Gerwitz et. al. 1996). For in vitro cell culture systems, the oligonucleotides can be simply added to the media and are taken up into the cells by passive diffusion and endocytosis. For more efficient uptake and delivery the antisense oligos can be incorporated in liposomes for lipofection or transferred by electroporation. Lipofection involves the complexing of the DNA/nucleic acid with cationic liposomes which adhere to the cell surface, fuse with the cell membrane and release the DNA/nucleic acid into the cytoplasm. Electroporation is a physical process that uses an electrical pulse to transiently permeabilize cell membranes, permitting the uptake of nucleotides and large DNA molecules. For the delivery of expression plasmids, the common transfection techniques, (calcium phosphate, electroporation, lipofection) are widely used. The use of viral vectors, such as retroviruses and adenoviruses, is becoming increasingly common (Skotzko et al. 1995). These vectors afford very efficient and even cell specific delivery of the oligonucleotides and are the method of choice for use in vivo experiments. However, the use of recombinant viruses is strictly controlled and requires stringent safety procedures as well as special permits before their use is allowed. The remainder of this chapter will deal with the practical use of two of the more common techniques used for oligonucleotide delivery in cell culture systems, lipofection and electroporation, as well as highlighting certain techniques useful in the molecular examination of the antisense effect. For the calcium phosphate transfection method, refer to the chapter by Daly et al.

Outline

A summary of the antisense methods in cell culture is shown in the flow chart (Fig. 29.1).

Fig. 29.1. Summary of the antisense methods

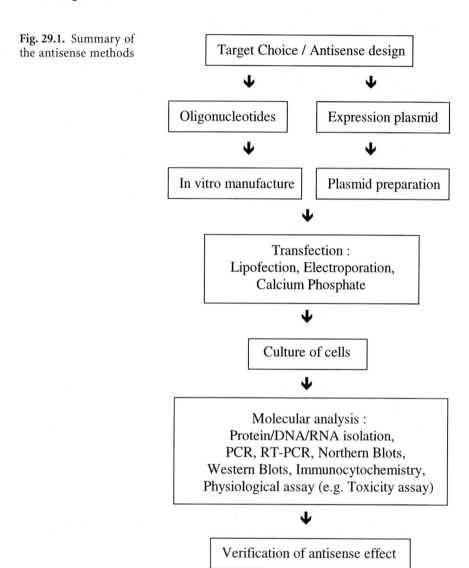

Materials

lipofection

- Transfection reagent (DOTAP; Boehringer Mannheim, cat. no.
  1 202 375)
- 20 mM Hepes buffer (cell culture grade): pH 7.4, sterile
- Culture medium with and without serum: sterile
- Antisense (oligonucleotides or plasmid)

electroporation

- Electroporation apparatus (Gene Pulser; Bio Rad, cat. no. 165 2077)
- Electroporation medium: (a) phosphate buffered saline: without Ca^{2+} or Mg^{2+}; (b) phosphate buffered sucrose: 272 mM sucrose; 7 mM sodium phosphate, pH 7.4; 1 mM $MgCl_2$; (c) cell culture medium (without serum)
- Electroporation cuvettes: 0.4 cm (Bio Rad, cat. no. 165 2088)
- Trypsin solution: sterile

All other equipment and reagents as described in referred chapters and articles.

Procedure

Lipofection

Note: Refer to DOTAP information sheet supplied by Boehringer Mannheim.

preparation of
the cells

1. The day before the transfection procedure, subculture the cells (which should be actively growing, but not 100 % confluent) according to standard procedures.

2. Replate the cells at a density of roughly 10^4 cells/ml of culture media (containing serum)

3. Incubate cells overnight at 37 °C.

preparation of
the transfection
reagent/nucleic
acid mixture

1. Dilute 5 µg nucleic acid to a concentration of 0.1 µg/µl in Hepes buffer (final volume 50 µl) in a sterile reaction tube.

2. In a separate sterile reaction tube mix 30 µl of transfection reagent (DOTAP) with Hepes buffer to a final volume of 100 µl.

3. Transfer the nucleic acid solution (50 µl) to the reaction tube already containing the DOTAP in Hepes buffer (100 µl) and carefully mix the transfection mixture by gently pipetting the mixture several times. **Do not vortex or centrifuge.**

4. Incubate at room temperature for 10–15 min.

transfection of the cells

1. Pipette up the DOTAP/nucleic acid mixture and add to the culture medium in the flask containing the cells. Mix by gentle swirling or gentle pipetting up and down to distribute evenly.

2. Incubate the cells with the DOTAP/nucleic acid mixture for 3–6 h at 37 °C.

3. Replace the medium with fresh culture medium and further incubate.

notes

- The above protocol is for the stable transfection, with an expression plasmid, of adherent cells growing in a $75 \, cm^2$ flask at a density of about 1×10^5 cells per flask. For other flask/container sizes and cell densities/types, adjust accordingly or refer to the manufacturers product information sheet (e.g., DOTAP, Boehringer Mannheim).

- The amount of nucleic acid and DOTAP reagent used depends on the transfection efficiency required and the cell type. It is suggested that the optimal conditions for a particular cell line be determined by examining the effects of variations in incubation time, DOTAP concentration and nucleic acid concentration on transfection efficiency (as measured by a suitable assay, e.g., β-Gal reporter assay).

- The amount of oligonucleotides used for transient transfection can vary greatly depending on the length and base content of the oligonucleotide (as this will affect the molecular weight and thus molarity of the antisense molecules) and also the required concentration which can vary from about 0.05 to 20 µM. Lower concentrations of oligonucleotides are required when using active transfection techniques, such as the above, as compared to relying on passive diffusion and endocytosis.

- For the selection of stable transfections, it is recommended that the cells be incubated for at least 24–48 h before the addition of geneticin or other selection agent. After this time the cells can also be cloned as described in the chapter by McBride et al.

- For a more detailed description of the procedure refer to the manufacturers product information sheet and protocol.

Electroporation

Note: Refer to the Bio-Rad Gene Pulser User Manual.

1. Grow cells to 70–80 % confluency in their normal culture medium containing serum.

2. Harvest the cells with trypsin (or by other suitable subculturing method), inactivate trypsin with small amount of serum and pellet the cells at 500 g for 5 min.

3. Wash the cells twice in the appropriate ice cold electroporation medium (PBS, phosphate buffered sucrose or serum free media), then resuspend in the same medium at a concentration of 0.5 to 1.0×10^7 cells/ml.

preparation of cells for electroporation

1. Add 2.5–40 µg of linearized expression vector (or roughly 0.1–100 µM of the oligonucleotides), in appropriate sterile nucleic acids buffer, to 0.8 ml of the cell suspension and transfer to the 0.4 cm electroporation cuvette. Mix the cells thoroughly by pipetting and then incubate on ice for 10 min prior to electroporation.

2. Place the cuvette containing the cells and the nucleic acids in the electroporation chamber, pulse once and then return to ice and incubate for an additional 10 min.

3. The following are the pulse settings for the different electroporation medium:

- Electroporation in PBS: 25 µF, voltage settings from 100 to 1600 V
- Electroporation in phosphate buffered sucrose: 25 µF, voltage settings from 100 to 1000 V
- Electroporation in serum-free culture medium: 960 µF, voltage settings from 250 to 450 V

4. Replate the cells with an appropriate volume of culture medium (with serum) and incubate for 24–48 h before beginning selection or cloning.

5. Stable transfectants selected with geneticin (G418), or other selection agents, should be grown for a number of weeks before any analysis is done to ensure the stability of the transfected plasmid.

6. Cells transiently transfected with oligonucleotides should be incubated for a sufficient amount of time to allow for down-regulation of

electroporation of cells

the protein by the antisense, before any molecular analysis is done. The half-life of the protein must therefore be taken into consideration.

notes
- Caution should be observed while using the electroporation apparatus as very high voltages can be obtained.

- As with the lipofection method, this electroporation procedure should be optimized for the cell line being used to determine the most effective electroporation medium, nucleic acid concentration and pulse settings to give the best transfection efficiency.

- Due to the severe nature of this transfection method, a high rate of cell death (up to 50 %) is quite common.

- Conditions and recommendations will vary between different electroporation apparatuses, therefore the manufacturers handbook should always be consulted.

Molecular Analysis of Cells

After the transfection has been performed and, in the case of stable transfection, selection completed, cells should be grown to a sufficient density to allow for their molecular characterization and analysis by the following methods:

- RNA isolation: refer to chapter by O'Doherty et al.
- Protein isolation: refer to chapter by Moran et al.
- Reverse transcription PCR: refer to chapter O'Doherty et al.
- Northern blot: refer to chapter by N. Daly and Clynes
- Immunocytochemistry/western blot: refer to chapter by Moran et al.

A physiological assay should also be carried out to establish if the normal function of the targeted gene product has been affected by the antisense molecule. For example, in the case of multiple drug resistance related genes (e.g., mdr-1, mrp, Lrp, Topo II), a drug toxicity assay would determine if the resistance of the cells has been affected by the downregulation of the gene. For details of in vitro toxicity assays, refer to the chapter by O'Connor et. al.

Results

Following the molecular analysis, the following results would be observed from a typical antisense gene transfection. Initial PCR reactions would confirm the presence of the antisense construct, at the DNA level, in the cell, while reverse transcriptase-polymerase chain reaction (RT-PCR) would confirm the expression of the antisense mRNA, if contained in an expression plasmid. RT-PCR may also give an indication as to the down-regulation of the mRNA of the target gene, but results may not be fully quantitative. Northern blots should show a measurable decrease in the mRNA levels of the gene indicating a down-regulation. To determine if this down-regulation at the DNA or RNA level is conserved at the protein level, immunocytochemistry or Western blotting would be needed. A reduction in the target protein levels from cell extracts in the western blot and a decrease in the expression and antibody presentation at a cellular level, as shown by immunocytochemistry, should be observed. If the target protein is related to some measurable physiological characteristic, e.g., drug resistance, then an appropriate assay should be performed to determine if any observed down-regulation of the protein levels is reflected at a functional level.

However, this is an ideal situation, and such clear cut results are rarely achieved without much difficulty and optimization of experimental conditions and parameters. The levels of down-regulation will vary from cell type to cell type, clone to clone and from one antisense molecule to another. Also the transfection efficiencies achieved will vary greatly between different cell lines and methods used. Once again, great care must be taken in the analysis of results to ensure their validity.

References

Askari F, McDonnell W (1996) Antisense-oligonucleotide therapy. New Eng J Med 334:316–318

Bennett C, Dean N, Ecker D, Monia B (1995) Pharmacology of antisense therapeutic agents. In: Agrawal S (ed) Methods in molecular medicine: antisense therapeutics. Human,Totowa, New Jersey, pp13–46

Bergan R, Connell Y, Fahmy B, Neckers L (1993) Electroporation enhances c-myc antisense oligonucleotide efficacy. Nucleic Acids Res 21(15):3567–3573

Bergan R, Graham M, Crooke M, Crooke S (1995) In vitro pharmokinetics of phosphorothioate antisense oligonucleotides. J Pharmacol Exp Ther 275:462–473

Crooke R, Graham M, Crooke M, Crooke S (1995) In vitro pharmokinetics of phosphorothioate antisense oligonucleotides. J Pharmacol Exp Ther 275:462–473

Editorial (1994) Problems in interpretation of data derived from in vitro and in vivo use of antisense oligodeoxynucleotides. Antisense Res Dev 4:67–69

Editorial (1995) Delivery of oligonucleotides and polynucleotides. J Drug Target 3:185–190

Gerwitz A, Stein C, Glazer P (1996) Facilitating oligonucleotide delivery: helping antisense deliver on its promise. Proc Natl Acad Sci USA 93:3161–3163

Ma L, Calvo F (1996) Recent status of the antisense oligonucleotide approaches in oncology. Fundam Clin Pharmacol 10:97–115

Scanlon K, Ohta Y, Ishida H, Kijma H, Ohkawa T, Kaminski A, Tsai J, Horng G, Kashani-Sabet M (1995) Oligonucleotide-mediated modulation of mammalian gene expression. FASEB J 9:1288–1296

Sharma H, Narayanan R (1995) The therapeutic potential of antisense oligonucleotides. BioEssays 17(12):1055–1063

Skotzko M, Wu L, Anderson W, Gordon E, Hall F (1995) Retroviral Vector-mediated gene transfer of antisense cyclin G1 (CYCG1) inhibits proliferation of human osteogenic sarcoma cells. Cancer Res 55:5493–5498

Stein C (1995) Does antisense exist. Nature Med 11:1119–1121

Wagner R (1994) Gene inhibition using antisense oligodeoxynucleotides. Nature 372:333–335

Wagner R (1995) The state of the art in antisense research. Nature Med 11:1116–1118

Zamecnik P, Stephenson M (1978) Inhibition of Rous sarcoma virus replication and cell transformation by a sepcific oligodeoxynucleotide. Proc Natl Acad Sci USA 75: 1:280–284

Zhang W (1996) Antisense oncogene and tumor suppressor gene therapy of cancer. J Mol Med 74:191–204

Suppliers

Boehringer Mannheim

Bio-Rad Laboratories Ltd.
Export Division
Bio-Rad House
Maylands Avenue
Hemel Hempstead, Hertfordshire, HP2 7TD
UK
(Tel: 44 442 232552, Fax: 44 442 259118)

Chapter 30

Transfection with Ribozymes to Investigate the Role of Specific Gene Expression

CARMEL DALY*, DARAGH BYRNE, and MARTIN CLYNES

Introduction

The transfer of cloned genes into mammalian cells has become a routine tool for studying gene structure and function and to identify regulatory sequences that control gene expression. Techniques have also been developed which specifically inhibit the expression of single genes already present in a cell. The ability to switch off or inhibit specific gene expression may be beneficial in the treatment of cancer or viral disease, since it is desirable to affect only the disease associated oncogenes or viral genes without disrupting expression of the other genes (for review see Sharma and Narayan 1995; Zhang 1996).

The word ribozyme is derived from the words ribonucleic acid (RNA) and enzyme. Ribozymes are RNA molecules which bind to a target RNA, catalyze cleavage and dissociate to enter further cycles of binding, cleavage and dissociation (Haseloff and Gerlach 1988; Cech 1990; Symons 1992). Hammerhead ribozymes have been most extensively studied and have been engineered in such a way that they can act in *trans* against other RNA molecules (Uhlenbeck 1987; Haseloff and Gerlach 1988). The *trans* acting hammerhead ribozyme developed by Haseloff and Gerlach (1988) consists of an antisense section (stems I and III) and a catalytic domain with a flanking stem II and loop section (Fig. 30.1). Substrates can have any sequence as long as the cleavage site contains NUX where N can be any nucleotide and X can be any nucleotide except G (Haseloff and Gerlach 1988; Ruffner et al. 1990; Perriman et al. 1992). Most efficient cleavage occurs after GUC triplets (Koizumi et al. 1988; Ruffner et al. 1990) with cleavage occurring 3' to the X nucleotide.

* *Correspondence to* Carmel Daly, National Cell and Tissue Culture Centre, Dublin City University, Glasnevin, Dublin 9, Ireland; phone +353–1–7045700; fax +353–1–7045484; e-mail 75021862@DCU.ie

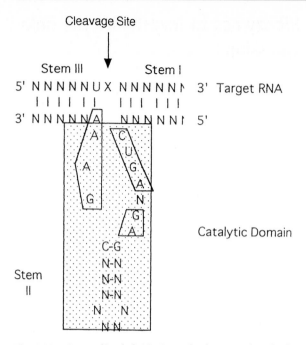

Fig. 30.1. Generalized depiction of a hammerhead ribozyme. The *Ns* represent any nucleotide and *X* represents A, C or U. The site of substrate cleavage is indicated by the *arrow*. Once cleavage has occurred the ribozyme can dissociate from the cleaved products and bind and cleave another substrate molecule. The catalytic domain of the hammerhead is *shaded* and the conserved nucleotides required for catalytic activity are indicated within the *boxes*. The hammerhead ribozyme hybridizes to sequences in the target RNA flanking the NUX cleavage site to form stems I and III. These stems determine the specificity of the ribozyme for its target. Nucleotides that form base pairs are indicated. Stem II forms part of the catalytic core

Ribozyme gene targets may be foreign as in a viral infection or may be a normal gene which has undergone mutation such as an activated proto-oncogene. There are numerous reports of the successful use of ribozymes, in particular hammerheads, for the inhibition of gene expression in cell culture (Sarver et al. 1990; Scanlon et al. 1991; Chen et al.1992; Kashani Sabet et al. 1992; Holm et al. 1994; Lange et al. 1993; for review see Stull and Szoka 1995) A prerequisite for the application of ribozymes is the efficient cleavage of large biologically active RNAs like mRNAs or viral RNAs. Large RNA molecules form complex secondary and tertiary structures which can interfere with ribozyme recognition and cleavage. Studies have shown that sequences do exist in target mRNAs that are cleaved efficiently and a search for them in a target sequence is worthwhile (Crisell et al. 1993; Bertrand et al. 1994; Palfner et al. 1995; Holm et al. 1994). For a discussion of the parameters involved in designing ribozymes see Christoffersen and Marr 1995 and Sullivan 1994. In order to design and optimize ribozymes (in terms of efficiency and specificity) it is essential to use cellular in addition to cell-free systems (Crisell et al. 1993; Bertrand et al. 1994, Bertram et al. 1995). In different cell lines the set of RNA binding proteins may differ resulting in altered access of a selected target sequence for the same oligo, in different cell lines.

The vector used to carry ribozyme constructs into cells can be either a plasmid, a retroviral vector or a viral vector (such as that from adenovirus). A number of methods are available to transfer DNA into mammalian cultured cells. In this chapter we will describe one of the most widely used methods, which involves using a calcium phosphate precipitation protocol for the introduction of DNA (plasmid), containing a ribozyme, into monolayers of adherent cells in culture. Using this method the DNA is mixed with a carefully buffered solution containing phosphate. Addition of calcium chloride forms a white precipitate of calcium phosphate and DNA. This precipitate is pipetted onto a monolayer of cells and left on the cells for several hours during which time many of the cells take up the precipitated DNA.

Methods which can be used for the selection of stable transfectants and for the analysis of the effect of ribozyme expression in these cells will be suggested. The technique described can be modified for the transfection of cells growing in suspension (Sambrook et al. 1989). For each different cell line used it is important to optimize the parameters for DNA transfection.

Outline

The steps involved in the transfection of a ribozyme construct into mammalian cells and subsequent analysis of clones containing the ribozyme are shown in Fig. 30.2.

Materials

– 37 °C Incubator
– Laminar flow cabinet

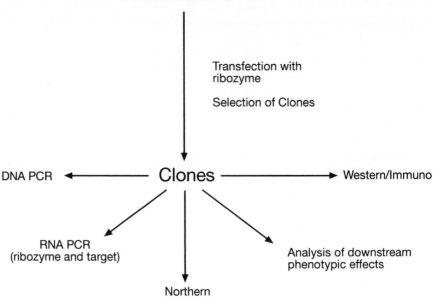

Fig. 30.2. Flow chart

- pZ523 Spin Columns (cat. no. 5305–523325, 5Prime→3Prime Inc., Boulder, Colarado)
- Exponentially growing eukaryotic cells
- Complete medium
- Serum free media
- Pure plasmid DNA
- 2 M $CaCl_2$, 10 % glycerol/1× Hepes buffered saline; TE, pH 8
- 10 mM Tris
- 1 mM EDTA
- Sterile water
- 2× HBS (Hepes buffered saline)
- PBS, pH 7.4
- 280 mM NaCl
- 0.01 M Phosphate buffer
- 50 mM Hepes
- 0.0027 M Potassium chloride
- 1.5 mM Na_2HPO_4
- 0.137 M Sodium chloride

Procedure

Transfection with Ribozymes

1. Twenty four hours before transfection the cells should be split into a 10 cm tissue culture plate. The optimum density at which cells should be set-up for transfection will vary with cell type. On the day of the transfection it is important that the cells are thoroughly separated (subconfluent) on the dish, as the ability to take up DNA is related to the surface area of the cell exposed to the medium.

Note: For optimal transfection efficiency it is necessary that the recipient cells are healthy and it is assumed that the conditions needed for optimal growth and passage of the cell line to be used will have already been determined. In our experience fungal contamination of transfection plates can be a problem. To minimize the risk of contamination, cells for transfection can be set up in cell culture flasks; transfection can be carried in these flasks and clones selected as described (see "Selection of Stable Transfectants").

2. Plasmid DNA to be transfected should be diluted to 1 μg/μl in TE. Place the required amount of DNA in a 3-ml tube and make up to 410 μl with sterile water. Store overnight at 4 °C.

Note: The gene to be transfected (in this case a ribozyme) is generally cloned into a mammalian expression vector; a variety of vectors are available from molecular biology reagent suppliers. The particular vector and promoter to be used depend on the cell line to be transfected and the gene to be targeted. The optimum amount of DNA to be used varies from 10 to 50 μg per 10 cm plate, depending on the cell line to be transfected. We routinely prepare plasmid DNA using pZ523 Spin Columns (5 Prime → 3 Prime, Boulder, Colorado).

3. The tube containing the DNA is incubated at 37 °C for 1 h before transfection; 480 μl 2× HBS is placed into a sterile tube and left at room temperature.

4. Using a vortex mixer with continual mixing, 60 μl 2 M $CaCl_2$ is added dropwise into the tube containing the DNA. The DNA-$CaCl_2$ mixture is added immediately into the 2× HBS. The DNA should also be added dropwise into the HBS with continual mixing using a vortex. The DNA-$CaCl_2$-HBS solution is left to precipitate in the laminar flow at room temperature for 30 min. Optimal time for precipitation may vary from cell type to cell type.

transfection

Note: Published procedures differ widely in the manner and rate of mixing of DNA/HBS/CaCl$_2$ and in the optimum precipitation time required. The object is to avoid the rapid formation of a coarse precipitate which would result in a decreased efficiency of transfection. Several factors other than the speed of mixing affect the size of the precipitate, including the concentration and size of the DNA and the exact pH of the buffer. If it is crucial to achieve the highest transfection efficiency, time should be spent optimizing these factors for each particular system.

5. Remove cells from the incubator and add the DNA-CaCl$_2$ mix dropwise onto the cells, swirling the plate gently to ensure even mixing. Return the plates to the incubator and incubate for 4–16 h under standard growth conditions.

Note: The amount of time that the precipitate should be left on the cells will vary with cell type.

6. After incubation the cells should be rinsed twice with 5 ml PBS and fed with complete medium. The plates should then be incubated under standard growth conditions until transfected cells are to be analyzed.

glycerol shock   Transfection efficiency in some cell lines is dramatically increased by "shocking" the cells with glycerol (DMSO can also be used; see Sambrook et al. 1989). Directly after removal of the DNA precipitate the cells are shocked. So at step 6 above proceed as follows:

6a. Remove media containing precipitate from the cells. Add 5 ml 10 % glycerol/1× HBS. Incubate for 3 min at room temperature. Remove the glycerol and rinse twice in 5 ml serum-free media. Refeed the cells with 10 ml complete media and incubate under standard growth conditions until the transfected cells are to be analyzed.

Note: The length of time that the cells are exposed to glycerol is critical as excessive exposure to glycerol will kill the cells. Different cell lines will show varying degrees of sensitivity to glycerol and each cell line being studied should be tested for optimal transfection efficiency with and without glycerol shocking, and for different time periods in glycerol.

Selection of Stable Transfectants

The selection conditions for the parental cell line should be determined, i.e., the minimum level of the selective drug that must be added to the medium to prevent cell growth.

Following 18–24 h of incubation in nonselective medium to allow expression of the transfected gene(s), the appropriate selective medium can be added. Mammalian cells can divide once or twice under selective conditions that will eventually kill them so it may be 24–48 h before the effects of a selective drug are visible on the cells. Selective medium should be changed every 2–4 days for 2–3 weeks to remove the debris of dead cells and to allow colonies of resistant (i.e., transfected) cells to grow. The cells are allowed to grow under selection for approximately ten doublings before individual colonies are picked and expanded into cell lines.

Note: The concentration of the selective drug can be increased gradually to the maximum recommended level. For example, for G418, selection can be started at 200 µg/ml, and over the course of 2–4 weeks the drug level can be increased to 800 µg/ml. If there is considerable cell kill at a particular drug concentration, the cells can be maintained at this concentration until they are growing well before increasing the concentration of the selective drug. We have found that a gradual increase in the selective drug concentration (rather than starting the selection at the maximum tolerated dose) promotes the development of stable transfectants.

Analysis of Ribozyme Clones

To determine the potential usefulness of specific ribozymes, their ability to cleave mRNA targets must be correlated with biochemical and physiological changes that result from target cleavage. Proof of ribozyme efficiency relies on demonstration of inhibition of targeted gene expression (using methods to show decreases in a target genes mRNA and protein) and any downstream phenotypic effects (such as decreased tumor cell growth, viral replication or drug resistance).

Analysis of RNA expression of the ribozyme itself, of its mRNA target and of any genes believed to be affected by the target can be studied by RT-PCR (see O'Driscoll et al. 1994 for description of RT-PCR method) or northern analysis (see chapter by N. Daly and M. Clynes in this volume). Since the ribozyme transcript may also act as antisense it is of interest to demonstrate that the cellular ribozyme retains the ability to cleave (as discussed by Symons 1992; Altman 1993). Detection of the cleavage products of the mRNA targeted has been elusive in many studies due to the rapid degradation of short RNAs by intracellular nucleases. One case where an intact hammerhead ribozyme mediated RNA cleavage product could be isolated and verified by RNA sequencing has been reported for *Xenopus* oocytes (Saxena and Ackerman 1990). The use of a disabled control ribozyme with mutations in the catalytic core (Kashani Sabet et al. 1992; Scanlon et al. 1994) also provides strong evidence that ribozyme mediated cleavage can contribute significantly to the inhibition of RNA function in mammalian cells. Several studies have detected cleavage products by PCR analysis using pairs of primers flanking the expected cleavage site (Sarver et al. 1990; Kashani Sabet et al. 1992). Recently detection of the cleaved fragment of *mdr1* mRNA was reported in human ovarian carcinoma cells by northern analysis (Scanlon et al. 1994). Other groups have shown that cellular extracts of ribozyme expressing cells and not control cells cleave target RNA in vitro (Chang et al. 1990; Scanlon et al. 1991).

Depending on the mRNA targeted various downstream phenotypic changes may be apparent. If a proto-oncogene such as *ras* is targeted cells may display altered growth characteristics and morphology compared to non-ribozyme containing cells (studied using [^{3}H]thymidine incorporation assays and colony formation in soft agar; Kashani Sabet et al. 1992). Western blotting or immunocytochemistry (see chapter by Moran et al in this volume for a description of these techniques) can be used to demonstrate that a reduction in the protein target has taken place compared to the untransfected parent cell line. If a gene believed to be involved in drug sensitivity is targeted, phenotypic changes such as alterations in drug sensitivity can be assayed using toxicity tests (Daly et al. 1996; Holm et al. 1994).

Results

Due to its high efficiency, transfection mediated by calcium phosphate is the method of choice for both transient expression of foreign DNA and also to establish cell lines that carry integrated copies of foreign DNA. Depending on the cell type up to 20 % of a population of cultured cells can be transfected at any one time using this method.

Having optimized parameters for CaPO$_4$ transfection approximately one in 10^4 cells in a transfection experiment will stably integrate DNA. Using a dominant selectable marker isolation of these stably transfected cells is possible.

Troubleshooting

There are two types of transfection that are routinely carried out in mammalian systems, transient and stable or permanent transfection.

Using a transient transfection assay, transcription or replication of the transfected gene can be analysed between 1 and 4 days after the introduction of the DNA. After this time expression of the foreign gene is lost. Alternatively many experiments require the formation of cell lines that contain the introduced genes integrated into chromosomal DNA resulting in stable or permanent transfection. Dominant selectable markers are used to permit isolation of stable transfectants. The gene to be studied is now generally cloned into a mammalian expression vector which also expresses a gene for a selectable marker. Several selection markers commonly used in mammalian cell culture are described in Ausubel et al. (1993). The choice of marker is determined by both the cell type to be transfected and the reason for carrying out the transfection.

Irrespective of the method used to introduce DNA into cells the efficiency of transient or stable transfection is determined largely by the cell type that is used. Every cell type has a characteristic set of requirements for optimal introduction of foreign DNA. Different lines of cultured cells vary by several orders of magnitude in their ability to take up and express exogenously added DNA. Transient assay systems using a reporter system (e.g., a plasmid coding for β-galactosidase (plasmid pCH110 eukaryotic assay vector; cat. no. 27–4508–01; Pharmacia Biochemicals) can be used to optimize transfection efficiencies. Transfections should be carried out and colonies stained and counted (Muller 1972); in this way the efficiency of a particular transfection method in a particular cell line can be ascertained. The first factor which should be optimized for a particular cell line is selection of the appropriate transfection protocol. It is important to compare the efficiencies of several different methods for each cell type being studied. In addition to $CaPO_4$ precipitation a number of other methods can be used to introduce DNA into cultured cells (Sambrook et al. 1989; Ausubel et al. 1993).

The efficiency with which cells take up DNA using $CaPO_4$ precipitation is extremely variable – the process is not difficult to do, it just does not always work even in the hands of people who routinely use it. Using $CaPO_4$ precipitation the most important factors that influence efficiency of DNA uptake are the pH of the HBS buffer, the amount of DNA in the precipitate, the length of time the precipitate is left on the cells and the use and duration of glycerol shock. One of the most common reason for transfection failure is use of a 2× HBS solution that is no longer at the appropriate pH. The optimum pH range for efficient transfection is extremely narrow (between 7.05 and 7.12, Graham and van der Eb 1973). The pH of this solution can change during storage and an old HBS solution may not work efficiently. It has also been reported that the efficiency of CaCl$_2$ solutions can deteriorate over time. Both solutions should be freshly prepared if transfection experiments are not working efficiently.

Total DNA concentration used in a particular experiment can have a dramatic effect on efficiency of DNA uptake. For some cell lines 10–15 µg of DNA added to a 10 cm dish results in excessive cell death and very little uptake of DNA, while for other cell types a high concentration of DNA in the precipitate is necessary to get any DNA at all into the cell. Efficiency is also dependent on the purity of the DNA. DNA prepared as described (see "Transfection"") should not be toxic to cells. If it is suspected that a particular plasmid prep is toxic, a control plasmid, one known not to be toxic to the particular cells, should be used to test for toxicity. If the plasmid DNA is toxic new DNA stocks should be prepared using an alternative protocol (for alternative methods for preparing plasmid DNA, see Sambrook et al. 1989; Ausubel et al. 1993).

The optimal length of time that the precipitate is left on the cells varies with cell type. Some cell types are efficiently transfected by leaving the DNA precipitate on for 16 h while other cell types cannot survive this length of exposure to DNA. Glycerol shock can improve the efficiency with which cells take up DNA; but this is very much dependent on individual cell lines. Each cell line should be transfected with and without a glycerol shock and the results monitored. The length of time the cells are exposed to glycerol can also be varied. For a more thorough discussion on optimization of transfection procedures see Ausubel et al. (1993).

Separation of transfected cells into clones (i.e., before selection) can be carried out at this point (i.e., the transfected cells should be split by a 1:15 dilution such that it is possible to identify isolated colonies) or after selection (i.e., feed in selective medium and then clone cells, as described in the chapter by McBride et al., this Vol.).

Having isolated a number of clones which grow in selective medium these clones can then be further analyzed. Presence of the plasmid carrying the gene of interest (in this case a ribozyme construct) in the cells can be confirmed by PCR using primers designed to flank the insert DNA in the plasmid (for example see Holm et al. 1994; Daly et al. 1996). Expression of the gene can be monitored using RT-PCR with specifically designed primers. A number of clones should be analyzed in this way. In our experience the clones expressing the highest levels of transfected ribozyme may not necessarily be the most efficient in terms of down-regulation of the target gene. It is important to analyze the expression of the target gene both in terms of mRNA expression (RT-PCR and northern blot analysis) and protein expression (western blot and immunocytochemistry) and also any other down-stream phenotypic changes.

References

Altman S (1993) RNA enzyme directed gene therapy. Proc Natl Acad Sci USA 90:10892–10900

Ausubel FM, Brent R, Kingston RE, Moore DD, Seidman JG, Smith JA, Struhl K (1993) Current protocols in molecular biology. Wiley, New York

Bertram J, Palfner K, Killian M, Brysch W, Schlingensiepen KH, Hiddemann W, Kneba M (1995) Reversal of multiple drug resistance in vitro by phosphorothioate oligonucleotides and ribozymes Anti Cancer Drugs 6:124–135

Bertrand E, Picket R, Grange T (1994) Can hammerhead ribozymes be efficient tools to inactivate gene function? Nucleic Acids Res 22(3):293–390

Cech TR (1990) Self splicing of group I introns. Annu Rev Biochem 59:543–568

Chang PS, Cantin EM, Zaia JA (1990) Ribozyme mediated site specific cleavage of the HIV-1 genome. Clin Biotechnol 2:23–31

Chen CJ, Banerjea AC, Harmison GG, Haglund K, Schubert M (1992) Multitarget ribozymes to cleave at up to 9 highly conserved HIV-1 env RNA regions inhibits HIV-1 replication – potential effectiveness against most presently sequenced HIV-1 isolates. Nucleic Acids Res 20:4581–4589

Christoffersen RE, Marr JJ (1995) Ribozymes as human therapeutic agents. J Med Chem 38:2023–2037

Crisell P, Thompson S, James W (1993) Inhibition of HIV-1 replication by ribozymes that show poor activity in vitro. Nucleic Acids Res 21:5251–5255

Daly C, Coyle S, McBride, O'Driscoll L, Daly N, Scanlon KJ, Clynes M (1996) *mdr1* mediated reversal of the multidrug resistant phenotype in human lung cell lines. Cytotechnology 19:199–205

Graham FL, van der Eb AJ (1973) A new technique for the assay of infectivity of human adenovirus 5 DNA. Virology 52:456

Haseloff J, Gerlach WL (1988) Simple RNA enzymes with new and highly specific endoribonuclease activity. Nature 334:585–591

Holm PS, Scanlon KJ, Dietel M (1994) Reversion of multidrug resistance in the p-glycoprotein positive human pancreatic cell line (EPP85–181RDB) by introduction of a hammerhead ribozyme. Br J Cancer 70:239–243

Kashani Sabet M, Funato T, Tone T, Jiao L, Wang W, Yoshida E, Hahfinn BI, Shitara T, Wu AM, Moreno JG, Traveek ST, Ahlering TE, Scanlon KJ (1992) Reversal of the malignant phenotype by an anti-*ras* ribozyme. Antisense Res Devel 2:3–15

Koizuni M, Iwai S, Ohtsuka (1988) Construction of a series of self cleaving RNA duplexes using a synthetic 21 mer. FEBS Letts 228:228–230

Lange W, Cantin EM, Finke J, Dolken G (1993) In vitro and in vivo effects of synthetic ribozymes targetted against BCR/Abl mRNA. Leukemia 7:1786–1794

Muller JH (1972) Experiments in molecular genetics. Cold Spring Harbor Laboratory, Cold Spring Harbor, New York, p 352

O'Driscoll L, Daly C, Saleh M, Clynes M (1994) The use of reverse-transcriptase polymerase cahin reaction (RT-PCR) to investigate specific gene expression in multidrug-resistant cells. Cytotechnology 12:289–314

Palfner K, Kneba M, Hiddemann W, Bertram J (1995) Improvement of hammerhead ribozymes cleaving *mdr1* mRNA. Biol Chem Hoppe Seyler 376:289–295

Perriman R, Delves A, Gerlach WL (1992) Extended target site specificity for a hammerhead ribozyme. Gene 113:157–163

Ruffner DE, Stormo GD, Uhlenbeck OC (1990) Sequence requirements of the hammerhead ribozyme RNA self cleavage reaction. Biochemistry 29:10695–10702

Sambrook J, Fritsch E, Maniatis (1989) Molecular cloning: a laboratory manual. Cold Spring Harbor Laboratory, Cold Spring Harbor, New York

Sarver N, Cantin EM, Chang PS, Zaia JA, Ladre PA, Stephens DA, Rossi JJ (1990) Ribozymes as potential HIV-1 therapeutic against. Science 247:1222–1225

Saxena SK, Ackerman EJ (1990) Ribozymes correctly cleave a model substrate and endogenous RNA in vivo. J Biol Chem 265:17106–17109

Scanlon KJ, Ishida H, Kashani Sabet M (1994) Ribozyme mediated reversal of the multidrug resistant phenotype. Proc Natl Acad Sci USA 91:11123–27

Scanlon KJ, Jiao L, Funato T, Wang W, Tone T, Rossi JJ, Kashani Sabet M (1991) Ribozyme mediated cleavage of *cfos* mRNA reduces gene expression of DNA synthesis enzymes and metallothionein. Proc Natl Acad Sci USA 88:10591–10595

Sharma HW, Narayanan R (1995) The therapeutic potential of antisense oligonucleotides. BioEssays 17(12):1055–1063

Stull RA, Szoka (Jr) FC (1995) Antigene, ribozyme and aptamer nucleic acid drugs: progress and prospects. Pharm Res 12:465–483

Sullivan SM (1994) Development of ribozymes for gene therapy. J Invest Dermatol 103:855–895

Symons RH (1992) Small catalytic RNAs. Annu Rev Biochem 61:641–671

Uhlenbeck OC (1987) A small catalytic oligoribonucleotide. Nature 328:596–600

Zhang WW (1996) Antisense oncogene and tumour suppressor gene therapy of cancer. J Mol Med 74:191–204

Methods for Studying the Transcription and DNase I Hypersensitive Sites of Genes in Cells in Culture

Noel Daly* and Martin Clynes

▦ Introduction

The study of gene regulation has expanded enormously over the last 20 years due to the availability of a wide range of new techniques. It is now central to molecular biology and is essential to an understanding of any gene or protein. Initially it was believed that transcription was driven by the interaction of transcription factors with conserved *cis*-elements in gene promoters. However, with the discovery of enhancer sequences and transcription factor binding sites within introns it was realized that gene transcription was a far more complex process. Gene expression is controlled (in part) by the binding of two classes of transcription factors to conserved regulatory elements. These are:

- General transcription factors, e.g., TFIID, which are used by the vast majority of RNA polymerase II transcribed genes

- Gene-specific transcription factors, e.g., AP-1, which are used specifically and in precise arrangements

In order that the binding of this array of transcription factors can proceed, the regularly spaced nucleosomes, which play a role in condensing chromatin structure, must be removed or displaced from around the protein-binding sites. This creates regions of DNA with open conformations within and around genes. This makes these regions particularly sensitive to digestion by nucleases such as DNase I and micrococcal nuclease. These DNase I hypersensitive sites can therefore be exploited to map the protein-binding sites within and around genes.

---

* *Correspondence to* Noel Daly, The National Cell and Tissue Culture Centre, Dublin City University, Glasnevin, Dublin 9, Ireland; phone +353–1–7045804; fax 353–1–7045804; e-mail 75029910@raven.dcu.ie

Many methods are now used to identify these protein-binding regions and to characterize the proteins that bind to these sites. This chapter will describe some of the techniques used to identify the DNase I hypersensitive sites of genes. Sambrook et al. (1989) and the series *Current Protocols in Molecular Biology* provide detailed descriptions of all the techniques that are currently used to study DNA-binding proteins.

Gene expression can often be induced, up-regulated or down-regulated by a range of compounds such as growth factors, differentiation-inducing agents, etc. In order to study these effects the level of mRNA must be quantified. This is covered in Section 31.1 of this chapter. Section 31.2 describes the mapping of DNase I hypersensitive sites. Section 31.3 deals with the nuclear transcription runoff procedure. This technique can provide a wealth of information on the regulation of a gene through altering the cellular environment by treatment with compounds such as mitogens, etc. Some of the techniques that may be used to study in great detail a protein-binding site and to characterize the protein(s) that bind(s) to the site are briefly discussed here.

The use of gene constructs containing reporter genes such as chloramphenicol acetyltransferase (CAT), β-galactosidase, and firefly luciferase in in vivo and in vitro assays can be very helpful for determining which regions of genes (particularly 5' and 3' untranslated regions, UTRs) are necessary for basal and full gene expression. These reporter genes can also be particularly valuable for identifying and studying enhancer sequences. They are very well described in *Current Protocols in Molecular Biology.*

Once a protein-binding region(s) has been identified there are several techniques which can be used to study the transcription factor(s) which bind to the site. DNA mobility shift assays can be used to verify the identified protein-binding site by incubating the sequence with a nuclear extract from the cell line in question. DNase I footprinting is frequently used to determine precisely the region(s) of a protein-binding site to which the protein binds. Methylation and uracil interference assays are two methods for, respectively, determining which purine (adenine and guanine) and thymine bases within a defined DNA binding site are important for protein binding reactions.

After the protein-binding site has been detected the size of the protein itself can be determined by covalently linking it to its recognition sequence (by ultraviolet light) and electrophoresing the DNA-protein complex on an SDS-polyacrylamide gel. The protein can subsequently be purified to homogeneity, partially sequenced and a cDNA clone coding for the protein isolated by screening an expression library with oligo-nucleotide probes. More details on all of these methods can be found in *Current Protocols in Molecular Biology*.

31.1
Quantification of mRNA Transcripts

There are several methods that can be used to analyze mRNA transcripts. Reverse transcriptase polymerase chain reaction (RT-PCR) is perhaps the simplest and quickest procedure to perform, particularly for multiple samples. This technique is described in the chapter by O'Doherty et al. Other procedures are S1 nuclease mapping, RNase mapping, primer extension, and RNA dot blots. However, for comparative and quantitative measurements of mRNA levels, we use Northern blot analysis as we have found that it is the most reliable technique for accurately quantifying mRNA levels.

Materials

- 10× MOPS buffer: 0.25 M MOPS; 0.05 M Na. acetate; 0.01 M EDTA.   **buffers**
  Adjust pH to 7.0 with NaOH and autoclave. The solution will turn yellow.
- Loading buffer: 50 % glycerol; 1 mg/ml xylene cyanol; 1 mg/ml bromphenol blue; 1 mM EDTA. There is no need to autoclave this solution.

Note: Unlike RNA isolation techniques we have found that it is not necessary to treat the buffers and equipment used in this protocol with diethylpyrocarbonate (DEPC).

- Prehybridization/hybridization buffer: 0.43 M Na.phosphate, pH 7.2; 7 % SDS; 1 % BSA; 20 mM EDTA. Filter sterilize (at the hybridization temperature) before use.

Note: Prepare 1 M sodium phosphate pH 7.2 by mixing 31.6 ml of 1 M NaH_2PO_4 and 68.4 ml of 1 M Na_2HPO_4.

▪ Procedure

Isolation of Total RNA and Poly(A)⁺ RNA

These procedures are described in the chapter by O'Doherty et al. They are also described in detail in Sambrook et al. (1989) and in *Current Protocols in Molecular Biology.*

Northern blot analysis of mRNA transcripts

This technique is used to determine the levels and sizes of mRNA in biological materials. The RNA is electrophoresed through a formaldehyde-agarose gel, transferred to a nylon membrane support and hybridized to a DNA or an RNA probe. The decision to use total RNA or poly(A)⁺ RNA depends on the level of the mRNA being studied. From 10 to 50 μg of total RNA and 1 μg–5 μg of poly(A)⁺ RNA are commonly used.

formaldehyde-agarose electrophoresis of RNA

1. For a 100 ml gel: dissolve the agarose in 73.4 ml water and cool to approximately 60 °C. In a fume hood add 10 ml 10× MOPS buffer, 16.6 ml formaldehyde, 5 μl ethidium bromide (10 mg/ml), mix well, and pour the gel.

2. The RNA to be electrophoresed is prepared in the following way: RNA (up to 1.75 μl), 10× MOPS (1.0 μl), formaldehyde (1.75 μl), formamide (5.0 μl), loading buffer (0.5 μl), H_2O to 10 μl. These amounts can be scaled up or down. Incubate the samples at 65 °C for 15 min, place on ice and load. Run the gel in 1× MOPS until the bromphenol blue band has run half to two-thirds way down the gel.

Note: The volume of sample loaded depends on the concentration of the RNA and on the amount of RNA required for a signal. For this reason it is helpful if the RNA is resuspended at a high concentration at the end of the RNA preparation.

3. Rinse the gel in deionized water for 30 min (three 10 min washes) and photograph it. Blot the gel overnight on to nylon membrane, e.g., Hybond N (Amersham, UK), by the standard capillary blotting procedure as described in Sambrook et al. (1989). In total RNA samples the large (28S) and small (18S) ribosomal RNA bands should be clear and sharp (Fig. 31.1). Poly(A)⁺ RNA will appear as a smear on the gel. RNA molecular weight markers should be included in the gel in order to determine the size of the detected band.

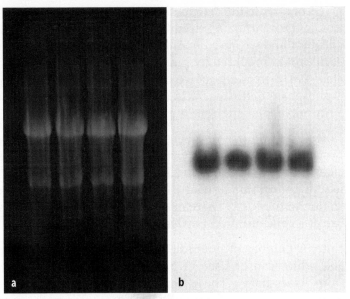

Fig. 31.1. Northern blot analysis of total RNA isolated from a drug-sensitive cell line (DLKP) and a number of cell lines resistant to the chemotherapeutic drug adriamycin using the procedure described by O' Doherty et al. *Lane 1* DLKP; *2* DLKPA; *3* DLKPA2B; *4,* DLKPA5F. Total RNA (10 µg) was electrophoresed on a 1.0 % formaldehyde-agarose gel, photographed, blotted, and hybridized to a glyceraldehyde-3-phosphate dehydrogenase (GAPDH) cDNA probe (obtained from the American Type Culture Collection) prepared by random primer labeling (Feinberg and Vogelstein 1983). The autoradiograph was exposed at −80 °C for several hours

4. Before removing the filter mark the position of each well and the orientation of the filter. Air dry briefly and bake at 80 °C for 2 h. Store the filter at 4 °C in a sealed bag. Total RNA should be clearly visible on the filter. Check the gel to ensure that the RNA has been fully transferred.

5. Prehybridize the membrane at 65 °C for 15–30 min. Boil the probe (if double-stranded DNA) for up to 5 min, add to the prehybridization buffer and hybridize overnight.

hybridization analysis

Note: The probe to be used for hybridization can be prepared by a number of procedures such as random primer labeling, nick translation, etc. For experiments in which high sensitivity is required an RNA probe prepared by in vitro transcription should be considered. These probes give higher sensitivity than double-stranded cDNA probes because there is no second strand to "compete" with the filter-bound RNA for hybridization to the hybridizing strand. However, for most genes cDNA probes are adequate and we have obtained good results for most experiments using such probes prepared by random primer labeling (Weinberg and Vogelstein 1983). In addition, we use a control probe such as glyceraldehyde-3-phosphate dehydrogenase (GAPDH) to ensure that equal amounts of RNA are loaded and that the RNA is not degraded (Fig. 31.1).

Note: The amounts of labeled cDNA probes used in the hybridization procedure can be very important. If too little is used there will be insufficient probe to hybridize to the filter-bound RNA. If too much probe is used hybridization of the two cDNA strands will be favored over hybridization to the RNA. Therefore, we label between 25–100 ng of cDNA and use at a concentration between 2 and 5 ng/ml.

Note: A number of nonradioactive detection procedures are now available which can be used in place of radioisotopes (e.g., the ECL system from Amersham). These have obvious advantages in terms of safety. Also, a large batch of nonradioactively labeled probe can be prepared in one reaction and then stored for months if necessary until required. However, these protocols are not as sensitive as those using radioisotopes and this should be borne in mind when choosing which protocols to use.

2. The posthybridization washes are performed at the hybridization temperature. Wash the filter once in $2\times$ SSC/0.1 % SDS for 5 min (to remove unhybridized probe) and twice in $0.5\times$ SSC/0.1 % SDS for 15 min each wash. The filter can be removed and autoradiographed at this stage. If higher stringencies are required wash the filter in either $0.2\times$ SSC/0.1 % SDS or $0.1x$ SSC/0.1 % SDS for two 15 min washes. It is important to keep the filter moist at all times if it is to be stripped and reused.

31.2
Mapping of DNase I Hypersensitive Sites

When a gene is transcribed or "active" the nucleosomes are removed around those regions to which the transcription factors must bind in order to drive transcription. These regions are therefore characterized by their open conformation which makes them particularly susceptible to DNase I digestion. These hypersensitive sites can be exploited to map the protein binding regions of active genes and in some cases inactive genes in which gene repression is due to the action of negative transcription factors. The technique is particularly useful for studying large regions of a gene in one experiment where other techniques would not be as feasible.

The procedure is described in detail by Wu (1989) and Bellard et al. (1989). Both references offer alternative methods of isolating nuclei from cells in culture and from tisssues and of performing the DNase I digestions. Both references provide a wealth of background information as well as troubleshooting guides when problems are encountered in this technique. In addition, Wu describes exonuclease III protection while Bellard et al. explain the use of micrococcal nuclease.

Materials

- NP-40 lysis buffer: 10 mM Tris-HCl, pH 7.4; 10 mM NaCl; 3 mM $MgCl_2$; **buffers** 0.5 % Nonidet P-40 (Sigma, N-6507). Autoclave then add NP-40.
- Nuclear storage buffer (Wu 1989): 15 mM Tris-HCl, pH 7.4; 60 mM KCl; 15 mM NaCl; 5 mM $MgCl_2$; 0.1 mM EGTA; 300 mM sucrose; 5 % glycerol; 0.5 mM DTT (Boehringer Mannheim, cat. no. 197 777); 0.1 mM PMSF (Boehringer Mannheim cat. no. 236 608). Autoclave. Add DTT and PMSF just before use. PMSF is prepared as a 0.1 M stock solution in isopropanol. Both DTT and PMSF are stored at −20 °C.
- Denaturing solution: 1.5 M NaCl; 0.5 M NaOH. The solution does not need to be autoclaved.
- Neutralizing solution: 1.5 M NaCl; 0.5 M Tris. Adjust pH to 7.5 with HCl. The solution does not need to be autoclaved.
- Solution I: 15 mM Tris-HCl, pH 7.5; 60 mM KCl; 15 mM NaCl; 5 mM $MgCl_2$; 0.5 mM EGTA; 300 mM sucrose; 0.5 mM β-mercaptoethanol; 0.2 % NP-40. **Note:** Add the β-mercaptoethanol and NP-40 after the other components have been autoclaved.
- Solution II: 50 mM Tris-HCl, pH 8.0; 20 mM EDTA; 1 % SDS

- Nylon membrane, e.g., Hybond N (Amersham, UK)   **equipment**
- Promega Prime-a-gene labeling system (cat. no. U1100)
- Autoradiography film, e.g., Kodak X-OMAT AR

▦ Procedure

Isolation of Nuclei from Cells in Culture

The particular protocol used to isolate the nuclei can vary depending on the type of cell line used, e.g., whether it grows attached or in suspension. Marzluff and Huang (1985) give an excellent background to the procedure as well as a comprehensive explanation of the problems that may be encountered. The procedure described here is that of Greenberg and Bender (1989), with some modifications, using roller bottles to provide enough cells to give an adequate yield of nuclei for multiple DNase I digests. For this procedure all steps are performed at 4 °C. The centrifuge, buffers, tubes, etc., must be precooled to 4 °C. It can be helpful to carry out the entire procedure in a coldroom.

isolation of nuclei

1a. For adherent cell lines: wash the cells twice in PBS. Trypsinize, centrifuge the cell suspension at 1000 rpm for 5 min, remove the supernatant and leave on ice for 5–10 min. Alternatively, wash the cells in ice-cold PBS and scrape them off using a cell scraper (e.g., Sigma C2677). Centrifuge at 1000 rpm for 5 min and remove the supernatant.

1b. For suspension cultures: centrifuge the cells at 1000 rpm for 5 min. Resuspend the pellet in 25 ml PBS and centrifuge again. Repeat this twice and proceed to step 2.

2. Resuspend the pellet in 10 ml NP-40 lysis buffer. Leave the cells on ice for 5 min. The sample should be examined using phase contrast microscopy to ensure that the cells have lysed completely and that the nuclei are intact. Centrifuge at 1000 rpm for 5 min. Remove the supernatant.

3. Resuspend the nuclear pellet in 10 ml NP-40 lysis buffer. Centrifuge at 1000 rpm for 5 min. Remove the supernatant and resuspend the pellet in 100–2000 µl nuclear storage buffer to give a concentration of between $1–5 \times 10^7$ nuclei/ml. Freeze the nuclei in liquid nitrogen and store at −80 °C.

DNase I Digestion

1. Remove the stored nuclei from −80 °C and supplement with 0.1 mM CaCl$_2$ at a concentration equivalent to 200 μg/ml of DNA. Set up several tubes with varying levels of DNase I (Boehringer Mannheim, cat. no. 776 785), from 0.5 to 6.0 μg/ml. Incubate at 30 °C for 10 min (Neznanov and Oshima 1993). Two controls should be included. DNase I is not added to either control and the first control is incubated under the same reaction conditions as the DNase I samples while the second control is retained on ice until the DNA isolation step. These controls will provide information about the endogenous levels of DNases.

2. Terminate the digestion by adding three volumes of 1 % SDS/20 mM EDTA. Add proteinase K to 200 μg/ml and incubate at 37 °C overnight. The sample should be appreciably less viscous after this step. **Note:** Use of three volumes of the SDS/EDTA solution aids in the proteinase K digestion and makes the phenol/chloroform extraction step easier.

3. Phenol/chloroform (50:50) extract once or twice (until there is no significant interface) and ethanol precipitate. Pellet the DNA by centrifugation in a microfuge for 15 min, wash twice with 70 % ethanol and resuspend in distilled water. Estimate the DNA concentration by spectrophotometry. An aliquot of each sample should be electrophoresed on an agarose gel to check the level of digestion (Fig. 31.2). **Note:** To remove the protein interface during the extractions it can help to use blue or yellow tips which have been cut to make the tips wider. This disturbs the interface less and makes removing the aqueous phase easier.

Detection of Hypersensitive Sites

To detect the hypersensitive sites the DNA is digested with a restriction enzyme, blotted on to nylon membrane, e.g., Hybond N (Amersham, UK), and hybridized with a labeled probe. The hybridization reaction is performed using a probe which hybridizes close to either end of a restriction fragment. The size(s) of the band(s) on the autoradiograph, when measured from the end of the restriction fragment closest to which the probe hybridizes, give an accurate estimate of the location of the protein binding site(s).

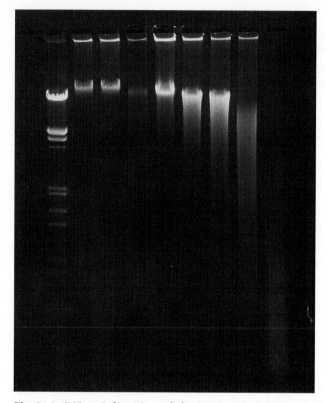

Fig. 31.2. DNase I digestion of chromatin. Nuclei were isolated from the lung ade-nocarcinoma cancer cell line A549 and digested with increasing concentrations of DNase I (as indicated) for 10 min at 30 °C. The DNA was isolated and an aliquot elec-trophoresed on a 1.0 % agarose gel. *Lane 1* bacteriophage lambda DNA *Eco*RI/*Hin*d III digest; *2* control without DNase I; *3* control without DNase I and incubated under the reaction conditions; *4* 0.5 mg/ml DNase I; *5* 1.0 µg/ml; *6* 2.0 mg/ml; *7* 3.0 µg/ml; 5.0 µg/ml; *9* 10.0 µg/ml

1. Digest 10–20 µg of the DNA samples overnight using 7–10 units of restriction enzyme per µg DNA. Electrophorese the digests on a 0.7 % to 1.0 % agarose gel overnight. **Note:** In initial experiments a restric-tion enzyme should be chosen which produces large fragments. This will help to identify the hypersensitive sites over a large area. Subse-quent experiments can then be carried out to map the hypersensitive sites in greater detail using enzymes which produce smaller frag-ments.

2. Photograph the gel under UV. Depurinate the DNA by treating the gel with 0.25 M HCl for 30 min. Wash once with deionized water. Treat the gel in denaturing solution for two 20 min washes on a shaking platform. Wash once with deionized water. Treat the gel in neutralizing solution for two 20 min washes. Wash once with deionized water.

3. Set up a capillary blot using 20× SSC as the transfer buffer, as shown in Sambrook et al. (1989), and leave overnight. Remove the absorbent paper and mark the orientation of the gel and the position of the wells with a pencil. Air dry the filter briefly and bake it at 80 °C for 2 h. Check the gel to ensure that the DNA has transferred completely to the filter.

4. Label the DNA probe using the random primer labeling procedure (Feinberg and Vogelstein 1983). Prehybridize the filter at 65 °C for 15–30 min (using the same buffer as that used for the Northern blot analysis), boil the probe for up to 5 min and add it to the prehybridization buffer. Incubate overnight. The posthybridization washes are the same as those used for the Northern blot analysis. Expose the filter to X-ray film and store at −80 °C.

Alternative Protocol

This protocol has the advantage of being significantly quicker than the procedure outlined above, as the DNase I digestion step is performed on the cells in culture (Stewart et al. 1991). It is used with cell lines that grow attached as a monolayer. It has the added advantage that, since the DNase I digestion is performed in situ, there is less chance that some chromatin proteins will be lost or that some chromatin structures will be altered. A good example of how it has been successfully used is provided by Parsa et al. (1996).

1. Set up the cells in 6-cm dishes (approximately 5×10^6 cells). Remove the culture media and cover the cells with 2 ml of solution I containing DNase I (200–1200 U/ml) (Boehringer Mannheim, cat. no. 104 159).

alternative digestion

2. Remove the solution after 3.5 min and lyse the cells with 1 ml solution II. Add proteinase K to 200 µg/ml and incubate overnight at 37 °C.

3. Phenol/chloroform (50:50) extract the lysate once or twice and precipitate. Centrifuge at 13 000 rpm for 15 min, rinse the pellet with

70 % ethanol and resuspend in distilled water. Estimate the DNA concentration by spectrophotometry. Proceed from step 1 of "Detection of Hypersensitive Sites."

31.3
Nuclear Runoff Transcription Assays

This technique is used to measure the transcription rate of a gene and to elucidate the mechanism of gene regulation by altering the cellular environment. This can be done by treating the cells with a range of compounds such as retinoic acid, growth factors, etc., and thereafter comparing the transcription rates of genes from treated and untreated cells. The procedure can be divided into three parts: (1) isolation of nuclei, (2) runoff transcription and isolation of RNA, and (3) quantitation of mRNA. The protocol we use is a variation of the procedure described by Greenberg and Bender (1989). We observe all the precautions used for the isolation of RNA in order to prevent RNA degradation.

■ Materials

equipment
- Nylon membrane, e.g., Hybond N (Amersham, UK)
- Dot blot apparatus, e.g. Hybridot (BRL)
- Autoradiography film, e.g. Kodak X-OMAT AR

buffers
- Glycerol storage buffer: 50 mM Tris-HCl, pH 7.5; 40 % glycerol; 5 mM $MgCl_2$; 0.1 mM EDTA
- High salt buffer: 10 mM Tris-HCl, pH 7.4; 50 mM $MgCl_2$; 2 mM $CaCl_2$; 0.5 M NaCl
- 2x Reaction buffer: 10 mM Tris-HCl, pH 8.0; 5 mM $MgCl_2$; 300 mM KCl
- 2x Reaction buffer (plus nucleotides): 1 ml 2X reaction buffer; 1 mM each of ATP, CTP, GTP; 5 mM DTT. **Note:** Prepare this solution just before use.
- Resuspension buffer (Marzluff and Huang 1985): 10 mM Tris-HCl, pH 7.5; 0.3 M NaCl; 0.1 % SDS; 1 mM EDTA
- Hybridization buffer (Bossard et al. 1993): 50 % formamide; 1 % glycine; 4 % SDS; 10x Denhardt's solution; 0.003 M EDTA; 0.54 M NaCl; 0.06 M NaH_2PO_4; 100 µg/ml polyadenylic acid (Boehringer Mannheim, cat. no. 108 626); 200 µg/ml salmon sperm DNA (Sigma D9156)

▓ Procedure

Isolation of Nuclei

Isolate nuclei as described in Section 31.2. However, for this protocol the nuclei are resuspended finally in a different glycerol storage buffer at a concentration of 5×10^7/ml. Freeze the nuclei in liquid nitrogen and store at $-80\,°C$. Marzluff and Huang (1985) provide alternative methods to isolate the nuclei and to perform the nuclear runoff assays. They also provide a wealth of background information for both procedures.

Runoff Transcription and Isolation of RNA

1. To 100 µl (5×10^6 nuclei) of thawed nuclei add 100 µl 2x reaction buffer (plus nucleotides) and 10 µl $[\alpha^{32}P]UTP$ (10 mCi/ml; 400 Ci/mmol) (Amersham, UK). Incubate at $30\,°C$ for 30 min. **Note:** In this step labeled mRNA transcripts are synthesized. It is important to use at least 5×10^6 nuclei. Below this number the incorporation of the labeled nucleotide is very low. It is also important to use the same number of nuclei for each sample.

2. Mix 20 µl of 1 mg/ml RNase-free DNase I (Boehringer Mannheim cat. no. 776 785) and 0.5 ml high salt buffer. Add 0.3 ml of this solution to nuclei. Incubate at $30\,°C$ for 5 min.

3. Add three volumes 1 % SDS/10 mM EDTA and 10 mg/ml proteinase K (Boehringer Mannheim) to 200 µg/ml and incubate at $50\,°C$ for 30 min. **Note:** These two steps degrade the DNA and proteins leaving the cellular RNA intact. Marzluff and Huang provide an alternative hot phenol-SDS procedure for isolating the RNA.

4. Extract the samples with phenol/chloroform/isoamyl alcohol (25:24:1) once or twice (until there is no significant interface) and ethanol precipitate overnight at $-20\,°C$.

5. Recover the RNA by centrifugation at $4\,°C$. Dissolve the RNA in resuspension buffer. Remove the unincorporated label by using a Sephadex G50 column. Ethanol precipitate, recover the RNA at $4\,°C$ and resuspend the pellet in water. Store at $-80\,°C$.

isolation of RNA

Quantitation of mRNA

quantitation

1. Linearize the plasmid DNA by digestion with an appropriate restriction enzyme. Ethanol precipitate the DNA and centrifuge at 13 000 rpm at 4 °C. Wash the pellet twice with 70 % ethanol and dry the pellet. Resuspend the DNA in 0.2 M NH_4OH, 2.0 M NaCl at a concentration of 50 µg/ml. Boil the DNA for up to 5 min, cool on ice and spot approximately 5 µg DNA on to nylon membrane using a dot blot apparatus (Marzluff and Huang 1985). Hybridize the RNA to the plasmid DNA at 42 °C for 48 h.

Note: When the experiments are being set up it is important that the same quantity of labeled mRNA transcripts (as determined by the counts per minute) are used for each sample to hybridize to the filter-bound DNA. This will ensure that comparisons of the hybridization signal strengths between the samples are meaningful. This of course presumes that the transcription rates are comparable for each sample. This may have to be verified. Two plasmid controls should be included. Firstly, a gene such as β-actin or GAPDH should be used to ensure that the transcription rates for each sample is approximately the same. Secondly, a nonrecombinant plasmid should be used as a negative control. It is important to keep the hybridization volume as low as possible.

2. All washes are carried out at the hybridization temperature. Wash the filters once in 2× SSC/0.1 % SDS for 5 min and twice in 0.5× SSC/ 0.1 % SDS for 15 min each wash, and autoradiograph the filters overnight.

▦ References

Bellard M, Dretzen G, Giangrande A, Ramain P (1989) Nuclease digestion of transcriptionally active chromatin. In: Wasserman PM, Kornberg RD (eds) Nucleosomes. Methods Enzymol 179:317–346

Bossard P, Parsa R, Decaux J-F, Iynedjian P, Girard J (1993) Glucose administration induces the premature expression of liver glucokinase gene in newborn rats. Eur J Biochem 215:883–892

Feinberg AP, Vogelstein B (1983) A technique for radiolabeling DNA restriction endonuclease fragments to high specific activity. Anal Biochem 132:6–13

Greenberg ME, Bender TP (1989) Identification of newly transcribed RNA. Nuclear runoff transcription in mammalian cells. In: Ausubel FM, Brent R, Kingston RE, Moore DD, Seidman JG, Smith JA, Struhl K (eds) Current protocols in molecular biology. Wiley, Chichester, pp 4.10.1–4.10.9

Marzluff MF, Huang RCC (1985) Transcription of RNA in isolated nuclei. In: Hames BD, Higgins SJ (eds) Transcription and translation: a practical approach. IRL, Oxford, pp 89–129

Neznanov NS, Oshima RG (1993) Cis regulation of the keratin 18 gene in transgenic mice. Mol Cell Biol 13:1815–1823

Parsa R, Decaux J-F, Bossard P, Robey BR, Magnuson MA, Granner DK, Girard J (1996) Induction of the glucokinase gene by insulin in cultured neonatal rat hepatocytes. Relationship with DNase-I hypersensitive sites and functional analysis of a putative insulin-response element. Eur J Biochem 236:214–221

Sambrook J, Fritsch EF, Maniatis T (1989) Molecular cloning: a laboratory manual, 2nd edn. Cold Spring Harbor Laboratory, New York

Stewart AF, Reik A, Schutz G (1991) A simpler and better method to cleave chromatin with DNase I for hypersensitive site analyses. Nucleic Acids Res 19:3157.

Wu C (1989) Analysis of hypersensitive sites in chromatin. In: Wasserman PM, Kornberg RD (eds) Nucleosomes. Methods Enzymol 170:269–289

Chapter 32

Analysis of Specific and Differentially Expressed mRNAs in Cell Culture

Toni O'Doherty*, Róisín Nic Amhlaoibh*, and Martin Clynes

Introduction

Analysis of mRNA levels of a gene is desirable in a broad range of research interests including cell development, differentiation, death, normal physiological function, and in many disease states in which the steady state level of many gene transcripts is altered. In some situations the gene of interest has been characterized and its cDNA sequence is known. However, in other cases the specific gene has not been identified and it is necessary to search for unknown mRNAs. In this chapter we discuss analysis of specific and differentially expressed mRNAs in cell culture including various methods for RNA extraction.

prevention of RNase contamination

- Glassware to be used should be baked at 180 °C for 8 h or more
- All chemicals should be weighed out onto baked tin-foil
- Sterile disposable plastics should be used where possible
- All solutions, except those containing Tris, which can be autoclaved should be treated with 0.1 % (v/v) diethylpyrocarbonate (DEPC) before autoclaving
- Disposable gloves should be worn at all times
- All procedures should be carried out under sterile conditions when feasible

Materials

buffers, solutions

Note: Many of the chemicals used in the following protocols are highly toxic and great care should be taken when they are in use.

* *Correspondence to* Toni O'Doherty, National Cell and Tissue Culture Centre, Dublin City University, Glasnevin, Dublin 9, Ireland; phone +353–1–7045728; fax +353–1–7045484; e-mail 93700466@tolka.dcu.ie, 75028577@raven.dcu.ie

- 4 M Guanidium thiocyanate: 50 g guanidium thiocyanate (Sigma, G-6639); 0.5 g N-lauroyl sarcosine (Sigma,, L-5125); 5 ml 1 M Na-citrate, pH 7. Bring to 100 ml with H_2O and check that pH is approximately 7. Filter through 0.45 µm filter and store at room temperature in the dark. Before use add: 700 µl/100 ml β-mercaptoethanol (Sigma,, M-6250); 330 µl/100 ml Antifoam A (30 %) (Sigma, A-5758).
- 5.7 M Cesium chloride: 95.8 g CsCl (Sigma, C-3032); 2.5 ml 1 M Na-citrate, pH 7. Bring to 100 ml with H_2O. Filter sterilize, DEPC-treat and autoclave. Store at room temperature.
- 3 M Sodium acetate, pH 5.2
- 2 M Sodium acetate, pH 4
- Ice-cold ethanol
- 1 M Na-citrate: 29.4 g Na-citrate (RDH, 32320). Bring to 80 ml with H_2O and pH to 7 with HCl. Bring to 100 ml and filter sterilize. Store at room temperature.
- Denaturing solution: 4 M guanidium thiocyanate; 25 mM sodium citrate, pH 7; 0.5 % N-laurylsarcosine. Prepare as a stock solution and aliquot. Heating at 60 °C is required to dissolve the N-laurylsarcosine. Add 2-mercaptoethanol to a final concentration of 0.1 M just prior to use.
- Lysis buffer: 10 mM Tris, pH 7.6; 10 mM NaCl: 2 mM EDTA, pH 8.0; 1 % sodium dodecyl sulfate (prepared from RNase free stock solutions)
- Water saturated phenol: Dissolve 50 g of phenol crystals in H_2O at 65 °C in a water bath. Then aspirate the water phase and store at 4 °C.
- Chloroform: isoamyl alcohol (24:1)
- Isopropanol
- Oligo dT cellulose
- High salt buffer: 10 mM Tris-Cl, pH 7.4; 0.4 M NaCl; 2 mM EDTA pH 8.0; 0.2 % sodium dodecyl sulfate (Prepared from Rnase-free stock solutions.)
- Low salt buffer: 10 mM Tris-Cl, pH 7.4; 0.1 M NaCl; 2 mM EDTA, pH 8.0; 0.2 % sodium dodecyl sulfate (prepared from RNase free stock solutions)
- Salt free buffer: 5 mM Tris-Cl, pH 7.4; 1 mM EDTA, pH 8.0; 0.2 % sodium dodecyl sulfate (prepared from Rnase-free stock solutions)
- Proteinase K: 10 mg/ml dissolved in DEPC-treated H_2O
- 5X Reverse transcription buffer: 125 mM Tris-Cl, pH 8.3; 188 mM KCl; 7.5 mM $MgCl_2$; 25 mM DTT
- 10× PCR amplification buffer: 100 mM Tris-Cl, pH 8.3; 500 mM KCl; 15 mM $MgCl_2$; 0.01 % gelatin
- Formamide loading buffer: 95 % formamide; 10mM EDTA, pH 8.0; 0.09 % xylene cyanol; 0.09 % bromphenol blue

- 10× TBE: 890 mM Tris base; 890 mM boric acid; 20 mM EDTA, pH 8.0
- 10× Phosphate buffered saline (PBS): 40 g/500 ml NaCl; KCl 1 g/500 ml; $Na_2HPO_4.7H_2O$ 5.75 g/500 ml; KH_2PO_4 1 g/500 ml
- 20× SSC: 3 M NaCl (87.66 g/500 ml); 0.3 M Na-citrate (44.12 g/500 ml)
- Pepsin/HCl: Immediately prior to use dissolve 5 mg pepsin in 2.17 ml DEPC-H_2O; 0.33 ml 1.5 M HCl is then added slowly to the pepsin solution.
- 0.4 % Paraformaldehyde: 1× PBS 100 ml; 0.4 g paraformaldehyde. Make up fresh and keep in dark.
- Prehybridization buffer: 50 % deionized formamide*; 2× SSC, 5 % dextran sulfate*; 0.3 % Triton X-100; 1×Denhardt's solution*, 150 μg/ml herring sperm DNA*; 150 μg/ml tRNA*; DEPC-H_2O (*all components are stored individually in aliquots at −20 °C)
- Hybridization buffer: As for prehybridization buffer but supplemented with the specific probe required (0.1–10 μg/ml)
- Buffer A: 100 mM Tris-HCl, pH 9.5 (6.055 g/500 ml); 100 mM NaCl (2.922 g/500 ml); 50 mM $MgCl_2$ (5.083 g/500 ml)
- NBT/BCIP color solution: 5 ml buffer A; 17.5 μl BCIP; NBT 16.88 μl
- 100× Denhardt's solution: 2 % (w/v) bovine serum albumin; 2 % (w/v) Ficoll 400; 2 % (w/v) polyvinylpyrollidone
- 20× TBS/BSA: 1 M Tris, pH 7.6; 3 M NaCl (87.66 g/500 ml); 40 mM $MgCl_2$; 2 % (w/v) BSA (10 g/500 ml)

32.1
RNA Extraction

▦ Procedure

total RNA extraction

1. Cells are grown to 80 % confluency in 175 cm^2 tissue culture flasks.

2. The medium is removed and the cells are rinsed twice in PBS.

3. The cells are lysed in 25 ml 4 M guanidium thiocyanate solution and subsequently homogenized for 1–2 min.

4. The viscous solution is then carefully layered on to 5.5 ml of 5.7 M cesium chloride solution in an ultracentrifuge tube.

5. The cell lysate is centrifuged at 26 000 rpm at 15 °C for 21–24 h and the RNA is pelleted to the bottom of the tube.

6. The guanidium thiocyanate solution and the "jelly-like" layer is removed by aspiration using a Pasteur pipette until all but 1 ml of the CsCl layer remains.

7. The bottom of the tube containing the RNA pellet is removed using a heated scalpel blade.

8. The tubes are inverted and allowed to drain and the RNA pellet is rinsed with 95 % ethanol at room temperature.

9. The pellet is then resuspended in DEPC-H$_2$O on ice and can be precipitated out of solution by the addition of 3 M Na-acetate, pH 5.2 (to a final concentration of 0.3 M) and 2 volumes of ice-cold ethanol (overnight at −20 °C or 30 min at −80 °C).

10. The RNA is pelleted at 4 °C in a centrifuge, the pellet is dried and resuspended in an appropriate volume of DEPC-H$_2$O.

RNA extracted is quantified spectrophotometrically at 260 nm and 280 nm. An optical density of 1 at 260 μm is equivalent to 40 mg/ml RNA. The A$_{260}$/A$_{280}$ ratio of pure RNA is approximately 2. Partially solubilized RNA has a ratio of < 1.6 (Ausubel et al. 1991). The yield of RNA from an average 175 cm^2 flask is 200 μg RNA using this protocol.

This protocol is based on the method of Chomczyski and Sacchi (1987). **rapid isolation** Cells are lysed in a denaturing solution containing 4 M guanidium thio- **of total RNA** cyanate. The lysate is treated with an acid phenol solution which retains most of the DNA and proteins in the organic phase, while the RNA remains in the aqueous phase. RNA purified using this method is suitable for all applications including Northerns, RT-PCR and differential display PCR. Many kits which are commercially available for RNA preparation use this single step method, i.e. Bio/RNA- Xcell from Bio/Gene Ltd.

1. Remove all medium from the cells and add 1 ml of denaturing solution per 10^7 cells. The cells should lyse immediately and a viscous supernatant should result.

2. Transfer the supernatant into a 50 ml tube and, if a number of flasks are used, pool the supernatants. Add 0.1 ml of 2 M sodium acetate, pH 4, and mix.

3. Add 1 ml of water-saturated phenol and gently mix. Leave on ice for 5 min.

4. Add 0.2 ml of chloroform/isoamyl alcohol (24:1). Mix thoroughly and then return to the ice for a further 10 min.

5. Centrifuge at 9000 g for 15 min. Transfer the upper aqueous phase to a fresh 50 ml tube.

6. Add 1 ml isopropanol (approximately 1 volume) to precipitate the RNA. Transfer to −20 °C for 30 min. Centrifuge at top speed in a microfuge for 15 min to pellet the RNA and aspirate the supernatant.

7. Resuspend the pellet in 0.3 ml denaturing solution. Add 0.3 ml 100 % isopropanol and place at −20 °C for 20 min.

8. Centrifuge at top speed in a microfuge for 15 min and aspirate the supernatant.

9. Wash the pellet well with 1 ml 70 % ethanol and centrifuge at top speed in a microfuge.

10. Discard the supernatant and dry the pellet at 65 °C.

11. Dissolve the pellet in 150 µl of DEPC-treated H_2O. Store the RNA at −80 °C for long-term storage or at −20 °C for short-term storage.

Isolation of mRNA

For some procedures it is necessary to isolate messenger RNA rather than total RNA from cells. This is particularly true when nonabundant messages are being studied. The method given here is based on the method of Sambrook et al. (1989).

1a. For every 2–3 75 cm^2 flasks used, suspend 0.1 g oligo dT cellulose in 0.1 N NaOH.

1b. Mix by gentle inversion of tube and then allow to settle for 10 min.

1c. Carefully aspirate the supernatant.

1d. Repeat steps 1a–1c.

1e. Wash the oligo dT cellulose twice with DEPC-treated H_2O.

1f. Resuspend the oligo dT cellulose in column loading buffer.

2. After removal of the culture medium, rinse the cells twice with 7 ml cold PBS. If suspension cells are used pellet the cells and then rinse with PBS.

3. To each 75 cm² flask add 600 µl of lysis buffer. Bring all flasks to this point before proceeding further.

4. Using a rubber policeman scrape the flask to remove all cells. Transfer the viscous lysate to a sterile tube and place on ice.

5. Using a #22 gauge needle and syringe shear the genomic DNA.

6. Add proteinase K to the pooled lysate to a final concentration of 200 µg/ml and incubate at 37 °C for 30 min.

7. Adjust the concentration of NaCl in the lysate to 0.4 M using a 5 M stock. The solution is 0.1 M NaCl already.

8. Add 10 ml oligo dT cellulose as prepared in step 1. Mix gently.

9. Place on an orbital shaker and rock gently for 2 h at room temperature. The oligo dT cellulose should be kept in suspension.

10. Centrifuge at 500 g for 5 min. Aspirate the supernatant.

11. Wash the oligo dT pellet with 20 ml high salt buffer, then pellet the oligo dT cellulose and aspirate the supernatant.

12. Add 10 ml of high salt buffer and add to a prepared column. (A siliconized Pasteur pipette plugged with sterile glass wool which has been DEPC-treated and autoclaved is fine).

13. Rinse any remaining oligo dT cellulose out of the tube using 5 ml high salt buffer and add to the column.

14. Wash the column three times with 5 ml high salt buffer.

15. Wash the column twice with 1 ml low salt buffer.

16. Place collection tubes under the column and add 500 µl salt-free buffer. Repeat twice. Collect all the eluant from the salt-free buffer in the same tube.

17. Add 3 M sodium acetate to a final concentration of 0.3 M and two volumes of ethanol. Precipitate overnight at −20 °C.

18. Centrifuge at 13 000 rpm in a microfuge for 15 min.

19. Wash the pellet with 1 ml 70 % ethanol. Centrifuge at 13 000 rpm for 10 min.

20. Aspirate the supernatant and heat at 65 °C to dry pellet.

21. Resuspend in 300 µl DEPC-treated H₂O.

32.2
Analysis of Differentially Expressed mRNAs in Animal Cells in Culture

The analysis of differentially expressed mRNAs is a useful tool in the study of development, differentiation and the effect of xenobiotics on cells in culture. Until 1992 the identification of differentially expressed

Fig. 32.1. Flow chart of the RNA differential display technique

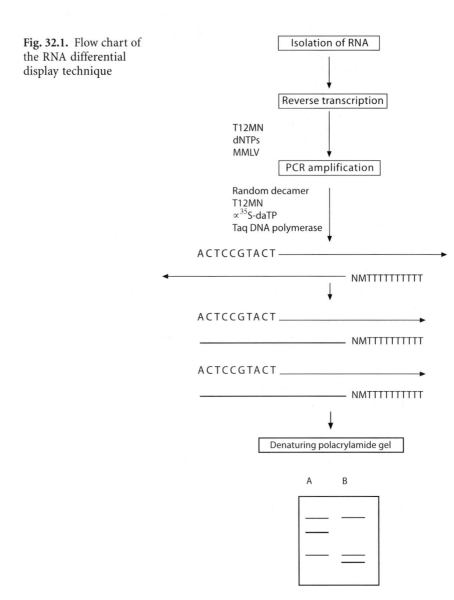

genes involved the generation of cDNA banks and a complex differential screening procedure (Sargent 1987). Pardee and Liang (1992) have developed a rapid procedure called differential display PCR (Fig. 32.1). Since that time several modifications have been published including Zhao et al. (1996, 1995), Guimarães et al. (1995), Callard et al. (1994), Li et al. (1994), Mou et al. (1994), Sokolov et al. (1994), Bauer et al. (1993) and Liang et al. (1993).

This technique involves the isolation of total RNA from samples of interest. This RNA is then used in a reverse transcription reaction with specially designed oligo-dT primers which contain a specific base followed by a random base and 12 T bases. This primer anchors to specific mRNAs within the total RNA. This reverse transcription reaction is then used as a template for PCR with a second 10-mer oligonucleotide which is arbitrary in sequence. A portion of this PCR reaction is then examined on a DNA sequencing gel. A radioactively labeled nucleotide incorporated in the reaction allows easy visualization of the DNA bands obtained using autoradiography. Differentially expressed mRNAs can be selected by differences seen in the banding pattern obtained. These bands can then be isolated from the dried DNA sequencing gel. After further amplification these bands can be used as probes for Northern analysis or subcloned for DNA sequencing.

Differential display can be a useful tool to isolate differentially expressed genes. However, the quality of the RNA used in the procedure is very important. It must be free from contaminating genomic DNA which can be amplified in the PCR reaction and give rise to high background. The combination of primer pairs is important, with many researchers reporting that while certain combinations work well others give no products. Another disadvantage of this technique is that it can be difficult to reproduce the results. Therefore it is important to set up reactions in duplicate and to repeat the process before selecting bands for purification, to ensure false positives are not selected.

As this protocol involves the use of a radioactive isotope ^{35}S, only those trained in the use of isotopes should perform it. Appropriate precautions should be taken to prevent contamination (Ausubel et al. 1991)

Materials

- Total cellular RNA
- Human placental RNase inhibitor (RNasin): 20–40 U/µl (Promega, cat. no. N2111)

- 3 M Sodium acetate, pH 5.2
- Diethylpyrocarbonate (DEPC)-treated H$_2$0 (Sigma, D5758)
- Moloney murine leukemia virus (MMLV) reverse transcriptase: 200 U/µl (Promega, M1701)
- dNTPs: 250 µM and 25 µM (Promega, U1240)
- α^{35}S-labeled dATP: >1000 Ci/mmol (Amersham, SJ1304)
- AmpliTaq DNA polymerase: 5 U/µl (Perkin Elmer, N8080160)
- Mineral oil (Sigma, M5904)
- Glycogen: 10 mg/ml
- Primers: degenerate anchored oligo dT primer set
- Random decamer: 2 µM
- Perkin Elmer DNA thermocycler
- PAGE sequencing apparatus
- 6 % Denaturing polyacrylamide gel
- 10× TBE

▓ Procedure

Differential Display PCR

reverse transcription of RNA

1. Thaw all components and place on ice.

2. For each sample of RNA to be examined set up four reverse transcription reactions each containing one of the four degenerate oligo dT primers. RNA should be freshly diluted using DEPC-treated H$_2$O just prior to use to a concentration of 0.5 µg/µl.

3. Into each reaction add the following:
   - 8.4 µl H$_2$O
   - 4.0 µl 5x reverse transcription buffer
   - 1.6 µl dNTP (250 µM)
   - 1.0 µl RNase inhibitor
   - 2.0 µl degenerate oligo dT primer (T12MN, 10 µM)
   - 2.0 µl total RNA
   - 19.0 µl total volume

Note: To prevent pipetting error a core mix should be made for all the RNA samples to be examined which can be added separately to the RNA templates.

4. Place the tubes in the thermocycler and program for: 5 min at 65 °C; 60 min at 37 °C; 5 min at 95 °C; 4 °C storage.

5. When the tubes have been at 37 °C for 10 min, 1 µl of MMLV reverse transcriptase should be added. Mix using a tip or by gently tapping the tube and quickly return to the thermocycler.

6. The reactions can be used immediately for PCR or they can be stored at −20 °C.

7. Set up the PCR reactions as follows using a core mix to avoid pipetting errors:
   – 2.0 µl reverse transcription reaction
   – 9.2 µl H₂O
   – 2.0 µl 10× PCR buffer
   – 2.0 µl arbitrary decamer (2 µM stock)
   – 2.0 µl degenerate oligo dT primer(T12MN, 10 µM)

PCR

8. Place tubes in the thermocycler and heat to 95 °C for 5 min.

9. Add a core mix containing the following:
   – 1.6 µl dNTP (25 µM stock)
   – 1.0 µl α^{35}S-labeled dATP($>$1000 Ci/mmole)
   – 0.2 µl Amplitaq (5 U/µl)
   – 20.0 µl final volume in each reaction

10. Mix well using a sterile tip or by gentle tapping.

11. Add 30 µl of mineral oil.

12. Program the thermocycler for 40 cycles as follows: 30 s at 95 °C; 30 s at 40 °C; 2 min at 72 °C. The samples should then be stored at 4 °C.

13. Samples are now ready for application to a 6 % denaturing polyacrylamide gel.

14. Prepare a 6 % denaturing PAGE gel in TBE. Allow to polymerize for 1–2 h.

PAGE electrophoresis

15. Prerun gel until it is 50 °C. Meanwhile, mix 3.5 µl of each sample with 2 µl formamide loading buffer and heat to 80 °C for 2 min to denature the products.

16. After washing urea from the wells, load the samples and run for approximately 2.5 h at 60 W until the xylene cyanol runs to within 15 cm from the end of the gel.

17. Transfer the gel carefully onto 3M paper and dry it on a gel dryer at 80 °C for 1 h. Do not fix in methanol/acetic acid as this makes it difficult to reamplify the products (Ausubel et al. 1991).

18. Using a needle align the dried gel and X-ray film before exposure.

19. After 24–72 h develop the film and realign to the dried gel. Using a needle punch holes around the band of interest and cut out using a clean scalpel blade. Carefully place the paper containing the band into a clean tube.

20. Add 100 µl sterile H_2O to each band and incubate at room temperature for 10 min.

21. After wrapping the tube in Parafilm, heat to 100 °C for 15 min.

22. Centrifuge the tube to collect the condensation and pellet paper debris.

23. Transfer the supernatant to a fresh tube and add 10 µl 3 M sodium acetate, 5 µl glycogen (10 mg/ml) and 450 µl 100 % ethanol. Place at −20 °C for 1 h.

24. Centrifuge at 13 000 rpm for 15 min. Wash the pellet with 300 µl ice-cold 85 % ethanol. Centrifuge at 13 000 rpm for a further 15 min and dry the pellet briefly at 65 °C.

25. Redissolve the pellet in 12 µl and use 4 µl as a template for reamplification. Store the remainder of the reaction at −20 °C.

reamplification 26. Reamplify using the primers used in the original PCR reaction. Set up the reaction as follows:

- 4.0 µl 10× PCR buffer
- 3.2 µl dNTP (250 µM stock)
- 4.0 µl cDNA template
- 20.4 µl H_2O
- 4.0 µl arbitrary decamer (2 µM stock)
- 4.0 µl degenerate oligo dT primer(T12MN, 10 µM)
- 0.4 µl Amplitaq (5 U/µl)
- 40.0 µl final volume in each reaction

27. Place tubes in the thermocycler and program as in step 12.

28. Analyze 10 µl on a 2.0 % agarose gel containing ethidium bromide.

29. If no product is visible on the agarose gel, use 5 µl of a 1:100 dilution of the PCR reaction as a template for a second round of amplification, keeping the conditions identical. It is important to confirm that the size of the reamplified PCR product is consistent with the size of

the original product isolated from the polyacrylamide sequencing gel. This product can now be used for Northern analysis (see chapter by N. Daly and Clynes). For subcloning of the PCR products we recommend either the eukaryotic TA cloning vector from Invitrogen (K3000–01), or Stratagene's PCR script cloning vector (#211188), both of which have been specifically designed for use in subcloning PCR products.

32.3
Analysis of Specific mRNA Expression

The mRNA levels of specific genes are generally studied using such techniques as Northern blot (Alwine et al. 1977; Thomas 1980), RNA slot/dot blot (Kafatos et al. 1979), RNase protection assay (Reyes and Wallace 1987), in situ hybridization (for review of method see Leitch et al. 1994) and RT-PCR (Wright and Wynford-Thomas 1990). Many of these techniques are insensitive and, although they do provide semiquantitative analysis of mRNA, a high quantity of purified polyA$^+$ RNA is necessary to produce a signal (i.e., 5–10 µg needed for Northerns; 1–10 µg for slot/dot blot).

The two most sensitive techniques used for specific mRNA detection are RT-PCR (Fig. 32.2) and in situ hybridization (Fig. 32.3). RT-PCR is 1000–10 000 times more sensitive than traditional blot techniques and can be used to detect multiple mRNA signals in as little as 1–1000 cells. In situ hybridization can detect 10–100 molecules of mRNA in a given cell and also reveals 3D information about transcript distribution in a

Fig. 32.2. Flow chart of the RT-PCR protocol

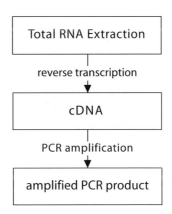

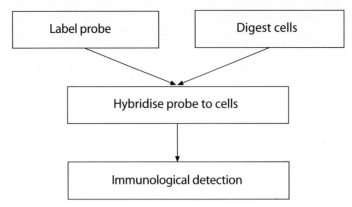

Fig. 32.3. The main steps involved in in situ hybridization

cell population). Despite the sensitivity of these techniques some serious disadvantages must be noted:

• In situ hybridization is technically difficult, time consuming and does not allow for quantification of transcripts.

• The exponential nature of PCR may lead to false positives and, although it is a semiquantitative technique, it is difficult to differentiate low-level changes.

Materials

– DIG RNA labeling kit (Boehringer-Mannheim)
– Alkaline phosphatase anti-DIG antibody (Boehringer-Mannheim)
– NBT (nitrobluetetrazolium)/BCIP (5-bromo-4-chloro-3-indolylphosphate) (Boehringer-Mannheim)
– Hybridization chamber
– Coplin jars
– Pepsin/HCl
– Paraformaldehyde: 0.4 %
– Prehybridization buffer
– Hybridization buffer
– Buffer A
– NBT/BCIP color solution

RT-PCR Analysis of Specific Genes

The basic principle of analyzing specific mRNA populations within a total RNA population requires the use of specific probes annealing to the cDNA in question and acting as a primer during the PCR amplification. To achieve this specificity the cDNA sequence of the specific gene must be known and the probe must be chosen so as to amplify only this gene from the total RNA population; a constitutive house-keeping gene should also be included in the amplification step if semiquantitative analysis of gene expression is required. For a detailed analysis of the RT-PCR protocol and step-by-step instructions on how to design PCR primers including which internal control should be used, the reader is referred to O'Driscoll et al. (1993).

1. Mix the following components together on ice: **RT reaction**
   - 1.0 µl oligo (dT) 12–18 primers (1 µg/µl)
   - 1.0 µl total RNA (1 µg/µl)
   - 3.0 µl H_2O

2. Heat at 70 °C for 10 min and then chill on ice to remove any RNA secondary structure formation and allow the oligo (dT) primers to bind to the polyA$^+$ tail on the mRNA.

3. The following is then added:
   - 4.0 µl 5× buffer
   - 2.0 µl DTT (dithiothreitol) (100 mM)
   - 1.0 µl RNasin (40 U/µl)
   - 1.0 µl dNTPs (10 mM each)
   - 1.0 µl MMLV-RT (200 U/µl)
   - 6.0 µl H_2O

4. Incubate for 1 h at 37 °C followed by 95 °C for 2 min to form cDNA from single-stranded RNA.

1. The following components are mixed on ice and heated to 94 °C for **PCR reaction**
   5 min:
   - 26.5 µl H_2O
   - 5.0 µl 10× buffer ($MgCl_2$ free)
   - 2.0 µl $MgCl_2$ (25 mM)
   - 0.5 µl internal control primer (a) (250 ng/µl)
   - 0.5 µl internal control primer (b) (250 ng/µl)
   - 1.0 µl target primer (a) (250 ng/µl)

- 1.0 µl target primer (b) (250 ng/µl)
- 5.0 µl cDNA (from room temperature reaction, step 20)

2. The following is then added to the above; 8.0 µl dNTP (1.25 mM); 0.5 µl Taq DNA polymerase

3. A drop of mineral oil is added to each tube and the cDNA is amplified using the following procedure:
   - 95 °C for 1.5 min (denatures double-stranded DNA)
   - 30 cycles:
     - 95 °C for 1.5 min (denature)
     - 55 °C for 1.0 min (anneal)
     - 72 °C for 3.0 min (extend)
   - 72 °C for 7.0 min (extend)
   - Hold at 4 °C.

4. The result of the RT-PCR is easily visualized and analyzed by agarose gel electrophoresis with ethidium bromide (intercalates with the cDNA forming a readily visible band under UV transillumination).

In Situ Hybridization

The following protocol uses a digoxigenin-labeled riboprobe to detect and localize specific mRNAs in cells. The protocol may need to be adapted for each particular gene under analysis, i.e., varying components in the hybridization buffers, altering the temperature and length of hybridization, and possibly changing the stringency of the washes. A detailed analysis of the effect each component used in the following protocol has on the hybridization conditions has been described by Leitch et al. (1994).

cell preparation   A choice of two procedures can be used to prepare slides of cells for use in in situ hybridization. The first involves direct spinning or dropping of cells on to slides and subsequent fixation; the other involves pelleting cells, embedding in an agarose plug, fixing and then paraffin-embedding the agarose plugs. Subsequently, sections of the cell block are cut on to slides.

The slides used for both procedures should be coated in an adhesive such as poly-L-lysine (Sigma) or APES (3-aminopropyltriethoxysilane) (Sigma).

The cells, prepared by either method, are routinely fixed in 10 % formalin.

The method here describes the procedure for the digoxigenin-labeling of the probe of interest by in vitro transcription from the SP6 and T7 promoter sites of an insert in the pGEM4 vector to produce directional riboprobes that are subsequently used for detection of mRNA in cells by in situ hybridization.

probe labeling

Oligonucleotide probes can also be used as can probes labeled with biotin or radioactively labeled probes.

To generate sense and antisense probes from the pGEM4 vector the plasmid must be linearized with an appropriate restriction enzyme and transcribed with the corresponding polymerase.

A transcription kit (Boehringer-Mannheim) is used to label the riboprobes according to the manufacturer's instructions. Incorporation of labeled nucleotides is measured using a dot-blot assay (as described in the manufacturer's protocols) and show yields of ~5 µg of RNA per µg of template cDNA. SP6 labeling efficiency was increased greatly when an incubation temperature of 40 °C was used in preference to the recommended 37 °C.

nonradioactive in situ hybridization

- If paraffin-embedded cell sections are in use then sections are washed twice in xylene for 10 min to remove the paraffin and rehydrated in 99 % ethanol and 95 % ethanol (two washes each for 2 min) and then incubated in DEPC-H_2O for 10 min at 37 °C.

- Nonembedded cells are rehydrated in 1× PBS.

1. Cells are digested in pepsin/HCl at 37 °C for 20 min and rinsed in 1× PBS at 37 °C.

2. Control slides are treated with RNaseA (100 µg/ml) for 30 min at 37 °C to destroy any RNA.

3. Postfixation in 0.4 % paraformaldehyde (in 1× PBS) is then carried out at 4 °C for 20 min and the slides are rinsed in DEPC-H_2O.

4. Slides are prehybridized in prehybridization buffer for 1 h at 37 °C in a humid chamber.

5. Hybridization buffer is added to slides under cover-slips. Hybridization is carried out at 95 °C for 15 min followed by 50 °C for 2 h.

6. To remove any unhybridized probe the slides are washed twice in solutions containing 30 % formamide and (a) 4× SSC, (b) 2× SSC and (c) 0.2× SSC for 5 min each at 50 °C.

7. To detect the hybridized probe, an alkaline phosphatase anti-DIG antibody is used. The slides are blocked first in 1× TBS/0.1 %BSA/ 0.1 %Triton for 25 min at room temperature with gentle shaking. The antibody is then added to the slides (1:500 in 1× TBS/0.1 %BSA) for 30 min at room temperature. To remove any unbound antibody the slides are then washed twice in 1× TBS/0.1 %BSA for 5 min each at room temperature with vigorous agitation.

8. The revealing step involves the formation of a purple precipitate at the site of hybridization due to the enzymatic action of alkaline phosphatase with NBT/BCIP. The slides are washed two to three times in buffer A, pH 9.5, and then color solution (NBT/BCIP) is added for 30–90 min in the dark (a longer time may be necessary if an mRNA expressed at low level is being analyzed). When the color develops the slides are washed under running tap water and allowed to dry.

The cells can be counter-stained with methylene green and the slides should be mounted with an aqueous mounting gel such as Glycergel (Dako).

Comments

- Parafilm can be used as cover-slips during hybridization.

- All hybridization steps must be carried out in a humid atmosphere. This is simply achieved by placing tissue soaked in 2× SSC at the base of a sealed hybridization chamber.

- The amount of solution required during hybridization can be reduced (thus lowering the overall cost of the procedure) by circling the area of the slide containing the cells with a Dako-pen (Dako) that forms a seal with the slide.

- For accurate analysis of mRNA expression by nonradioactive in situ hybridization, a number of controls must be set up with the positive slide, namely, hybridization with: (a) a sense probe, (b) a control probe (gene not present in cells under investigation), (c) an oligo-dT or β-actin probe (to assess mRNA quality throughout the cell sample), (d) pretreatment with RNase and (e) hybridization buffer without labeled probe.

■ References

Alwine JC, Kemp DJ, Stark GR (1977) Method for detection of specific RNAs in agarose gels by transfer to diazobenzyloxymethyl-paper and hybridisation with DNA
probes. Proc Natl Acad Sci USA 74:5350–5354

Ausubel FM, Brent R, Kingston RE, Moore DD, Seidman JG, Smith JA, Struhl K (eds)
(1987) Current protocols in molecular biology. Greene, Wiley-Interscience, New
York

Callard D, Lescure B, Mazzolini L (1994) A method for the elimination of false positives generated by the mRNA Differential display technique. BioTechniques
16:1096–1103

Chomczynski P, Sacchi N (1987) Single-step method of RNA isolation by acid guanidium thiocyanate-phenol-chloroform extraction. Anal Biochem 162:156–159

Guimarães MJ, Lee F, Zlotnik A, McClanahan T (1995) Differential display by PCR:
novel findings and applications. Nucleic Acids Res 23:1832–33

Kafatos FC, Weldon JC, Efstratiadis A (1979) Determination of nucleic acid sequence
homologies and relative concentrations by a dot hybridisation procedure. Nucleics
Acids Res 7:1541–1552

Leitch AR, Schwarzacher T, Jackson D, Leitch IJ (1994) In situ hybridization. BIOS,
Oxford

Li F, Barnathan ES, Karikó K (1994) Rapid method for screening and cloning cDNAs
generated in differential mRNA display: application of Northern blot for affinity
capturing of cDNAs. Nucleic Acids Res 22:1764–1765

Liang P, Pardee A (1992) Differential display of eukaryotic messanger RNA by means
of the polymerase chain reaction. Science 257:967–971

Mou L, Miller H, Li J, Wang E, Chalifour L (1994) Improvements to the Differential
display method for gene analysis. Biochem Biophys Res Comm 199:564–569

O'Driscoll L, Daly C, Saleh M, Clynes M (1993) The use of reverse transcriptase-
polymerase chain reaction (RT-PCR) to investigate specific gene expression in
multidrug-resistant cells. Cytotechnology 12:289–314

Reyes AA, Wallace RB (1987) Mapping of RNA using S1 nuclease and synthetic oligonucleotides. Method Enzymol 154:87–101

Sambrook J, Fritsch EF, Maniatis T (1989) Molecular cloning: a laboratory manual.
Cold Spring Harbor Laboratory, Cold Spring Harbor, New York

Sargent T (1987) The isolation of differentially expressed genes. In: Abelson J, Simon
M (eds) Methods Enzymol 152: 423–432

Sokolov BP, Prockop DJ (1994) A rapid and simple PCR-based method for the isolation of cDNAs from differentially expressed genes. Nucleic Acids Res 22:4009–15

Thomas PS (1980) Hybridisation of denatured RNA and small DNA fragments transferred to nitrocellulose. Proc Natl Acad Sci USA 77:5201–5205

Wright PA, Wynford-Thomas D (1990) The polymerase chain reaction, miracle or
mirage? A critical review of its uses and limitations in diagnosis and research.
J Pathol 162:99–117

Zhao S, Ooi SL, Pardee A (1995) New primer strategy improves precision of differential display. Biotechniques 18:842–850

Zhao S, Ooi SL, Yang F, Pardee A (1996) Three methods for identification of true positive cloned cDNA fragments in differential display. Biotechniques 20:400–404

Chapter 33

Human Gene Therapy: Review and Outlook

GERARD J. McGARRITY*

33.1
Introduction

The first clinical trial for gene transfer in humans was approved in the United States in 1989. In this trial, genetically marked lymphocytes that had been removed from a melanoma biopsy were transduced with a retroviral vector containing the gene conferring resistance to the neomycin analogue G-418, propagated ex vivo and returned to the patient. The purpose of this gene marking study was to determine if the G-418 resistant lymphocytes honed specifically to melanoma and to determine the lifespan of the marked lymphocytes. Results of various gene marker trials have been published (Dunbar et al. 1994; Brenner 1996). Marker trials have demonstrated some homing to tumors, and marked lymphocytes have been detectable for years past re-infusion.

The first therapeutic gene therapy trial occurred on September 14, 1990 and consisted of lymphocytes transduced with a retroviral vector encoding the gene for adenosine deaminase (ADA). The transduced lymphocytes were reinfused back into a child with severe combined immune deficiency (SCID) due to ADA deficiency. This procedure was also performed ex vivo, meaning the gene was inserted into cells during expansion of the autologous lymphocytes in the laboratory prior to reinfusion into the patient. Results from this trial have been published (Blaese et al. 1995); significant increases in blood ADA levels and other immune parameters were found. Interpretation of clinical effectiveness is compromised by the fact that the treated children were continued on enzyme replacement therapy during gene therapy due to clinical and regulatory

* Gerard J. McGarrity, Genetic Therapy, Inc., 938 Clopper Road, Gaithersburg, MD, 20878, USA; phone +1–301–208–2403; fax +1–301–208–2412; e-mail gerard.mcgarrity@pharma.novartis.com

concerns. In addition, three newborns with ADA deficiency have also been treated with gene therapy 3–4 days after birth (Kohn et al. 1995). This trial used the same retroviral vector to transduce CD34+ cells from cord blood. In addition to the ADA gene, both trials also used a a gene encoding for resistance to G-418, a neomycin analogue in a bicistronic vector. The newborn trial demonstrated the presence of the transgenes in different hematopoietic cell lineages.

From these beginnings in 1989 and 1990, more than 200 clinical trials have been approved by regulatory authorities around the world. More than 3000 patients have been treated worldwide. These trials have been for a wide diversity of clinical indications, including cancer, genetic diseases, and AIDS. They have also utilized different methods to deliver genes to target cells.

This chapter will review the basic gene delivery systems, or vectors, and summarize some relevant clinical findings gathered from published studies.

33.2
Methods

The term vector refers to instruments, viral and nonviral, that deliver genes to target cells. To date, the most commonly employed vector has been the retroviral system, accounting for approximately 70 % of vectors used in clinical trials. Retroviral vectors are replication incompetent and typically are derived from the Moloney strain of murine leukemia virus. This retrovirus is among the simplest genomically and consist of three major genes: *gag*, *pol*, and *env*. The *gag* gene encodes a total of four different proteins; *pol* encodes enzymes including polymerase and integrase. The *env* gene encodes for the viral envelope. The genes are driven or promoted by sequences at the 5' and 3' ends known as long terminal repeats (LTRs).

Retroviral Vectors

Retroviral vectors are made by removal of the three functional genes, *gag*, *pol*, and *env*, from the virus and the insertion of the transgene(s) of interest. Retroviral vectors can accommodate up to 8.5 kilobases (kb) of sequence. One (monocistronic) or two (bicistronic) genes can be added. Typically, the first gene is promoted by the 5' LTR. The second gene can follow an internal promoter, often a viral promoter such as respiratory syncytial virus (RSV) or cytomegalovirus (CMV). Alternately, an internal ribosome entry site (IRES) sequence derived from a virus, e.g., polio, can be placed between the two transgenes for efficient splicing and expression. The second transgene using an internal promoter typically has less gene expression in bicistronic vectors.

Retroviral vectors are made in genetically engineered cell lines known as packaging cells. The vectors express only the product of the transgene. Packaging cells contain the structural genes encoding for *gag*, *pol*, and *env*, and provide these in *trans* to the vector. Packaging cells without vector generate empty vector particles. Packaging lines have been derived from mouse, canine and human cells. When packaging cells are transfected with vector DNA, they are referred to as producer cells. The vector DNA also contain sequences known as psi, or the packaging signal, immediately 3' to the 5' LTR and extending into *gag*. Psi sequences assemble the internal RNA into the vector particle. The objective of vector construction is to minimize the potential for recombination between vector and package sequences which would result in the generation of a replication competent retrovirus (RCR). Removal of homologous sequences between vector and producer cell is essential to minimize this recombination. Another objective of vector construction is to maximize the level of gene expression by such measures as clone selection and/or insertion of specific promoters or enhancers. More details on packaging cell construction and characterization are available (Miller 1990; Miller and Buttimorer 1986). Titers of retroviral vectors are lower than those of adeno vectors and typically are on the order of 10^5–10^6/ml. Downstream concentration can yield titers higher than 10^9/ml.

Replication incompetent retroviral vectors transduce only replicating cells. Therefore, they are effective for ex vivo gene therapy in which cells are undergoing rapid expansion. These cells include tumor cells, skin fibroblasts, lymphocytes, and bone marrow, among others. They are not suited for in vivo or in situ transduction of non-replicating cells. Retroviral vectors are rapidly inactivated by human serum probably due to antibody directed to galactosyl epitope. Also, retroviral vectors randomly integrate into the genome of the cells they transduce. Since genomic integration is random, the possibility exists that insertional mutagenesis could result. This could lead to the death of the cell, oncogene activation, or inactivation of tumor supressor genes. While these possibilities exist, Anderson has calculated that the increase in mutagenic potential as related to vector genomic insertion is insignificant (Anderson 1993; Anderson et al. 1993).

More than 2000 patients have received retroviral vector mediated gene transfer/therapy. No adverse event due to the retroviral vector has been reported. The number of patients, while increasing, is still modest, and more observations are needed before risk factors can be more fully described. Many of the patients that received retroviral vectors had limited life expectancies, such as patients with cancer or AIDS.

Safety studies have been performed with retroviral vectors in mice and nonhuman primates. Cornetta et al. (1991) have reported no adverse events after intravenous inoculation of replicating vectors (RCR). This report confirmed and extended earlier reports. However, Donahue et al. (1992) reported that administration of large numbers of RCR in the presence of nonreplicating vectors to nonhuman primates that were severely immunosuppressed due to whole body irradiation and/or 5-fluorouracil resulted in lymphomas in three out of ten animals. These tumors were first detected 180–210 days following inoculation of transduced autologous CD34+ cells. These results demonstrate that exposure of severely immunosuppressed patients to significant numbers of RCR may potentially result in malignancy, based on these studies in nonhuman primates. Although the estimated numbers of RCR in these animals have been high, described as "overwhelming retroviremia," the potential risk for patients is clear. However, these results clearly identified a threshhold for potential adverse effects in patients, especially those having severe immune suppression as part of their therapy or disease, such as recipients of allogeneic bone marrow transplants. In response to this finding and for general safety concerns, regulatory authorities have required manufacturers of retroviral vectors to perform specific and sensitive assays to assay for RCR in product lots. Beginning in 1993, the Food and Drug Administration (FDA) in the United States also recommended that all patients receiving retroviral vectors in clinical trials be regularly screened for RCR by sensitive and efficient assays for evidence of vector and RCR in lymphocytes. These are performed monthly during treatment, quarterly for the next year and annually thereafter.

At the present time, retroviral vectors represent the most widely characterized and used gene delivery system. It is estimated that approximately 70 % of patients who have received gene therapy have been treated by retroviral vectors.

Adenoviral Vectors

These vectors are derived from adenoviruses which contain double-stranded DNA and have approximately 36 kb of sequence. Vectors derived from serogroups 2 and 5 have been used in clinical trials. To generate replication incompetent adenoviral systems, genes have been deleted – most often E1, E2, and E3 – either alone or in combination. The transgene(s) is (are) inserted in the space provided by the deletion(s). Replication incompetence is usually accomplished by complete or partial removal of early adenogenes. This typically involves the E1 gene (E1a/E1b), which directly or indirectly regulates other transcriptional units, and the E3 genes, which encode a protein that aids the infected cells to evade the host immune surveillance. Elimination of these genes permits insertion of approximately 7 kb of the inserted transgene. In later vector constructs, more sequence of vector is accommodated by further viral deletions, e.g., E4. Packaging cells which contain the deleted sequences and those required for replication allow production of the vector. The adeno system differs from retroviral vectors in several aspects. Packaging cells are available, but true continuous producer cells do not exist. The vector is generated by transfecting vector DNA into the packaging cell and analyzing the generated plaques for titer and transgene expression and selection. Once obtained, the cloned vector is inoculated into the packaging cells to generate more adenoviral vector, killing the packaging cells in the process. Therefore, in adeno systems, single batches of product are collected and purified, generating high titers, on the order of 10^{11}–10^{12}. Human cell lines 293 and A549 have been employed as the matrix to construct adeno vector packaging cells. A review of adeno vectors for use in gene therapy has been published (Shenk 1995).

Adenoviral vectors can transduce nonreplicating cells, making them capable of in vivo/in situ administration. Adenoviral vectors do not integrate, but reside in the nucleus as episomes. Adenoviral vector transduced cells have shorter periods of transgene expression than retroviral vectors. However, adenoviral vectors can be prepared in high titer following concentration and partial purification from transduced cell culture supernatants.

The majority of the population of developed countries has antibodies to adenoviruses, including serogroups 2 and 5. Therefore, even initial administration of these vectors to patients represents an immunological boost and can lead to local inflammation. This has been reported in early clinical trials using adeno vector in cystic fibrosis (CF) patients. This first generation adeno vector deleted only E1 sequences leaving remaining adeno gene products to initiate immune response in the recipient. To reduce the potential of inflammatory and anamnestic responses to adeno vectors, elimination of adeno E3 sequences has been achieved, but it remains to be seen whether this will reduce immune responses to vector backbone. MHC class I restricted CD8+ cytotoxic T lymphocytes are activated in response to newly synthesized antigens, resulting in destruction of vector transduced cells. Further elimination of sequences will allow additional sequences, e.g., E-4, of transgenes to be inserted, if necessary, and hopefully minimize immune response. If immune responses cannot be controlled by further gene deletion or possibly immune suppression, the utility of adenovectors will be minimized, especially for those diseases requiring multiple administrations.

Adeno-Associated Virus Vectors

This double-stranded DNA containing vector is derived from adeno associated virus (AAV) which has several interesting characteristics. AAV can only replicate in the presence of adeno and related viruses. The parental virus has a genome of approximately 4.7 kb. AAV also has a tendency to integrate into a specific site, 19Q, in the human genome. The helper viruses provide functions in *trans* for AAV replication and encapsulation. This site specific tendency would be useful to eliminate whatever low risk may be associated with random integration. This tendency, however, is lost when the virus is converted into a vector. Attempts are being made to identify the sequences that tend to direct AAV to integrate in a site specific fashion. This would be helpful not only in AAV but in other vector systems in which long-term gene expression is desired.

To date, AAV vectors have been used in clinical trials for CF and for Gaucher disease. Present production methods and quality control methods are not efficient and need to be further developed (Fisher et al. 1996). Continuous expression of the AAV rep protein can produce significant toxicity.

Nonviral Vectors: Liposome Encapsulation

Wolff et al. (1990) demonstrated gene expression following inoculation of marker DNA into skeletal muscle using a RSV promoted plasmid. Initial levels of gene expression that involved encapsulation of the plasmid into cationic liposomes originally resulted in low levels of transfection, on the order of 0.02 μg of DNA expression/10^6 cells. A limitation of liposome-mediated gene transfer is the low concentration of DNA-liposome complexes employed compared to other vector systems. Previous attempts to increase the DNA-liposome concentration resulted in aggregation and toxicities in animals and were not suitable for human studies.

A number of modifications have been made to the type of cytofectins that have been employed. Felgner has updated some of the newer developments in liposome construction (Stephan et al. 1996; Felgner 1996). He reported that GAP-DLRIE results in a 3300-fold increase in DNA expression in skeletal muscle. GAP DLRIE/DOPE has also been shown to have significant levels of gene expression. Incorporation of sequences that increase DNA expression has produced significant improvement in DNA-liposome complexes and expression of gene product. These include the use of improved promoters, cleavable introns, terminators, and poly A signals. Polymerases have also been effective.

Significant progress has been made in the design, effectiveness and stability of synthetic gene delivery systems. A variety of approaches involving cationic and other liposomes, "stealth" liposomes and other nonprotein systems offer potential to transfer genes in an efficient, nonimmunogenic fashion that can, in the long run, avoid the potential side effects of viral-mediated gene delivery.

Future Directions

Significant improvements have been made in vector design since the initial clinical trials in 1989–1990. These improvements will continue in both viral and non-viral systems. What is likely to occur is that sequences will continue to be removed from viral vectors, further simplifying them, to further increase safety. With nonviral delivery systems, components will continue to be added to further increase the level and the location of transgene expression. Other viral systems have also been investigated, including SV-40, herpes, lentiviruses, and canary pox.

It is likely that an efficient and safe hybrid of viral and nonviral gene delivery systems will result. Hopefully, the resultant systems will combine the advantages of high gene expression for the desired length of time with the safety of true nonreplicating systems, including lack of immune response to the vector backbone. These will have to be manufactured in a cost-efficient manner than can reduce cost of goods and provide long-term stability and ease of use.

33.3
Clinical Trials: A Brief Overview

At the end of 1996, more than 150 clinical trials had been approved by the Recombinant DNA Advisory Committee (RAC) of the US National Institutes of Health (NIH). Approval by the RAC does not imply approval by the appropriate US regulatory agency, the FDA, or that the trials have been initiated. However, this database supplies important information as to what clinical protocols and what vector systems are being contemplated by investigators and gene therapy companies.

Gene therapy clinical trials may be ex vivo or in situ/in vivo. In ex vivo gene therapy, cells are removed from the patient, propagated in the laboratory (ex vivo) transduced with a vector, typically retroviral, and returned to the patient. To insure that the maximum number of transduced cells are reinfused, bicistronic vectors are frequently used. The first transgene is the therapeutic gene of interest; the second is a selectable marker, most often a gene that confers resistance to the neomycin analogue G-418. This gene is typically designated neo^r. The use of a selectable marker enables selection and quantitation of the number of transduced cells returned to the patient.

To obtain a pure population of transduced cells containing neo^r, the cells are grown in the presence of G-418. Only cells containing the neo^r gene and expressing that gene will survive. Nerve growth factor-receptor (NGF-R) has also been used as a marker. An advantage of this latter approach is that selections can take hours, not days, as in neo selection. Other markers that can be selected in hours rather than days have advantages in clinical protocols in which autologous cells can be reinfused into the patient as soon as practical. It does however, require good manufacturing practices-produced monoclonal antibody to the NGF-R for human studies.

To date, all ex vivo protocols have involved retroviral vectors. Target cells have included lymphocytes, tumor cells, bone marrow and skin fibroblasts. Retroviral vectors are suited for these rapidly proliferating cells.

Cancer

Most tumor types have been used as targets for cancer gene therapy. These include: melanoma, renal cell carcinoma, colorectal cancer, breast, ovarian, prostatic and head and neck cancers, neuroblastoma, glioblastoma, small cell and non-small cell lung cancers, astrocytoma, mesotheloma, lymphoma, multiple myeloma and hepatocellular carcinoma. Results of many preclinical studies and some clinical studies have been published. These have been reviewed (McGarrity and Chiang 1997).

Several types of approaches have been executed in cancer gene therapy. These are:

- Use of cytokines and growth factors

- Suicide genes

- Tumor suppressor genes

- Histocombatibility genes to confer "foreigness" to tumors

The transgenes for cytokines used in gene therapy clinical trials are extensive and have included: interleukin (IL)-1β, IL-2, IL-4, IL-7, IL-12, insulin-like growth factor, interferon-γ, granulocyte/macrophage colony-stimulating factor (GM-CSF), and tumor necrosis factor (TNF)-α, among others. Suicide genes have included thymidine kinase (TK) derival from herpes simplex and cytosine deaminase from *Escherichia coli*. In TK trials, vectors encoding TK are transduced into replicating tumor cells. These cells are subsequently killed by administration of gancyclovir, an approved therapeutic. TK trials have been designed for glioblastoma multiforme, ovarian cancer, and mesothelioma.

The therapeutic rationale in cytokine gene therapy trials is to exert a stimulation on the presentation and effective processing of putative tumor antigens to the immune system. Studies in animals and in humans have shown tumors are immunogenic (Kaposi's sarcoma seems to be an exception), but for some reason an immunological response is not made to neoplastic cells. Apparently, a block exists in antigen presentation and/or antigen processing. Somehow, tumors are tolerogenic. Suggestive data have been generated that tumors can down-regulate MHC class expression and produce TGF-β that can suppress immune functions. High concentrations of cytokine in the microenvironment of the tumor are designed to remove this (these) blocks. One advantage of gene therapy is the ability to deliver high concentrations of cytokine to the tumor microenvironment. The ultimate challenge is to generate a therapy that can control local and metastatic tumors.

If immune modulation of tumor cells presents a rational therapeutic target, how can this modulation be optimally exploited? That is the critical question clinical investigators are struggling to answer. While most clinical trials are in early phase I/II, definitive answers await the results of carefully designed controlled, prospective phase III trials. It is unclear if a combination of cytokines will be needed to effect clinically significant tumor killing or whether immunogene therapy can serve as an adjunct to conventional chemo- and radiotherapies.

Duration of gene expression is less problematic in cancer trials. Presumably, expression for periods on the order of one month, perhaps less, may be sufficient. Introduction of tumor antigen into antigen presenting cells, especially dendritic cells, are especially promising. Data are needed to determine if longer transgene expression occurs in other tissues and, if so, what are the long-term consequences. Incorporation of suicide genes that can kill transduced cells could abrogate potential side effects of uncontrolled cytokine/growth factor expression.

Genetic Diseases

A different therapeutic objective exists in gene therapy for genetic diseases. Therapy must last for life. Gene expression from a vector must remain at high levels for long periods or the vector must be readministered at regular intervals. Readministration with adeno vector presents a particular problem since a strong inflammatory response is mounted to the vector backbone. Alternate possibilities include mild immunosuppression or further modification of the vector to reduce immunogeneity. It would be expected that readministration of any viral vector would result in strong immunological responses.

The ADA trials were described earlier in this chapter. The largest number of trials in genetic disease has been for CF. These involve the CF transmembrane regulator (CFTR), the defective gene in this disease. Trials using adenoviral vectors, AAV and plasmids have been conducted. Significant, but not therapeutic, gene expression has been demonstrated in the nasal epithelium and respiratory alveoli. The significant problem remains the duration of transgene expression and immunological responses to the adeno vector backbone. Life-long clinical efficiency still remains a goal. Life expectancy for CF patients in the United States is approximately 30 years. Similar problems exist for the genetic diseases, such as hemophilia.

Trials for a variety of other genetic diseases are also in progress. These include: Fanconi's anemia, chronic myelogenous disease, Gaucher disease, Hunter's and Hurler's syndromes. The objective in these and other disease targets may be slowing the progression of the disease or reversal of the symptoms, i.e., actual cure. As stated, a critical factor for gene therapy of genetic disease is, of course, long-term gene expression. The length of gene expression will be influenced by the type and efficiency of vector used and the target cell for transgene delivery. Delivery of the vector to stem cells should result in longer term gene expression in all stem cell progeny, especially if the vector DNA is integrated. However, this would have to be verified by assays for the gene product and eventually for clinical efficiency.

AIDS

A variety of therapeutic approaches have been used to develop an effective approach to AIDS prevention and cure. In gene therapy trials, approaches have included immunological, transducing P-120 antigen into lymphocytes in a vaccine, use of new mutants and *tar* (transactivating response element of HIV) anti-sense constructs.

The concept of intracellular immunization, first described by Baltimore, attempts to block transcription of HIV gene products upon release from the nucleus. A critical success factor for this approach is getting the vector into stem cells so that the stable source of HIV resistant progeny can be generated.

A variety of protocols are in progress or in planning. The gene target varies in different clinical protocols. These have included Gp120, *rev* and *tar* anti sense. The trials for Gp120 have included both classical and intracellular vaccination.

Future Directions

In the early 1990s, many claims and promises were made for the potential of gene therapy. What was not appreciated by many outside (and inside) the field was that much work was required, not only on molecular biology and improvements in gene delivery, but that an entire field of product development had to be initiated. This is similar to what was done in other areas of biologics, including recombinant protein and monoclonal antibodies. Assays had to be developed, manufacturing methods had to be devised and scaled to large volumes, and crisp, controlled, perspective clinical trials had to be designed.

In the years since the first therapeutic clinical trial was initiated in September 1990, much progress has been made. This has included improvements in vector design and performance. A continuing database on clinical experiences, including safety profiles, introduction of serum-free medium, growth of producer cells in large scale with downstream processing and lyophilization of vectors, has been achieved. These experiences provide the basis for vigorously pursued research in a growing number of biotech and pharmaceutical companies that are investing resources for commercial development in this field.

It is difficult to project what the future may hold. Clearly more clever and more diverse vector systems will be introduced into the clinic. These will likely involve the use of tissue specific or tumor specific promoters, gene switches that can regulate gene expression as well as vectors that can effectively transduce cell targets that have proven difficult till now, such as CD34 cells from bone marrow.

What is needed more than anything is the scientific proof, obtained from controlled, prospective trials, that gene therapy can unequivocally improve human health by curing disease or slowing disease progression.

References

Anderson WF (1993) What about these monkeys that got T-cell lymphoma? Hum Gen Ther 4:1–2

Anderson WF, McGarrity GJ, Moen RC (1993) Report to the NIH Recombinant DNA Advisory Committee on murine replication competent retroviruses (RCR) assays. Hum Gen Ther 4:311–321

Blaese RM, Culver KW, Miller AD et al. (1995) T-lymphocyte directed gene therapy for ADA-SCID: initial trial results after 4 years. Science 270:475–480

Brenner MK (1996) Gene marking. Hum Gen Ther 7:1927–1936

Cornetta K, Morgan RA, Anderson WF (1991) Safety issues related to retroviral-mediated gene transfer. Hum Gen Ther 2:5–14

Donahue RE, Kessler SW, Bodine D et. al. (1992) Helper virus induced T-cell lymphoma in non-human primates after retroviral mediated gene transfer. J Exp Med 176:1125–1135

Dunbar CE, Emmons RVB (1994) Gene transfer into hematopoeitic progenitor and stem cells; progress and problems. Stem Cells 12:563–576

Felgner PL (1996) Improvements in cationic liposomes for in vivo gene transfer. Hum Gen Ther 7:1791–1793

Fisher KJ, Gao GP, Weitzman MD, DeMattio R, Burda JF, Wilson JM (1996) Transduction with recombinant adeno-associated virus for gene therapy is limited by leading strand synthesis. J Virol 70:520–532

Kohn DB, Weinberg KI, Nolta JA, et. al. (1995) Engraftment of gene-modified umbilical cord blood cells in neonates with adenosine deaminase deficiency. Nat Med 1:1017–1023

McGarrity GJ, Chiang YL (1997) Gene therapy of local tumors. In: Lotze M, Rubin R (ed) Regional therapy of advanced cancer. Lippincott-Raven, pp 375–390

Miller AD (1990) Retroviral packaging cells. Hum Gen Ther 1:5–14

Miller AD, Buttimore C (1986) Redesign of retropackaging cell lines to avoid recombination leading to helper virus production. Mol Cell Biol 6:2895–2902

Shenk T (1995) Group C adenoviruses as vectors for gene therapy. Viral Vectors 89–107

Stephan DJ, Yang Z-Y, Shan H, Simari RD, Wheeler CJ, Felgner PL, Gordon D, Nabel GJ, Nabel EG (1996) A new cationic liposome DNA complex enhances the efficiency of arterial gene transfer in vivo. Hum Gen Ther 7:1803–1812

Wolff JA, Malone RW, Williams P, Chong W, Ascadi G, Jariu P, Felgner PL (1990) Direct gene transfer into mouse muscle in vivo. Science 247:1465–1468

Subject Index